D1620376

Edited by
Marino Xanthos

Functional Fillers for Plastics

Related Titles

Faulkner, Edwin B. / Schwartz, Russell J. (eds.)

High Performance Pigments

Second, Completely Revised and Enlarged Edition

2009

ISBN: 978-3-527-31405-8

Mathers, Robert T. / Meier, Michael A. R. (eds.)

Green Polymerization Methods

Renewable Starting Materials, Catalysis and Waste Reduction

2010

ISBN: 978-3-527-32625-9

Janssen, Leon / Moscicki, Leszek (eds.)

Thermoplastic Starch

A Green Material for Various Industries

2009

ISBN: 978-3-527-32528-3

Gujrati, Purushottam D. / Leonov, Arkadii I. (eds.)

Modeling and Simulation in Polymers

2010

ISBN: 978-3-527-32415-6

Matyjaszewski, Krzysztof / Müller, Axel H. E. (eds.)

Controlled and Living Polymerizations

From Mechanisms to Applications

2009

ISBN: 978-3-527-32492-7

Elias, Hans-Georg

Macromolecules

Series: Macromolecules (Volume 1-4)

2009

ISBN: 978-3-527-31171-2

Martín, Nazario / Giacalone, Francesco (eds.)

Fullerene Polymers

Synthesis, Properties and Applications

2009

ISBN: 978-3-527-32282-4

Edited by
Marino Xanthos

Functional Fillers for Plastics

Second, updated and enlarged edition

WILEY-VCH Verlag GmbH & Co. KGaA

The Editor

Dr. Marino Xanthos
New Jersey Inst. of Technology
Dept. of Chemical Engineering
Tiernan Hall, Univ. Heights
Newark, NJ 07102
USA

Library of Congress Card No.: applied for

British Library Cataloguing-in-Publication Data
A catalogue record for this book is available from the British Library.

Bibliographic information published by the Deutsche Nationalbibliothek
The Deutsche Nationalbibliothek lists this publication in the Deutsche Nationalbibliografie; detailed bibliographic data are available on the Internet at http://dnb.d-nb.de.

Cover Design Grafik-Design Schulz, Fußgönnheim
Typesetting Thomson Digital, Noida, India
Printing and Binding Strauss GmbH, Mörlenbach

Printed in the Federal Republic of Germany
Printed on acid-free paper

ISBN: 978-3-527-32361-6

Contents

Functional Fillers for Plastics: Second, updated and enlarged edition. Edited by Marino Xanthos

ISBN: 978-3-527-32361-6

Preface to the 2nd Edition

The organization of this edition follows that of the earlier one. Part I (Chapters 1–3) provides an introduction to the types and properties of functional fillers and their polymer composites, including melt mixing aspects. Part II (Chapters 4–6) discusses different types of surface modifiers and coupling agents for fillers. Part III discusses in detail individual types of man-made, natural, in-situ generated and mineral fillers, and their functions. High aspect ratio fillers (Chapters 7–10), low aspect ratio fillers (Chapters 11–16), and a variety of specialty fillers for specific applications (Chapters 17–24) are included in Part III. Each filler chapter, typically, contains information on: (a) production methods, (b) structure and properties, (c) suppliers, (d) cost/availability, (e) environmental/toxicity considerations, and (f) applications based on primary and secondary functions.

This edition differs from the previous one in several aspects. The most important is that new authors with considerable industrial experience bring their own perspective in the revision of several chapters. They include: Dr. B. Borup on silane coupling agents (Chapter 4), Ms. S. Robinson on wollastonite (Chapter 14), Dr. Y. Khanna on calcium carbonate (Chapter 16), Dr. A. Kron in Chapter 21, which is now significantly expanded to include organic spherical fillers, and Dr. C. DeArmitt whose new Chapter 23 on polyhedral oligomeric silsesquioxanes (POSS) brings the total number of chapters in this edition to twenty four.

All chapters have been updated and, some, significantly expanded vs. the first edition. Two chapters, however, due to unavailabilty of the authors remain essentially the versions that appeared in the 2005 edition of this book, with only minor modifications in the References section. Information on the rapidly growing field of nanosized fillers can now be found, not only in Chapters 9 and 10, but also in many other chapters of the book (see, for example, Chapters 1, 2, 3, 5, 12, 17, 18, 20, 22, 24).

I am indebted to users and reviewers of the first edition for their comments and suggestions. We implemented, to the extent possible, the suggestion to include titles of articles cited in the end-of chapter references. Although this has obviously resulted in an overall increase of the number of book pages, it is my belief that it will be beneficial to readers who need more detailed information on any given topic.

Functional Fillers for Plastics: Second, updated and enlarged edition. Edited by Marino Xanthos

ISBN: 978-3-527-32361-6

The topic of functional fillers for plastics is a multidisciplinary field, covering important materials of current and future everyday experience. I trust that this volume will be of continuing use to practicing engineers and scientists in the polymer field, a companion to university-level graduate textbooks in polymer engineering and science, and a basis for short courses to industry on types, properties and application of fillers and polymer composites. Finally, I am grateful to my industrial and academic colleagues for responding enthusiastically to my invitation to contribute in a revised version of this volume.

Newark, NJ, USA,
November 2009

Marino Xanthos

Preface to the 1st Edition

It is generally accepted that growth in plastics consumption and the development of new and specialized applications are related to advances in the field of multicomponent, multiphase polymer systems. These include composites, blends and alloys and foams. Fillers are essential components of multiphase composite structures; they usually form the minor dispersed phase in a polymeric matrix.

Increased interest in the use of discontinuous fillers as a means to reduce the price of molding compounds begun about 30 years ago when increasing oil prices made necessary the replacement of expensive polymers with less costly additives. When such additives had also a beneficial effect on certain mechanical properties (mostly modulus and strength) they were also known as reinforcing fillers. Since that time, there has been a considerable effort to extend the uses (and functions) of existing fillers by: a) particle size and shape optimization, b) developing value- added materials through surface treatments and c) developing efficient methods for their incorporation in plastics.

The term "filler" is very broad and encompasses a very wide range of materials. We arbitrarily define in this book as fillers a variety of natural or synthetic solid particulates (inorganic, organic) that may be irregular, acicular, fibrous or flakey and are used in most cases in reasonably large volume loadings in plastics, mostly thermoplastics. Continuous fibers or ribbons are not included. Elastomers are also not included in this definition as well as many specialty additives that are used at low concentrations (e.g. pigments, lubricants, catalysts, etc).

Among the best known handbooks on fillers that appeared in the English language in the past 25 years are the detailed works edited by Katz and Milewski (1978, 1987) and Zweifel (2001) and the monograph compiled by Wypych (2000). This present volume is not intended to be a handbook listing individual fillers according to their generic chemical structure or name but rather a comprehensive and up-to-date presentation, in a unified fashion, of structure/property/processing relationships in thermoplastic composites containing discontinuous fillers that would help the identification of new markets and applications. For convenience, fillers are grouped according to their primary functions that include modification of: a) mechanical properties, b) flame retardancy, c) electrical and magnetic properties, d) surface properties and e) processability. For each filler there is always a series of additional functions. Examples include degradability enhancement, bioactivity,

Functional Fillers for Plastics: Second, updated and enlarged edition. Edited by Marino Xanthos

ISBN: 978-3-527-32361-6

radiation absorption, damping enhancement, enhancement of dimensional stability, reduced permeability and reduced density.

Functional Fillers has been the focus of International Conferences such as those organized annually by Intertech Corp. in N. America and Europe over the past 10 years and attended by the editor and several contributing authors of this book and the biannual "Eurofillers" Conference. Judging from the interest generated from these conferences it became clear that there is a need for a volume that would capture the current technologies applicable to "commodity" fillers and compare them with new technologies and emerging applications that would reflect the multifunctional character of new or modified existing fillers. Examples of advances in the latter category include nanoplatelets of high aspect ratio produced by exfoliation of organoclays, nanoscale metal oxides, carbon nanotubes, ultrafine talc, TiO_2 and hydroxyapatite particles, ceramers and ormosils, new rheology modifiers and adhesion promoters, and increased usage of natural fibers. This volume is expected to address the needs of engineers, scientists and technologists involved in the industrially important sector of polymers additives and composites.

The book is divided into three main parts:

Part I, entitled *Polymers and Fillers,* contains a general introduction to polymer composites, a review of the parameters affecting mechanical and rheological properties of polymers containing functional fillers and an overview of mixing and compounding equipment along with methods of filler incorporation in molten and liquid polymers.

Part II focuses on the use of *Surface Modifiers and Coupling Agents* to enhance the performance of functional fillers and includes sections on silanes, titanates, functionalized polymers and miscellaneous low molecular weight reactive additives.

Part III on *Fillers and their functions* describes in a systematic manner the most important inorganic and organic functional fillers with examples of existing and emerging applications in plastics. Fillers have been grouped into seven families, each family representing the primary function of the filler. The families and the corresponding fillers covered in this book are:

- *High Aspect Ratio Mechanical Property Modifiers* with detailed description of glass fibers, mica flakes, nanoclays, carbon nanotubes/nanofibers and carbon fibers, and natural fibers, Chapters 7–11.
- *Low Aspect Ratio Mechanical Property Modifiers* with detailed description of talc, kaolin, wollastonite, wood flour, and calcium carbonate, Chapters 12–16.
- *Fire Retardants* with emphasis on metal hydroxides but also inclusion of antimony oxide, ammonium polyphosphate, borate salts and low melting temperature glasses, Chapter 17.
- *Electrical and Magnetic Property Modifiers* with emphasis on carbon black but also inclusion of metal particles and various magnetic fillers, Chapter 18.
- *Surface Property Modifiers* with further division into: a) solid lubricants/tribological additives that include molybdenite, graphite, PTFE and boron nitride and b) antiblocking fillers such as silica, Chapter 19.
- *Processing Aids* including rheological modifiers such as MgO and fumed silica, and process stabilizers such as hydrotalcites, Chapter 20.

Under *Specialty Fillers* (Chapters 21–23) a variety of multifunctional inorganics are discussed. They include: a) *glass and ceramic spheres* with primary functionalities as rheology modifiers and enhancers of dimensional stability (solid spheres) or weight reduction (hollow spheres), b) a variety of phosphate, carbonate and silicate calcium salts and specialty glasses that show *bioactivity* in tissue engineering applications and, c) *in-situ* generated fillers such as *organic-inorganic hybrids* with important functions as mechanical property or surface modifiers, depending on the system.

For commercially available fillers, contributing authors to chapters of Part III have been asked to broadly adhere to a uniform pattern of information that would include: a) production methods of the respective filler, b) its structure and properties, c) a list of major suppliers, information on availability and prices, d) a discussion of environmental/toxicity issues including applicable exposure limits proposed by regulatory agencies, and e) a concluding section on applications that considers both primary and secondary functions of each filler and presents specific data on properties and information on current and emerging markets.

Many authors used government and company websites as sources for updated Material Safety Data Sheets (MSDS) and information on the threshold limit values (TLV) for the airborne concentration of filler dusts in the workplace. Reliable information on possible risks to human health or the environment is extremely important to current and potential users of existing fillers, or new fillers of different origin and different particle size/shape characteristics. It should be recognized that health issues have been responsible in the past for the withdrawal from certain plastics markets of natural and synthetic fibrous fillers with unique properties such a as chrysotile asbestos, microfibers, whiskers and the recently mandated very low content of crystalline silica in mineral fillers.

In presenting the different topics of this book, efforts were made to produce selfcontained chapters in terms of cited references, abbreviations and symbols. Although this may have resulted in duplication of information, it should be useful to readers interested in only certain chapters of the book. All Tables, Figures and Equations are labeled in terms of the chapter where they first appear to facilitate cross-referencing.

The authors who contributed to this book all have significant credentials in the field of fillers and reinforcements for plastics and represent industry, academe, consulting and R&D organizations. Their different backgrounds and insight reflect on their chapter contents, style of presentation and the emphasis placed on the information presented. I would like to thank my colleagues from the Polymer Processing Institute (Drs. Davidson, Patel, Todd), from industry (Drs. Ashton, Dey, Mack and Weissenbach, and Messrs. Duca, Kamena and Monte), from academic institutions (Drs. Flaris, Iqbal, Mascia and Rothon), from US Government laboratories (Drs. Clemons and Caulfield), and our doctoral students at the New Jersey Institute of Technology (Ms. Chouzouri and Mr. Goyal) for their hard work and excellent cooperation and patience during the long editing process. Special thanks are due to my friends and coworkers for many years in the field of polymer composites, Dr. Leno Mascia of Loughborough University, UK and Dr. Ulku Yilmazer of Middle East Technical University, Turkey for reading Chapters 1 and 2 and offering many useful suggestions for their improvement. Finally, many thanks to Dr. Michael Jaffe of NJIT for his input on Chapter 22 and to a great number of colleagues and graduate students that I have been associated with over a 30-year industrial and academic career in polymer modification and multicomponent polymer systems. They were instrumental in helping me summarize the concepts and information presented in this book.

This book was largely completed during a sabbatical leave of the editor/contributing author from the New Jersey Institute of Technology in 2003–2004.

Fort Lee, NJ, USA, November 2004 *Marino Xanthos*

List of Contributors

Henry C. Ashton
Schneller LLC
6019 Powdermill Road
Kent, OH 44240
USA

Björn Borup
Evonik Degussa GmbH
Weissfrauenstr. 9
60287 Frankfurt a.M.
Germany

Georgia Chouzouri
319 Livingston Court
Livingston
NJ 07039
USA

Craig M. Clemons
USDA Forest Service
Forest Products Laboratory
One Gifford Pinchot Drive
Madison, WI 53705-2398
USA

Theodore Davidson
109 Poe Road
Princeton, NJ 08540-4121
USA

Chris DeArmitt
Phantom Plastics
1821 Rambling Rose Lane
Hattiesburg, MS 39402
USA

Subir K. Dey
Sonoco
Mail Code P21
1 North Second Street
Hartsville, SC 29501
USA

Joseph Duca
Engelhard Corp.
101 Wood Avenue
Iselin, NJ 08830
USA

Vicki Flaris
University of New York
Bronx Community College of the City
University Avenue and W. 181st Street
Bronx, NY 10453-3102
USA

Functional Fillers for Plastics: Second, updated and enlarged edition. Edited by Marino Xanthos

ISBN: 978-3-527-32361-6

Amit Goyal
New Jersey Institute of Technology
Department of Chemistry and
Environmental Science
Newark, NJ 07102
USA

and

Exelus Inc.
110 Dorsa Avenue
Livingston, NJ 07039
USA

Zafar Iqbal
New Jersey Institute of Technology
Department of Chemistry and
Environmental Science
Newark, NJ 07102
USA

Karl Kamena
Southern Clay Products Inc.
1212 Church Street
Gonzales, TX 78629
USA

Yash P. Khanna
New Jersey Institute of Technology
Otto H. York Department of Chemical,
Biological and Pharmaceutical
Engineering
Tiernan Hall
University Heights
Newark, NJ 07102
USA

Anna Kron
Expancel, Eka Chemicals AB
Akzo Nobel
Box 13000
SE-850 13 Sundsvall
Sweden

Leno Mascia
Loughborough University
Department of Materials
Loughborough LE11 3TU
UK

Salvatore J. Monte
Kenrich Petrochemicals, Inc.
140 East 22nd Street
P.O. Box 32
Bayonne, NJ 07002-0032
USA

Subhash H. Patel
New Jersey Institute of Technology
Polymer Processing Institute
University Heights
Newark, NJ 07102
USA

Sara M. Robinson
Sara Robinson Consultants
3162 Lonesome Mountain Road
Charlottesville
VA 22911
USA

Roger N. Rothon
Rothon Consultants/Manchester
Metropolitan University
3 Orchard Croft
Guilden Sutton
Chester CH3 7SL
UK

David B. Todd
New Jersey Institute of Technology
Polymer Processing Institute
GITC Building
Newark, NJ 07102
USA

Kerstin Weissenbach
Evonik Degussa GmbH
Untere Kanalstraße 3
79618 Rheinfelden
Germany

Marino Xanthos
New Jersey Institute of Technology
Otto H. York Department of Chemical,
Biological and Pharmaceutical
Engineering
Tiernan Hall
University Heights
Newark, NJ 07102
USA

List of Symbols

a, b	unit vectors of the hexagonal lattice
A	absorbance
A	constant
A	surface area
A, B	Nielsen's equation parameters
c	concentration of absorbing species
c	specific heat
C	chiral vector
C	cost of composite
C_i	cost of the individual component
d	fiber or platelet diameter
D	fractal dimension
E	tensile modulus
E	Young's modulus
E_i	modulus of inorganic phase
E_o	modulus of organic phase
G	shear modulus
h	thermal diffusivity
I	intensity of light
I_0	intensity of the incident light
k	thermal conductivity
K	modulus efficiency parameter
K	stress transfer efficiency factor
K'	orientation parameter
K_e	Einstein coefficient
l	path length of light
L	length
M	moisture content
n	degree of polymerization
n, m	integers
N	number of constituent gases

Functional Fillers for Plastics: Second, updated and enlarged edition. Edited by Marino Xanthos

ISBN: 978-3-527-32361-6

P_c	composite permeability
P_m	matrix permeability
t_f	film thickness
T_g	glass transition temperature
T_m	melting temperature
u	reinforcement parameter
v_f, v_m	volumes of the individual components
V	volume
V_0	dry volume
V_1	wood volume
V_c	critical volume fraction
V_f	volume fraction of filler
V_m	volume fraction of matrix
W_a	work of adhesion
α	absorption coefficient
α	aspect ratio
α	coefficient of thermal expansion
χ	packing parameter
ε	molar absorptivity
$\varepsilon_{ifracture}$	strain value
ϕ_{max}	maximum packing factor
Φ	interaction parameter
γ	surface tension
γ_c	critical surface tension
Γ	fracture surface energy
μ	viscosity
ν	Poisson's ratio
θ	contact angle
ϱ	density
σ	stress
$\sigma_{bnc(fracture)}$	strength of nanocomposite
τ	interface or matrix shear strength
ξ	Halpin–Tsai parameter
Ψ	Nielsen's equation parameters

Part One
Polymers and Fillers

Functional Fillers for Plastics: Second, updated and enlarged edition. Edited by Marino Xanthos

ISBN: 978-3-527-32361-6

1
Polymers and Polymer Composites

Marino Xanthos

1.1
Thermoplastics and Thermosets

Almost 85% of the polymers produced worldwide are thermoplastics [1]. They can be divided into two broad classes, amorphous and crystalline, depending on the type of their characteristic transition temperature. Amorphous thermoplastics are characterized by their glass transition temperature T_g, a temperature above which the modulus rapidly decreases and the polymer exhibits liquid-like properties; amorphous thermoplastics are normally processed at temperatures well above their T_g. Glass transition temperatures may be as low as 65 °C for polyvinyl chloride (PVC) and as high as 295 °C for polyamideimide (PAI) [1]. Crystalline thermoplastics or, more correctly, semicrystalline thermoplastics can have different degrees of crystallinity ranging from 20 to 90%; they are normally processed above the melting temperature T_m of the crystalline phase and the T_g of the coexisting amorphous phase. Melting temperatures can be as high as 365 °C for polyetherketone (PEK), as low as 110 °C for low-density polyethylene (LDPE), and even lower for ethylene–vinyl acetate (EVA) copolymers [1]. Upon cooling, crystallization must occur quickly, preferably within a few seconds. Additional crystallization often takes place after cooling and during the first few hours following melt processing.

Over 70% of the total production of thermoplastics is accounted for by the large-volume, low-cost commodity resins: polyethylenes (PE) of different densities, isotactic polypropylene (PP), polystyrene (PS), and PVC. Next in performance and in cost are acrylics, acrylonitrile–butadiene–styrene (ABS) terpolymers, and high-impact polystyrene (HIPS). Engineering plastics such as acetals, polyamides, polycarbonate, polyesters, polyphenylene oxide, and blends thereof are increasingly used in high-performance applications. Specialty polymers such as liquid-crystal polymers, polysulfones, polyimides, polyphenylene sulfide, polyetherketones, and fluoropolymers are well established in advanced technology areas because of their high T_g or T_m (290–350 °C).

Common thermosetting resins are unsaturated polyesters, phenolic resins, amino resins, urea/formaldehyde resins, polyurethanes, epoxy resins, and silicones. Less

Functional Fillers for Plastics: Second, updated and enlarged edition. Edited by Marino Xanthos

ISBN: 978-3-527-32361-6

common thermosets employed in specialized applications are polybismaleimides, polyimides, and polybenzimidazoles. Thermosetting resins are usually low-viscosity liquids or low molecular weight solids that are formulated with suitable additives known as cross-linking agents to induce curing and with fillers or fibrous reinforcements to enhance both properties and thermal and dimensional stability. It has been frequently stated that in view of their excessive brittleness many thermosets would have been nearly useless had they not been combined with fillers and reinforcing fibers.

1.2
Processing of Thermoplastics and Thermosets

The operation by which solid or liquid polymers are converted to finished products is generally known as polymer processing. Polymer processing consists of several steps [2]:

1) *Preshaping* operations involving all or some of the following individual operations:
   - handling of particulate solids (particle packing, agglomeration, gravitational flow, compaction, and others);
   - melting or heat softening;
   - pressurization and pumping of the polymer melt;
   - mixing for melt homogenization or dispersion of additives;
   - devolatilization and stripping of residual monomers, solvents, contaminants, and moisture.

   The common goal of the above operations is to deliver thermoplastics or cross-linkable thermosets in a deformable fluid state that will allow them to be shaped by a die or mold; thereafter, they can be solidified by cooling below T_g or T_m (thermoplastics) or by a chemical reaction (thermosets).
2) *Shaping* operations during which "structuring" occurs (morphology development and molecular orientation that modify and improve physical and mechanical properties). Principal shaping methods include die forming, molding, casting, calendering, and coating.
3) *Postshaping* operations, such as decorating, fastening, bonding, sealing, welding, dyeing, printing, and metallizing.

Following the explosive development of thermoplastics after World War II, many improvements and new developments have led to today's diversity of polymer processing machines and technologies. Some processes are unique to thermoplastics; some are applicable only to thermosets and cross-linkable thermoplastics, whereas others, after certain modifications, can be applied to both thermoplastics and thermosets. Table 1.1, adapted from Ref. 3, summarizes the principal processing/shaping methods. For thermoplastics, extrusion is the most popular with approximately 50% of all the commodity thermoplastics being used in extrusion process equipment to produce profiles, pipe and tubing, films, sheets, wires, and cables. Injection molding follows as the next most popular processing method, accounting for

Table 1.1 Principal processing methods for thermoplastics and thermosets.

Thermoplastics	Thermosets/cross-linkable thermoplastics
Extrusion	Compression molding, transfer molding, casting
Pipe, tubing, sheet, cast film, profile	Injection molding, resin injection molding
Blown film	Polyurethane foam molding
Coextrusion, extrusion coating	Open mold reinforced plastics
Wire and cable coating	Lay-up
Foam extrusion	Spray-up
Extrusion blow molding	Filament winding
Injection molding, injection blow molding, resin injection molding (RIM)	Closed mold reinforced plastics
	Pultrusion
Foam molding	Resin transfer molding (RTM)
Structural	
Expandable bead	
Thermoforming	
Vacuum	
Pressure forming	
Rotational molding, calendering	

about 15% of all the commodity thermoplastics processed. Other common methods include blow molding, rotomolding, thermoforming, and calendering.

The range of processes that may be used for fabricating a plastic product is determined by the scale of production, the cost of the machine and the mold, and the capabilities and limitations of the individual processes. For example, complex and precise shapes can be achieved by injection molding, hollow objects by blow molding or rotational molding, and continuous lengths by extrusion. Processing methods for thermosets, particularly those related to reinforced thermosets involving liquid polymers, are often quite different from those employed for thermoplastics.

Increased polymer consumption over the past 20 years has not only stimulated machinery sales but also led to a parallel growth in the usage of a large variety of liquid and solid modifiers including fillers and reinforcements [4]. Significant advances have been made to accommodate such additives by improving the efficiency of polymer mixing/compounding equipment. Thermoplastic resin compounders combine the polymer(s) with the modifiers in high-intensity batch mixers and continuous extruders (mostly twin-screw extruders), and the material is then pumped into a pelletizer to produce the feed for subsequent shaping operations (see Chapter 3). Thermosetting resin suppliers compound heat-sensitive resins with fillers, additives, and/or pigments in a variety of mixers to produce molding compounds in such forms as powder, granules, and pastes to be fed into the molding equipment.

1.3
Polymer Composites

Modification of organic polymers through the incorporation of additives yields, with few exceptions, multiphase systems containing the additive embedded in a continuous polymeric matrix. The resulting mixtures are characterized by unique microstructures or macrostructures that are responsible for their properties. The primary reasons for using additives are

- Property modification or enhancement.
- Overall cost reduction.
- Improving and controlling of processing characteristics.

In addition to polymer composites that are introduced in this chapter, important types of modified polymer systems include polymer–polymer blends and polymeric foams.

1.3.1 Types and Components of Polymer Composites

Polymer composites are mixtures of polymers with inorganic or organic additives having certain geometries (fibers, flakes, spheres, and particulates). Thus, they consist of two or more components and two or more phases. The additives may be continuous, for example, long fibers or ribbons; these are embedded in the polymer in regular geometric arrangements that extend throughout the dimensions of the product. Familiar examples are the well-known fiber-based thermoset laminates that are usually classified as high-performance polymer composites, or as *macrocomposites* based on the length of the fibers or ribbons. On the other hand, additives may be discontinuous (short), for example, short fibers (say, $<3\,\text{cm}$ in length), flakes, platelets, spheres, or irregulars (millimeter to micrometer size); fibers and flakes are usually dispersed in different orientations and multiple geometric patterns throughout the continuous matrix forming *microcomposites*. Such systems are usually based on a thermoplastic matrix and are classified as lower performance polymer composites compared to their counterparts with continuous additives. When the fibers, platelets, or spheres as the dispersed phase are of nanoscale dimensions (see, for example, hydrotalcite nanoplatelets in Figure 1.1), the materials are known as *nanocomposites*. They differ from microcomposites in that they contain a significant number of interfaces available for interactions between the intermixed phases [5]. As a result of their unique properties, nanocomposites have a great potential for advanced applications. Microcomposites and, to a lesser extent, nanocomposites form the topic of this book.

Composites may also be classified based on the origin (*natural* versus *synthetic*) of the matrix or filler. Nature uses composites for all her hard materials. These are complex structures consisting of continuous or discontinuous fibrous or particulate material embedded in an organic matrix acting as glue. Wood is a composite of fibrous cellulose and lignin. Bone is a composite of collagen and other proteins and calcium phosphate salts. Spider silk consists of organic nanocrystals in an organic amorphous matrix. The shells of mollusks (Figure 1.2) are made of layers of hard mineral separated by a protein binder [6]. A similar platy structure providing a tortuous path for vapors and liquids can be obtained in a microcomposite containing mica flakes embedded in a synthetic thermoset polymeric matrix (Figure 1.3).

Figure 1.1 Scanning electron micrograph of hydrotalcite, a synthetic anionic nanoclay. (Courtesy of Dr. T.G. Gopakumar, Polymer Processing Institute.)

Composites can also be classified on the basis of the intended application. For example, one can distinguish between two types of *biocomposites*. Biocomposites for ecological applications are combinations of natural fibers or particulates with polymer matrices from both nonrenewable and renewable resources and are characterized by environmental degradability. Biocomposites for biomedical applications are combinations of biostable or degradable polymers with inert or bioactive

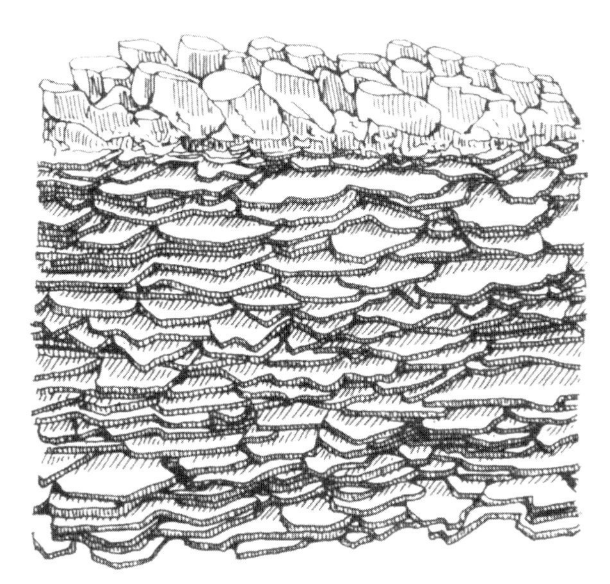

Figure 1.2 Natural microcomposite composite: the shell of a mollusk made of layers of calcium salts separated by protein. (Reprinted with permission from Ref. [6].)

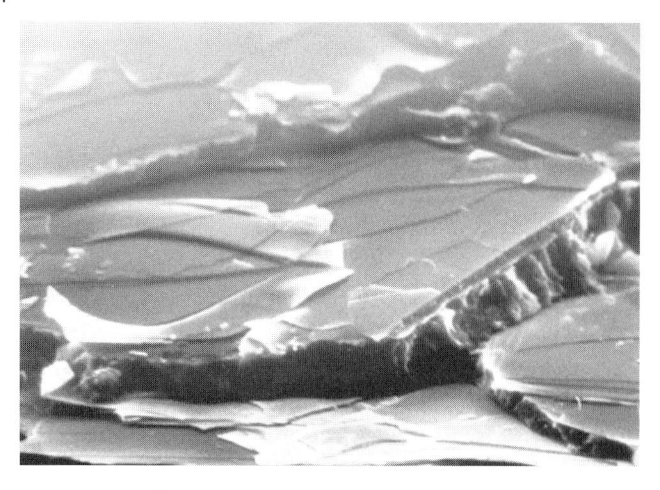

Figure 1.3 Synthetic microcomposite: scanning electron micrograph photo of cross section of fractured mica thermoset composite showing mica flakes with a thickness of about 2.5 μm separated by a much thicker polymer layer.

fillers intended for use in orthopedics, bone regeneration, or tissue engineering applications.

Additives for polymer composites have been variously classified as reinforcements, fillers, or reinforcing fillers. Reinforcements, being much stiffer and stronger than the polymer, usually increase its modulus and strength. Thus, mechanical property modification may be considered as their primary function, although their presence may significantly affect thermal expansion, transparency, thermal stability, and so on. *Continuous* composites contain long fiber or ribbon reinforcements mostly in thermosetting matrices; when prearranged in certain geometric patterns, they may become the major component of the composite (they can constitute as much as 70% by volume in oriented composites). For *discontinuous* composites, the directional reinforcing agents (short fibers or flakes) are arranged in the composite in different orientations and multiple geometric patterns, which are dictated by the selected processing and shaping methods, most often extrusion or injection molding. In this case, the content of the additive does not usually exceed 30–40% by volume. It should be noted, however, that manufacturing methods for continuous oriented fiber thermoplastic composites are available resulting in much higher fiber contents, as used in high-performance engineering polymers [7]. In this book, the term reinforcement will be mostly used for long, continuous fibers or ribbons, whereas the term filler, performance filler, or functional filler will mostly refer to short, discontinuous fibers, flakes, platelets, or particulates.

1.3.2 Parameters Affecting Properties of Composites

In general, parameters affecting the properties of polymer composites, whether continuous or discontinuous, include

- the properties of the additives (inherent properties, size, and shape);
- the composition;
- the interaction of components at the phase boundaries, which is also associated with the presence of a thick interface, also known as the interphase; this is often considered as a separate phase, controlling adhesion between the components;
- the method of fabrication.

With regard to methods of fabrication, all processes in Table 1.1 that are applicable to unfilled, unmodified thermoplastics can also be used for discontinuous systems (with the exception of expandable bead molding). In addition to thermoforming, hot stamping of reinforced thermoplastic sheets mostly containing randomly oriented continuous or discontinuous fibers is used for the production of large semistructural parts. Fillers can also be used in the thermoset processes in Table 1.1, often in combination with the primary continuous fiber reinforcement. The content and inherent properties of the additive, as well as its physical/chemical interactions with the matrix, are important parameters controlling the processability of the composite.

1.3.3
Effects of Fillers/Reinforcements: Functions

Traditionally, most fillers were considered as additives, which, because of their unfavorable geometrical features, surface area, or surface chemical composition, could only moderately increase the modulus of the polymer, whereas strength (tensile, flexural) remained unchanged or even decreased. Their major contribution was in lowering the cost of materials by replacing the most expensive polymer; other possible economic advantages were faster molding cycles as a result of increased thermal conductivity and fewer rejected parts due to warpage. Depending on the type of filler, other polymer properties could be affected; for example, melt viscosity could be significantly increased through the incorporation of fibrous materials. On the other hand, mold shrinkage and thermal expansion would be reduced, a common effect of most inorganic fillers.

The term reinforcing filler has been coined to describe discontinuous additives, the form, shape, and/or surface chemistry of which have been suitably modified with the objective of improving the mechanical properties of the polymer, particularly strength. Inorganic reinforcing fillers are stiffer than the matrix and deform less, causing an overall reduction in the matrix strain, especially in the vicinity of the particle as a result of the particle–matrix interface. As shown in Figure 1.4, the fiber "pinches" the polymer in its vicinity, reducing strain and increasing stiffness [8]. Reinforcing fillers are characterized by relatively high aspect ratio α, defined as the ratio of length over diameter for a fiber or the ratio of diameter over thickness for platelets and flakes. For spheres, which have minimal reinforcing capacity, the aspect ratio is unity. A useful parameter for characterizing the effectiveness of a filler is the ratio of its surface area A to its volume V, which needs to be as high as possible for effective reinforcement. Figure 1.5 (from Ref. 8) shows that maximizing A/V and

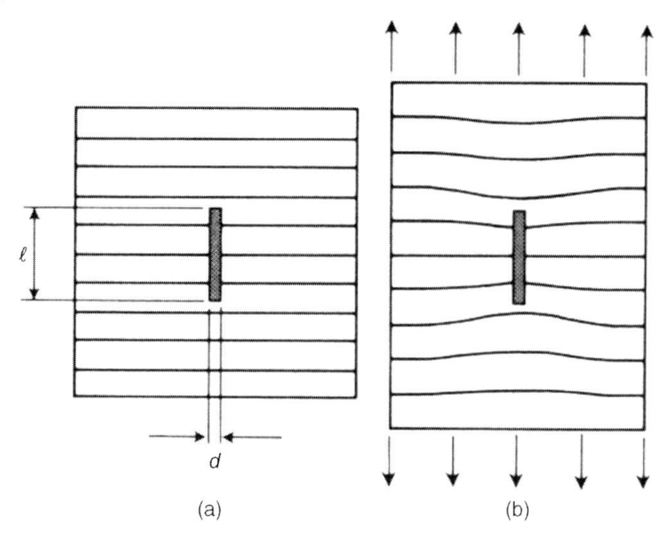

Figure 1.4 A cylindrical reinforcing fiber in a polymer matrix: (a) in the undeformed state; (b) under a tensile load. (Reprinted with permission from Ref. [8].)

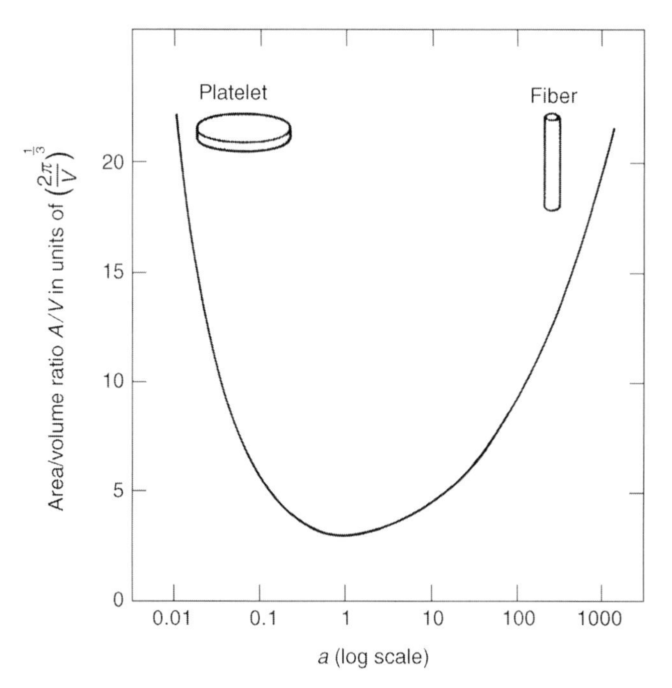

Figure 1.5 Surface area-to-volume ratio A/V of a cylindrical particle plotted versus aspect ratio $\alpha = l/d$. (Reprinted with permission from Ref. [8].)

particle–matrix interaction through the interface requires $\alpha \gg 1$ for fibers and $1/\alpha \ll 1$ for platelets.

In developing reinforcing fillers, the aims of process or material modifications are to increase the aspect ratio of the particles and to improve their compatibility and interfacial adhesion with the chemically dissimilar polymer matrix. Such modifications may not only enhance and optimize the primary function of the filler (in this case, its use as a mechanical property modifier) but also introduce or enhance additional functions. New functions attained by substitution or modification of existing fillers, thus broadening their range of applications, are illustrated in the examples below.

As described by Heinold [9], the first generation of fillers soon after the commercialization of polypropylene included talc platelets and asbestos fibers for their beneficial effects on stiffness and heat resistance. The search for an asbestos replacement due to health issues led to calcium carbonate particles and mica flakes as the second-generation fillers. Mica was found to be more effective than talc for increasing stiffness and heat resistance, whereas calcium carbonate was less effective in increasing stiffness but increased the impact resistance of PP homopolymers. Surface modification of mica with coupling agents to enhance adhesion and stearate modification of the calcium carbonate to assist dispersion were found to enhance these functions and introduce additional benefits such as improved processability, a means of controlling color, and reduced long-term heat aging. Other fillers imparted entirely different functions. For example, barium sulfate enhances sound absorption, wollastonite enhances scratch resistance, solid glass spheres add dimensional stability and increase hardness, hollow glass spheres lower density, and combinations of glass fibers with particulate fillers provide unique properties that cannot be attained with single fillers.

An additional example of families of fillers imparting distinct new properties is given by the pearlescent pigments produced by platelet core–shell technologies [10]. These comprise platelets of mica, silica, alumina, or glass substrates coated with films of oxide nanoparticles, for example, TiO_2, Fe_2O_3, Fe_3O_4, and Cr_2O_3 (Figure 1.6). In addition to conventional decorative applications, new functional applications such as solar heat reflection, laser marking of plastics, and electrical conductivity are possible through the proper selection of substrate/coating combinations.

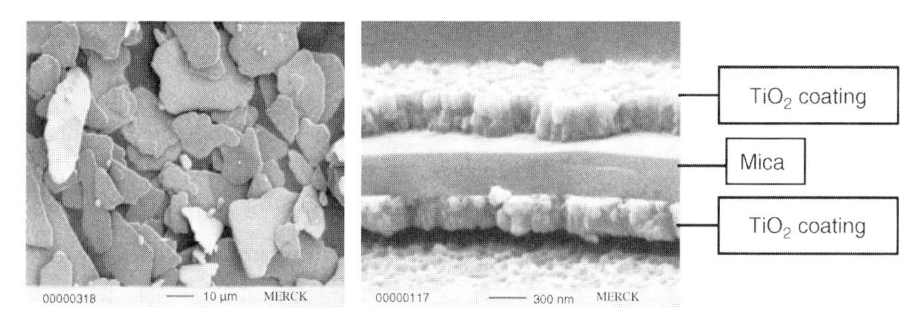

Figure 1.6 Scanning electron micrograph of mica flakes and a cross section of an anatase/mica pigment particle [10].

1.3.4
Rules of Mixtures for Composites

Rule of mixtures equations (often modified according to the type, shape, and orientation of the reinforcement/filler) are commonly used to describe certain properties of the composites. For example

1) Concentrations are usually expressed by volume, as volume fractions of filler V_f and matrix V_m, obtained from the volumes ν_f and ν_m of the individual components:

$$V_f = \nu_f/(\nu_f + \nu_m), \tag{1.1}$$

$$V_m = \nu_m/(\nu_f + \nu_m), \tag{1.2}$$

$$V_f + V_m = 1. \tag{1.3}$$

2) Volume fractions are also used to predict a theoretical density of the composite ϱ, based on the respective densities of the components and assuming a total absence of voids:

$$\varrho = V_f\, \varrho_f + (1-V_f)\varrho_m. \tag{1.4}$$

3) The total cost per unit weight of the composite C can also be calculated from the volume fractions and the costs of the individual components and the cost of compounding per unit weight of the composite C_i [7]:

$$C = V_f\, \varrho_f/\varrho\, C_f + (1-V_f)\, \varrho_m/\varrho\, C_m + C_i. \tag{1.5}$$

After introducing incorporation costs, the cost of the composite may be higher or lower than that of the unfilled polymer. For low-cost commodity plastics, the term filler (implying cost reduction) may be a misnomer since manufacturing costs may offset the lower cost of most mineral fillers. For higher cost specialty high-temperature thermoplastics, the final cost of, for example, glass fiber-reinforced polyetherimide is usually less than that of the unmodified polymer.

Rule of mixtures equations are also used to describe mechanical, thermal, and other properties as shown in Chapter 2.

1.3.5
Functional Fillers

1.3.5.1 Classification and Types

The term filler is very broad and encompasses a very wide range of materials. In this book, we arbitrarily define as fillers a variety of solid particulate materials (inorganic, organic) that may be irregular, acicular, fibrous, or plate-like in shape and that are used in reasonably large volume loadings in plastics. Pigments and elastomeric matrices are normally not included in this definition.

There is a significant diversity in chemical structures, forms, shapes, sizes, and inherent properties of the various inorganic and organic compounds that are used as

Table 1.2 Chemical families of fillers for plastics.

Chemical family	Examples
Inorganics	
Oxides	Glass (fibers, spheres, hollow spheres, and flakes), MgO, SiO_2, Sb_2O_3, Al_2O_3, and ZnO
Hydroxides	$Al(OH)_3$ and $Mg(OH)_2$
Salts	$CaCO_3$, $BaSO_4$, $CaSO_4$, phosphates, and hydrotalcite
Silicates	Talc, mica, kaolin, wollastonite, montmorillonite, feldspar, and asbestos
Metals	Boron and steel
Organics	
Carbon, graphite	Carbon fibers, graphite fibers and flakes, carbon nanotubes, and carbon black
Natural polymers	Cellulose fibers, wood flour and fibers, flax, cotton, sisal, and starch
Synthetic polymers	Polyamide, polyester, aramid, and polyvinyl alcohol fibers

fillers. They are usually rigid materials, immiscible with the matrix in both molten and solid states, and, as such, form distinct dispersed morphologies. Their common characteristic is that they are used at relatively high concentrations (>5% by volume), although some surface modifiers and processing aids are used at lower concentrations. Fillers may be classified as inorganic or organic substances and further subdivided according to chemical family (Table 1.2) or according to their shape and size or aspect ratio (Table 1.3). Wypych [11] reported more than 70 types of particulates or flakes and more than 15 types of fibers of natural or synthetic origin that have been used or evaluated as fillers in thermoplastics and thermosets. The most commonly used particulate fillers are industrial minerals, such as talc, calcium carbonate, mica, kaolin, wollastonite, feldspar, and aluminum hydroxide. The most commonly used fibrous fillers are glass fibers and, recently, a variety of natural fibers. Carbon black has long been considered a nanofiller. More recent additions, rapidly moving to commercial markets, are nanoclays such as montmorillonite and hydrotalcite, a variety of oxides, and nanofibers such as single- or multiple-wall carbon nanotubes. Graphene sheets and halloysite nanotubes are potential additives in

Table 1.3 Particle morphology of fillers.

Shape	Aspect ratio	Example
Cube	1	Feldspar and calcite
Sphere	1	Glass spheres
Block	1–4	Quartz, calcite, silica, and barite
Plate	4–30	Kaolin, talc, and hydrous alumina
Flake	50–200 + +	Mica, graphite, and montmorillonite nanoclays
Fiber	20–200 + +	Wollastonite, glass fibers, carbon nanotubes, wood fibers, asbestos fibers, and carbon fibers

advanced nanocomposites; the first are single layers of carbon atoms tightly packed in a honeycomb structure [12], whereas the second are naturally occurring nanotubes produced by surface weathering of aluminosilicate minerals [13].

A more convenient scheme, first proposed by Mascia [14] for plastic additives, is to classify fillers according to their specific function, such as their ability to modify mechanical, electrical, or thermal properties, flame retardancy, processing characteristics, solvent permeability, or simply formulation costs. Fillers, however, are multifunctional and may be characterized by a primary function and a plethora of additional functions (see Table 1.4). The scheme adopted in this book involves classification of fillers according to five primary functions, as follows:

- mechanical property modifiers (and further subdivision according to aspect ratio);
- fire retardants;
- electrical and magnetic property modifiers;
- surface property modifiers;
- processing aids.

Additional functions may include degradability enhancement, barrier characteristics, antiaging characteristics, bioactivity, radiation absorption, warpage minimization, and so on. Such attributes for certain fillers will be identified in subsequent chapters of the book.

1.3.5.2 Applications, Trends, and Challenges

Global demand for fillers/reinforcing fillers including calcium carbonate, aluminum trihydrate, talc, kaolin, mica, wollastonite, glass fiber, aramid fiber, carbon fiber, and carbon black for the plastics industry has been estimated to be about 15 million tons [15]. Primary end-use markets are building/construction and transportation, followed by appliances and consumer products; furniture, industrial/machinery, electrical/electronics, and packaging comprise smaller market segments. Flexural modulus and heat resistance are the two critical properties of plastics that are enhanced by the inclusion of performance minerals. Automotive exterior parts, construction materials, outdoor furniture, and appliance components are examples of applications benefiting from enhanced flexural modulus. Automotive interior and underhood parts, electrical connectors, and microwaveable containers are examples of applications requiring high-temperature resistance. Environmental acceptance and improved sustainability of automotive parts are attributes of composites containing natural fibers. Life cycle assessment studies [16] tend to support the position that natural fiber composites are environmentally superior to glass fiber composites.

Recent statistics (2007) estimate the U.S. demand for fillers and extender minerals to a total of 3.2 million tons per annum [17]. Annual growth rates are estimated to be 2–3% with much higher rates for fire retardant fillers such as aluminum hydroxide (5.5–7%). Data (not including glass products and natural fibers but including TiO_2 and organoclays) indicate the highest demand for ground calcium carbonate followed by TiO_2 and aluminum trihydrate. Talc, kaolin, mica, wollastonite, silica, barites, and organoclays have a much smaller share of the market.

Table 1.4 Fillers and their functions.

Primary function	Examples of fillers	Additional functions	Examples of fillers
Modification of mechanical properties	High aspect ratio: glass fibers, mica, nanoclays, carbon nanotubes, carbon/graphite fibers, and aramid/synthetic/natural fibers Low aspect ratio: talc, $CaCO_3$, kaolin, wood flour, wollastonite, and glass spheres	Control of permeability	Reduced permeability: impermeable plate-like fillers: mica, talc, nanoclays, glass flakes Enhanced permeability: stress concentrators for inducing porosity: $CaCO_3$ and dispersed polymers
Enhancement of fire retardancy	Hydrated fillers: $Al(OH)_3$ and $Mg(OH)_2$	Bioactivity	Bone regeneration: hydroxyapatite, tricalcium phosphate, and silicate glasses
Modification of electrical and magnetic properties	Conductive, nonconductive, and ferromagnetic: metals, carbon fiber, carbon black, and mica	Degradability	Organic fillers: starch and cellulosic fibers
Modification of surface properties	Antiblock, lubricating: silica, $CaCO_3$, PTFE, MoS_2, and graphite	Radiation absorption	Metal particles, lead oxide, and leaded glass
Enhancement of processability	Thixotropic, antisag, thickeners, and acid scavengers: colloidal silica, bentonite, and hydrotalcite	Improved dimensional stability	Isotropic shrinkage and reduced warpage: particulate fillers, glass beads, and mica
		Modification of optical properties	Nucleators, clarifiers, and iridescent pigments: fine particulates and mica/pigment hybrids
		Control of damping	Flake fillers, glass, and $BaSO_4$

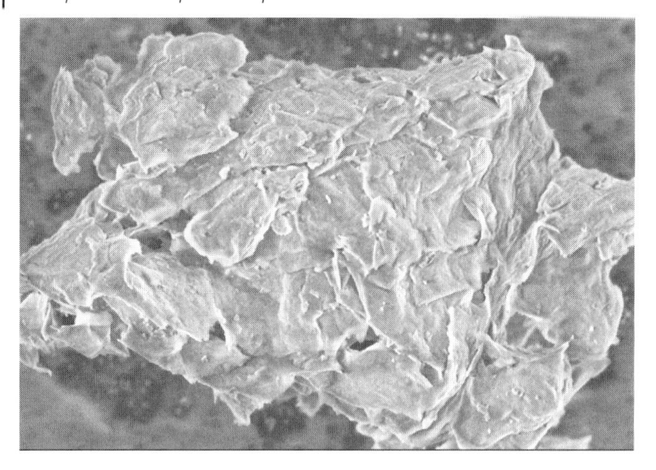

Figure 1.7 Scanning electron micrograph of montmorillonite agglomerate (a natural cationic clay) prior to its dispersion into high aspect ratio nanoplatelets. 7000×. (Courtesy of Dr. S. Kim, Polymer Processing Institute.)

There are a significant number of technological advances that will undoubtedly contribute to the additional growth in the usage of certain functional fillers: for example, for wood-filled plastics, introduction of specially configured counter-rotating twin-screw extruders with vent zones to remove moisture [18]; for mica and talc, development of new grinding technologies to retain the lamellarity and aspect ratio of the plate-like fillers [19] and for other minerals to produce ultrafine particles by special grinding methods [20]; for deagglomeration, dispersion, and exfoliation of agglomerated nanoclays (see, for example, Figure 1.7), particularly in high-temperature thermoplastic matrices, clay modification with additives such as ionic liquids [21] having higher thermal stability than the existing alkylammonium modifiers; in addition, melt compounding in extruders with improved screw configuration, and optional ultrasonic assistance or the use of supercritical fluids; for carbon nanotube composites, appropriate interfacial modification to improve dispersion and adhesion and minimize deagglomeration; also, equipment/process modification to ensure the desired orientation and maintain the high aspect ratio.

Some new exciting application areas for composites containing cationic or anionic nanoclays, nanooxides, carbon nanotubes, ultrafine TiO_2, talc, and synthetic hydroxyapatite are

1) structural materials with improved mechanical, thermal, and barrier properties, electrical conductivity, and flame retardancy;
2) high-performance materials with improved UV absorption and scratch resistance;
3) barrier packaging for reduced oxygen degradation;
4) multifunctional fillers that could release in a controlled manner corrosion sensing additives, corrosion inhibitors, insecticides, active pharmaceutical ingredients, and so on;
5) bioactive materials for tissue engineering applications.

Concerns have been raised regarding the safety of certain nanomaterials in a variety of products, since their inhalation toxicology has not been fully evaluated yet and few data exist on dermal or oral exposures [22, 23]. The environmental, health, and safety (EHS) issues may be related to real risks, perceptual risks, and/or the lack of clear regulations. Current practices by some material suppliers are to supply precompounded masterbatches of nanoclays or nanofibers, slurries of nanoparticles, or high bulk density dispersible powders. Note that there is a significant amount of ongoing work in government, industrial, and academic laboratories that seeks to identify and address potential EHS risks of nanomaterials. Recent advances are described in Refs [24, 25].

References

1 Xanthos, M. and Todd, D.B. (1996) Plastics processing, in *Kirk-Othmer Encyclopedia of Chemical Technology*, vol. 19, 4th edn, John Wiley & Sons, Inc., New York, pp. 290–316.

2 Tadmor, Z. and Gogos, C.G. (2006) *Principles of Polymer Processing*, 2nd edn, John Wiley & Sons, Inc., Hoboken, NJ, pp. 14–17.

3 Xanthos, M. (2000) Chapter 19: Polymer processing, in *Applied Polymer Chemistry: 21st Century* (eds C.E. Carraher and C.D. Craver), Elsevier, Oxford, UK, pp. 355–371.

4 Xanthos, M. (1994) Chapter 14: The physical and chemical nature of plastics additives, in *Mixing and Compounding of Polymers: Theory and Practice* (eds I. Manas-Zloczower and Z. Tadmor), Carl Hanser Verlag, Munich, pp. 471–492.

5 Ajayan, P.M., Schadler, L.S., and Braun, P.V. (eds) (2003) *Nanocomposite Science and Technology*, Wiley-VCH Verlag GmbH, Weinheim, pp. 77–144.

6 Vogel, S. (1998) *Cats' Paws and Catapults*, W.W. Norton & Co., New York, pp. 123–124.

7 Raghupathi, N. (1990) Chapter 7: Long fiber thermoplastic composites, in *Composite Materials Technology* (eds P.K. Mallick and S. Newman), Hanser Publishers, Munich, pp. 237–264.

8 McCrum, N.G., Buckley, C.P., and Bucknall, C.B. (1997) *Principles of Polymer Engineering*, 2nd edn, Oxford University Press, New York, pp. 242–245.

9 Heinold, R. (1995) Broadening polypropylene capabilities with functional fillers. Proceedings of the Functional Fillers 95, Intertech Corp., Houston, TX.

10 Pfaff, G. (2002) Chapter 7: Special effect pigments, in *High Performance Pigments* (ed. H.M. Smith), Wiley-VCH Verlag GmbH, Weinheim, pp. 77–101.

11 Wypych, G. (2000) *Handbook of Fillers*, ChemTec Publishing, Toronto, Canada.

12 Jacoby, M. (2009) Graphene: carbon as thin as can be. *C&EN*, **87** (9), 14–20.

13 Lvov, Y. and Price, R. (2009) Halloysite Nanotubes. Accessed May 2009 at http://www.sigmaaldrich.com/materials-science/nanomaterials/nanoclay-building/halloysite-nanotubes.html.

14 Mascia, L. (1974) *The Role of Additives in Plastics*, Edward Arnold, London, UK.

15 Mahajan, S. (2003) Proceedings of the Functional Fillers for Plastics, Intertech Corp., Atlanta, GA, October 2003.

16 Joshi, S.V., Drzal, L.T., Mohanty, A.K., and Arora, S. (2004) Are natural fiber composites environmentally superior to glass fiber reinforced composites? *Composites Part A*, **35**, 371–376.

17 Blum, H.R. (2008) Functional fillers: a solution towards polymer sustainability & renewability. Proceedings of the Functional Fillers for Plastics, PIRA Intertech Corp., Atlanta, GA, September 2008.

18 Wood, K.E. (2007) Wood-filled composites jump off the deck. *Compos. Technol.*, **13.6**, 25–29.

19 Roth, J. (2008) Influencing functional properties by different grinding methods. Proceedings of the Functional Fillers for

Plastics, PIRA Intertech Corp., Atlanta, GA, September 2008.

20 Holzinger, T. and Hobenberger, W. (2003) *Ind. Miner.*, **443**, 85–88.

21 Ha, J.U. and Xanthos, M. (2009) Functionalization of nanoclays with ionic liquids for polypropylene composites. *Polym. Compos.*, **30** (5), 534–542.

22 Cheetham, A.K. and Grubstein, P.S.H. (2003) *Nanotoday*, 16–19.

23 Warheit, D.B. (2004) *Mater. Today*, 32–35.

24 Wetzel, M.D. (2008) Environmental, health and safety issues and approaches for the processing of polymer nanocomposites. Proceedings of the 66th SPE ANTEC, vol. 54, pp. 247–251.

25 Hussain, S.M. *et al.* (2009) Toxicity evaluation for safe use of nanomaterials: recent advancements and technical challenges. *Adv. Mater.*, **21** (126), 1549–1559.

2
Modification of Polymer Properties with Functional Fillers

Marino Xanthos

2.1
Introduction

Parameters affecting the performance of polymer composites containing functional fillers are related to

1) the characteristics of the filler itself, including its geometry (particle shape, particle size and size distribution, and aspect ratio), its surface area and porosity, and its physical, mechanical, chemical, thermal, optical, electrical, and other properties. Relevant concepts introduced in Chapter 1 are further discussed in this chapter and also in other chapters dealing with specific fillers and surface modifiers;
2) the type and extent of interactions at the phase boundaries, which affect adhesion and stress transfer from the matrix to the filler. Interfacial interactions are also related to surface characteristics of the filler, such as surface tension and surface reactivity. These are parameters that control its wetting and dispersion characteristics. The importance of the interface is also emphasized in Chapters 4–6;
3) the method of filler incorporation into the polymer melt (discussed in Chapter 3) and its distribution in the final product part; processing/structure/property relationships are briefly discussed in this chapter and are elaborated in other chapters covering specific fillers.

Given the overall importance of mechanical properties, this chapter focuses on parameters controlling such properties as related to the filler, the filler/polymer interface, and the method of fabrication. Concepts presented below may also be applicable to the modification of other polymer properties (e.g., permeability, thermal expansion) through the addition of functional fillers.

Functional Fillers for Plastics: Second, updated and enlarged edition. Edited by Marino Xanthos

ISBN: 978-3-527-32361-6

2.2 The Importance of the Interface

Interactions at phase boundaries affect not only the mechanical behavior but also the rheology and processing characteristics, environmental resistance, sorption and diffusion, and many other properties of composites. The strength (tensile, flexural) of the composite and its retention at higher temperatures, after prolonged times and under adverse environmental conditions, are particularly affected by interfacial adhesion. The principal sources of information presented in this section are Refs [1–6].

The extent of adhesion at the polymer/filler interface may be related to various parameters of adsorption and wetting. Factors related to the adsorption of the polymer onto the filler are types of interfacial forces (secondary, primary bonds), molecular orientation/conformation at the interface, and polymer mobility. Contact angle, surface tension, and substrate critical surface tension are among factors related to wetting.

For a drop of liquid in equilibrium on a solid surface, Young's equation relates interfacial tensions at the solid/vapor interface γ_1, liquid/vapor interface γ_2, and solid/liquid interface γ_{12}, with the contact angle θ, which is a measure of the degree of wetting and takes a value of zero for ideal wetting

$$\gamma_1 = \gamma_{12} + \gamma_2 \cos \theta \tag{2.1}$$

Critical surface tension γ_c equals the surface tension of a liquid that exhibits zero contact angle on the solid. Any liquid (melt) with a surface tension less than that of the solid's critical surface tension will wet the surface. Uncoated inorganic fillers may have very high surface tension $\gamma > 200\,\text{mJ/m}^2$, whereas polymers such as polystyrene and polyethylene have lower surface tension $\gamma < 50\,\text{mJ/m}^2$. Thus, polymer melts will spread on the high-energy surfaces of fillers, unless the γ_c value of the filler is reduced by absorbed water layers ($\gamma = 21.8\,\text{mJ/m}^2$), by contamination with low surface tension impurities, or by surface irregularities. This will result in incomplete wetting and void formation at the interface.

The need to minimize contact angle to maximize the work of adhesion W_a is shown by the following Young–Dupré equations:

$$W_a = \gamma_1 + \gamma_2 - \gamma_{12} \tag{2.2}$$

$$W_a = \gamma_2 (1 + \cos \theta) \tag{2.3}$$

The need to minimize unfavorable interfacial interactions by minimizing the interfacial tension γ_{12} can also be inferred from the following simplified Good–Girifalco equations:

$$\gamma_{12} = \gamma_1 + \gamma_2 - 2\Phi\,(\gamma_1 \gamma_2)^{0.5} \tag{2.4}$$

$$W_a = 2\Phi\,(\gamma_1 \gamma_2)^{0.5} \tag{2.5}$$

where Φ is the interaction parameter that depends on polarity. Polarity is defined as the ratio of the polar component of the surface tension to the total surface tension. Φ

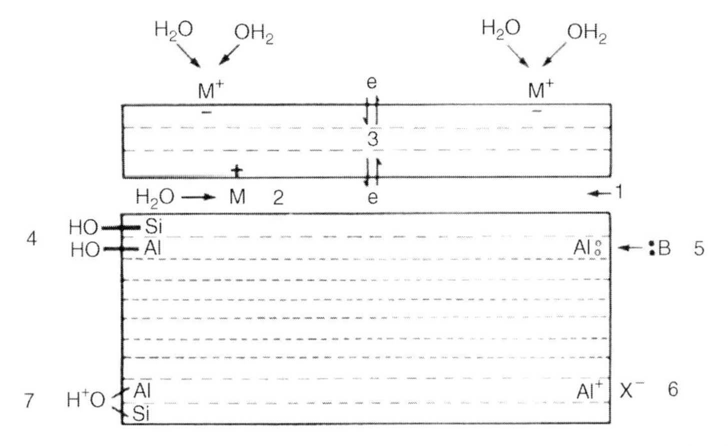

Figure 2.1 Structure and reactivity of a hypothetical silicate mineral (1) interlamellar spaces; (2) exchangeable cations (acidic potential); (3) variable valence species; (4) reactive hydroxylic species; (5) Lewis acids; (6) anion-exchange sites; (7) bridged hydroxyl groups (Reproduced from Ref. [7]).

is maximal when the polarities are equal (approaching unity) and is minimal (approximately zero) when polarities are totally mismatched. It follows that finite γ_{12} and low W_a are the result of disparity between polarities (e.g., between a nonpolar polyolefin and a hydrophilic polar filler surface). Surface modification of fillers can reduce γ_{12}, modify the γ_c of the filler and reduce polarity differences.

Surface modification of fibrous or nonfibrous fillers through the introduction of new functional groups, or the modification of existing ones, may be accomplished by oxidation, thermal treatment, plasma treatment, vapor deposition, and ion exchange or through additives that may react or interact with both the filler and the polymer matrix. Figure 2.1 shows the structure of a hypothetical mineral and the availability of multiple sites for interactions or reactions directly with the polymeric matrix or through additives such as coupling agents [7]. Surface modification of the mineral for improved adhesion can, in effect, convert an ordinary filler into a value-added filler with multiple functionalities. Surface modification is further covered in detail in Chapters 4–6 of this book.

2.3 Modification of Mechanical Properties

2.3.1 General

Modification of mechanical properties and, in particular, the enhancement of modulus and strength is undoubtedly one of the most compelling reasons for incorporating functional fillers in thermoplastics. Appropriate selection of a filler based on its size and shape, modulus and strength, and density is of paramount importance in order to establish its potential reinforcing capacity and to provide guidelines for its method of incorporation into the polymer. For directional fillers with a certain aspect ratio (e.g., short fibers and flakes or platelets) embedded in thermoplastic matrices, the load is transferred from matrix to fibers or flakes by a shear stress and the ends of fibers or flakes do not bear a load. As a result, the properties of the resulting composites are inferior to those of equivalent composites containing continuous fibers or ribbons. Although in some cases thermoplastic composites containing continuous fiber or ribbons have been produced, the methods suitable for the production of short-fiber or flake composites are those that are normally used for the processing of unfilled thermoplastics (e.g., extrusion, injection molding, and blow molding). Rigid filler particles may break during such operations with a concomitant reduction in aspect ratio.

In Table 2.1, average values for modulus and strength of commercially available continuous and discontinuous inorganic fibers/ribbons/flakes having different densities are compared. Data have been obtained from a variety of sources [1, 8–15] and often may be subject to variation, considering measurement difficulties, particularly for short fillers of different origins. In particular, the strength values may be viewed with caution since they depend on the method of testing and the effect of flaws and edges. Data for typical polymer matrices are also included, as well as data for whiskers or single-crystal platelets, which are considered to be virtually flaw-free and, therefore, have extremely high strengths. Metallic wires have relatively higher diameters and are typically used as continuous reinforcements. The densities of particulate mineral fillers (e.g., calcium carbonate, silica, talc, kaolin, wollastonite, and aluminum hydroxide) range from 2.4 to 2.75 g/cm^3. The corresponding Young's modulus values have been quoted as ranging from 25 to 35 GPa [14, 16]. Data are often expressed in terms of specific properties (modulus or strength over density). It is obvious, therefore, that for a specific application requiring high stiffness and strength combined with light weight, the choice of filler with the optimal specific properties would be desirable.

In the following sections, theoretical and empirical treatments that have been used to describe composite modulus and strength are presented for continuous fillers (aspect ratio approaching infinity), discontinuous directional fillers (finite aspect ratio, >1) and particulates (aspect ratio taken as unity). An attempt is made to demonstrate the principles governing the mechanical behavior of polymer composites through model systems rather than real molded parts having variable filler orientation and distribution. For the ease of analysis, stresses are applied only in tension since the situation in flexure or compression becomes significantly more complicated in multiphase, multicomponent systems. The principal sources of information presented in the following sections on modulus and strength are Refs [8, 10, 12, 14, 15, 17–19]. Parameters controlling other mechanical properties are also briefly covered.

2.3.2 Modulus of Fiber and Lamellar Composites

2.3.2.1 Continuous Reinforcements

Uniaxially oriented fiber systems with fibers randomly spaced when viewed from an end cross section are anisotropic materials. For continuous ("long") fiber composites

Table 2.1 Comparison of commercially available high aspect ratio fibers, ribbons, and platelets.

Filler	Density (g/cm^3)	Tensile axial modulus, GPa (average value)	Tensile axial strength, MPa (average value)
Inorganic fibers			
E-glass fibers	2.54	76	1500
S-glass fibers	2.49	86	1900
Asbestos (chrysotile) fibers	2.5	160	2000
Boron	2.57	400	3600
Organic fibers			
Carbon fibers	1.79–1.86	230–340 (7–13)[a)]	3200–2500
Carbon nanotubes	1.2	1000–1700	180 000
Aramid fibers	1.45	124 (5)[a)]	2800
Polyester (terylene)	1.38	1.2	600
Nylon fibers	1.14	2.9	800
UHMWPE (Spectra 900)	0.97	117	2600
Natural fibers			
Sisal fibers (*A. sisalana*)	1.5	16.7	507
Jute fibers (*C. capsularis*)	—	24.1	900
Flax fibers (*Lin usitatissimum*)	1.52	110	900
Cotton fibers	1.50	1.1	350
Wood fibers (aver. tropical hardwoods)	0.6	13.5	400
Wood fibers (Kraft)	1.0	72	900
Metallic wires			
Steel wire	7.9	210	2390
Tungsten	19.3	407	2890
Whiskers			
Silicon nitride	3.2	350–380	5000–7000
Silicon carbide	3.2	480	20 000
Aluminum oxide	4.0	700–1500	10 000–20 000
Ribbons, flakes, platelets			
Glass ribbons	2.47–3.84	59–78	Up to 21 000[b)]
SiC platelets	3.2	480	10 000
AlB_2 platelets	2.7	500	6000
Mica flakes	2.7–2.9	175	3000[c)]
Exfoliated silicate nanoclay platelets	2.8–3.0	170	Up to 1000
Exfoliated graphite platelets	2.0	1000	10 000–20 000
Polymers (excluding elastomers)	0.90–1.35	0.2–3.3	8.5–95

a) Anisotropic fibers; values in parentheses are related to the radial direction.
b) Extrinsic property depending on manufacturing process.
c) Maximum value with perfect edges; in practice, the strength of small flakes can be as low as 850 MPa.

containing all fibers aligned in one direction, subjected to a tensile stress applied along the fiber axis, the total stress σ_{cL} equals the weighted sum of stresses in the fibers σ_f and matrix σ_m:

$$\sigma_{cL} = V_f\sigma_f + (1 - V_f)\sigma_m \tag{2.6}$$

where V_f is the volume fraction of filler.

Table 2.2 summarizes the most commonly used predictive equations for the longitudinal tensile modulus E_{cL} (parallel to the fiber axis) and the transverse tensile modulus E_{cT} (perpendicular to the fiber axis) of the aligned fiber composite. Such predictions have been verified in a plethora of experimental systems. Derivation of the longitudinal modulus, Eq. (2.7), assumes an elastic fiber and matrix, equal Poisson's ratios, and good adhesion and isostrain conditions, that is, strains in fibers and matrix equal to the strain in the composite. Derivation of the transverse modulus, Eq. (2.8), assumes isostress conditions and is based on a simple model lumping all fibers together in a band normal to the tensile stress applied perpendicularly to the fibers (along the directions perpendicular to the fiber axis) [19]. Table 2.2 also contains predictive equations for the modulus of aligned composites tested at intermediate angles to fibers' axes, Eq. (2.9). Modulus values are also given for random planar and 3D orientations (Eqs. (2.10) and (2.11), respectively).

With respect to continuous lamellar composites containing ribbons or tapes, isotropy in the plane is essentially provided at large ribbon aspect ratios (width to thickness) without any angle dependency [12]. The corresponding Eq. (2.12) is similar to Eq. (2.7) for fibers. To summarize the information in Table 2.2:

- The main parameters affecting composite modulus E_c are the fiber or ribbon modulus, the volume fractions, and the angle of application of the stress relative to the reinforcement axis.
- The highest modulus in the fiber composites is obtained in the longitudinal case, at an application angle of 0°. Longitudinal and transverse moduli provide the upper and lower bounds of modulus versus fiber concentration curves. In the extreme case of fiber misalignment (90°), the fibers play only a minor role in determining the overall stiffness of the composite.
- Composite modulus decreases rapidly with increasing orientation angle. Experiments confirm that even a few degrees of misalignment can significantly reduce the modulus.
- Fibers that are randomly oriented in the plane or in space may provide isotropy but at the expense of the overall composite modulus
- Isotropy in a plane may be achieved with aligned ribbons; the modulus value in this case is approximately equal to that of an oriented continuous fiber composite tested in the longitudinal direction.

2.3.2.2 **Discontinuous Reinforcements**

Prediction of the modulus of a short-fiber composite needs to take into account end effects since isostrain conditions are not satisfied at the fiber ends. Stress builds up along each fiber from zero at its end to a maximum at its center. As shown in Figure 1.4, at the interface, the matrix is severely sheared at the fiber ends.

For aligned short fibers subject to a stress along the fiber axis, the stress borne by the fibers is no longer σ_f as defined in Eq. (2.6), but it assumes a lower mean value that takes into account the effect of fiber ends and the rate of stress

Table 2.2 Comparison of modulus equations for continuous fiber and ribbon composites.

Tensile modulus	Continuous fibers	Eq. no.	Continuous ribbons (tapes) of high aspect ratio	Eq. no.
Longitudinal (0°)	$E_{cL} = V_f E_f + (1 - V_f) E_m$	(2.7)	$E_{cL} \approx E_{cT} = V_f E_f + (1 - V_f) E_m$	(2.12)
Transverse (90°)	$1/E_{cT} = V_f / E_f + [(1 - V_f)/E_m]$	(2.8)	$E_{cL} \approx E_{cT}$ Assume isotropy in plane	(2.12)
Intermediate angles (θ)	$E_{cL}/E_{c\theta} = \cos^4\theta + \sin^4\theta E_{cL}/$ $E_{cT} + \cos^2\theta \sin^2\theta (E_{cL}/G_{cLT} - 2\nu_{cLT})$	(2.9)[a]	$E_{cL} \approx E_{cT}$ Assume isotropy in plane	(2.12)
Random orientation in plane	$E_c = 3/8 E_{cL} + 5/8 E_{cT}$	(2.10)	$E_{cL} \approx E_{cT}$ Assume isotropy in plane	(2.12)
Random 3D orientation	$E_c = 1/5 E_{cL} + 4/5 E_{cT}$	(2.11)	—	

a) G_{cLT} and ν_{cLT} are longitudinal–transverse shear modulus and Poisson's ratio, respectively, of the composite.

build-up along the length toward the central regions. Eq. (2.7) can be written in a modified form as Eq. (2.13) (see Table 2.3), where K is an efficiency parameter that approaches unity for very long fibers. Values of K calculated by different authors for single fibers and by assuming random overlap are given by Eqs. (2.14) and (2.15). In these equations, the parameter u includes constants such as fiber length and diameter, fiber volume fraction, fiber modulus, and matrix shear modulus. It has been shown [8, 14, 15] that for optimal efficiency in stress transfer to the fibers, the value of u should be as high as possible; this implies that the aspect ratio and the ratio G_m/E_f should also be as high as possible. Equation (2.13) may be modified as Eqs. (2.16) and (2.17) to account for the effect of variation in fiber orientation, which is a natural consequence of melt processing.

An interpolation procedure applied by Halpin and Tsai [17, 18] has led to general expressions for the moduli of composites, as given by Eqs. (2.18) and (2.19). Note that for $\xi = 0$, Eq. (2.18) reduces to that for the lower limit, Eq. (2.8), and for ξ = infinity, it becomes equal to the upper limit for continuous composites, Eq. (2.7). By empirical curve fitting, the value of $\xi = 2(l/d)$ has been shown to predict the tensile modulus of aligned short-fiber composites in the direction of the fibers, and the value of $\xi = 0.5$ can be used for the transverse modulus. Other mathematical relationships for modulus calculations of composites with discontinuous fillers include the Takayanagi and the Mori–Tanaka equations [20].

Equation (2.13) has also been used to predict the modulus of flake (platelet) composites containing planar oriented reinforcement for uniform arrays of flakes, Eq. (2.14), and for random overlap, Eq. (2.15) [10, 12, 14, 19]. Equations for the parameter u are somewhat different from those used for fibers, but they still contain the important parameters affecting the modulus of the composite, that is, aspect ratio, volume fraction, and flake/matrix modulus ratio. Equation (2.18) has also been used to predict the modulus of platelet-reinforced plastics [17, 21].

In summary, general parameters affecting the modulus of a rigid polymeric material containing discontinuous directional reinforcements are the moduli of the reinforcement and the matrix, the ratio G_m/E_f, the volume concentration of the reinforcement, the reinforcement orientation, and its size (aspect ratio). Values of initial modulus (tangent) are usually not affected by the extent of interfacial adhesion; however, secant moduli, measured at higher strains, are usually higher in the case of good adhesion. In Figure 2.2, E_c/E_m (the modulus enhancement factor) is plotted as a function of fiber or flake aspect ratio; the curves show the effects of aspect ratio and orientation angle [18]. Theoretical plots of this type have been confirmed in many experimental systems. Thus, it can be easily inferred that efficient reinforcement occurs when volume fraction and aspect ratio are fairly large, and there is a minimum misalignment of fibers or flakes from the axis or plane of application of the stress. Practical limitations are the limited volume fraction that may be attained depending on the manufacturing method, the reduced aspect ratio during processing, and the complex orientation characteristics of injection molded parts.

Table 2.3 Comparison of modulus equations for short fiber, flake, and particulate composites.

Tensile modulus	Short fibers	Eq. no.	Flakes	Eq. no.	Particulates	Eq. no.
Longitudinal	$E_{cL} = KV_f E_f + (1-V_f)E_m$ where $K \leq 1$	(2.13)[a]	$E_{cL} = KV_f E_f + (1-V_f)E_m$ where $K \leq 1$	(2.13)[a]	$E_c/E_m = (1+ABV_f)/(1-B\chi V_f)$	(2.20)[d]
Longitudinal	$E_c/E_m = (1+\xi\eta V_f)/(1-\eta V_f)$ for $\xi = 2(l/d)$	(2.18)[b]	$E_c/E_m = (1+\xi\eta V_f)/(1-\eta V_f)$ for $\xi = 2(\alpha)$	(2.18)[b]	Assume 3D isotropy	
Transverse	$E_c/E_m = (1+\xi\eta V_f)/(1-\eta V_f)$ for $\xi = 0.5$	(2.18)[b]	$E_{cL} \approx E_{cT}$ Assume isotropy in plane		Assume 3D isotropy	
Intermediate angles (θ)	$E_c = K' KV_f E_f + (1-V_f)E_m$	(2.16)[c]	$E_{cL} \approx E_{cT}$ Assume isotropy in plane		Assume 3D isotropy	

a) For single fibers

$$K = 1-(\tanh u/u) \quad (2.14)$$

or, by assuming random overlap,

$$K = 1-[\ln(u+1)/u]. \quad (2.15)$$

b) $$\eta = (E_f/E_m-1)/(E_f/E_m+\xi) \quad (2.19)$$

and ξ adjustable parameter related to aspect ratio

c) $$K' = \sum_{i=1}^{k} \alpha_i \cos^4 \theta_i, \quad (2.17)$$

where α_i is the fraction of the fibers oriented at an angle θ to the direction in which the value of E_c is required and k is the number of intervals of angle defining the orientation distribution (Ref. 10, p. 53).

d) Where $A = K_E - 1$ with K_E geometric factor,

$$B = (E_f/E_m-1)/(E_f/E_m+A) \quad (2.21)$$

and

$$\chi = 1 + V_f(1/\phi_{max}^2 - 1/\phi_{max}). \quad (2.22)$$

where ϕ_{max} maximum packing fraction of the filler (0.74 for hexagonal close packing, 0.524 for simple cubic packing, etc.).

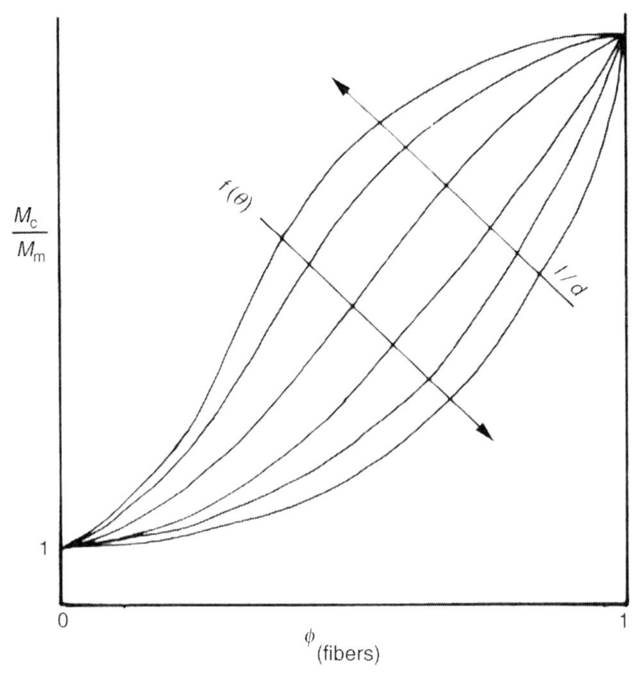

Figure 2.2 Modulus enhancement factor for reinforced thermoplastics as a function of fiber volume fraction, orientation angle, and aspect ratio. (Adapted from Ref. [18].)

2.3.3 Modulus of Composites Incorporating Particulates

Parameters affecting the modulus of rigid plastic matrices (Poisson's ratio <0.5) containing particles with aspect ratio close to unity are primarily the ratio of the moduli of the two phases, and the volume fraction of the filler. Other parameters are the geometry, packing characteristics, degree of agglomeration of the filler, and its size and size distribution. The stiff behavior of particulate composites can be represented by Eqs. (2.20)–(2.22) included in Table 2.3, which were devised by Kerner and further modified by Nielsen and coworkers [1, 17]. The theories predict that the elastic moduli of rigid particulate composites are lower than those of composites with directional fillers but higher than those of the composites containing elastomeric particles or voids (foams) as shown in Figure 2.3 [22]. In this figure, the modulus increase obtained with rigid particulate fillers with an aspect ratio of approximate unity (system 4) is compared with the modulus increases obtained with long fibers in the longitudinal and transverse directions (systems 7 and 2, respectively), and with short fibers, which are either aligned (system 6) or randomly oriented (system 5).

The theories suggest that the elastic moduli of composites containing particulates with an aspect ratio of unity should be independent of the dimensions of the filler and dependent only on the relative moduli of filler and matrix, their volume fractions, and

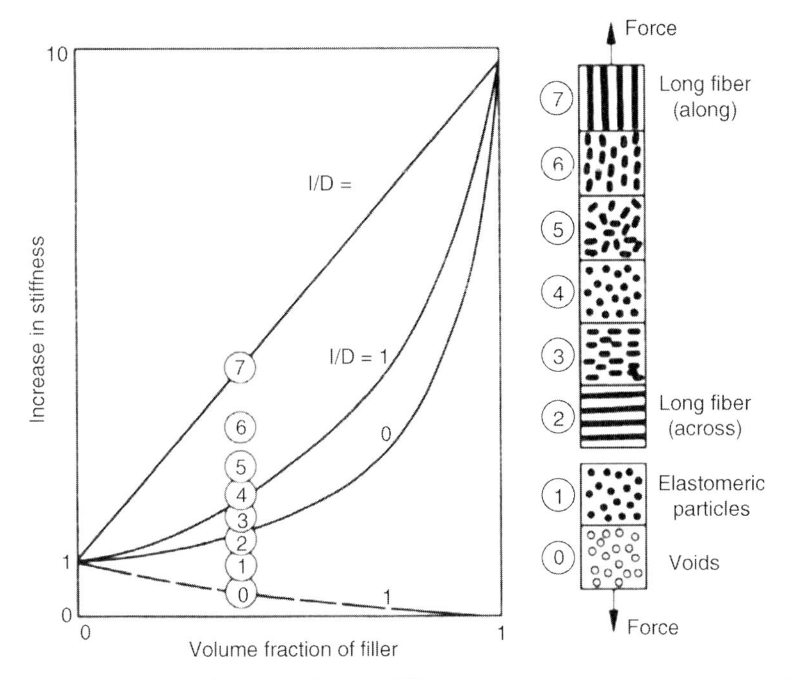

Figure 2.3 Relation between stiffness and filler type and orientation in polymeric materials. (Reproduced with permission from Ref. [21].)

geometric factors. Good adhesion between the organic and inorganic phases is assumed, even though they usually have vastly different coefficients of thermal expansion. Thus, even in the case of poor adhesion, the theoretical equations are valid because there may not be any relative motion at the interface as a result of a squeezing force on the filler surface imposed during cooling down from the fabrication temperature. Although the theories predict no dependence of the composite modulus on the filler dimensions, experiments show that an increase in modulus is obtained when the particle size is decreased and that there is a corresponding decrease in modulus when the particle size distribution is shifted to higher values. These discrepancies can be attributed to increased particle surface area and surface energy, and increased maximum packing fractions as discussed in Ref. [17].

2.3.4 Strength of Fiber and Lamellar Composites

2.3.4.1 Continuous Reinforcements

Table 2.4 provides the most important equations proposed to model the tensile strength of composites containing continuous fibers and ribbons (microtapes) with large aspect ratio as a function of degree of interfacial adhesion and reinforcement orientation.

Table 2.4 Comparison of strength equations for continuous fiber and ribbon composites.

Tensile strength	Continuous fibers	Eq. no.	Continuous ribbons (tapes) of high aspect ratio	Eq. no.
Longitudinal (0°) – perfect adhesion	$\sigma_{cL} = \sigma_{fu} V_f + \sigma'_m (1 - V_f)$	(2.23)	$\sigma_{cL} = \sigma_{fu} V_f + \sigma'_m (1 - V_f)$	(2.23)
Longitudinal (0°) – no adhesion or very low V_f	$\sigma_{cL} = \sigma_{mu}(1 - V_f)$	(2.24)	$\sigma_{cL} = \sigma_{mu}(1 - V_f)$	(2.24)
Longitudinal (0°) – intermediate adhesion	$\sigma_{cL} = K\sigma_{fu} V_f + \sigma'_m (1 - V_f)$ where $0 < K < 1$	(2.25)	$\sigma_{cL} = K\sigma_{fu} V_f + \sigma'_m (1 - V_f)$ where $0 < K < 1$	(2.25)
Transverse (90°)	σ_{cT} = bond or matrix strength		$\sigma_{cL} \approx \sigma_{cT}$ Assume isotropy in plane for perfect adhesion	
Intermediate angles (θ)	$1/\sigma_{c\theta}^2 = \cos^4\theta/\sigma_{cL} + \sin^4\theta/\sigma_{cT} + \cos^2\theta\,\sin^2\theta(1/\tau^2 - 1/\sigma_{cL}^2)$	(2.26)[a]	$\sigma_{cL} \approx \sigma_{cT}$ Assume isotropy in plane for perfect adhesion	

a) 0–5° longitudinal tensile failure; 5–45° shear failure; 45–90° transverse tensile failure.

In summary,

- Parameters affecting composite strength σ_c are the ultimate strength of the fibers or ribbons σ_{fu}, the ultimate strength of the matrix σ_{mu}, or the stress borne by the matrix, σ'_m, when the fibers or ribbons fail, volume fractions, the shear strength of the matrix or bond strength τ, and the angle between the direction of stress and the fiber axis.
- Maximal strength in the fiber composites is obtained in the longitudinal case, that is, at an angle of 0° in the case of perfect adhesion. Under these conditions, the composite efficiently utilizes the strong reinforcement, and the overall failure is by fiber or ribbon fracture. The strength rapidly decreases and the mode of failure changes with increasing angle. At 90°, failure of the fiber composites occurs by failure either in the matrix or at the interface. For example, the longitudinal tensile strength of a unidirectional 50 vol% glass fiber–unsaturated polyester composite is 700 MPa; this is significantly higher than its transverse strength of 20 MPa, which is closer to the strength of the matrix [23].
- Fibers that are randomly oriented in the plane may provide isotropy but at the expense of the overall composite strength.
- Isotropy in a plane may be achieved with aligned, large aspect ratio ribbons; strength values in this case are approximately equal to those of an oriented continuous fiber composite tested in the longitudinal direction.

2.3.4.2 Discontinuous Reinforcements

In this case, the matrix deforms more than the filler at the fiber ends and shear stresses are set up at the interface. Failure of the composite may occur either by fiber fracture or bond failure (fiber pullout), depending on the filler aspect ratio. A critical fiber length L_{cr} defines the transition point between the two modes of failure. Similar concepts apply to flakes (platelets), for which a critical platelet diameter d_{cr} defines the transition point. Table 2.5 summarizes common predictive equations used to describe the tensile strength of composites containing short fibers or platelets as a function of orientation, aspect ratio, volume fraction, adhesion, and strength of matrix and fillers. In these equations, the effects of adjacent fibers or flakes, and of the presence of edges in irregularly shaped flakes, are ignored. The critical aspect ratio α_{cr}, determining the transition from fiber or platelet fracture to failure by debonding or shear failure of the matrix at lower stresses, is defined for fibers (Eqs. (2.27) and (2.28)) as

$$\alpha_{cr} = \sigma_{fu}/2\tau \tag{2.29}$$

and for platelets (Eqs. (2.32) and (2.33))

$$\alpha_{cr} = \sigma_{fu}/\tau \tag{2.34}$$

where τ is interface or matrix shear strength and σ_{fu} is the fiber or platelet strength.

In summary,

- The fiber or platelet aspect ratio has to be above the critical value for an efficient utilization of the reinforcing properties of the filler. In addition to the aspect ratio,

Table 2.5 Comparison of strength equations for short fiber, flake, and particulate composites.

Tensile strength	Short fibers	Eq. no.	Flakes	Eq. no.	Particulates	Eq. no.
Longitudinal for fiber $L > L_{cr}$ or for platelet $\alpha > \alpha_{cr}$	$\sigma_{cL} = \sigma_{fu}(1-L_{cr}/2L)V_f + \sigma'_m(1-V_f)$	(2.27)[a]	$\sigma_{cL} = \sigma_{fu}(1-\alpha_{cr}/\alpha)V_f + \sigma'_m(1-V_f)$	(2.32)[a]	$\sigma_c = \sigma_{mu}(1-a V_f^b + cV_f^d)$	(2.35)[c]
					or $\sigma_c = \sigma_{mu}(1-1.21V_f^{2/3})$	(2.36)
					or $\sigma_c = \lambda\sigma_{mu} - KV_f$	(2.37)[d]
Longitudinal for fiber $L < L_{cr}$ or for platelet $\alpha < \alpha_{cr}$	$\sigma'_{cL} = \tau L/dV_f + \sigma_{mu}(1-V_f)$	(2.28)[b]	$\sigma'_{cL} = \tau\alpha/2V_f + \sigma_{mu}(1-V_f)$	(2.33)[b]	3D isotropy	
Transverse	Bond or matrix strength		σ_{cL} or σ'_{cL}		3D isotropy	
Random orientation in plane for fiber $L > L_{cr}$ or for platelet $\alpha > \alpha_{cr}$	$\sigma_c = \sigma_{fu}V_f/2(1-L_{cr}/L)$	(2.30)	σ_{cL}		3D isotropy	
Random orientation in plane if $L < L_c$ or $\alpha < \alpha_{cr}$	$\sigma_c = \sigma_{fu}V_f l/4L_{cr}$	(2.31)	σ'_{cL}		3D isotropy	

a) Failure by reinforcement fracture.
b) Failure by reinforcement debonding (pullout) or shear failure of the matrix.
c) a, b, and c constants depending on particle size and adhesion.
d) λ is the stress concentration factor; K is a constant depending on adhesion.

other parameters affecting composite strength are the ultimate strength of the filler, the ultimate strength of the matrix σ_{mu}, or the stress borne by the matrix, σ'_{m}, when the filler fails, the volume fractions, bond strength or matrix shear strength, and the angle of application of the stress relative to the fiber axis.

- The strength values of composites containing short fiber or platelet-type fillers are lower than those of their counterparts with continuous reinforcements.
- Maximal strength for fiber composites is obtained in the longitudinal case, at an angle of 0°, with good adhesion, and at aspect ratios well above the critical value. Composite strength decreases with increasing angle of application of stress. In the case of randomly oriented long fibers ($L > L_{cr}$), the (isotropic) composite strength is much lower than the longitudinal strength of an oriented fiber-filled composite.
- The strength of flake-containing composites is isotropic in the plane of the oriented flakes and much lower perpendicular to the flake plane axis. The highest strength for flake-containing composites is obtained at any angle of stress applied parallel to the flake surface, with good adhesion, and with aspect ratios well above the critical value, when composite failure is by flake fracture.

Experimental data on the tensile strength of a variety of polymer/filler combinations compiled by Wypych [9] indicate the complexity of the parameters affecting strength in real systems.

2.3.5 Strength of Composites Containing Particulates

Table 2.5 contains a general equation, Eq. (2.35), for the effect of particulate fillers on the tensile strength of a polymer, a common modification of this equation by Nicolais and Narkis, (Eq. (2.36)), and an additional equation proposed by Piggott and Leidner, (Eq. (2.37)) [9, 17, 19]. These equations predict that failure is either by matrix failure or by loss of adhesion without utilization of the inherent strength of the particulate. Experimental results for a variety of nondirectional filler particles show that in most cases, tensile strength decreases with increasing volume fraction; relatively higher values, however, are obtained in the case of improved adhesion [9].

2.3.6 Toughness Considerations

In addition to the effects on modulus and strength, the use of fillers can, also in most cases, improve polymer toughness, that is, its ability to resist crack propagation, normally expressed as the area under the stress/strain curve. Impact strength, a common measure of toughness at high strain rates or dynamic conditions is the most important property for real application conditions.

In short-fiber composites, areas around fiber ends, areas of poor adhesion, and regions of fiber–fiber contacts may reduce the resistance to crack initiation by acting as stress concentrators. However, fibers may also reduce crack propagation by diverting cracks around the fibers or by bridging cracks. Materials with high-impact strength spread the absorbed energy throughout as large a volume as possible to prevent brittle failure. Dissipation of energy in short-fiber composites, and hence an increase in impact strength, may be accomplished by (a) mechanical friction, as in the case of pullout of the fibers from the matrix that prevents localization of the stresses or (b) by controlled debonding of the fibers, which disperses the region of stress concentration through a larger volume and tends to stop the crack propagation. The energy dissipated in forming a unit amount of new surface Γ by pullout depends on fiber properties such as strength, length, critical length, and volume fraction:

$$\Gamma = [\sigma_{fu} L_{cr}^2 V_f]/12L \quad \text{for } L > L_{cr}, \tag{2.38}$$

$$\Gamma = [\sigma_{fu} L^2 V_f]/12L_{cr} \quad \text{for } L < L_{cr}. \tag{2.39}$$

The energy dissipated in forming a unit amount of new surface, Γ, by debonding is given by

$$\Gamma = [\sigma_{fu}^2 L_d V_f]/4E_f \tag{2.40}$$

where L_d is the debonded fiber length [17].

These theoretical equations for maximum energy dissipation have been correlated with maximum impact strength for impact loads applied parallel to short fibers having length close to their critical length and in the case of less than perfect adhesion [17]. Transverse impact strength is generally lower than longitudinal, since the aforementioned crack propagation toughening mechanisms are inoperative while crack initiation factors still exist. Similar equations have been derived for round platelets and flakes [19]. In practice, the use of flakes has, in most cases, shown to adversely affect the impact strength, possibly because of significant stress concentration effects associated with shape irregularities and sharp corners.

Control of aspect ratio and bond strength appear to be the most important parameters for impact strength enhancement through the use of fibers, particularly in the case of brittle matrices. Maximizing the impact strength is usually accomplished at the expense of the longitudinal tensile strength, which is greatest for longer fibers well bonded to the matrix. The effect of nondirectional fillers on toughness is correlated in a complex manner with the particular testing method used and to the particle size, shape, concentration and rigidity of the filler, the nature of the interface, and the specific type of polymer matrix. Studies on a variety of rigid particulate fillers ranging in size from 0.8 to 30 μm and added in a rigid matrix show a peak in falling dart impact strength at about 2 μm; studies on $CaCO_3$-filled polypropylene confirm the beneficial effect of less than perfect adhesion on impact strength, as was observed in fibers also [9]. Impact strength is largely determined by dewetting and formation of narrow zones of highly deformed, crazed polymer as a result of stress application. In brittle rigid polymers, rigid spherical fillers with higher modulus than the matrix act as crack initiators, promoting crack propagation and lowering impact strength. In tough, rigid polymers, toughness may be raised as a result of enhanced crazing [17]. A compilation of impact strength data for particulate and fibrous fillers [9] in a variety of polymers suggests that for certain systems correlations corresponding to reality may be difficult to establish and that one type of test may contradict the results of another.

2.3.7
Temperature and Time Effects

The mechanical behavior of polymer composites is not only defined by short-term properties such as stiffness, strength, and toughness but also long-term properties such as creep, stress relaxation, and fatigue. All such properties, which are affected by temperature and type of environment, can be modified through the addition of fillers.

In general, most fillers increase the heat distortion temperature (HDT) of plastics as a result of increasing modulus and reducing high-temperature creep. Thermoelastic properties such as coefficient of thermal expansion (CTE) are also affected by the presence of fillers and have been modeled through a variety of equations derived from the rule of mixtures [8]. For directional fillers, this property is strongly orientation-dependent, and because of the difference between the CTE of the filler and that of the matrix, internal stresses may lead to undesirable warpage.

In general, fillers also decrease creep and creep rate of polymers as long as there is no serious debonding of the particles. At high strains and long times, when debonding may occur, creep and creep rate may dramatically increase and, accordingly, time for rupture may be significantly decreased. Increased adhesion usually minimizes these effects. Similarly, in many glass fiber-reinforced plastics subjected to cyclic tensile stresses, debonding may occur after a few cycles even at moderate stress levels. Debonding may then be followed by resin cracking, severe localized damage, and eventually, fatigue failure. Plastics containing fibers with higher modulus than glass (e.g., carbon) are found to be less vulnerable to fatigue [24].

2.3.8
Other Properties

2.3.8.1 Tribological Properties

Polymers in the unfilled state often need an enhancement of properties to adequately perform in tribological situations. This can be accomplished by either reducing coefficient of friction against metals through internal lubricants, for example, PTFE, silicones (see also Chapter 19) or by increasing the resistance of the polymer to the combined normal and tangential stresses through the use of reinforcements. Carbon, aramid, glass fibers, and glass spheres decrease the wear factor of unfilled Nylon 6,6, although not all these reinforcements will necessarily lower the coefficient of friction. In fact, the coefficient of friction may increase with the addition of certain rigid fillers [25]. In MMT nanoclay Nylon 6,6 composites, it is shown that the coefficient of friction is lower than that of the unfilled polymer depending on loadings and sliding velocity. Wear factor decreases with clay concentration up to 5% and increases significantly thereafter [26]. It should be noted that the rule of mixtures calculations can also be used to predict the coefficient of friction of unidirectional continuous fiber composites [25].

2.3.8.2 Permeability

A variety of equations for predicting the reduction in permeability of polymeric films containing an impermeable dispersed phase have been reviewed in Ref. [27]. Examples include the following:

1) A simplified version of the Maxwell model that includes only volume fraction, ϕ_d, of dispersed spherical fillers given by

$$\frac{P_m}{P_c} = \left[\frac{(1+\phi_d/2)}{(1-\phi_d)}\right], \tag{2.41}$$

where P_c is the permeability of the composite and P_m is the permeability of the matrix.

2) A simple tortuous 2D model developed by Nielsen to depict the effect of the size and aspect ratio α of platelet fillers with orientation perpendicular to the diffusion path on the barrier properties of the polymer composite; related Eqs. (8.1) and (8.2) are found in Chapter 8.
3) Equations accounting for the effects of deviation from planar orientation of flakes on film permeability in the presence of a second impermeable crystalline phase [21]), and in the case of randomly oriented layered clay platelets [28, 29].

2.4 Effects of Fillers on Processing Characteristics of Polymers

2.4.1 General

Unfilled polymers behave as non-Newtonian liquids during melt processing. Melt rheology has been extensively investigated over the past 20 years and is documented in a large number of texts. The significant effect of dispersed particulates, fiber, or flakes on polymer melt rheology and melt elasticity are directly related to processability with respect to both mixing (compounding) and shaping operations. For directional fillers, understanding the flow-induced orientation and the possibility of particle segregation to the region of the highest fluid velocity are of paramount importance in controlling the microstructure of the final product and its properties. Principal sources of information presented in this section are Refs [1, 2, 10, 17, 22, 31–33].

2.4.2 Melt Rheology of Filled Polymers

Filler effects on viscosity and elasticity depend on several parameters including concentration, size, shape, and aspect ratio of the filler; interactions with the polymer; shear rate, the presence of agglomerates, fiber/flake alignment, and surface treatment.

2.4.2.1 **Concentration and Shear Rate**

In general, shear and elongational viscosities increase with increasing filler volume fraction. The effect on shear viscosity is more pronounced at low shear rates; "yield" effects due to the formation of structured networks are often encountered at low shear rates and at high loadings of submicron particles [10, 22]. At processing shear rates above $10\,s^{-1}$, the network structure is destroyed and the rheological nature of the matrix dominates [30]. High shear rates tend to orient fibers and flakes to different degrees depending on their size, rigidity, concentration, and interactions with the matrix. The increase in viscosity relative to the unfilled matrix becomes less pronounced at higher shear rates, as shown in Figure 2.4 [10]. Much larger deviations from Newtonian behavior than for the corresponding polymer matrix are observed in filled polymer melts. Velocity profiles in circular and slit channels become very flat, due to a decrease in the power law index, and plug-like flow behavior is observed [1, 31]. Increasing the amount of fillers, regardless of shape, reduces melt elasticity, as shown by reduced extrudate (die) swell and concomitant effects on normal stress differences [32]. Reduced melt elasticity has significant practical effects for extrusion and injection molding, including depressed melt fracture.

2.4.2.2 **Filler Size and Shape**

Particle size effects may be negligible at high shear rates since for filled systems high shear viscosity is often governed by the matrix characteristics and low shear viscosity

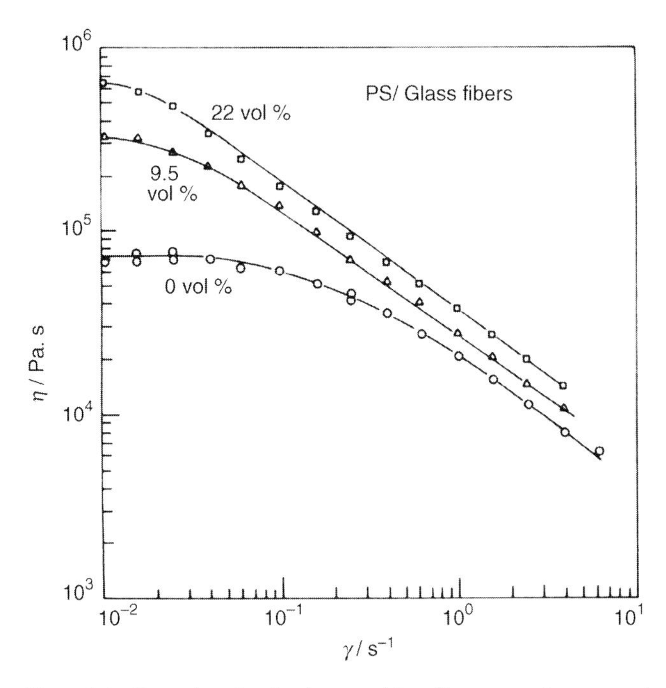

Figure 2.4 Shear viscosity of polystyrene/glass fiber suspension versus shear rate at 180 °C. (Reproduced with permission from Ref. [10], Chapter 5.)

is governed by the filler. Low specific area fillers such as large particulate fillers and low aspect ratio fibers or flakes have less interactions with the polymer and yield lower viscosities than higher surface area, higher aspect ratio fillers. High shear stresses tend to not only break filler agglomerates but also cause additional reduction of fiber length or flake diameter, with a concomitant effect on viscosity.

2.4.2.3 Filler Surface Treatments

Interfacial agents that tend to wet or lubricate the filler surface (titanates, stearates, etc.) tend to reduce viscosity. This may result from attenuated interparticle forces and less tendency to flocculation since polymer molecules may slip between treated filler particles encountering less frictional resistance (see Figure 2.5) [10]. Reduced particle–particle interactions may lead to further flow orientation of fibers and flakes and to a further decrease in viscosity at high shear rates. However, an increase in viscosity may occur when the surface treatment causes strong adhesion of the filler to the polymer.

Several equations have been proposed to predict the ratio of the viscosity of the composite to that of the unfilled matrix, μ_c/μ_m, and to explain viscosity effects as a

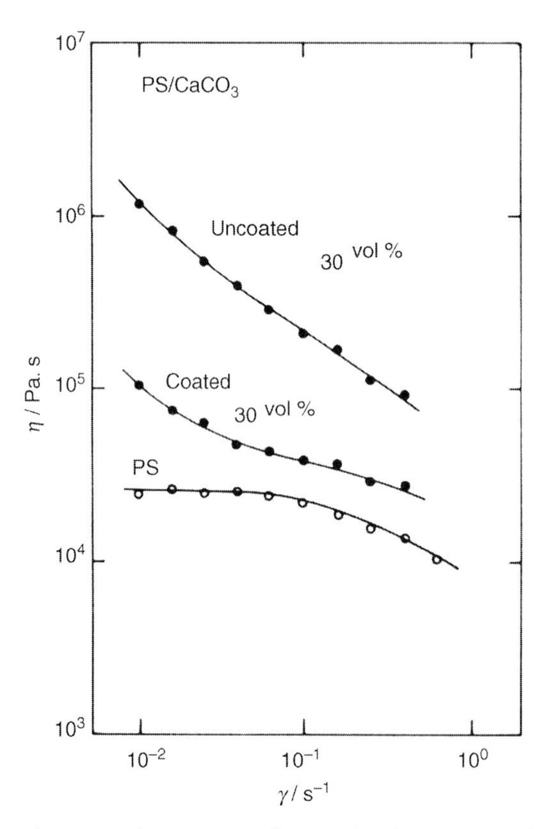

Figure 2.5 Shear viscosity of uncoated and stearate-coated calcium carbonate-filled polystyrene melt versus shear rate. (Reproduced with permission from Ref. [10], Chapter 5.)

function of volume fraction, shape factors, aspect ratio, packing characteristics, interaction parameters, and non-Newtonian or yield parameters. Examples are as follows:

1) The Mooney equation [2, 31] that is valid for the entire concentration range

$$\ln(\mu_c/\mu_m) = K_e V_f/(1-V_f/\phi_{max}), \quad (2.42)$$

where ϕ_{max} is the maximum packing factor, defined as true volume of filler/apparent volume occupied by filler, K_e is a geometric parameter known as the Einstein coefficient (see also Eq. (3.4)), which depends on aspect ratio and degree of agglomeration, and for rods also on degree of orientation, which, in turn, depends on shear rate.

2) The Nielsen equation [17]

$$\mu_c/\mu_m = (1+ABV_f)/(1-B\Psi V_f), \quad (2.43)$$

where A, B, and Ψ are functions of component properties, packing characteristics, and aspect ratio, respectively.

In practical terms, the high viscosities obtained through the addition of fibers and flakes can be reduced through wetting agents, through filler orientation, and/or by reduction of aspect ratio. Reduction of aspect ratio during processing may be undesirable but is common for fragile fibers and large flakes (glass, mica); in contrast, organic flexible fibers may orient and bend without fracture. These effects will lead to an increase in ϕ_{max} and a decrease in K_e in Eq. (2.42) with a concomitant decrease in the relative viscosity μ_c/μ_m [31].

2.4.3
Processing/Structure/Property Relationships

Primary filler properties controlling the morphology and properties of plastic products are geometry, concentration, density, modulus, strength, and surface chemistry. Additional filler properties related to processing (compounding and shaping) are as follows:

- *Hardness,* which is usually expressed on the Mohs' scale ranges from 1 for talc to 10 for diamond. Soft fillers (e.g., talc, calcite with Mohs' hardness 3) are preferred against hard fillers (e.g., silica with Mohs' hardness 7) that tend to cause excessive wear of processing equipment.
- *Thermal properties* such as thermal conductivity that for most mineral fillers is about an order of magnitude higher than that of thermoplastics; specific heat, which is typically about half that of polymers; and coefficient of thermal expansion (see Section 2.3.7) that is lower than that of polymers. The net effect is that most fillers (nonfibrous) usually produce a faster rate of cooling in injection molding, lower volume shrinkage, and promote less warpage and shorter cycle times. Fibrous fillers, however, may cause differential shrinkage with an increased tendency to warp as a result of orientation. Combinations of

fibers with flakes (planar orientation) or spheres (no orientation) tend to minimize warpage.

- *Thermal stability* (up to 300 °C for high-temperature thermoplastics), which is required during processing so that there is minimum weight loss or structural changes.
- *Moisture absorption* that needs to be minimized since it may affect the quality of the compound or the stability of hydrolytically sensitive matrices such as nylons or polyesters.

Quantitative predictions of the effects of fillers on the properties of the final product are difficult to make, considering that they also depend on the method of manufacture, which controls the dispersion and orientation of the filler and its distribution in the final part. Short-fiber- and flake-filled thermoplastics are usually anisotropic products with variable aspect ratio distribution and orientation varying across the thickness of a molded part. The situation becomes more complex if one considers anisotropy, not only in the macroscopic composite but also in the matrix (as a result of molecular orientation) and in the filler itself (e.g., graphite and aramid fibers and mica flakes have directional properties). Thus, thermoplastic composites are not always amenable to rigorous analytical treatments, in contrast to continuous thermoset composites, which usually have controlled macrostructures and reinforcement orientation [8, 17].

Morphological features resulting from orientation of directional fibers and flakes in complex flow fields are directly related to the part properties. Attempts have been made to model and predict orientation distributions in fiber and flake composites [10, 34, 35] in extruders and injection molders equipped with ordinary or special molds. In injection molding, mold-filling patterns and filler orientation depend, among other factors, on mold geometry and cavity thickness, type and position of gate, injection speed, and rheology of matrix material [8, 10]. Typically, three distinct regions with different fiber orientations can be identified: (a) a skin originating from the expanding melt front, (b) an intermediate layer, in which the fibers are oriented parallel to the flow direction, and (c) a core layer, with fiber orientation transverse to the flow direction. An example of complex fiber orientation in a glass-reinforced polypropylene, parallel to the flow direction, is shown in Figure 2.6 [8]. For flakes such as mica or talc, flake orientation during extrusion, injection molding, or blow molding is also predominantly parallel to the flow direction, with a region of misalignment in the core [2, 20]. Such morphologies can be modified through the application of a shear force to the melt as it cools (e.g., SCORIM™), which has a marked effect on the physical properties [24].

Improper location of the gate or multiple gates may produce a zone where the two melt fronts meet (weldline). For unfilled polymers, this region usually has inferior properties compared to other locations within the part; mechanical weakness becomes more pronounced in the presence of high aspect ratio fibers and flakes, which may not interpenetrate and instead lie in their most unfavorable orientation for effective reinforcement. A more moderate detrimental effect on properties has been observed with low aspect ratio fillers. In general, fibrous fillers cause up to 50% loss in

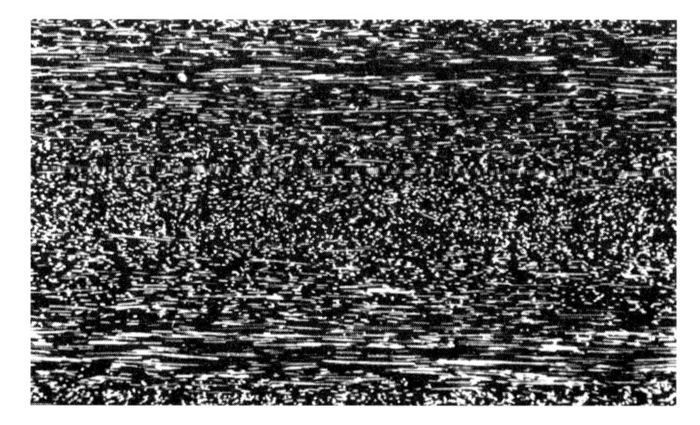

Figure 2.6 Section parallel to the flow direction through a glass fiber-reinforced polypropylene injection molding showing longitudinal orientation near the mold surface and transverse orientation in the core region. (Reproduced from Ref. [8] with permission of Oxford University Press.)

tensile yield strength near weld lines, plate-like fillers up to 30%, and cubic fillers up to 15% [16].

For semicrystalline polymers, fillers may affect crystallinity, size of crystallites, and direction of crystal growth. Filler surface may provide a large number of nucleation sites, although this also depends on surface functional groups and surface treatments. In certain polymers, fillers may promote transcrystallinity, which can improve adhesion and other properties [10].

References

1 Sheldon, R.P. (1982) Chapters 1, 3, 4, 5, in *Composite Polymeric Materials*, Applied Science Publishers, Barking, Essex, England.

2 Xanthos, M. (2009) Chapter 20: compatibilizers (theory and practice), in *Mixing and Compounding of Polymers*, 2nd edn (ed. I. Manas-Zloczower), Carl Hanser Verlag, Munich, Germany.

3 Wu, S. (1982) *Polymer Interface and Adhesion*, Marcel Dekker, Inc., New York.

4 Yosomiya, R. *et al.* (1990) Chapters 1, 2 and 3, in *Adhesion and Bonding in Composites*, Marcel Dekker Inc., New York.

5 Xanthos, M. (1988) Interfacial Agents for Multiphase Polymer Systems: Recent Advances *Polym. Eng. Sci.*, **28**, 1392.

6 Pukanszky, B. and Fekete, E. (1999) Adhesion and surface modification, *Advances in Polymer Science*, vol. 139, Springer-Verlag, Berlin, Heidelberg, pp. 121–153.

7 Solomon, D.H. and Hawthorne, D.G. (1983) *Chemistry of Pigments and Fillers*, John Wiley & Sons, Inc., New York, p. 6.

8 McCrum, N.G., Buckley, C.P., and Bucknall, C.B. (1997) Chapters 6 and 8, in *Principles of Polymer Engineering*, 2nd edn, Oxford, New York.

9 Wypych, G. (2000) *Handbook of Fillers*, ChemTec Publishing, Toronto, Canada.

10 Clegg, D.W. and Collyer, A.A. (eds) (1986) *Mechanical Properties of Thermoplastics*, Elsevier Applied Science Publishers, Barking, Essex, England.

11 Callister, W., Jr. (2003) Chapter 1 and App. B, in *Materials Science and Engineering: An Introduction*, 6th edn, John Wiley & Sons, Inc., New York.

12 Milewski, J.V. and Katz, H.S. (eds) (1987) Chapters 4 and 5 *Handbook of Reinforcement for Plastics*, Van Nostrand Reinhold Co., New York.
13 Drzal, L.T. (2003) Proceedings of the Functional Fillers for Plastics Conference, Intertech Corp., Atlanta, GA, October 2003, paper 7.
14 Katz, H.S. and Milewski, J.V. (eds) (1978) Chapters 2, 3 and 20, in *Handbook of Fillers and Reinforcements for Plastics*, Van Nostrand Reinhold Co., New York.
15 Kelly, A. (1966) Chapter V and App. A, in *Strong Solids*, Clarendon Press, Oxford, UK.
16 Hohenberger, W. (2001) Chapter 17, in *Plastics Additives Handbook* (ed. H. Zweifel), Hanser Publishers, Munich, Germany.
17 Nielsen, L.E. and Landel, R.F. (1994) Chapters 7 and 8, in *Mechanical Properties of Polymers and Composites*, 2nd edn, Marcel Dekker, Inc., New York.
18 Mascia, L. (1982) Chapter 4, in *Thermoplastics: Materials Engineering*, Applied Science Publishers, Barking, Essex, UK.
19 Piggott, M.R. (1980) Chapters 4, 5, 6, 8, in *Load Bearing Fibre Composites*, Pergamon Press, Inc., Oxford, UK.
20 Sperling, L.H. (2006) *Introduction to Physical Polymer Science*, 4th edn, John Wiley & Sons, Inc., Hoboken, NJ, pp. 698–706.
21 Xanthos, M., Faridi, N., and Li, Y. (1998) Processing/structure relationships of mica-filled polyethylene films with low oxygen permeability. *Int. Polym. Process.*, **13** (1), 58.
22 Osswald, T.A. and Menges, G. (1995) Chapter 8, in *Materials Science of Polymers for Engineers*, Hanser Publishers, Munich, Germany.
23 Callister, W.D., Jr. and Rethwisch, D.G. (2008) *Fundamentals of Materials Science and Engineering*, 3rd edn, John Wiley & Sons, Inc., Hoboken, NJ, p. 632.
24 Chanda, M. and Roy, S.K. (2007) *Plastics Technology Handbook*, CRC Press, Boca Raton, Florida, pp. 3–68.
25 Gerdeen, J.C., Lord, H.W., and Rorrer, R.A.L. (2006) *Engineering Design with Polymers and Composites*, CRC Press, Boca Raton, Florida, pp. 284–292.
26 Mu, B., Wang, Q., Wang, T., Wang, H., and Jian, L. (2009) The friction and wear properties of clay filled PA66. *Polym. Eng. Sci.*, **48** (1), 203.
27 Gopakumar, T.G., Patel, N.S., and Xanthos, M. (2006) Effect of nanofilllers on the properties of flexible protective coatings. *Polym. Comp.*, **27** (4), 368–380.
28 Gusev, A.A. and Lusti, H.R. (2001) Rational Design of nanocomposites for barrier applications. *Adv. Mater.*, **13**, (21), 1641–1643.
29 Gusev, A.A., Guseva, O. and Lusti, H.R. (2004) The influence of platelet disorientation on the barrier properties of composites: a numerical study. *Model. Simul. Mater. Sci. Eng.*, **12**, 1201–1207.
30 Tadmor, Z. and Gogos, C.G. (2006) *Principles of Polymer Processing*, 2nd edn, John Wiley & Sons, Inc., Hoboken, NJ, pp. 638–643.
31 Fisa, B. (1990) Injection molding of thermoplastic composites, *Composite Materials Technology* (eds P.K. Mallick and S. Newmann), Hanser Publishers, Munich, Germany, pp. 265–320.
32 Han, C.D. (1981) Chapter 3, in *Multiphase Flow in Polymer Processing*, Academic Press, New York.
33 Hornsby, P.R. (1999) Rheology, compounding and processing of filled thermoplastics, *Advances in Polymer Science*, vol. **139**, Springer-Verlag, Berlin, pp. 155–217.
34 Advani, S.G. and Tucker, C.L. (1987) The use of tensors to describe and predict fiber orientation in short fiber composites. *J. Rheol.*, **31**, 751.
35 Wang, J., Silva, C.A., Viana, J.C., van Hattum, F.W.J., Cunha, A.M., and Tucker, C.L., III (2008) Prediction of fiber orientation in a rotating compressing and expanding mold. *Polym. Eng. Sci.*, **48** (7), 1405–1413.

3
Mixing of Fillers with Plastics

David B. Todd

3.1
Introduction

Deriving the maximum benefit from the incorporation of fillers into polymers depends upon achieving a uniform distribution of well wet-out individual particles and/or fibers.

Filler incorporation may be performed in either batch or continuous equipment. For small production requirements or for very viscous products such as some filled elastomers, double-armed sigma blade mixers (Figure 3.1), Banbury mixers (Figure 3.2), or two-roll mills may still be used. Batch mixers allow sequence control of the addition of ingredients to obtain the final desired product without the need for multiple feeders. Disadvantages of batch mixers may be difficulty in emptying and extra processing required for the end-product shaping.

Bulk molding compounds, wherein chopped fiberglass and/or sisal is incorporated into a plastic (usually thermosetting) matrix along with other fillers, are frequently prepared in double-arm mixers (Figure 3.1). The 180° single-curve blade (Figure 3.3b) with relatively large blade-to-housing clearance was developed with the aim of achieving less fiber degradation and more complete discharge without the tendency of the material to wrap around the center wing sections of the sigma blades (Figure 3.3a).

For most applications of filler incorporation into plastics, most processors now use extruders, either single-screw or twin-screw, to achieve the desired compounds. The primary task to be accomplished in extrusion compounding generally include the following steps:

- Possible pretreatment of the ingredients
- Metering and feeding the ingredients
- Melting of solid-fed polymers
- Breakup of agglomerates
- Providing uniform distribution of filler
- Venting
- Developing pressure for discharge

Functional Fillers for Plastics: Second, updated and enlarged edition. Edited by Marino Xanthos

ISBN: 978-3-527-32361-6

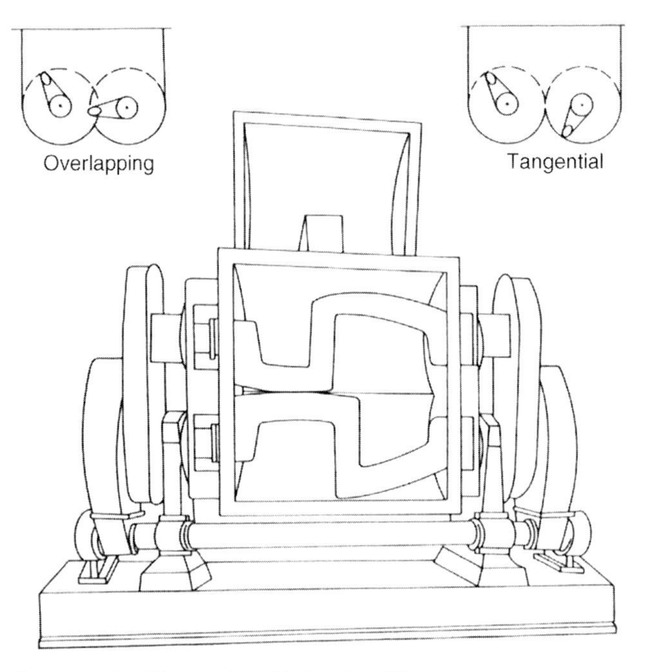

Figure 3.1 Double-arm sigma blade mixer [18].

Further aims are

- avoiding excessive screw and barrel wear;
- minimizing energy consumption.

There may be significant differences between single- and twin-screw extruders in how they achieve the above functions.

3.2
Pretreatment of Fillers

Frequently, either the polymer or the filler(s) may contain too much moisture such that predrying is required, otherwise unremoved moisture may

- interfere with the bonding between the filler and the polymer;
- contribute to the degradation of the polymer;
- create undesired bubble formation in the product.

Drying may be accomplished by the passage of low-humidity hot air through batch mixers such as ribbon blenders, flow mixers, or high intensity mixers. Polymers such as nylons and polyesters require drying under vacuum to avoid degradation by hydrolysis. Grulke [1] outlines some of the process-engineering considerations involved in the slow diffusion process of removing moisture from solids.

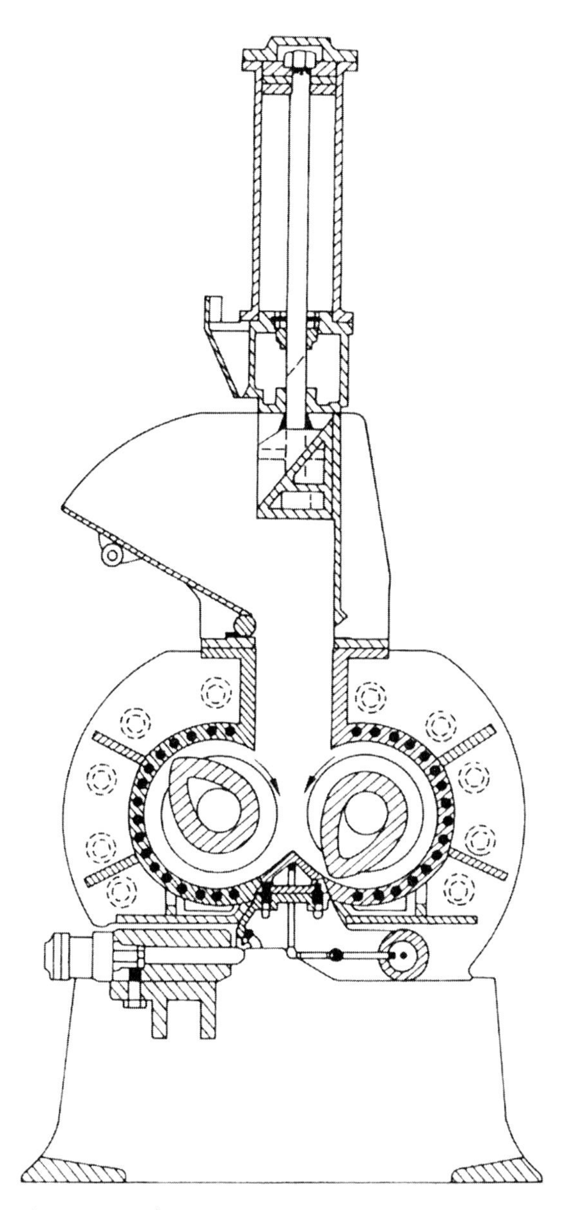

Figure 3.2 Banbury mixer [18].

3.3
Feeding

Generally, single-screw extruders (SSEs), such as that shown in Figure 3.4, are flood fed; that is, the rate of feed addition is controlled by the extruder screw speed.

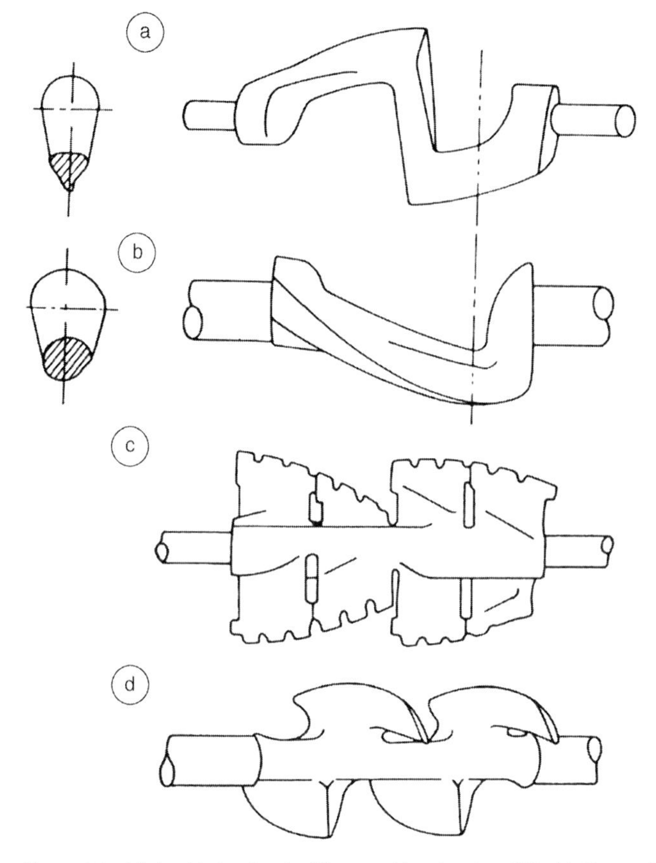

Figure 3.3 Mixing blades for double-armed batch mixer [18]. (a) Sigma, (b) 180° single curve, (c) multiwing overlap, and (d) double naben.

Dry preblending of multiple solid ingredients can be performed in simple ribbon or plow mixers. With low levels of addition of some nonabrasive fillers, the polymer and filler can be dry blended and charged to the feed hopper. Feeding large quantities of powdered fillers through a downstream feedport in an SSE is difficult and will likely require a crammer feeder and provision for venting the air accompanying low bulk density fillers.

The primary advantages of using twin-screw extruders (TSEs) for filler incorporation arise from

- greater volumetric feeding capacity (especially important with low bulk density powders);
- the possibility of eliminating preblending operations;
- flexibility in porting for downstream feeding and venting;
- independence of feed rate and screw speed;
- greater flexibility in controlling the mixing action.

Thrust Bearing Assembly
Hopper
Throat
Barrel
Insulated Barrel Guard
Powder Seal
Water Cooled Heaters
Screw
Barrel Support
Belt and Sheave Assembly
Motor
Reducer
Base
Closed Loop Cooling System

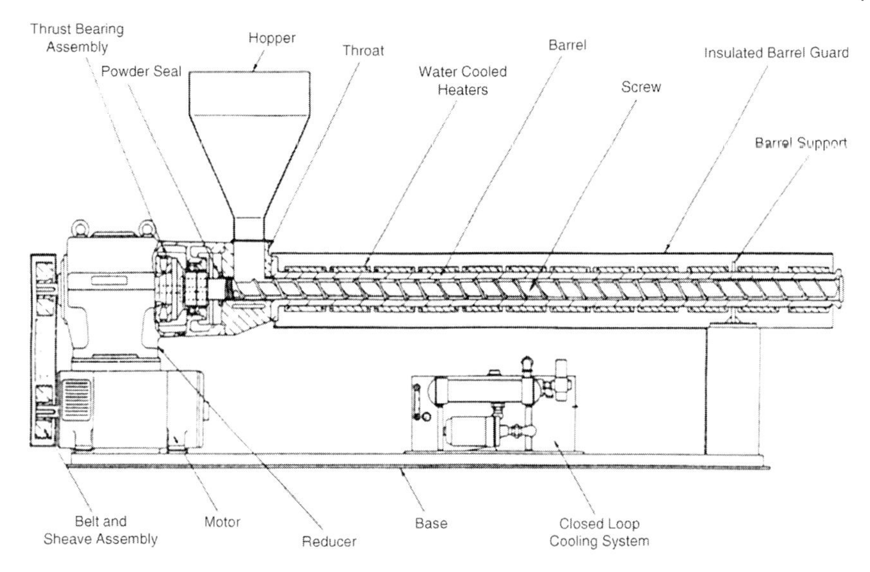

Figure 3.4 Typical single-screw extruder [18].

Twin-screw extruders are classified as tangential or intermeshing and counter-rotating or corotating (Figure 3.5). The most prevalent is the corotating intermeshing type (Figure 3.6), available from dozens of manufacturers. Most suppliers have learnt how to solve the problems associated with incorporation of high

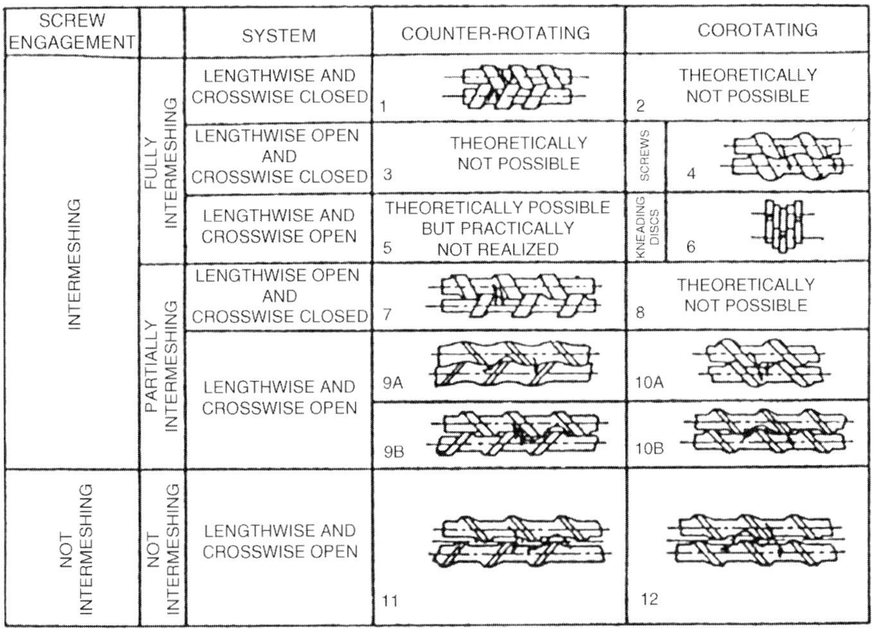

SCREW ENGAGEMENT
SYSTEM
COUNTER-ROTATING
COROTATING
INTERMESHING
FULLY INTERMESHING
LENGTHWISE AND CROSSWISE CLOSED
1
2
THEORETICALLY NOT POSSIBLE
LENGTHWISE OPEN AND CROSSWISE CLOSED
3
THEORETICALLY NOT POSSIBLE
SCREWS
4
LENGTHWISE AND CROSSWISE OPEN
5
THEORETICALLY POSSIBLE BUT PRACTICALLY NOT REALIZED
KNEADING DISCS
6
PARTIALLY INTERMESHING
LENGTHWISE OPEN AND CROSSWISE CLOSED
7
8
THEORETICALLY NOT POSSIBLE
LENGTHWISE AND CROSSWISE OPEN
9A
10A
9B
10B
NOT INTERMESHING
NOT INTERMESHING
LENGTHWISE AND CROSSWISE OPEN
11
12

Figure 3.5 Classification of twin-screw extruders [18].

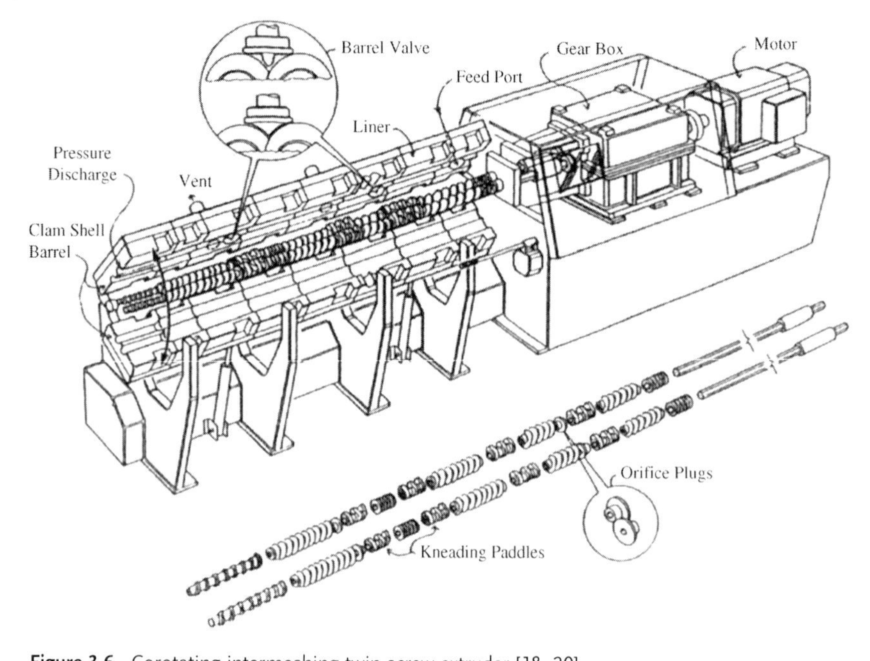

Figure 3.6 Corotating intermeshing twin-screw extruder [18, 20].

loadings of fillers, and some offer special features specifically addressing these problems.

Although the number of screw starts can vary from single to triple, the most common is double, as shown in Figure 3.7. The flight tips of one screw wipe the

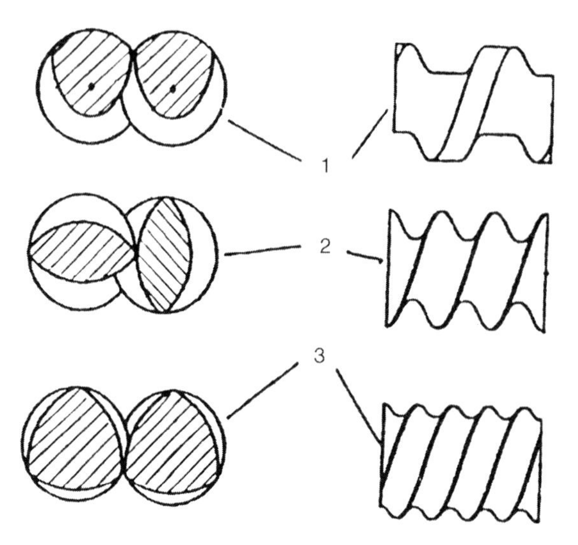

Figure 3.7 Single, double, and triple start corotating intermeshing screws [18, 20].

channels of the opposite screw. A deep intermesh implies a good conveying capacity, but too deep an intermesh means a smaller root diameter, and therefore less torque capability, which can also limit capacity. The optimum operating conditions occur when all available extruder power is used just as the feed intake has reached its limit.

The increasingly faster screw speeds (to 1200 rpm) and shorter residence times (<20 s) now being offered by TSE manufacturers place a greater precision requirement for the feeders. Generally, loss-in-weight feeders will be required for the desired product uniformity [2], as back-mixing in extruders is so minimal that feed irregularities are not dampened out during the brief passage through the extruder.

Metering and feeding of fibers is particularly difficult because of the tendency of the fibers to clump together. Some combination of a weight belt feeder in turn feeding a side entering crammer feeder is probably the best solution for materials such as fiberglass [3]. With such a feeding system, a large volume of air will also be pumped into the extruder, and it is desirable to provide for upstream venting of the starved screw section into which the filler is being fed. Typical barrel sections combining side entry plus venting are shown in Figure 3.8.

Melting of a polymer in the presence of abrasive fillers is undesirable for two reasons. The large forces encountered during melting can result in agglomeration (briquetting) of powder fillers and will also cause much more abrasive wear of screws and barrels. The mechanism for the formation of agglomerates in single-screw extruders has been described by Gale [4], as shown in Figure 3.9.

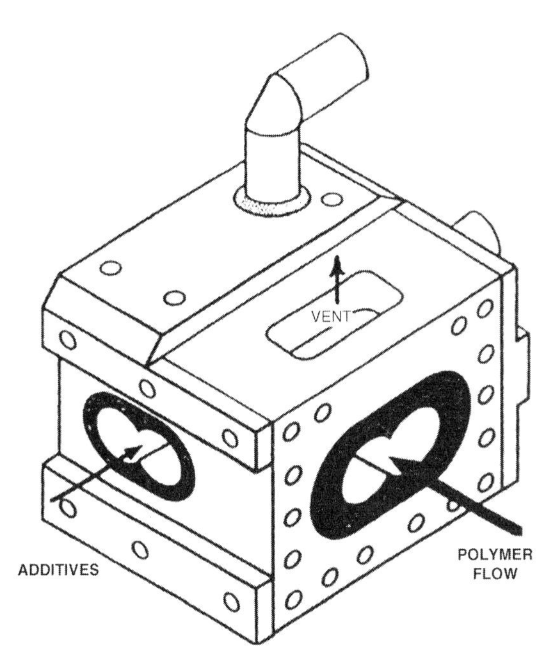

Figure 3.8 Barrel section for side entry feeding plus venting [20].

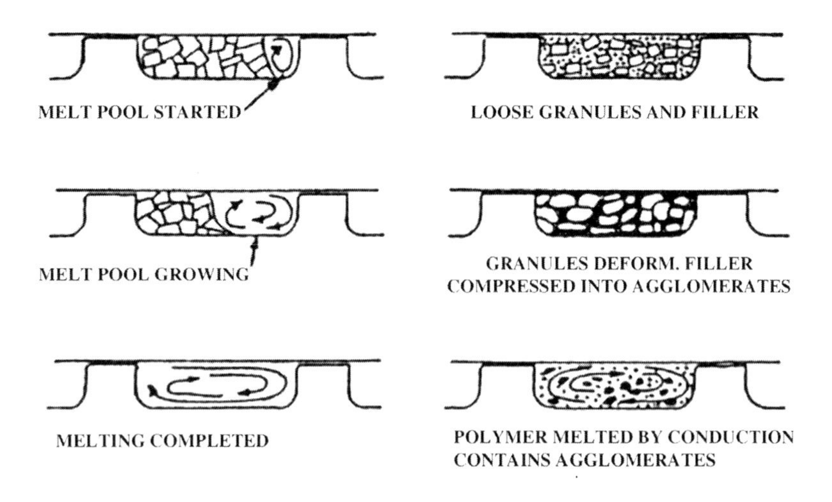

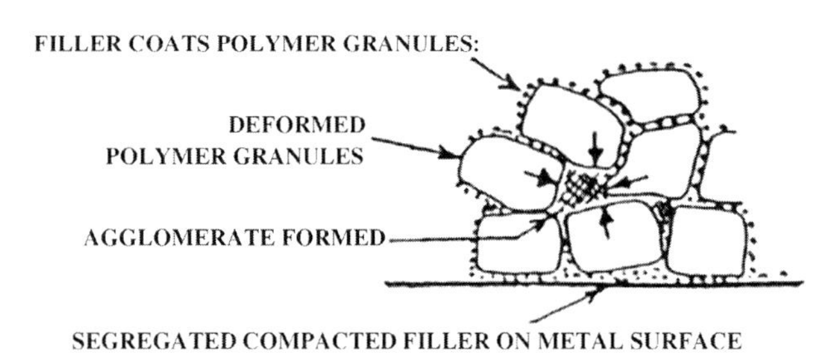

Figure 3.9 Mechanism of agglomerate formation [4].

The best way of overcoming the problems of agglomerates is simply avoiding their formation, generally by feeding the filler downstream after the polymer is fully melted. The agglomeration effect accompanying polymer melting in the presence of solid fillers has clearly been shown by Rogers *et al.* [5] who reported that joint feeding of 15 wt% $CaCO_3$ and 85 wt% polystyrene pellets led to significant agglomerate formation (diameter $d > 20\,\mu m$) of initially 0.7 μm diameter $CaCO_3$ powder. The potential for concomitant agglomerate formation with pellet melting is particularly high in kneading block sections of intermeshing TSEs because of the high pressures that can be developed in the compression/expansion cycles of corotating TSEs or during the calendering impact of counter-rotating TSEs. Simultaneous pellet melting and filler addition can cause excessive screw and barrel wear.

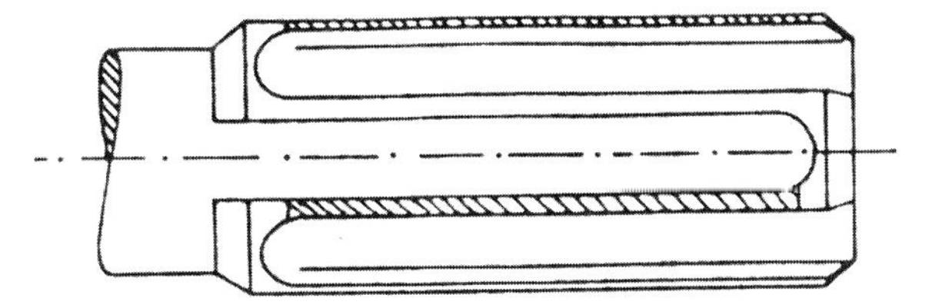

Figure 3.10 Maddock (Union Carbide) mixing element.

3.4 Melting

The generally accepted melting model for SSEs is the one described by Tadmor and Gogos [6] wherein the passive solid bed moves axially downstream being gradually compressed in the transition sections such that a melt film forms at the barrel wall by a combination of direct heat transfer through this wall and friction of the unmelted solids against it.

Initial melting in TSEs is usually dominated by plastic energy dissipation [7], which is the unrecovered energy of dynamic compression of the solid pellets in the kneading block section of the extruder. After the initial melting produces a slurry, which predominantly consists of some unmelted solid polymer pellets floating in a "sea" of the continuous melt phase, completion of the melting process occurs by heat transfer at the melt–polymer interface, with the energy supplied mainly by viscous energy dissipation and to a lesser extent by barrel heat transfer.

In SSEs, it is often prudent to utilize a Maddock (Union Carbide) mixing element (Figure 3.10). This acts as a crude filter to prevent passage of the unmelted floaters and retaining them until melting reduces their size to less than that of the clearance gaps. In TSEs, the same filtering effect can be achieved with intermeshing blister rings.

3.5 Solids Introduction and Mixing

After the screw configuration, in which the polymer is melted, a starved section is provided into which the filler solids can be introduced. Preferably, as previously mentioned in Section 3.3, provision should also be made for venting the air accompanying the powdered solids (e.g., in Figure 3.8). If oxygen is likely to cause degradation of the base resin, it will be necessary to purge the solids feeder with nitrogen. The lubrication supplied by the molten polymer and the low-pressure requirement of the initial incorporation will minimize screw and barrel wear. A typical configuration and pressure profile for incorporating fillers in a TSE is shown in Figure 3.11 [8].

The mixing requirement in filler incorporation is usually a combination of dispersive and distributive actions, as illustrated in Figure 3.12. Dispersive mixing

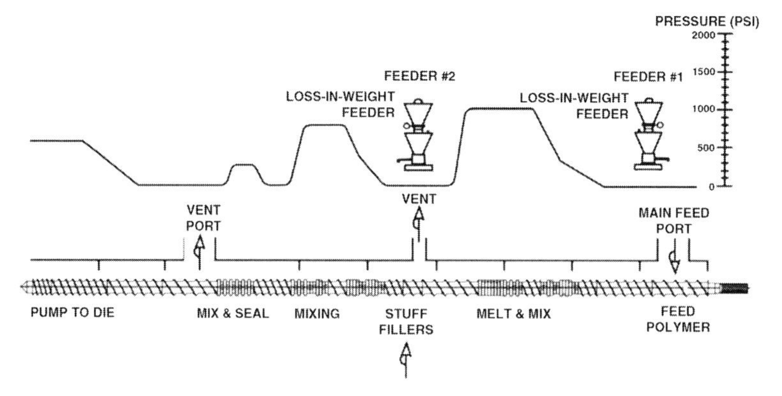

Figure 3.11 TSE configuration and pressure profile for filler incorporation [8].

is the breaking up of agglomerates or the unraveling of fiber bundles. Distributive mixing is achieving equal spatial distribution of the filler throughout the plastic matrix.

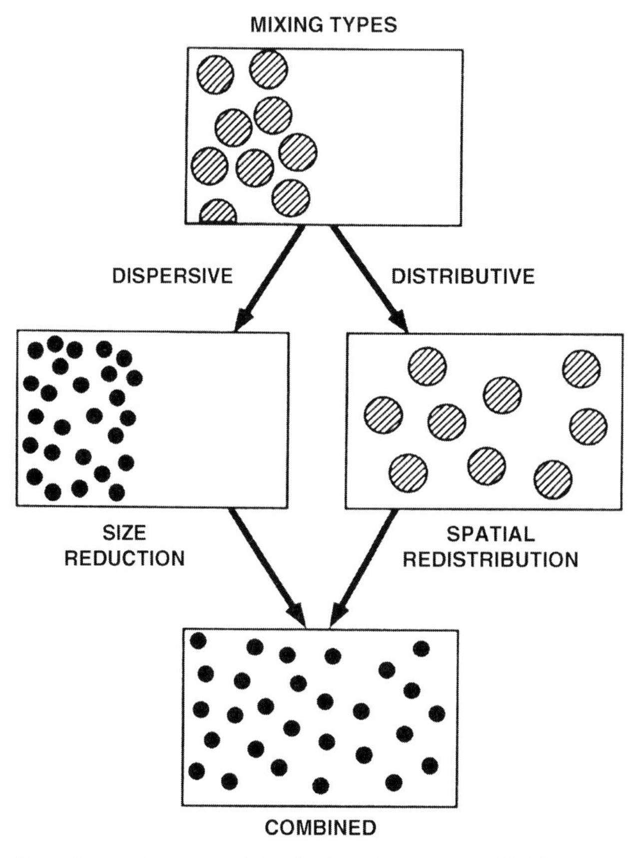

Figure 3.12 Dispersive and distributive mixing aspects [18, 20].

Screws by themselves do not provide much mixing. In an SSE, the requirement for developing pressure to form the extrudate leads to some cross-channel mixing.

Shearing forces will be required to break up agglomerates. The level of these shearing forces will depend upon the nature of the filler. For example, intense shearing is usually required when dispersing carbon black. Even more severe shearing will be needed to disperse montmorillonite clay into nanosized platelets.

Incorporation of fiberglass is a special case. The glass fibers are supplied in chopped bundles that have been treated with sizings and suitable coupling agents to increase the adhesion between polymer and glass. These bundles need to be "unwrapped" and wetted out, ideally with a minimum of fiber breakage. Dispersion of glass fibers in TSEs is further discussed in Chapter 7.

It is always wise to use some sort of mixing enhancer to improve distributive mixing rather than to rely on the mixing normally achieved with screws alone. Figure 3.13 illustrates some of the SSE mixing enhancers that have been employed. In general, all the enhancers improve the distribution of filler by a series of divisions and recombinations. Simple shear mixing should be interrupted by a

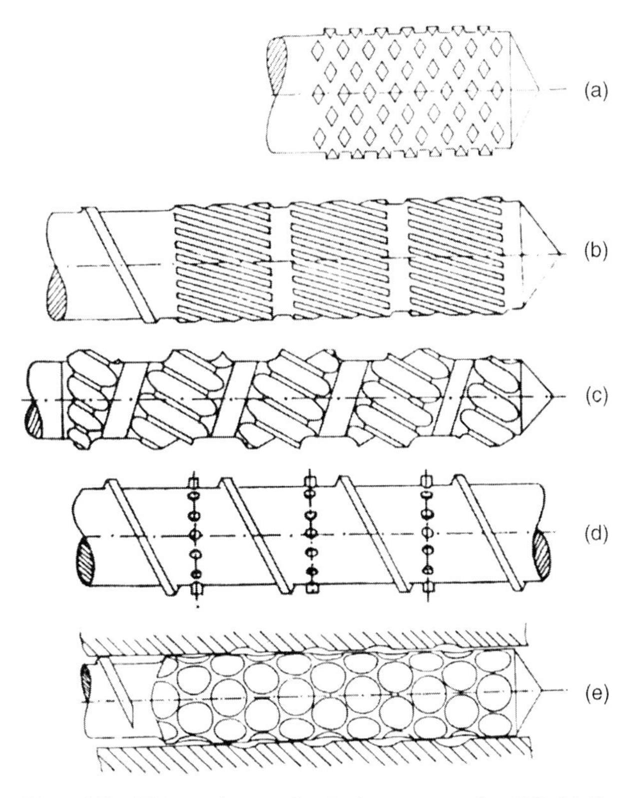

Figure 3.13 Mixing enhancers for single-screw extruders [20]. (a) Pineapple, (b) Dulmage, (c) Saxton, (d) pin, and (e) cavity transfer.

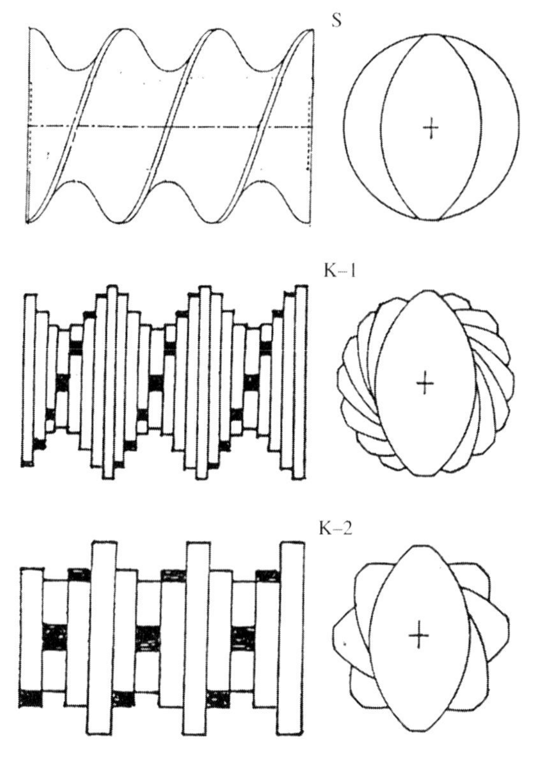

Figure 3.14 Comparison of kneading elements and screws [18].

combination of cutting and turning actions. Gale [9] illustrated the homogeneity achieved with nine different mixers along with the power consumption for each mixing configuration. Gale showed that the *Cavity Transfer Mixer* was very effective in achieving homogeneity with relatively low-pressure drop and low-temperature rise.

As in the case of SSEs, TSE screws themselves do not provide much mixing. The bulk of the melting and mixing arises from the interplay of kneading paddles, which are the offset straight elements with the same cross section as the screws. Kneading paddle arrays have some of the same conveying characteristics as screws (Figure 3.14), but they also provide unique mixing actions not available in single-screw extruders. The dispersion face (Figure 3.15) can provide the high-shearing action desired for agglomerate breakup. Excellent distribution is achieved by the expansion/compression action, as shown in Figure 3.16. Each quarter-turn produces an expanding and compression of the process parts of the cross section, and these squeezing actions cause extensional flow and redistribution of the polymer–filler mixture as it passes down the extruder. In addition to kneading paddles, toothed gear elements (Figure 3.17) can be effectively used to provide enhanced distributive mixing action.

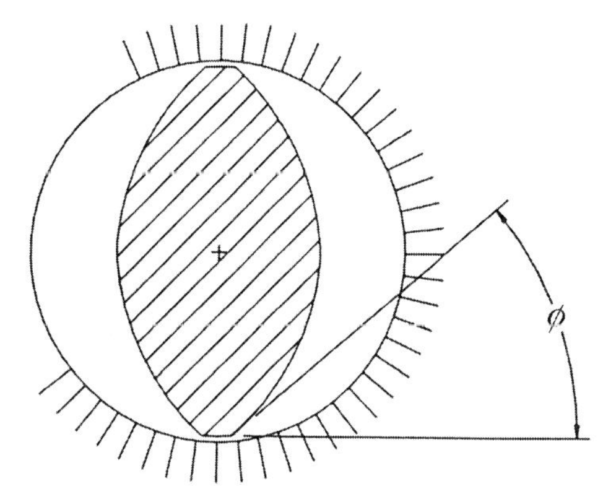

Figure 3.15 Dispersion face of a kneading paddle [18, 20].

3.6 Venting

Removal of air accompanying low bulk density fillers is essential, as noted in Section 3.3. Off-gassing of processing aids associated with some fillers, such as coatings on fiberglass, will also require venting. If a vacuum is required for adequate devolatilizing, the extruder screw(s) must have a melt seal upstream of the vent port zone to prevent sucking air or unincorporated powder out the vent port. The melt seal can be achieved with short sections of reverse screws, reverse-kneading blocks, or blister rings. The screws in the vent-port region need to be designed to be operating starved, preferably less than half-full, since the mixture may be foaming as it passes the melt seal.

The degree of fill (f) in the screws is approximately equal to the ratio of net flow (Q) to drag flow (Q_d). The drag flow per revolution is equal to one half of the volume contained in the open cross section (α) over the lead length (z):

$$Q_d/N = \alpha z/2 \tag{3.1}$$

$$f = Q/Q_d \tag{3.2}$$

For fully intermeshing corotating bilobe TSEs, the open cross section (α) can be calculated from the screw diameter (D) and the channel depth (h) [11]:

$$\alpha = 3.08hD \tag{3.3}$$

3.7 Pressure Generation

The pressure required for shaping the product (sheet profile or pellets) depends upon the flow rate, the aperture geometry, and the viscosity of the filled polymer at the exit shear rate. In general, the viscosity of a polymer–filler mixture (η_c) increases as the

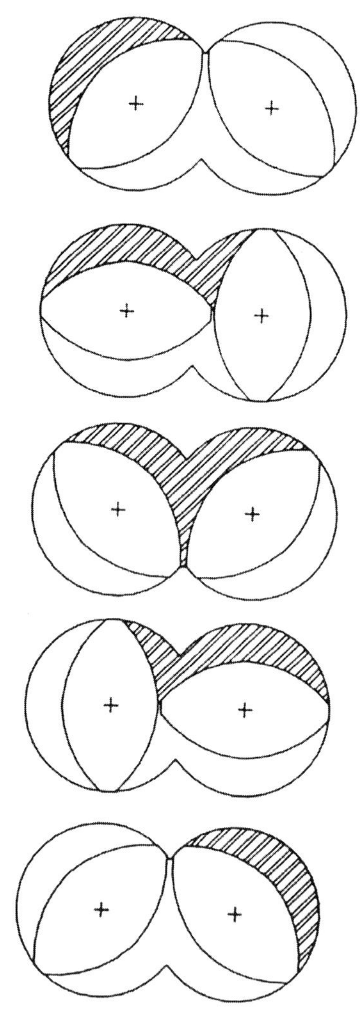

Figure 3.16 Expansion/compression squeezing action of corotating intermeshing bilobal paddles [18, 20].

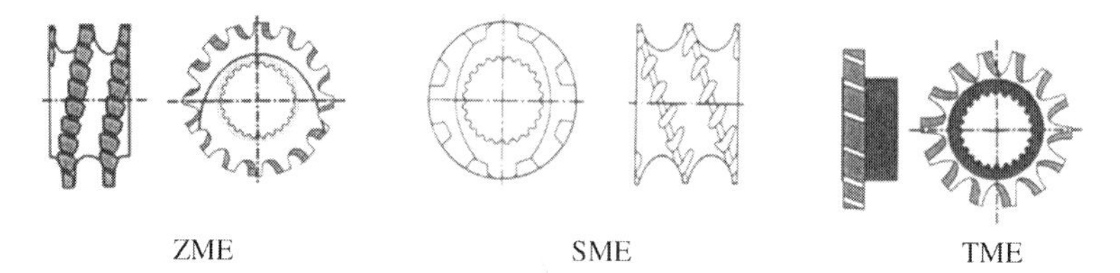

Figure 3.17 Toothed gear mixing elements [18, 20].

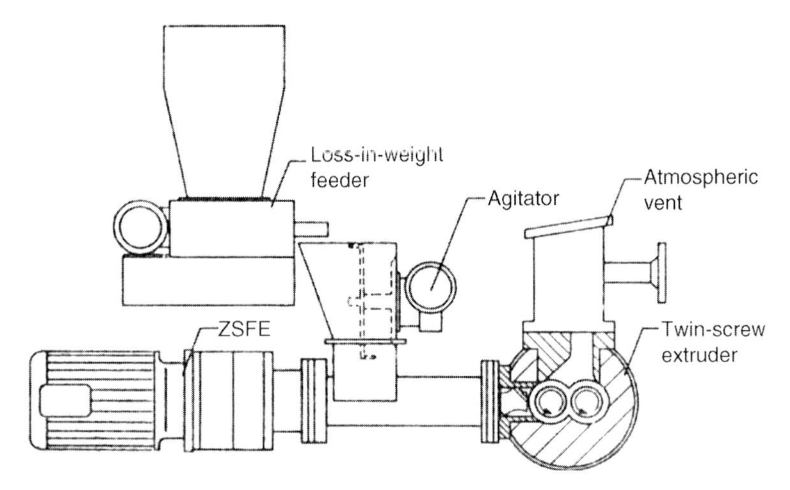

Figure 3.18 Horizontal side feed extruder [15, 20].

volume fraction concentration (V_f) of filler is increased. Because of the irregular shapes of most of the fillers, the viscosity of the mixture will increase more than that of uniformly sized spheres predicted by Einstein [10]:

$$\eta_c/\eta_m = 1 + 2.5 V_f \tag{3.4}$$

where η_m is the matrix viscosity.

In some platelet-type fillers such as talc and mica, the platelets can align with the streamlines at higher shear rates, and the viscosity increase is less than that predicted by the above equation (see Chapter 2).

3.8 Process Examples

Valsamis and Canedo [12] have illustrated how the use of different mixing elements in a Farrel FTX 80-twin-screw extruder affected glass breakage and compound properties in a nylon/fiberglass mixture. Wall [13] studied a tandem arrangement of melting nylon 6,6 and subsequent addition of 4.5 mm long glass fibers in a short TSE followed by pressure development for pelletizing in a more slowly rotating SSE. In the ensuing injection molding process, remelting of the fiber-filled pellets reduced the fiber length from about 500 to 300 μm. Grillo *et al.* [14] studied the effect of TSE mixing elements and die design on glass fiber length retention and ultimate physical properties with glass filled styrene-maleic anhydride copolymers.

Using equipment, such as that shown in Figure 3.18, Mack [15] has investigated the incorporation of high talc loadings into polypropylene. Although the true talc particle density is around 2.7 g/cm^3, the bulk density was only about 0.2 g/cm^3,

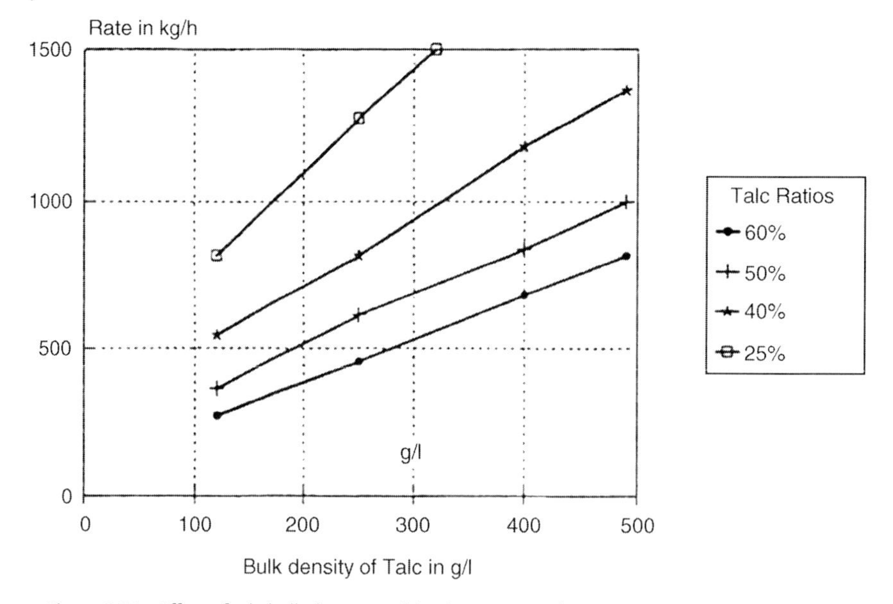

Figure 3.19 Effect of talc bulk density and loading on extruder capacity [15, 20].

and may well be as low as 0.05 g/cm^3 when aerated. Thus, a talc stream being forced into such an extruder may contain 95% air.

The effect of talc bulk density and talc loading on extruder capacity for a Berstorff ZE90-A TSE is shown in Figure 3.19. Increasing the talc ratio made the mixing process more difficult and decreased the output rate. Mack also showed that the melt viscosity of the polymer could affect the limits of filler incorporation. For example, the level of incorporation of calcium carbonate (0.34 g/cm^3 bulk density) was shown to be higher with less viscous polypropylene (Figure 3.20) because of easier wetting-out of the filler.

Deep-flighted TSEs may reach torque-limited capacity before the feeding rate limitation. Because the heat capacities of most fillers are only about half that of typical polymers, the capacity of such an extruder may actually increase as the filler loading is increased.

Andersen [16] has provided some guidelines for preparing nanocomposite polymer compounds, such as clay/polypropylene/maleated polypropylene, in TSEs. Clay delamination was found to be improved by adding all the ingredients in the feed port, as the stresses were the greatest as the polypropylene was melting, and there was no tendency to form conglomerates as occurs with $CaCO_3$. Kapfer and Schneider [17] studied the formation of highly filled compounds in both a TSE and a screw kneader wherein the interrupted screw oscillates as well as rotates, and the screw channels are wiped by rows of teeth protruding inward from the barrel wall. They concluded that the TSE would be preferred when the polymer is not susceptible to shear degradation, whereas the reciprocating single screw offers some benefits when fillers or fibers are susceptible to easy breakage. Deep channels

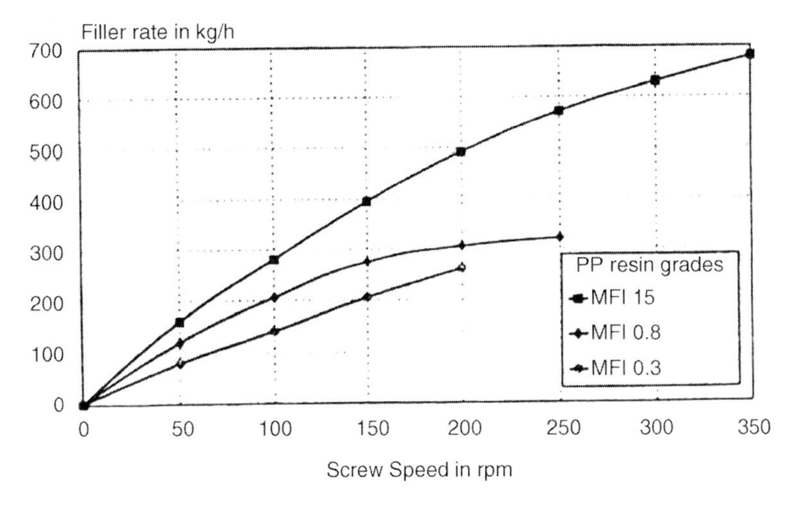

Figure 3.20 Effect of polymer viscosity on extruder capacity [15, 20].

are preferred with both types of equipment to facilitate easier intake of larger volumes of filler.

3.9 Further Information

Virtually all aspects of industrial mixing are described in the recent *North American Mixing Forum's Handbook* [18]. The *SPE Guide on Extrusion Technology and Troubleshooting* [19] contains chapters particularly relevant to processing of polymeric systems. Compounding operations as perceived from the equipment manufacturer's viewpoint are detailed in "Plastics Compounding: Equipment and Processing" [10].

References

1 Grulke, E.A. (1994) *Polymer Process Engineering*, Prentice Hall, Englewood Cliffs, NJ.
2 Welsch, R. (2003) Using loss-in-weight feeders in compounding. *Plast. Addit. Compound.*, **5**, 40–45.
3 Häuptli, A. (2003) Direct processing of long fiber reinforced plastics: selecting a feeding system. *Plast. Addit. Compound.*, **5**, 36–39.
4 Gale, M. (1997) Compounding with single screw extruders. *Adv. Polym. Technol.*, **16** (4), 251–262.
5 Rogers, M.J., Koelling, K.W., Read, M.D., and Spalding, M.A. (2001) The effect of three-lobe, off-set kneading blocks on the dispersion of calcium carbonate in polystyrene resins. Proceedings of the 59th SPE ANTEC, 47, pp. 129–133.
6 Tadmor, Z. and Gogos, C.G. (2006) *Principles of Polymer Processing*, 2nd edn, John Wiley & Sons, Inc., Hoboken, NJ.
7 Qian, B., Todd, D.B., and Gogos, C.G. (2003) Plastic energy dissipation and its role in heating/melting of single component polymers and

multicomponent polymer blends. *Adv. Polym. Technol.*, **22** (2), 1–11.
8 Martin, C. (2008) Twin screw extrusion advances for compounding, devolatilization and direct extrusion. Proceedings of the Plastics Extrusion Asia Conference, Bangkok, Thailand, March 17–18.
9 Gale, G.M. (1991) Mixing of solid and liquid additives into polymers using single screw extruders. Proceedings of the 49th SPE ANTEC, 37, pp. 95–98.
10 Todd, D.B. (ed.) (1998) *Plastics Compounding: Equipment and Processing*, Hanser Publishers, Munich.
11 Einstein, A. (1906) A new method of determining molecular dimensions. *Ann. Phys.*, **19**, 289; (1911), Corrections to my work: a new determination of molecular dimensions, **34**, 591.
12 Valsamis, L.N. and Canedo, E.L. (1997) Compounding glass-reinforced plastics using novel mixing sections. *Plast. Eng.*, **53** (4), 37–39.
13 Wall, D. (1987) The processing of fiber reinforced thermoplastics using co-rotating twin screw extruders. Proceedings of the 45th SPE ANTEC, 33, pp. 778–781.
14 Grillo, J., Andersen, P.G., and Papazoglou, E. (1991) Die designs for compounding fiberglass strand on co-rotating twin screw extruders. Proceedings of the 49th SPE ANTEC, 37, pp. 122–127.
15 Mack, M. (1997) Compounding highly filled polyolefins. *Plast. Eng.*, **53** (4), 33–35.
16 Andersen, P.G. (2002) Twin screw guidelines for compounding nanocomposites. Proceedings of the 60th SPE ANTEC, pp. 48–53.
17 Kapfer, K. and Schneider, W. (2003) Production of compounds with high filler or fiber loading on screw extruders. Proceedings of the 60th SPE ANTEC, 49, pp. 246–250.
18 Paul, E.L., Antiemo-Obeng, V.A., and Kresta, S.M. (eds) (2004) *Handbook of Industrial Mixing: Science and Practice*, John Wiley & Sons, Inc., Hoboken, NJ.
19 Vlachopoulos, J. and Wagner, J.R. (eds) (2001) *SPE Guide on Extrusion Technology and Troubleshooting*, Society of Plastics Engineers, Brookfield, CT.
20 Todd, D.B. (2000) Improving incorporation of fillers in plastics. *Adv. Polym. Technol.*, **19** (1), 54.

Part Two
Surface Modifiers and Coupling Agents

Functional Fillers for Plastics: Second, updated and enlarged edition. Edited by Marino Xanthos

ISBN: 978-3-527-32361-6

4
Silane Coupling Agents

Björn Borup and Kerstin Weissenbach

4.1
Introduction

After the paper and coatings markets, the plastics industry is the third largest outlet for white minerals in Europe, as well as in North America and Asia. These minerals include a number of natural products such as talc, kaolin, wollastonite, and ground calcium carbonate, as well as synthetic products such as precipitated calcium carbonate, aluminum trihydroxide $Al(OH)_3$ (ATH), magnesium dihydroxide $Mg(OH)_2$ (MDH), glass fibers, synthetic silicas, and silicates. A significant amount of all thermoplastics, thermosets, and elastomers worldwide are compounded and reinforced with fillers and fibers. A major need for silanes as coupling agents arose in the 1940s when glass fibers were first used as reinforcement in unsaturated polyester (UP) resins. The first commercial silanes appeared in the mid-1950s. Since then, they have become the most common and widely used coupling agents. Silanes offer tremendous advantages for virtually all market segments involving polymer/filler interactions and, as a result, are in widespread use in modifying the surface of fillers [1–6].

Surface modification of fillers with silanes may generate the following performance benefits:

- Improved dimensional stability
- Modified surface characteristics (water repellency or hydrophobicity)
- Improved wet-out between resin and filler
- Decreased water vapor transmission
- Controlled rheological properties (higher loadings with no viscosity increase)
- Improved filler dispersion (no filler agglomerates)
- Improved mechanical properties and high retention under adverse environments
- Improved electrical properties

The world market for silanes used in filler treatment is approximately divided as follows: Europe, 25%; Asia, 13%; Americas, 60%; rest of the world, 2%.

Functional Fillers for Plastics: Second, updated and enlarged edition. Edited by Marino Xanthos

ISBN: 978-3-527-32361-6

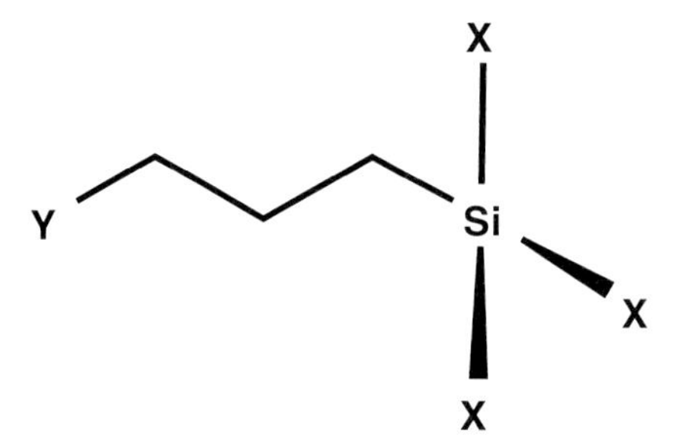

Figure 4.1 General structure of organosilanes.

Common silanes have the general formulas Y-$(CH_2)_3Si(X)_3$ (Figure 4.1) and Y-$(CH_2)_2Si(CH_3)(X)_2$. The silicon functional group X is a hydrolyzable group chosen to react with surface hydroxyl groups of the filler to produce a stable bond and is usually halogen or alkoxy. The silane coupling agents in commercial use are generally alkoxy based and bear one organic group attached to the silicon center, the general formula being Y-$(CH_2)_3Si(OR)_3$. The organofunctional group Y is bound tightly to the silicon via a short carbon chain and links with the polymer. This group has to ensure maximum compatibility with the resin system. Bonding to the polymer usually takes place by chemical reactions (grafting, addition, or substitution) or by physiochemical interactions (hydrogen bonding, acid–base interactions, interpenetrating polymer network (entanglements), or electrostatic attraction) [7–11]. The group Y may be nonfunctional or functional (reactive); examples of the latter are vinyl, amino, methacryl, epoxy, mercapto, and so on. Most silanes are colorless or slightly yellowish, low-viscosity liquids. Both nonfunctional and functional organosilanes, discussed below, are important commercial filler treatments.

4.2
Production and Structures of Monomeric Silanes

Chlorosilanes, as raw materials for organofunctional silanes, are technically produced by reaction of HCl with silicon metal in a fixed-bed reactor or fluidized bed. Products derived from this process are tetrachlorosilane ($SiCl_4$) and trichlorosilane ($HSiCl_3$). Tetrachlorosilane is used to manufacture optical fibers, high-purity silicon, silicic acid esters, and fumed silica. Trichlorosilane is, in addition to other uses, the raw material for organofunctional and alkyl silanes. Commercially available organofunctional silanes are produced by hydrosilylation reactions of olefins with hydrogen-containing silanes. Esterification then leads to the standard commercially available organofunctional silanes (Scheme 4.1).

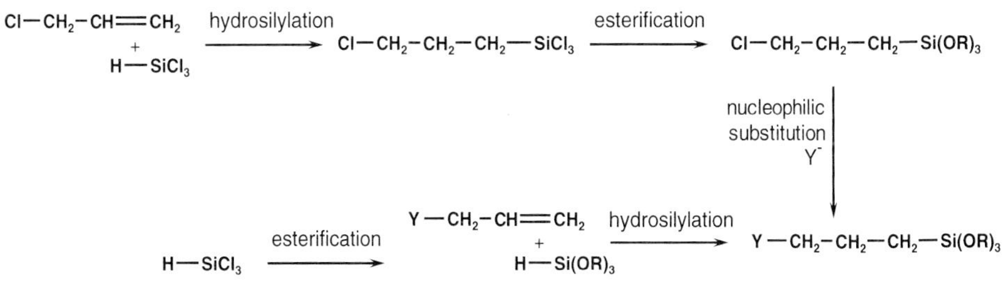

$$Cl{-}CH_2{-}CH{=}CH_2 + H{-}SiCl_3 \xrightarrow{\text{hydrosilylation}} Cl{-}CH_2{-}CH_2{-}CH_2{-}SiCl_3 \xrightarrow{\text{esterification}} Cl{-}CH_2{-}CH_2{-}CH_2{-}Si(OR)_3$$

$$Cl{-}CH_2{-}CH_2{-}CH_2{-}Si(OR)_3 \xrightarrow[\text{substitution } Y^-]{\text{nucleophilic}} Y{-}CH_2{-}CH_2{-}CH_2{-}Si(OR)_3$$

$$H{-}SiCl_3 \xrightarrow{\text{esterification}} H{-}Si(OR)_3;\quad Y{-}CH_2{-}CH{=}CH_2 + H{-}Si(OR)_3 \xrightarrow{\text{hydrosilylation}} Y{-}CH_2{-}CH_2{-}CH_2{-}Si(OR)_3$$

Scheme 4.1 Production of organofunctional silanes.

4.3 Silane Chemistry

Silanes are adhesion promoters that unite the different phases present in a composite material. These phases are typically organic resins (e.g., ethylene/vinyl acetate copolymer, EVA) and inorganic fillers (e.g., ATH, quartz, cristobalite, and clays) or fibrous reinforcements. Silanes form "molecular bridges" to create strong, stable, water- and chemical-resistant bonds between two otherwise weakly bonded surfaces. The properties and effects of silanes are defined by their molecular structure. The silicon at the center is combined with the organofunctional group Y and the silicon functional alkoxy groups, OR. The organofunctional group Y interacts with the polymer by chemical reactions and/or by physiochemical interactions. The silicon functional groups OR, usually alkoxy groups, can be hydrolyzed at the first stage of application, liberating the corresponding alcohol. Continuous reaction with water or moisture results in the elimination of all OR groups in the form of alcohol and their replacement with hydroxyl units forming silanols (see Scheme 4.2).

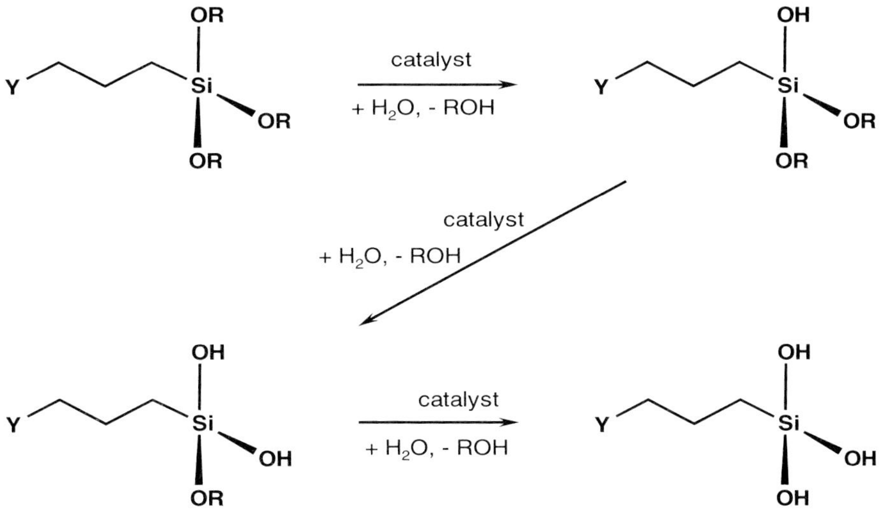

$$Y{-}CH_2CH_2CH_2{-}Si(OR)_3 \xrightarrow[+\,H_2O,\ -\,ROH]{\text{catalyst}} Y{-}CH_2CH_2CH_2{-}Si(OH)(OR)_2 \xrightarrow[+\,H_2O,\ -\,ROH]{\text{catalyst}} Y{-}CH_2CH_2CH_2{-}Si(OH)_2(OR) \xrightarrow[+\,H_2O,\ -\,ROH]{\text{catalyst}} Y{-}CH_2CH_2CH_2{-}Si(OH)_3$$

Scheme 4.2 Hydrolysis of trialkoxysilanes.

In the text that follows, the term "silanols" is used as a simplified description of the monosilanols, silanediols, and silanetriols in the case of trialkoxysilanes. Usually, the hydrolysis of the first OR group is the rate-controlling reaction step. The resultant silanol intermediates (Si−OH) react with active OH groups of the inorganic substrate, building up stable covalent (Si−O−substrate) bonds [12].

4.4 Types of Silanes

4.4.1 Waterborne Silane Systems

Organofunctional silanes can be applied to fillers neat or in aqueous or aqueous/alcohol solutions. If the silane is added to an aqueous filler slurry, the silane has to be water soluble. Some silanes will be water soluble after hydrolysis and the formation of silanols. These Si-OH functions prove to be very reactive in establishing a covalent bond between the filler and the silane. It has long been postulated that monomeric silanetriols are exclusively responsible for the activity of the silane and that these monomeric units are stable for a period of few hours to few days in aqueous solutions. However, it was proven by ^{29}Si-NMR spectroscopy that oligomeric silanes are active as well. Finally, the activity of the aqueous silane solution decreases as a result of cross-linking to give insoluble, polymeric siloxanes (gel structures); this is evident from a substantial decrease in adhesion. Knowledge of the silanol concentration and the degree of oligomerization are, therefore, important when using aqueous solutions [13].

For many filler treatment methods, the silanol form of silanes is desirable since silanols have greater solubility and reactivity than their alkoxysilane precursors. This proves to be a challenge for devising storage-stable water-borne silane systems that contain silanols. But using a special manufacturing process originally introduced by Hüls under the trade name Dynasylan® HYDROSIL has afforded a number of water-borne oligomeric silane systems with several different functionalities that are commercially available. These systems have a high concentration of active silanol groups that are stable in water for periods up to 1 year and are considered free of volatile organic compounds (VOCs) [14]. They are particularly useful in wet processes such as grinding and milling where VOCs are undesirable and the mineral can be treated *in situ* in the aqueous process slurry. An example of such a water-borne silane system is shown in Figure 4.2.

4.4.2 Oligomeric Silanes

Silane hydrolysis and condensation take place on the surface of the mineral filler, thus, forming oligomeric silane structures. Oligomeric silanes (Figure 4.3) are commercially available under the Dynasylan® trade name. They are low-viscosity

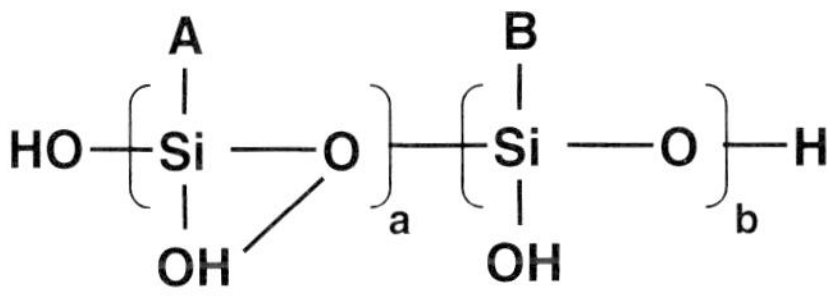

Figure 4.2 Example of a water-borne oligomeric siloxane.

liquids with high boiling and flash points releasing a significantly reduced amount of alcohol [15]. Easy handling and safe processing in standard extruders, kneaders, Banbury mixers, or cokneaders are assured as a result of the reduced VOCs. Besides low VOC evolution during application, oligomers provide additional benefits such as better wettability of the filler and formation of a more homogeneous, defect-free silane layer on the filler surface. Furthermore, reduced amounts of silane are required to achieve the same final properties. Different types of oligomers are available ranging from homooligomers to different types of cooligomers, the latter combining the benefits and properties of both silane monomers.

4.5 Silane Hydrolysis

Alkoxysilanes are activated through hydrolysis, the rate of which depends both on the pH and on the type of organofunctional and silicon functional groups. The silicon functional group significantly influences the hydrolysis rate. Reactivity is as follows: propoxy $\ll$ ethoxy $<$ methoxy. Typically, a large excess of water is used as reactant; under these conditions, it was found that hydrolysis of alkoxysilanes is a (pseudo) first-order reaction.

The next important parameter influencing the reaction rate is the pH of the silane hydrolysis medium. At high and very low pH values, the rate of hydrolysis is higher than that at neutral pH, at which silanes are most stable. For example, the rate of reaction of a monomeric trialkoxysilane in acetic acid solution increases by a factor of

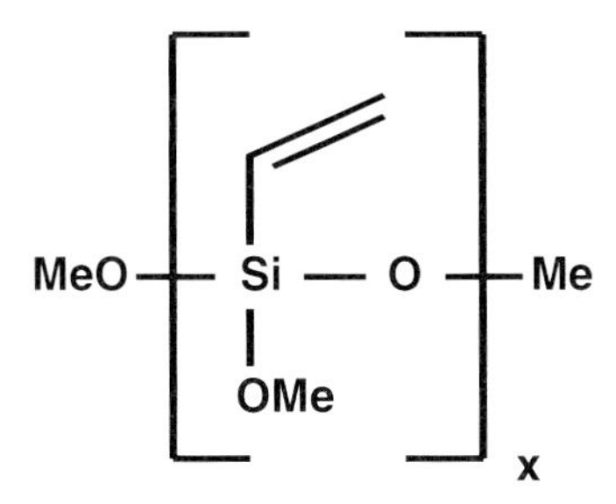

Figure 4.3 Chemical structure of a vinyl oligomeric silane.

10 when the pH is reduced from 4 to 3. This effect is even more marked when changing from neutral to acidic conditions (pH 3) where the factor is about 25–50 depending on the method of mixing. As an example, hydrolysis of 3-methacryloxypropyltrimethoxysilane proceeds within a few minutes in a low-pH environment as a result of the catalytic action of H^+ ions.

The strong dependency of reaction rate on pH is understandable if one takes into account the nucleophilic substitution reaction mechanism for the hydrolysis of alkoxysilanes in acid media since H^+ ions directly affect the rate-determining step of the reaction. The nature of the acid present also affects the hydrolysis behavior. A change from inorganic (e.g., HCl, H_2SO_4, or H_3PO_4) to organic acids (e.g., acetic acid, formic acid, and citric acid) has a significant effect on the hydrolysis rate. The experimental results for the hydrolysis of trialkoxysilanes, in particular, the nonpolar, long-chain or branched alkylsilanes, indicate a type of a "micelle formation," where the silane molecules are surrounded by a phase boundary layer comparable with structures found in surfactant chemistry. In the above model, attack of water molecules at the silicon atom is rendered more difficult by the formation of a phase boundary layer. Accordingly, an inorganic, fully dissociated acid finds it more difficult to develop its catalytic potential in this "nonpolar capsule," whereas the dual character of acetic acid encourages the transfer of water molecules to the reaction center. When comparing hydrolysis rate constants, it can be shown that variation of the structure of the substituents in the alkyltrialkoxysilane is clearly an important factor [13, 16].

The effect of pH on the stability of the formed silanols is different from the effect of pH on the stability of alkoxysilanes. Silanols are most stable around a pH of 3 and their reactivity is higher at a pH lower than 1.5 or higher than 4.5 (Figure 4.4). Silanols will condense to form oligomers and, finally, two- and three-dimensional networks.

When considering silane hydrolysis and condensation, different reactivities at different pH ranges can be expected. At very low pH, silanes will hydrolyze very quickly. The formed silanols are relatively stable and, with time, will form coordinated networks. At neutral pH, silanes will hydrolyze very slowly to silanols that are unstable and will condense. Thus, in both cases, there is a slow reaction in the

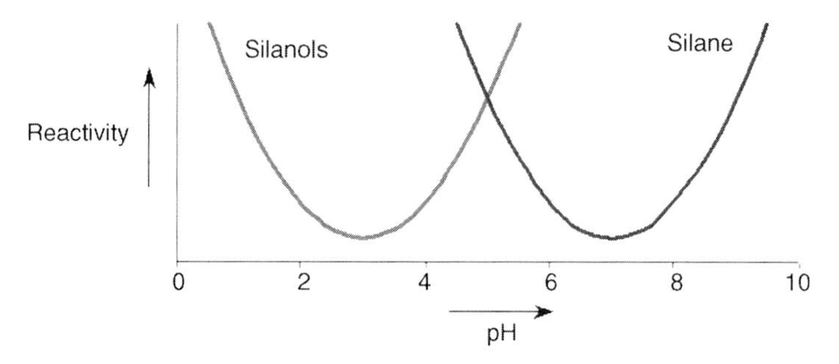

Figure 4.4 Reactivity of silanes and silanols.

transition from silanes to Si–O–Si networks. At pH higher than eight, silanes are highly reactive and form silanols very quickly. These silanols are very unstable and condense rapidly to uncoordinated Si–O–Si networks. The buildup of Si–O–Si networks cannot be controlled, and the uniform coating of the filler surface becomes more difficult, resulting in thicker, uncoordinated layers.

Hydrolysis is also influenced by the type of organic substituents in trialkoxysilanes. As the polarity of the organic substituent changes by increasing the length of the nonpolar chain, for example, in long-chain alkylsilanes, hydrolysis rate decreases. This behavior can be explained by the lower solubility of the nonpolar silanes in the aqueous reaction system and the associated formation of micelle structures. Incorporating polar moieties (functionalities other than alkyl) generally increases susceptibility to hydrolysis. However, it is not possible to determine whether the increased rate is directly linked to the functionality in the substituent or is merely a consequence of better solubility. pH and/or the use of a catalyst also has a decisive effect on the hydrolysis behavior of trialkoxysilanes. Without exception, in all the studied functional trialkoxysilanes, complete hydrolysis of the alkoxy substituents to the corresponding silanols takes place within a period ranging from few minutes to few hours, depending on the nature of the functional group. Reactivity toward hydrolysis increases in the following order of substituents: alkyl < vinyl ≈ methacryloxy < mercapto < epoxy < amino. In the case of epoxysilanes, such as 3-glycidyloxypropyltrimethoxysilane, the epoxy ring also opens, further increasing the rate of hydrolysis.

4.6
Reactivity of Silanes Toward the Filler

Organosilanes rely on the reaction with surface hydroxyl groups to produce a stable covalent bond and a stable layer on the filler surface [7, 8, 17]. They are, thus, most effective on fillers with high concentrations of reactive hydroxyls and a sufficient amount of residual surface water. Silica, silicates (including glass), oxides, and hydroxides are most reactive toward silanes. Silanes are generally not as effective on materials such as sulfates and carbonates, although encapsulation with, for example, silica, often done through condensation of silicic acid esters, can achieve stable silane modification even on these surfaces.

Fixation of the silanol on the filler surface usually proceeds in the first step via hydrogen bonding with the surface OH groups. Hydrogen-bonded silanes are still mobile on the filler surface (the hydrogen bond is thought to be reversible) until a water molecule is split off and eliminated to form a covalent bond [silane–O–filler] with the filler surface. In theory, the silane (and in the reactions that follow) the oligomers form a monolayer on the filler surface (see Scheme 4.3). In reality, the ability of trialkoxysilanes to condense with themselves to produce various three-dimensional networks makes the concept of monolayer coverage based on simple surface reaction of dubious value when considering this type of molecule. Some of the silanols (highly reactive species) will form oligomers prior to the reaction with the

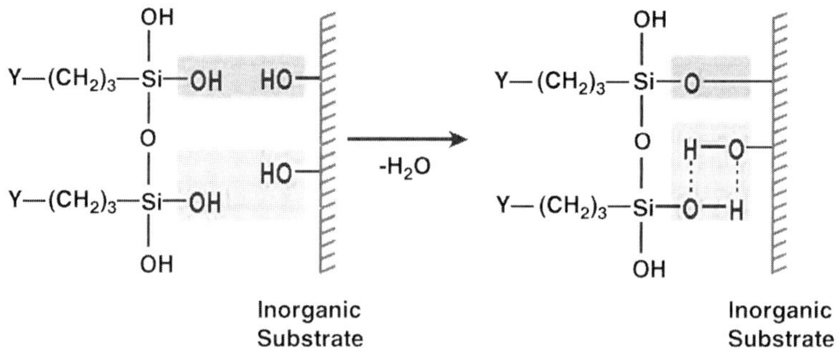

Scheme 4.3 Mode of reaction between silanol and inorganic surface.

filler surface. Oligomer formation is usually beneficial to the performance of the silane on the filler surface [18].

Elucidation of the exact nature of the surface layers and their relationship with the coating conditions has proven to be difficult. The current understanding is that silane layers on mineral surfaces are thicker than the postulated theoretical silane monolayer (see Figure 4.5). Such layers are very complex and depend on the nature of the coating conditions, the type of the mineral surface and the chemistry of the reactive functionalities present. A kind of ladder structure has been postulated to form on some fillers [19a]. In general, it is known that silane layers, as they are normally formed (and before the incorporation of the treated mineral in a composite), consist of a mixture of chemisorbed and physiosorbed material. The physiosorbed silane is readily removable by solvent washing, whereas the chemisorbed silane is not extractable.

4.7 Combining Silanes and Mineral Fillers

Several filler surface pretreatment procedures are commercially used; the neat silane (dry method I) or silane solution (slurry method II) are added uniformly onto the filler

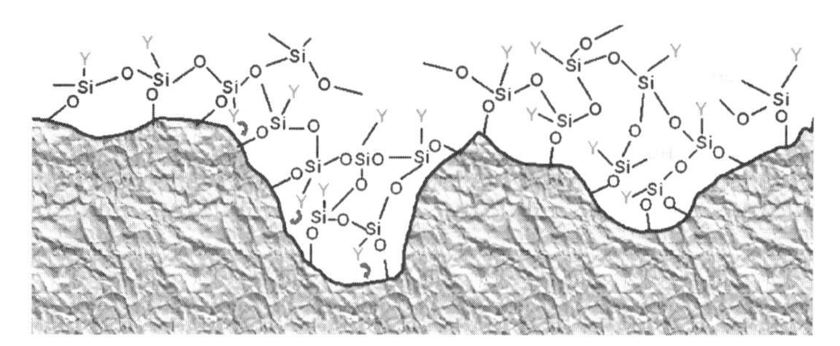

Figure 4.5 Overview of silanized surface.

with sufficient mixing or tumbling followed by heat treatment. Other methods include integral silane blend or *in situ* coating (method III) and the use of silane concentrates (method IV).

4.7.1 Method I

In the dry procedure, the silane is sprayed onto well-agitated filler. In order to obtain maximum efficiency, uniform silane dispersion is essential through the shear rates provided by the mixing equipment, for example, kneaders, Banbury, Hauschild, Primax, and Plowshare mixers, two-roll mills, or extruders [19b]. Most important commercial silane coating processes are continuous and have high-throughput rates. Silane addition control, dwell time, and exact temperature control within the system are essential. All parameters need adjustment depending on the type of silane employed.

Silanes need, in general, a certain time and temperature to react with the filler surface. Therefore, the treated filler is heated (commonly at about 30–150 °C) after the addition of the silane to remove reaction by-products (especially the alcohol resulting from hydrolysis), solvents, and water, and bond the silane completely and permanently to the filler surface. Explosive limits and concentrations for the evolved alcohol should be considered and special collection systems can be installed to reduce the risk of explosions.

In an industrial setting, dwell times of only 2–3 min are common. Since the main covalent reaction is completed after about 15–30 min, an additional heating step is recommended to assure fixation of the silane on the mineral. In some cases, a catalyst can help "activate" a slowly reacting silane or mineral [20]. Typical silane loadings are between 0.7 and 2 wt% relative to the filler and depend on filler surface area and chemistry and application procedure.

4.7.2 Method II

The slurry procedure of filler treatment is limited to alkoxysilanes and waterborne silane systems. Fillers that react with water, obviously, cannot be treated by this method. The slurry procedure is especially suited for commercial treatment when the filler is handled as slurry during manufacturing. Treating solutions may be aqueous, mixtures of alcohol and water, or a variety of polar and nonpolar solvents. Typically, low concentrations of the silane (up to 5%) are dissolved by hydrolysis. Alternatively, the silane can be applied as an emulsion. The silane solution or emulsion can be applied by spraying, dipping, or immersion. Removal of water, solvents, and reaction by-products requires additional steps such as setting, dehydration, and finally, drying. Silane loadings are comparable to those in method I.

4.7.3
Method III

The neat silane can also be added during compounding and is expected to migrate to the filler surface. The *in situ* treatment procedure provides the opportunity to coat freshly formed filler surfaces, for example, during silica/rubber compounding.

The undiluted silane is added directly to the polymer before or together with the filler. It is essential that the resin does not prematurely react with the silane as otherwise the coupling efficiency will be reduced. Typical compounding equipment consists of internal mixers, kneaders, Banbury mixers, two-roll mills, or extruders. The integral blend technique is widely used in resin/filler systems because of its great simplicity and possible cost advantages. This is mainly due to the one-step process and the lower raw material costs (untreated mineral plus silane compared to pretreated mineral) despite the fact that more silane is needed to achieve comparable performance in the finished composite.

4.7.4
Method IV

The silane can also be added as a dry concentrate (wax dispersion, dry liquid, or masterbatch). Here, the silane is adsorbed at very high levels onto suitable carriers and then blended with the polymer and filler during compounding. The use of "solid" silanes leads to very effective dispersion even with simple production equipment. In addition, an easy and safer handling method is assured. Silane loadings are comparable to those in the *in situ* method.

4.8
Insights into the Silylated Filler Surfaces

Any silane surface modification needs to prove its value through a practical test. Proper selection of mineral, silane, and production parameters will lead to optimum properties of the composite [21]. But still, there exists a desire to "see" the silane on the filler surface, to visualize the silane layers or, at least, to see the difference between a silane-modified and an unmodified mineral surface [22, 23]. In general, surface analysis depends on the type of organofunctional group of the silane. To avoid the analysis of physiosorbed instead of chemically bound silanes, the treated mineral can be eluted with an excess of solvent in which the respective silane is soluble.

4.8.1
Spectroscopy

4.8.1.1 FTIR/Raman Spectroscopy

IR spectroscopy is often referred to in technical literature as a method for the analysis of treated filler surfaces. It enables the detection of C–H groups as well as other organofunctional groups, for example, N–H, via their specific bands. This method is, in most cases, semiquantitative [24, 25]. Similar results may be obtained from Raman spectroscopy.

4.8.1.2 MAS–NMR Spectroscopy

Characterization of modified surfaces is also possible via MAS (magic angle spinning)–NMR (nuclear magnetic resonance) spectroscopy. Bound and physiochemically absorbed silane can be distinguished by this method [26]. Comparative ^{1}H-NMR spectra of 2-aminoethyl-3-aminopropyltrimethoxysilane (Dynasylan® DAMO) and 2-aminoethyl-3-aminopropylmethyldimethoxysilane (Dynasylan® 1411) are shown in Figures 4.6 and 4.7.

4.8.1.3 Auger Electron Spectroscopy

Auger electron spectroscopy (AES) activates the sample through electrons producing ionization of atoms on the outer silane layer. During refilling of this hole by an outer electron, the energy released can be transferred to a third electron that leaves the solid and can be detected. By this method, two-dimensional silane-coated surfaces can be analyzed [27]. Although it may be assumed that comparable processes will take place on the surfaces of mineral fillers, AES cannot be used for analysis since mineral fillers are three-dimensional.

Figure 4.8 shows the results of AES surface analysis of a control E-glass sample. The sample was rinsed with ethanol prior to the analysis and, as a result, the elemental composition consists of only silicon and oxygen (SiO_2) after a few seconds

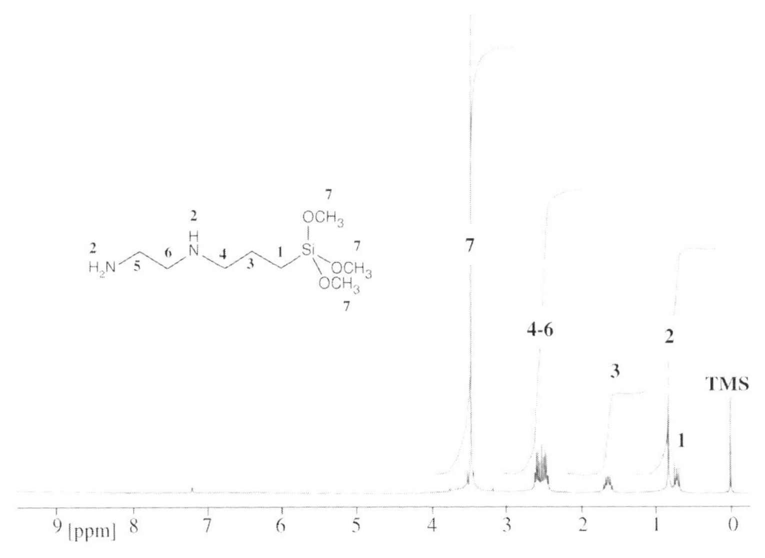

Figure 4.6 ^{1}H-NMR spectrum of 2-aminoethyl-3-aminopropyltrimethoxysilane (Dynasylan® DAMO).

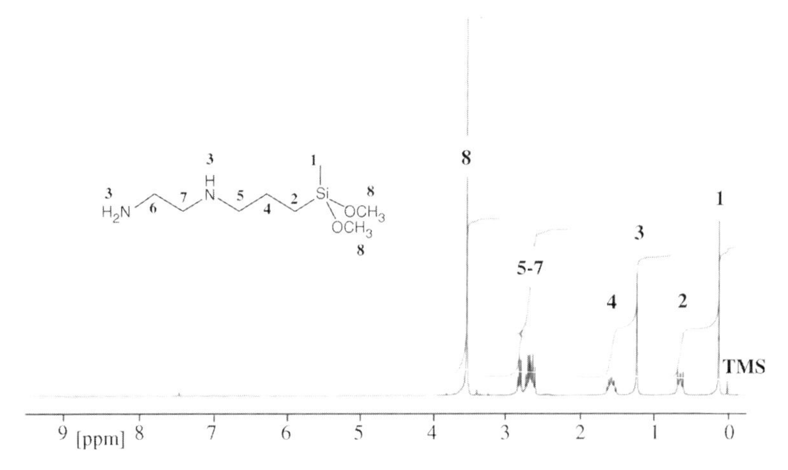

Figure 4.7 ^{1}H-NMR spectrum of 2-aminoethyl-3-aminopropylmethyldimethoxysilane (Dynasylan® 1411).

of sputtering. For the AES depth profiling, the ethanol-rinsed plate E-glass was dipped for 5 min in a 1 wt% aqueous cationic aminosilane solution and then dried for 1 h at room temperature.

Figure 4.9 shows the two-dimensional visualization of a monomeric and an oligomeric cationic aminosilane (Dynasylan® 1161) on E-glass. The lack of homogeneity of the monomeric cationic aminosilane surface layer is confirmed by the shady SEM (scanning electron microscopy) image (Figure 4.10). By contrast, the oligomeric cationic aminosilane leads to a much more homogeneous surface.

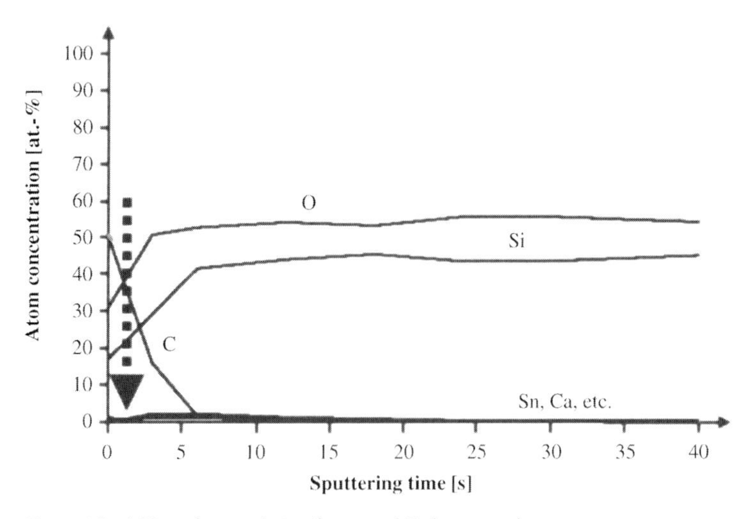

Figure 4.8 AES surface analysis of a control E-glass sample.

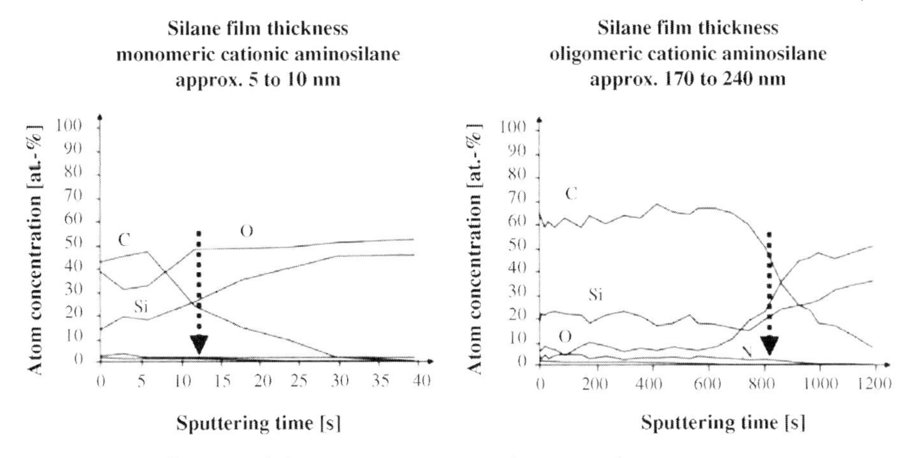

Figure 4.9 AES line scan of oligomeric cationic aminosilanes on E-glass.

AES depth profiling shows that the monomeric cationic aminosilane leads to an average silane layer thickness of approximately 10 nm, whereas the oligomeric cationic aminosilane forms much thicker silane layers with an average thickness of approximately 200 nm. The AES line scans (Figure 4.11) demonstrate the very homogeneous layer of the oligomeric cationic aminosilane. By contrast, the monomeric cationic aminosilane leads to pinholes in the silane coverage and isolated domains. The data show that oligomeric silanes wet surfaces much better than monomeric silanes and lead to a more homogeneous silane distribution on surfaces.

There are other spectroscopy methods also available for characterizing silane-coated surfaces such as SSIMS (static secondary ion mass spectrometry), AFM

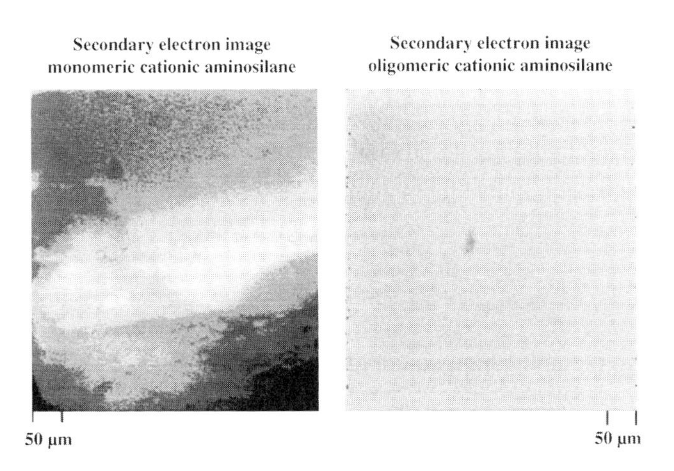

Figure 4.10 SEM of monomeric and oligomeric aminosilane layers on E-glass.

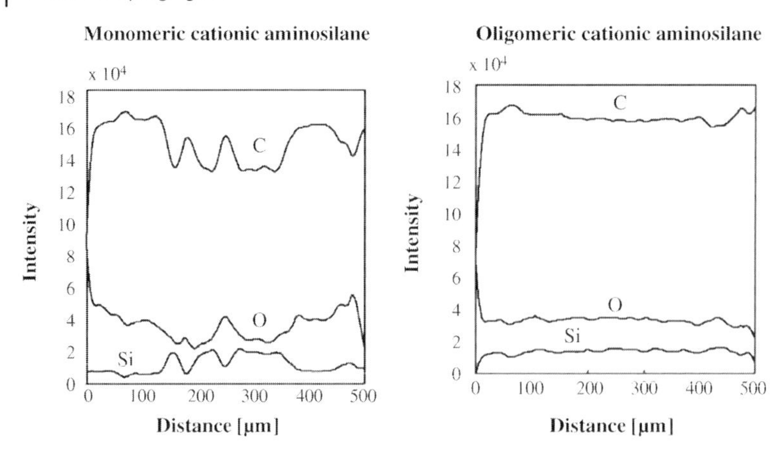

Figure 4.11 AES line scans for monomeric and oligomeric aminosilane layers on E-glass.

(atomic force microscopy), ESCA (electron spectroscopy for chemical analysis), and EDX (energy dispersive X-ray analysis) [28–33].

4.8.2 Pyrolysis–Gas Chromatography

Silane loading and the nature of the silane can be determined quantitatively via pyrolysis–gas chromatography (Py–GC). The samples are pyrolyzed and the volatilized parts of the silane (especially those originating from the organofunctional group) can subsequently be determined by GC. Typical test conditions for the octyltriethoxysilane/TiO_2 system shown in Figure 4.12 are pyrolysis at 700 °C, helium carrier gas, nonpolar column, and flame ionization detector [28].

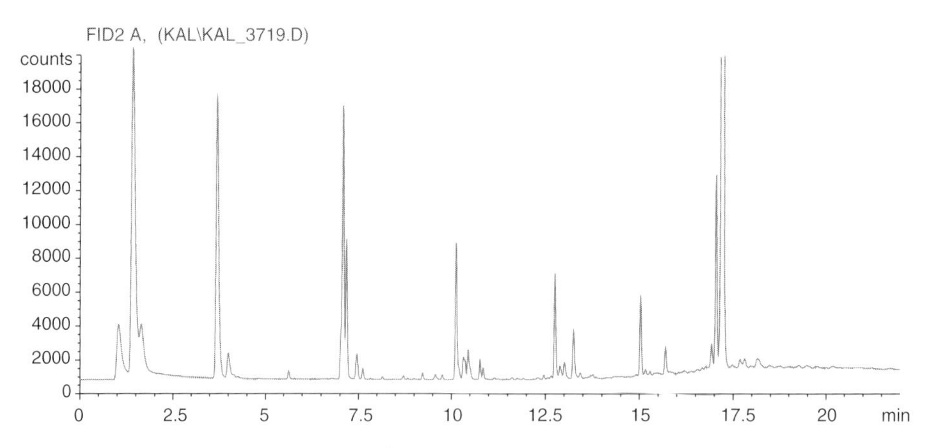

Figure 4.12 Gas chromatogram of pyrolyzed octyltriethoxysilane/TiO_2 system.

4.8.3 Carbon Analysis

Carbon elementary analysis on noncarbon-containing minerals such as synthetic TiO_2 is a very helpful method to determine the carbon content introduced by a silane surface treatment.

4.8.4 Colorimetric Tests

Some specific organofunctional silanes can be analyzed via colorimetry with suitable reagents. Examples include epoxysilane with bromothymol blue/thiosulfate solution; methacryloxy and vinylsilanes with permanganate salts; amino/diamino silanes with ninhydrine [33]. The results are semiquantitative within the same system of mineral and silane; in all other cases, they are only of qualitative nature.

4.8.5 Acid–Base Titration

Conventional acid–base titration can also be applied to slurries of treated fillers if the silane changes the pH of the mineral. For example, application of an aminosilane to an acidic mineral will change its surface pH. Such changes can be monitored by titration in the presence of conventional indicators. The results are semiquantitative within the same system of mineral and silane; in all other cases, they are only of qualitative nature.

4.8.6 Analytical Tests for Hydrophobicity

An elegant method to determine the hydrophobicity of a treated filler is to determine the methanol wetting. In this method several different mixtures of methanol with water are prepared (from 1:9 to 9:1). Treated filler placed into these mixtures and is centrifuged. The amount of sediment is determined giving an indication of filler wetting by the medium. For highly hydrophobic fillers the amount of sediment should be negligible until higher methanol concentrations are used [33]a.

A further method is to place a drop of water on heap of silane treated hydrophobic filler. The time required for the water to sink into the filler is a quick indication of the hydrophobicity the treatment.

4.8.7 Silane/Colorant Combined Surface Modification

Incorporation of a suitable colorant in the silane solution can show the distribution of the silane on the treated mineral (e.g., Rhodamine B in isobutyltriethoxysilane). It is important to note that the colorant will not block any of the active sides on the mineral surface and that it spreads equally on the surface and the silane. The normally invisible silane coating can, thus, be visualized; a proper treatment will then lead to a homogeneously colored surface.

4.9
Selection of Silanes

A list of silanes providing the best property enhancements in a variety of filled polymer systems is shown in Table 4.1. The list, based on experience, is not exhaustive and variations are possible. Major suppliers of silanes include Evonik-Degussa GmbH, Dow-Corning, Momentive, Shin-Etsu, and Wacker.

Table 4.1 Types of polymers and recommended silanes.

Polymer	Silane functionality	Examples of commercial silanes
Acrylic	Acrylate, methacrylate, vinyl	3-Methacryloxypropyltrimethoxysilane
Butyl rubber	Diamino	2-Aminoethyl-3-aminopropyltrimethoxysilane, aqueous diaminosiloxane oligomer
EVA	Amino, vinyl	3-Aminopropyltriethoxysilane, special amino-silane blends, aqueous amino/alkylsiloxane oligomer, vinyltriethoxysilane
Neoprene	Mercapto, diamino	3-Mercaptopropyltrimethoxysilane, 2-aminoethyl-3-aminopropyltrimethoxysilane, aqueous diaminosiloxane oligomer
Nitrile rubber	Mercapto	3-Mercaptopropyltrimethoxysilane
Polyamide	Amino, secondary amino	3-Aminopropyltriethoxysilane, aqueous amino/alkylsiloxane oligomer, *N*-(*n*-butyl-)-3-aminopropyltrimethoxysilane
Unsaturated polyester	Methacrylate, polyether	3-Methacryloxypropyltrimethoxysilane, polyetherfunctional trimethoxysilane
Polyester thermoplastic	Amino, epoxy	3-Aminopropyltriethoxysilane aqueous aminosiloxane oligomer, 3-glycidyloxypropyltrimethoxysilane
Polyolefin	Vinyl, alkyl	Vinyltriethoxysilane, hexadecyltrimethoxysilane, alkylsiloxane oligomer
Polypropylene maleated	Amino, secondary amino	Special aminosilane blends, *N*-(*n*-butyl)-3-aminopropyltrimethoxysilane
PTFE	Fluoroalkyl	Fluoroalkyltriethoxysilane, aqueous fluoroalkyl siloxane oligomer
EPR, EPDM	Vinyl, sulfur, mercapto, thiocyanato	Vinyltriethoxysilane, vinyltrimethoxysilane, vinyltris(2-methoxyethoxy)silane, vinyl/alkyl-functional siloxane oligomer
SBR	Sulfur, mercapto	Bis(triethoxysilylpropyl)polysulfane, 3-mercaptopropyltrimethoxysilane
Water dispersion	Polyether	Polyetherfunctional trimethoxysilane

Table 4.2 Comparison of 300% modulus of differently cured filled EPDM systems containing different silanes.

Silane concentration (phr)	300% Modulus (MPa) (sulfur cured EPDM, talc filled)	300% Modulus (MPa) (peroxide-cured EPDM, clay filled)
Control, no silane	490	420
Vinyl (Dynasylan® VTMO)	430	1110
Mercapto (Dynasylan® MTMO)	790	1200
Amino (Dynasylan® AMEO)	790	1440
Methacryloxy (Dynasylan® MEMO)	Not determined	1660

The following example on vulcanized EPDM filled systems shows the importance of the correct selection of the silane functionality [4, 34]. The surface reactivity of many nonblack fillers generally precludes strong bonding with a polymer matrix and, in some cases, leads to poor compatibility and dispersion, and thus careful consideration to the choice of the organofunctional group of the silane should be made to optimize reactivity. The importance of selecting the right coupling agent with an organofunctional group complementary to the curative/polymer system can be seen in Table 4.2. Comparative modulus values are shown for two filled EPDM systems, cured with sulfur and peroxide, respectively. In the formulation, one part per hundred parts of resin (phr) of each indicated silane was introduced during compounding of 100 phr filler in a two-roll mill, and the composite properties were determined by standard (DIN 53504) methods.

In sulfur-cured EPDM systems, the primary amino- and mercaptofunctional silanes can participate in the cure mechanism to a far greater extent than the vinylsilane as shown by the higher modulus values. In the peroxide-cured EPDM system, all silanes promote significant improvements in modulus but in different degrees depending on their relative reactivity. The methacryloxy functional silane

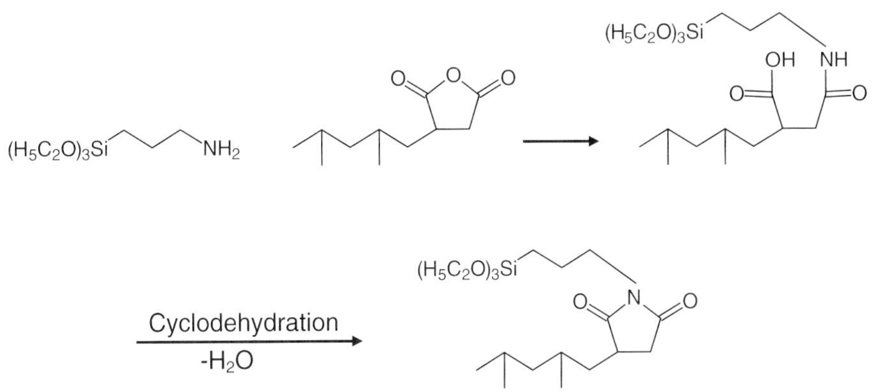

Scheme 4.4 Reaction of aminosilane with maleated polypropylene.

is considerably more effective than the vinylsilane, which is reasonably predicted from the relative reactivity of the double-bond moiety (methacryl versus vinyl). In the example, the aminosilane also provides a high level of filler–polymer interfacial bonding, whereas the mercaptosilane is relatively less effective than the methacryloxysilane.

An example of a different reaction of an amino-functional silane with a modified polymer is the reaction of the amino groups with maleated polypropylene (PP) to form an imide (Scheme 4.4). This reaction is extensively used to couple treated glass fibers to polyolefins (see Chapter 7).

4.10 Applications of Specific Silanes

4.10.1 Vinylsilanes

Commercial vinylsilanes usually have the vinyl group directly attached to the silicon atom. Common hydrolyzable groups are methoxy, ethoxy, or 2-methoxyethoxy. The vinyl functionality is used in polymers that are cross-linked by a free-radical process (peroxide cure), but it is, however, not sufficiently reactive for all systems, and methacryloxy functionality is sometimes preferred as shown in Table 4.2 [35]. Vinylsilanes because of their overall cost/performance advantages have become the industry standard for EPR and EPDM and wire and cable applications.

A particularly important area of application of vinylsilanes is in ATH-filled EVA. The first generation of halogen-free flame-retardant (HFFR) materials possessed excellent fire and smoke properties but were mechanically weak and slow to process compared to the PVC compounds that they were replacing. Today, the majority of thermoplastic HFFR cable materials are made of ATH/EVA and occupy a rapidly growing and specialized area of cable production. Vinylsilane adhesion promoters make possible the high loading levels of ATH required for effective flame retardation, the improvement in melt processability of the highly filled EVAs and the enhancement of the mechanical properties of the finished product. Oligomeric vinylsilanes, in addition, provide a significant reduction in alcohol emissions (and hence lower VOC emissions) upon reaction with moisture. Correct use of oligomeric vinylsilanes can also reduce compound viscosity and produce a smooth, defect-free surface.

Hydrated fillers such as ATH achieve their flame-retarding characteristics by endothermically decomposing with the release of water close to the temperature at which the polymers themselves decompose (see Chapter 17). They do not have the smoke and corrosive gas problems associated with other types of flame retardants. In order to produce an acceptable HFFR compound at very high ATH loadings of 60–65 wt%, the ATH particle size and shape have to be carefully controlled. Experience suggests that large and thick ATH particles with a low surface area are

Table 4.3 Comparison of EVA/ATH HFFR containing vinylsilanes.

	Vinylsilane (phr)	Peroxide (phr)	Tensile strength (MPa)	Elongation at break (%)
Oligomeric methoxy-based vinylsilane (Dynasylan® 6490)	1.5	No peroxide	12.3	210
	1.6	0.03	16.8	213
	1.6	0.04	17.4	203
	1.6	0.05	Scorch	Scorch
	1.85	0.03	16.9	213
	1.85	0.05	17.6	210
	2.1	0.03	16.4	212
	2.1	0.04	17.4	204
	2.1	0.05	17.9	199
Monomeric methoxy-based vinylsilane (Dynasylan® VTMO)	2.5	0.03	16.4	192

required for effective flame retardation. When using vinylsilanes in HFFR materials, a small amount of peroxide is required to obtain good coupling.

Tensile properties of different EVA/ATH formulations are summarized in Table 4.3. The basic formulation contained 160 phr of ATH, 1 phr stabilizer, and variable amounts of monomeric and oligomeric silanes and peroxide. A corotating twin-screw extruder was used to produce sheets for the tests. Silane content is based on filler, the silane was preblended with the EVA, dicumyl peroxide (DCP) and Irganox 1010 (phenolic stabilizer, Ciba) were used as peroxide and stabilizer, respectively. A control without silane is not included since it leads to scorch.

For oligomeric silanes, the absence of peroxide results in poor tensile strength. In the presence of peroxide, however, the overall picture changes dramatically. As the ATH couples to the EVA, tensile strength increases and water pickup is reduced both by increasing the cross-link density and by rendering the compound hydrophobic. Elongation at break is not significantly affected by the presence of peroxide. Oligomeric vinylsilanes perform better than monomeric vinylsilanes, even at the lower concentration of 1.6 phr. At such low silane levels, the vinylsilane/peroxide ratio has to be monitored, and thus, the peroxide level should not exceed 0.04 phr because of risk of scorch. As demonstrated in Table 4.3, increased silane levels significantly reduce the risk of scorch. The properties achieved with the use of oligomeric vinylsilanes clearly outmatch those of commonly employed monomeric vinylsilanes, even at lower concentrations.

4.10.2 Aminosilanes

4.10.2.1 General

Amino-functional silanes are widely used for the surface treatment of fillers such as wollastonite and calcined clay. Commercial silanes are usually based on the primary

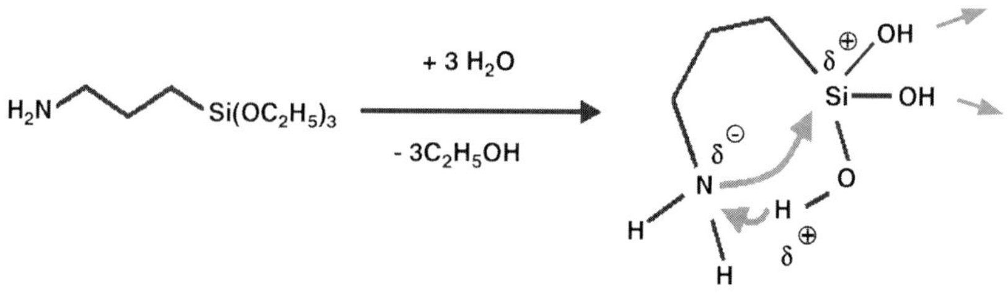

$$H_2N\text{-}CH_2CH_2CH_2\text{-}Si(OC_2H_5)_3 \xrightarrow[-\ 3C_2H_5OH]{+\ 3\ H_2O}$$

Scheme 4.5 Autocatalytic hydrolysis of η-aminopropyltriethoxysilane.

3-aminopropyl functionality. They also have a wide versatility being used in epoxies, phenolics, polyamides, thermoplastic polyesters, and elastomers. Unlike most other silanes, their aqueous solutions are quite stable as a result of hydrogen bonding between the silanol groups and the primary amine. An internal five- or six-membered ring (see Scheme 4.5) is formed.

The reactivity of the primary amino group has made elucidation of the nature of surface layers resulting from its adsorption on the filler very difficult. The amino group itself may absorb strongly on a variety of surfaces and has also been shown to be very prone to bicarbonate salt formation with atmospheric carbon dioxide. Modern analytical procedures are needed to elucidate some of the important features of these coatings [22, 23].

4.10.2.2 Calcined Clay-Filled Polyamides

Impact strength is very important for mineral-filled polyamides. It is directly related to the filler loading level and decreases with increasing loadings. In the absence of fine and uniform dispersion of the filler particles, agglomerates acting as stress concentrators will provide sites where impact failure originates. Optimizing filler dispersion, thus, minimizes the formation of filler agglomerates and yields more homogeneous materials with improved impact properties. Choosing the right aminosilane therefore optimizes the properties of the final compound [36].

The polyamide 6,6 selected in the example below was a general purpose, lubricated material. The filler used was a fine calcined clay that was surface modified with 1 wt% aminosilane by employing a standardized laboratory surface treatment process. The polyamide 6,6 and the silane-treated filler were predried (24 h at 80 °C) prior to compounding. Compounds were prepared at 40 wt% filler level in a corotating twin-screw extruder. All injection molded test samples were conditioned (24 h, 23 °C, 50% RH) prior to testing as per DIN 50014 Procedure. Table 4.4 summarizes the properties of the nylon 6,6/clay compounds using two different silanes. Aminosilane treatment of calcined clay dramatically improves the properties of the filled polyamide 6,6. Data are not shown for nylon compounds with 40 wt% untreated clay since without silane these led to scorch. By using the *N*-(*n*-butyl)-3-aminopropyltrimethoxysilane, the impact strength of the final compound can be further improved. Good dispersion of the calcined clay in the polymer phase leads to low compound viscosities (as measured by MFI) and better processability. The *N*-(*n*-butyl)-3-aminopropyltrimethoxysilane further reduces compound viscosity and results in higher

Table 4.4 Properties of nylon 6,6/clay compounds using two different aminosilanes.

Property	Primary γ-aminopropyltriethoxysilane (Dynasylan® AMEO)	Secondary *N*-(*n*-butyl-)-3-aminopropyltrimethoxysilane (Dynasylan® 1189)
Ultimate tensile strength (MPa)[a]	72.0	69.1
Ultimate tensile strength (MPa)[b]	36.5	34.7
Flexural modulus (GPa)	3.7	3.5
Charpy impact strength, unnotched (kJ/m^2)	32.2	41.7
Izod impact strength, unnotched (kJ/m^2)	19.5	30.1
Charpy impact strength, notched (kJ/m^2)	4.0	4.0
Izod impact strength, notched (kJ/m^2)	4.1	4.0
Melt flow index (g/10 min)[c]	43	55

a) Measurement after 7 days/90 °C immersion in water.
b) Measurement at 275 °C and 5 kg.
c) DIN ISO 1133 (method B), temperature 275 °C, preheating time: 1 min, load: 5 kg.

melt flow rates, improved dispersion, fewer agglomerates, and a smooth, defect-free surface. Treatment with this nonpolar silane also results in reduced filler moisture pickup.

4.10.2.3 ATH-Filled EVA

In addition to vinylsilanes, ATH- and MDH-filled ethylene vinylacetate copolymers can be effectively coupled to aminosilanes [37]. Fine and uniform ATH dispersion in the polymer leads to low HFFR compound viscosities (MFI) and high tear strength. The use of 3-aminopropyltriethoxysilane became an industrial standard for the production of reliable thermoplastic HFFR compounds. New primary/secondary aminosilane blends provide further improvements in the final HFFR compound [38]. Aminosilanes are also found in thermoplastic EVA, nonperoxide cross-linked.

Properties of different EVA/ATH formulations are summarized in Table 4.5. The basic formulation contained 160 phr of ATH, 1 phr stabilizer (Irganox 1010, Ciba), and 1.5 phr aminosilane based on filler preblended with EVA. Compounds were made in a corotating twin-screw extruder and test specimens were produced from extruded sheets.

4.10.2.4 MDH-Filled Polypropylene

In filled polypropylene cables, low water pickup is important. Increased water pickup leads to a significant deterioration in electrical properties. Often maleated PP is introduced into the compound to improve the mechanical properties of the mineral-filled cable system, but this leads to an increase in the water pickup. The use of silanes in conjunction with maleated PP can markedly improve the mechanical properties

Table 4.5 Comparison of EVA/ATH HFFR composites containing aminosilanes.

Silane	Ultimate tensile strength (N/mm^2)	Elongation at break (%)	Tear strength (N/mm)	Water uptake (14 days at 70 °C) (mg/cm^3)
3-Aminopropyltriethoxysilane (Dynasylan® AMEO)	16.3	200	10.2	4.02
Special aminosilane blend (Dynasylan® SIVO 214)	16.6	210	11.6	3.75

and also reduce the water pickup. It might be expected that aminosilanes are too hydrophilic for this task, but with the use of specially designed silanes, and particularly aminosilane blends, outstanding results can be achieved [39, 40].

PP and maleated PP, used in the compounds of Table 4.6, are types normally used for cable applications. The MDH was a fine precipitated grade modified with 1 wt% aminosilane by employing method I (see Section 4.7). Both filler and resin were predried (24 h at 80 °C) prior to compounding. Compounds were prepared using 185 phr aminosilane modified filler on a corotating twin-screw extruder. All test samples were made from extruded tapes. The water pickup was measured after 14 days storage in distilled water. Both the tensile and the elongation at break were measured after 7 days of storage (distilled or salt water, 3% NaCl). Table 4.6 displays

Table 4.6 Comparison of PP composites containing aminosilane-treated MDH.

Property	Without silane (control)	Secondary *N*-(*n*-butyl-)-3-aminopropyltrimethoxysilane (Dynasylan® 1189)	Special aminosilane blend (Dynasylan® SIVO 214)
Melt flow ratio (21.6 kg at 230 °C)	3.8	23.2	11.5
Water pickup 14 days at 70 °C	0.45	0.13	0.17
Tensile strength (MPa) (without water storage)[a)]	10.5	9.4	11.2
Tensile strength (MPa) (water storage)	8.9	8.9	10.8
Tensile strength (MPa) (water storage with 3% NaCl)[a)]	9.1	9.0	10.8
Elongation at break (%) (without water storage)	67	390	348
Elongation at break (%) (water storage)[a)]	31	406	259
Elongation at break (%) (water storage with 3% NaCl)[a)]	29	333	259

a) The tests were performed after 7 days at 70 °C storage in the respective media, aqueous 3% sodium chloride solution (NaCl) or distilled water.

how the use of the appropriate silane significantly improves the elongation at break without reducing the tensile strength.

4.10.3 Methacryloxysilanes

4.10.3.1 General

Methacryloxysilanes provide more reactive forms of unsaturation than vinylsilanes and are used extensively in free-radical curing formulations where the extrareactivity is beneficial. Commercial products usually contain 3-methacryloxypropyl groups. The presence of a carbonyl group in the molecule leads to a tendency for it to orient flat on the filler surface under certain conditions [20].

Thermosetting filler systems prepared from unsaturated polyesters, methyl methacrylate (MMA), vinyl esters, epoxy (EP), and phenolic and furan resins are used in many applications. After pretreatment of the filler, the organofunctional silanes cause a strong reduction in viscosity and enhancement of mechanical properties, such as flexural and impact strength. This is especially evident after exposure to moisture. When applied onto the filler, the methacryloxysilane acts immediately in acrylic casting systems by reducing the viscosity to 24% of the initial value; by contrast, the use of the silane as an additive brings about a viscosity reduction to only 76% of the initial value even after 24 h storage time.

As an example of the effect of methacryloxysilane in a filled UP system [34], a formulation containing 63% quartz flour was selected. Pretreatment of the filler with 1% silane results in 38% lower viscosity versus the non-silane-containing composite. The organofunctional silane can cause higher retention of mechanical properties versus the non-silane-containing system as shown by flexural strength data after 6 h in boiling water. Maximum effectiveness is achieved by introducing the silane via filler pretreatment.

4.10.3.2 Filled PMMA Resin Systems

The effect of the methacryloxy functional silane in a cristobalite-filled PMMA resin can be visualized via SEM on cryofractured samples that contain 100 phr filler and 0 or 1 phr silane (Figure 4.13). A gap between the polymer matrix and the filler particle indicating debonding is visible in the absence of silane. In the presence of silane, no gap can be detected and the composite breaks in the polymer phase as a result of the improved adhesion. Similar effects can be observed in highly filled ATH/PMMA systems. For example, in a system containing 60 wt% ATH, the viscosity drops to a third of the original value if 0.5% silane is applied on the filler. Flexural strength and impact strength are higher by 22% and 35%, respectively, than the values for the non-silane-containing composite.

4.10.3.3 Silica-Filled UV-Cured Acrylates for Scratch-Resistant Coatings

Pyrogenic silica is normally used in amounts smaller than 1 wt% in order to control the resin rheological behavior. However, for scratch-resistant coating systems the

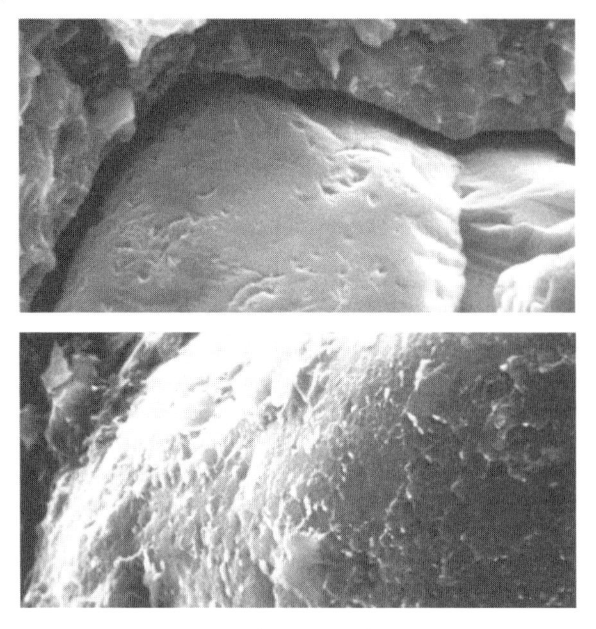

Figure 4.13 Comparison of cristobalite filled PMMA; above untreated, below treated with 1% methacryloxysilane.

silica content needs to be increased; nonetheless, the coating system must still remain at low viscosity. These high-performance fillers are, thus, pretreated with 3-methacryloxypropyltrimethoxysilane [41]. The silanes can also be used *in situ* during the formulation of the coating system [24, 42–45].[1)]

4.10.4
Epoxysilanes

Epoxy resins filled with particulate inorganic fillers provide insulation properties in many electronic applications. In addition to decreasing cost, these fillers serve to increase hardness, act as a heat sink for the exothermic curing reaction, decrease shrinkage during curing, and improve other properties, particularly retention of mechanical and electrical properties after extensive exposure to water [34]. Using 3-glycidyloxypropyltrimethoxysilane (Dynasylan® GLYMO), the viscosity of an epoxy containing 60 wt% quartz flour coated with 1 wt% silane drops to 80% of the value of the untreated filler composite; after 6 h in boiling water, flexural strength and impact strength are higher by 210 and 250%, respectively, than the values for the untreated material.

1) Note: For *in situ* treated pyrogenic silica in UV-cured coating systems, it is advisable to use vinylsilane oligomers such as Dynaslyan® 6490.

4.10.5 Sulfur-Containing Silanes

An important application of these silanes is in tires where silicas and silanes are used to reduce rolling resistance. For tires, the silane is added during rubber compounding, neat or in the form of a dry liquid reacting with the silanol groups of the silica. The rubber-active group of the silane (tetrasulfane, disulfane, thiocyanato, or mercapto group) has a strong tendency to form rubber-to-filler bonds during curing of the rubber compound [46].

4.10.6 Alkylsilanes

Alkylsilanes are often used for fillers in polyolefins. Such fillers are added either for reinforcement or for coloring. In comparison to other treatment methods, silanes have the advantage that they remain on the surface of the filler without migrating to the surface of the plastic film causing difficulty with further printing or coating, as is often the case for silicone-type modification. It has been shown that oligomeric alkylsiloxanes are especially effective in reducing the viscosity of the overall system and thus allowing high loading levels [47]. Furthermore, such alkylsiloxane oligomers have a lower VOC content.

4.10.7 Fluoroalkylfunctional Silanes

Fluoroalkylfunctional silanes are special as they provide not only hydrophobicity but also oleophobicity. Thus, they can be used only in special plastic formulations that are compatible with such fluoroalkyl groups. They are commonly used to treat fillers used in PTFE (polytetrafluoroethylene) for compatibilization and to improve the mechanical strength of the final composite. They have been especially useful as water-borne oligomeric fluoroalkyl siloxane systems, as the trialkoxyfluoroalkylsilane does not readily react with water and thus cannot react easily with the filler surface.

4.10.8 Polyetherfunctional Silanes

In special applications, it is desirable to increase the hydrophilic properties of fillers and pigments for easy dispersion in water. With polyetherfunctional silanes, it is possible to obtain, for example, self-dispersing iron oxide pigments [38]. Furthermore, the polyether silanes display a very significant reduction in the viscosity of unsaturated polyester systems [47] when high filler loadings are required.

4.11 Trends and Developments

Future trends for functional, surface-treated minerals for plastic applications are associated with the use of silanes that reduce water pickup particularly for polyamide applications. In recent years, special silanes have been developed for this task. Furthermore, the use of silanes with reduced VOC emission, such as oligomeric siloxanes or aqueous oligomeric siloxane systems is increasing significantly. Low-volatility silanes are of high interest in the industry allowing higher treatment temperatures and shorter dwell times (increased throughput). In this regard, the development of oligomeric siloxane systems has been a breakthrough. Also, the combination of silane functionalities in one silane molecule (e.g., combined amino and alkyl functionalities, alkyl/vinyl, etc.) is of interest for easier handling and better properties of the end product.

References

1 Ishida, H. (1993) Controlled interphases in glass fiber and particulate reinforced polymers: structure of silane coupling agents in solutions and on substrates. *Appl. Sci.*, **230**, 169–199.

2 Skudelny, D. (1987) Silanisierte füllstoffe und ihre einsatzgebiete. *Kunststoffe*, **77** (11), 1153–1156.

3 Ramney, M.W., Berger, S.E., and Marsden, J.G. (1972) Silane coupling agents in particulate mineral-filled composites. Proceedings of the Annual Reinforced Plastics/Composites Institute, Society of Plastic Industry (SPI), 21D, pp. 1–22.

4 Marsden, J.G. (1970) Silicone coupling agents and primers for thermosets, thermoplasts, and elastomers. *Appl. Polym. Symp.*, **14**, 107–120.

5 Marsden, J.G. and Ziemianski, L.P. (1979) Organofunctional silanes – functions, applications and advantages. *Br. Polym. J.*, **11** (4), 199–205.

6 Arkles, B. (1977) Tailoring surfaces with silanes. *Chemtech*, **7**, 766–778.

7 Plueddemann, E.P. (1980) Chemistry of silane coupling agents. *Macromol. Monogr.*, **7**, 31–53.

8 Plueddemann, E.P. (1978) Silane coupling agents. *Addit. Plast.*, **1**, 123–167.

9 Ramney, M.W., Berger, S.E., and Marsden, J.G. (1974) Silane coupling agents in particulate mineral filled composites. *Compos. Mater.*, **6**, 131–172.

10 Atkins, K.E., Gentry, R.R., Gandy, R.C., Berger, S.E., and Schwarz, E.G. (1978) Silane treated alumina trihydrate: a new formulating tool for flame retardant polyester fiber reinforced plastics. *Polym. Eng. Sci.*, **18** (2), 73–77.

11 Plueddemann, E.P. and Stark, G.L. (1977) Role of coupling agents in surface modification of fillers. *Mod. Plast.*, **54** (8), 76–78.

12 Plueddemann, E.P. (1980) Chemistry of silane coupling agents. *Midland Macromol. Monogr.*, **7**, 31–53.

13 Beari, F., Brand, M., Jenkner, M., Lehnert, R., Metternich, H.J., Monkiewicz, J., and Siesler, H.W. (2001) Organofunctional alkoxysilanes in dilute aqueous solution: new accounts on the dynamic structure mutability. *J. Organomet. Chem.*, **625**, 208–216.

14 Arkles, B., Steinmetz, J.R., Zazyczny, J., and Zolotnitsky, M. (1991) Stable, water-borne silane coupling agents. Proceedings of the 46th Annual Reinforced Plastics/Composites Institute, Society of Plastic Industry (SPI), Washington, DC.

15 Mack, H. (2004) Silane oligomers: a class of their own, *Silanes and Other Coupling Agents III* (ed. K.L. Mittal) VSP, Boston, pp. 11–20.

16 Brand, M., Frings, A., Jenkner, P.K., Lehnert, R., Metternich, H.J., Monkiewicz, J., and Schramm, J. (1999) NMR-spektroskopische Untersuchungen zur Hydrolyse von funktionellen Trialkoxysilanen. *Z. Naturforsch.*, **54b**, 155–164.

17 Rosen, M.R. (1978) From treating solution to filler surface and beyond: the life history of a silane coupling agent. *J. Coat. Tech.*, **50** (644), 70–82.

18 Giessler, S. and Jenkner, P.K. (2003) Silanes for easy-to-clean surfaces and glass fiber reinforced plastics. Proceedings of the Glass Processing Days, Tambpere, Finland, July 2003, pp. 1–3.

19 Bauer, F., Gläsel, H.-J., Decker, U., Ernst, H., Freyer, A., Hartmann, E., Sauerland, V., and Mehnert, R. (2003) Trialkoxysilane grafting onto nanoparticles for the preparation of clear coat polyacrylate systems with excellent scratch performance. *Prog. Org. Coat.*, **47**, 147–153; (b) Degussa Technical Brochure (2007) Silanes for Mineral Fillers and Pigments.

20 Hanisch, H., Steinmetz, J., and Peeters, H. (1987) Improved organosilane systems for highly filled acrylics. *Plast. Compound.*, **5**, 25–30.

21 Harding, P.H. and Berg, J.C. (1997) The role of adhesion in the mechanical properties of polymer composites. *J. Adhes. Sci. Tech.*, **11** (4), 471–493.

22 Ishida, H. (1984) A review of recent progress in the studies of molecular and microstructure of coupling agents and their functions in composites, coatings, and adhesive joints. *Polym. Compos.*, **5**, 101–123.

23 Ishida, H. (1985) Structural gradient in the silane coupling agent layers and its influence on the mechanical and physical properties. *Polym. Sci. Technol.*, **27**, 25–50.

24 Bauer, F., Ernst, H., Decker, U., Findeisen, M., Gläsel, H.-J., Langguth, H., Hartmann, E., Mehnert, R., and Peuker, C. (2000) Preparation of scratch and abrasion resistant polymeric nanocomposites by monomer grafting onto nanoparticles, 1. FTIR and multinuclear NMR spectroscopy to the characterization of methacryl grafting. *Macromol. Chem. Phys.*, **201**, 2654–2659.

25 Feresenbet, E., Raghavan, D., and Holmes, G.A. (2003) The influence of silane coupling agent composition on the surface characterization of fiber and on fiber matrix interfacial shear strength. *J. Adhes.*, **79**, 643–665.

26 Görl, U., Münzenberg, J., Luginsland, D., and Müller, A. (1999) Investigations on the reaction silica/organosilane and organosilane/polymer. *KGK Kautschuk Gummi Kunststoffe*, **52** (9), 588–598.

27 Hussain, A. and Pflugbeil, C. (1994) Neue Analysen bringen Oberflächen näher. *Kleben Dichten Adhaesion*, **38**, 22–25.

28 Rotzsche, H. and Ditscheid, K.-P. (1996) Verfahren zur untersuchung silanbehandelter, anorganischer materialien, EP0741293 A2.

29 Trifonova-Van Haeringen, D., Schönherr, H., Vancso, G.J., van der Does, L., Noordermeer, J.M.W., and Janssen, P.J.P. (1999) Atomic force microscopy of elastomers: morphology, distribution of filler particles, and adhesion using chemically modified tips. *Rubber Chem. Techn.*, **72** (5), 862–875.

30 Garbassi, F., Occhiello, E., Bastioli, C., and Romano, G. (1987) A quantitative and qualitative assessment of the bonding of 3-methacryloxypropyltrimethoxysilane to filler surfaces using XPS and SSIMS (FABMS) techniques. *J. Colloid Interface Sci.*, **117** (1), 258–270.

31 Bartella, J. (1997) Oberflächenanalyse an dünnen Schichten. VDI (Verein Deutscher Ingenieure) Bildungswerk, BW 6998.

32 Albers, P. and Lechner, U. (1991) Zur chemischen Fixierung von Silanen auf Feststoffoberflächen. *Kunststoffe*, **81** (5), 420–423.

33 Hartwig, A. (1997) Oberflächenanalyse für die qualitätssicherung durch farbreaktionen. *J. Oberflaechentechnol.*, **5**, 46–51.

34 Nargiello, M., Mertsch, R., Michael, G., Linares, M., Leder, G., and Hill, S. (2008) Nano-Structured-Particles to Enhance Green Coatings, Proceedings of the Radtech Symposium, Chicago, IL, USA.

35 Wang, G., Jiang, P., Zhu, Z., and Yin, J. (2002) Structure–property relationships

of LLDPE – highly filled with aluminum hydroxide. *J. Appl. Polym. Sci.*, **85** (12), 2485–2490.

36 Giessler, S. and Mack, H. (2003) Organofunctional silanes – molecular bridges for glass fibre reinforced polyamides. *Reinf. Plast.*, **22**, 28–32.

37 Schofield, W.C.E., Hurst, S.J., Lees, G.C., Liauw, C.M., and Rothon, R.N. (1998) Influence of surface modification of magnesium hydroxide on the processing and mechanical properties of composites of magnesium hydroxide and an ethylene vinyl acetate copolymer. *Compos. Interface*, **5** (6), 515–528.

38 Mack, H. (1999) HFFR cable materials: trends in silane coupling technology. 48th International Wire & Cable Symposium Proceedings, Atlantic City, NJ, USA, pp. 401–405.

39 Schlosser, T. (2004) Innovative silanes for polypropylene. Proceedings of the Cables 2004, Cologne, Germany.

40 Borup, B. and Ioannidis, A. (2006) Polypropylene, maleic acid anhydride, and silanes: the winning combination. Proceedings of the Cables 2006, Cologne, Germany.

41 Frahn, S., Valter, V., and Leder, G. (2001) Modified silica for UV-coatings. *Eur. Coat. J.*, **10**, 24–30.

42 Borup, B., Edelmann, R., and Mehnert, R. (2003) Silanes and inorganic particles: the winning combination for scratch and abrasion resistant coatings. *Eur. Coat. J.*, **6**, 21–27.

43 Borup, B., Edelmann, R., and Mehnert, R. (2002) A new role for silanes in coatings. Silanes: versatile ingredients in hybrid coating systems. Proceedings of the Annual Meeting Technical Program of the FSCT, International Coatings Exhibition, New Orleans, LA, USA, pp. 402–410.

44 Gläsel, H.-J., Bauer, F., Ernst, H., Findeisen, M., Hartmann, E., Langguth, H., Mehnert, R., and Schubert, R. (2000) Preparation of scratch and abrasion resistant polymeric nanocomposites by monomer grafting onto nanoparticles, 2. Characterization of radiation-cured polymeric nanocomposites. *Macromol. Chem. Phys.*, **201**, 2765–2770.

45 Bauer, F., Sauerland, V., Gläsel, H.-J., Ernst, H., Findeisen, M., Hartmann, E., Langguth, H., Marquardt, B., and Mehnert, R. (2002) Preparation of scratch and abrasion resistant polymeric nanocomposites by monomer grafting onto nanoparticles, 3. Effect of filler particles and grafting agents. *Macromol. Mater. Eng.*, **287**, 546–552.

46 Hunsche, H., Görl, U., Müller, A., Knaack, M., and Göbel, Th. (1997) Investigations concerning the reaction of silica/organosilane and organosilane/polymer. *KGK Kautschuk Gummi Kunststoffe*, **50**, 881–889.

47 Weissenbach, K. and Simões, D. (2007) Max it out: silane modification at its best. Proceedings of the Pira Intertech Conference on Functional Fillers, September 2007, Charleston, SC, USA.

5
Titanate Coupling Agents

Salvatore J. Monte

5.1
Introduction

Titanate coupling agents impart increased functionality to fillers in plastics. The different ways that these additives work in filled polymers can be explained by breaking down the various mechanisms of the titanate (or zirconate) molecule into six distinct functions. Filler pretreatment and *in situ* reactive compounding with titanates and zirconates to effect coupling, catalysis, and heteroatom functionality in the polymer melt are also discussed.

Esters of titanium or zirconium couple or chemically bridge two dissimilar species such as an inorganic filler/organic particulate/fiber and an organic polymer through proton coordination. This permits coupling to both nonhydroxyl bearing, and therefore non-silane reactive, inorganic substrates such as $CaCO_3$ and boron nitride and organic substrates such as carbon black (CB) and nitramines without the need of water of condensation as with silanes. The thermally stable quaternary carbon structure of the neoalkoxy organometallics permits *in situ* reactions to take place in the thermoplastic melt. In addition, the coupling of monolayers of a phosphato or a pyrophosphato heteroatom titanate or zirconate imparts synergistic intumescence to nonhalogenated flame retardants such as $Mg(OH)_2$ and aluminum trihydrate (ATH); flame retardance function to fillers such as $CaCO_3$; control of the burn rate and burn rate exponent of aluminum powder rocket fuels; and extinguishment of the flame spread of spalls of polymer-bound nitramines used in propellants and explosives. It is also believed that the organometallic monolayer-covered filler surface becomes a catalysis support bed for "repolymerization" of the surrounding polymer phase, thus allowing fillers to act as mechanical property improvers. Furthermore, the *in situ* monomolecular deposition of titanate on the surface of a particulate, such as a nanofiller, renders the particulate hydrophobic and organophilic. Under melt compounding shear conditions, the titanate assists in the removal of air voids and moisture from the particle surface, resulting in complete dispersion and formation of a true continuous phase, thus optimizing filler performance.

Functional Fillers for Plastics: Second, updated and enlarged edition. Edited by Marino Xanthos

ISBN: 978-3-527-32361-6

Minor amounts of thermally stable neoalkoxy titanate and zirconate additives may provide a means for postreactor, *in situ* metallocene-like "repolymerization" catalysis of a filled or unfilled polymer during the plasticization phase. This may result in the creation of metallocene-like (titanocene or zirconocene) behavior associated with effects such as increased composite strain to failure resulting in increased impact toughness or enhanced polymer foamability. Other effects to be discussed below with specific examples are related to enhanced processability, reduced polymer chain scission, shortened polymer recrystallization time, and compatibilization of dissimilar polymers.

There is a significant body of published information on titanium- and zirconium-based coupling agents. During the period 1974–2008, over 2000 patents and technical papers appeared. Some detailed historical documentation with more than 500 figures and tables is provided by the author in Ref. [1]. References [2–21] provide some of the 360 ACS, CAS abstracted technical papers and conference presentations by the author. Table 5.1 provides a chemical description of the coupling agents discussed in this chapter, along with an alpha-numeric code for the titanates and zirconates invented by the author. The alpha-numeric code is used often alone in this chapter for the sake of brevity. Table 5.2 indicates the designation of just a few of more than 50 commercial titanates and zirconates available from various vendors such as Kenrich Petrochemicals, Inc., E.I. du Pont de Nemours & Company, Synetix-Johnson Matthey, Nippon Soda, and Ajinomoto Fine Techno Co. (Kenrich licensee). Chemical structures of two common titanates (KR TTS and LICA 38) are shown in Figure 5.1.

5.2
The Six Functions of the Titanate Molecule

Organosilanes have long been used to enhance the chemical bonding of a variety of thermoset resins with siliceous surfaces and more recently of thermoplastics. However, Plueddemann observed [22] that organosilanes are essentially nonfunctional as bonding agents when employing carbon black, $CaCO_3$, boron nitride, graphite, aramid, or other organic-derived fibers.

A discussion of the six functional sites of a titanate (or zirconate) compared to a silane (see structures in Table 5.1 and Chapter 4) is useful to explain their performance differences and each may be represented as follows:

Titanate	Silane
(1) (2) (3)(4)(5)(6) $(RO\text{-})\text{-}_n\,Ti\text{-}(\text{-}O\;X\;R'\;Y)_{4-n}$	(1) (5) $(RO\text{-})\text{-}_3\,Si\text{-}(\text{-}R'\;Y)$

where

1) RO = hydrolyzable group/substrate reactive group with surface hydroxyls or protons.

Table 5.1 Coupling agent chemical description – alpha-numeric code.

Code	Nomenclature
Silanes[a)]	
A-187	3-Glycidoxypropyl, trimethoxysilane
A-1100	3-Aminopropyl, trimethoxysilane
Titanates[b)]	
KR TTS	Titanium IV 2-propanolato, tris isooctadecanoato-O
KR 7	Titanium IV bis-2-methyl-2-propenoato-O, isooctadecanoato-O 2-propanolato
KR 9S	Titanium IV 2-propanolato, tris(dodecyl)benzenesulfonato-O
KR 33CS	Titanium IV, tris(2-methyl)-2-propenoato-O, methoxydiglycolylato
KR 38S	Titanium IV 2-propanolato, tris(dioctyl)pyrophosphato-O
KR 41B	Titanium IV tetrakis-2-propanolato, adduct 2 mol (dioctyl) hydrogen phosphite
KR 46B	Titanium IV tetrakis octanolato adduct 2 mol (ditridecyl)hydrogen phosphite
KR 55	Titanium IV tetrakis (bis-2-propenolatomethyl)-1-butanolato adduct 2 mol (ditridecyl)hydrogen phosphite
KR 112	Titanium IV oxoethylene-diolato, bis(dioctyl)phosphato-O
KR 138S	Titanium IV bis(dioctyl)pyrophosphato-O, oxoethylenediolato, (adduct), (dioctyl) (hydrogen)phosphite-O
KR 238S	Titanium IV ethylenediolato, bis(dioctyl)pyrophosphato-O
LICA 01	Titanium IV 2,2(bis 2-propenolatomethyl)butanolato, tris neodecanoato-O
LICA 09	Titanium IV 2,2(bis-2-propenolatomethyl)butanolato, tris(dodecyl)benzene-sulfonato-O
LICA 12	Titanium IV 2,2(bis-2-propenolatomethyl)butanolato, tris(dioctyl)phosphato-O
LICA 38	Titanium IV 2,2(bis-2-propenolatomethyl)butanolato, tris(dioctyl)pyropho-sphato-O
LICA 38ENP	Titanium IV 2,2(bis-2-propenolatomethyl)butanolato, tris(dioctyl)pyropho-sphato-O: ethoxylated nonyl phenol – 1 : 1
LICA 38J	Titanium IV (bis-2-propenolatomethyl)-1-butanolato, bis(dioctyl)pyropho-sphato-O, (adduct) 3 mol *N,N*-dimethylamino-alkyl propenoamide
LICA 44	Titanium IV 2,2(bis-2-propenolatomethyl), tris 2-ethylenediamino)ethylato
LICA 97	Titanium IV 2,2(bis-2-propenolatomethyl)butanolato, tris(3-amino)phenylato
KS N100	Combined mononeoalkoxy titanates
KS N60WE	60% Active combined mononeoalkoxy titanates containing water emulsifiers
Zirconates[b)]	
KZ 55	Zirconium IV tetrakis(2,2-bis-propenolatomethyl)butanolato, adduct 2 mol bis (tridecyl)hydrogen phosphite
NZ 12	Zirconium IV 2,2(bis-2-propenolatomethyl)butanolato, tris (dioctyl)phospha-to-O
NZ 37	Zirconium IV bis-2,2(bis-2-propenolatomethyl)butanolato, bis(*para*-amino benzoato-O)
NZ 38	Zirconium IV 2,2(bis-2-propenolatomethyl)butanolato, tris (dioctyl)pyropho-sphato-O
NZ 39	Zirconium IV 2,2(bis-2-propenolatomethyl)butanolato, tris-2-propenoato-O
NZ 44	Zirconium IV 2,2(bis-2-propenolatomethyl), tris(2-ethylenediamino)ethylato
NZ 97	Zirconium IV (2,2-bis propenolatomethyl), tris(3-amino)phenylato
KS MZ100	Combined trineoalkoxy zirconates
KS MZ60WE	60% Active combined trineoalkoxy zirconates containing water emulsifiers

a) OSi Specialties, GE Silicones.
b) Kenrich Petrochemicals, Inc.

Table 5.2 Titanate and zirconate coupling agents[a] form designation.

Liquid form	
KR® # = Liquid coupling agent, 100% active monoalkoxy, chelate and coordinate titanate	
LICA® # = Liquid coupling agent, 100% active neoalkoxy titanate	
NZ® # = Liquid coupling agent, 100% active neoalkoxy zirconate	
Example: LICA 12 = 100% liquid neoalkoxy titanate	
Powder masterbatch form	
CAPOW® L® #/Carrier = Coupling agent powder	
Example: CAPOW L 12/H = 65% LICA 12/35% PPG Hi Sil 233 silica carrier	
Pellet masterbatch form	
CAPS® L® #/Binder = Coupling agent pellet system, 20% active neoalkoxy titanate/binder	
Example: CAPS L 12/L = 20% LICA 12/10% silica/70% LLDPE binder	
Water-soluble quat form	Quat blend (QB) part ratio
Designation	KR 238M:LICA 38J:NZ 38J
QB 012	0:1:2
QB 521	5:2:1

a) Kenrich Petrochemicals, Inc.

2) Ti (Zr), Si = tetravalent titanium, zirconium, or silicon. The Ti–O (or Zr–O) bond is capable of disassociation allowing transesterification, transalkylation, and other catalyzed reactions such as "repolymerization," while the Si–C bond is more stable and thus unreactive.
3) X = binder functional groups such as phosphato, pyrophosphato, sulfonyl, carboxyl, and so on that may impart intumescence, burn rate control, anticorrosion, quaternization sites, disassociation rate/electron transfer control, and so on.
4) R′ = thermoplastics-specific functional groups such as aliphatic and nonpolar isopropyl, butyl, octyl, isostearoyl groups; naphthenic and mildly polar dodecylbenzyl groups; or aromatic benzyl, cumyl, or phenyl groups.
5) Y = thermoset (but also thermoplastic)-specific functional groups such as acrylyl, methacrylyl, mercapto, amino, and so on.

KR TTS

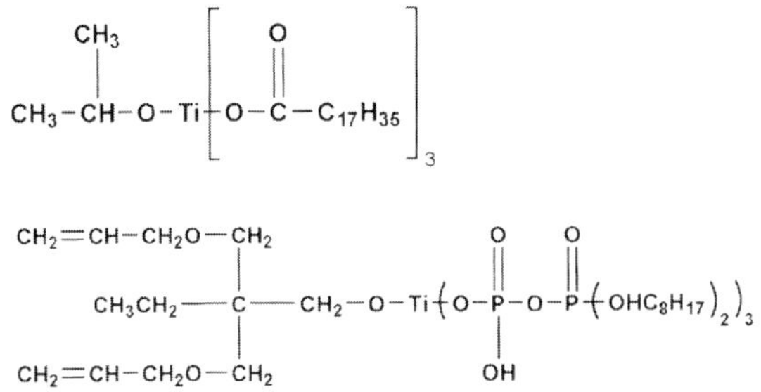

$$CH_3-CH(CH_3)-O-Ti\left[O-C(=O)-C_{17}H_{35}\right]_3$$

$$(CH_2=CH-CH_2O-CH_2)_2(CH_3CH_2)C-CH_2-O-Ti\left(O-P(=O)(OH)-O-P(=O)\left(OHC_8H_{17}\right)_2\right)_3$$

LICA 38

Figure 5.1 Chemical structures of two commercial titanate coupling agents.

6) $4-n$ = mono-, di-, or triorganofunctionality. Hybrid titanate (zirconate) coupling agents, such as those containing one mole each of a carboxyl (function 3) and aliphatic isostearoyl (function 4) ligand and two moles of carboxyl (function 3) and acrylyl (function 5) ligands, are possible.

Therefore, function (1) relates to filler/fiber substrate reaction mechanisms, while functions (2) to (6) are polymer/curative reactive.

5.2.1 Effects of Function (1)

The functional site (1) of the titanate molecule is associated with coupling, dispersion, adhesion, and hydrophobicity effects. These effects are also related to the method of application of the titanate on the filler surface as discussed below.

5.2.1.1 Coupling

In its simplest terms, the titanate function (1) mechanism may be classed as proton-reactive through solvolysis (monoalkoxy) or coordination (neoalkoxy) without the need of water of condensation, while the silane function (1) mechanism may be classed as hydroxyl-reactive through a silanol–siloxane mechanism requiring water of condensation. The silane's silanol–siloxane water of condensation mechanism limits its reactions to temperatures below 100 °C, thereby reducing the possibility of *in situ* reaction in the thermoplastic or elastomer melt above 100 °C as is possible with titanates. In addition, a variety of particulate fillers such as carbonates, sulfates, nitrides, nitrates, carbon, boron and metal powders used in thermoplastics, thermosets, and cross-linked elastomers do not have surface silane-reactive hydroxyl groups, while almost all three-dimensional particulates and species have surface protons, thereby apparently making titanates universally more reactive.

5.2.1.2 Dispersion

Dispersion of fillers results from the application of electrochemical and mechanical forces to the interface of the inorganic filler/polymer so as to cause complete deagglomeration to the attrited or original particle size in an organic phase, complete elimination of air voids and water, and creation of a true continuous inorganic/organic composition. The coupling of the titanate to the inorganic/organic substrate in monolayers allows elimination of air voids, enhanced hydrophobicity, and a complete continuous phase for stress/strain transfer. Figure 5.2 shows the "before and after" effect, envisaged by the author, of a titanate monolayer on agglomerated fillers.

C20 aliphatic mineral oil can be used as a low molecular weight model for polyolefins. Since it is nonpolar and, thus, a poor medium for dispersion of most polar fillers, coupling agent effects can be more easily measured. Figure 5.3 shows the effect of 0.5% isopropyl triisostearoyl titanate (KR TTS; see Table 5.1) on the dispersion of $CaCO_3$ in a nonpolar mineral oil. The deagglomeration effect is apparent. Significant viscosity reductions have been observed through the application

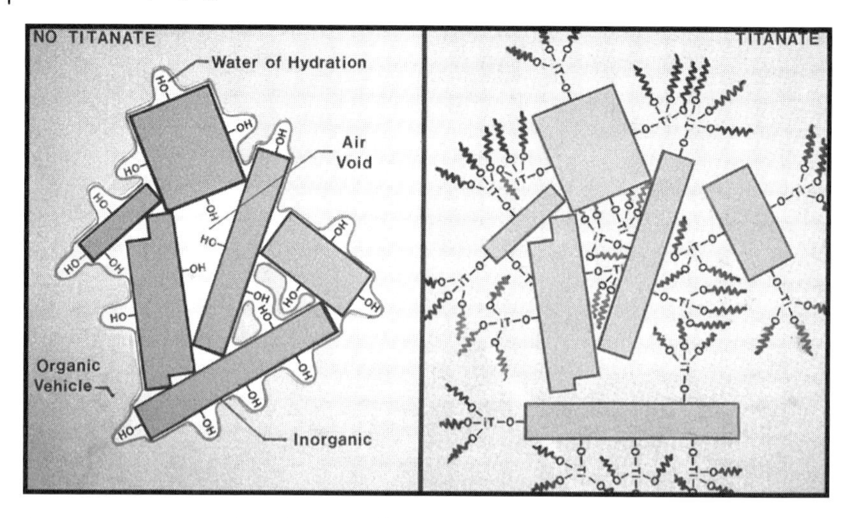

Figure 5.2 Illustration of the dispersion effect of coupling a titanate monolayer on an agglomerated inorganic (left, no titanate) in an organic phase, thereby creating a continuous inorganic/organic phase (right, with titanate) by deagglomeration and subsequent elimination of air and water from the interface.

of the same titanate (at 0.5–3 wt%) on numerous fillers such as 2.5 μm $CaCO_3$ (70 wt% filler), or clay (30 wt% filler) in mineral oil, and 40 wt% TiO_2 coated with 0.4% titanate in dioctyl phthalate plasticizer. Figure 5.4 shows the shift in the critical pigment volume concentration point (CPVC) of $CaCO_3$-filled mineral oil using 0.5 wt% KR TTS. The CPVC is defined as the point at which addition of more filler to an organic phase will cause incomplete wetting due to insufficient organic binder being available to wet the additional inorganic filler surface. The shift in the CPVC as a result of "coupling" (function 1) may be extended from the mineral oil model to filled thermoplastic and thermoset systems allowing higher loading to equivalent oil

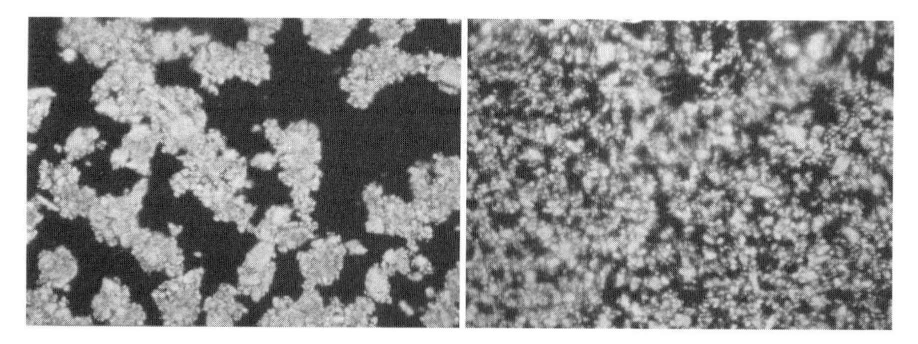

Figure 5.3 Left: a micrograph of a suspension of $CaCO_3$ (untreated)/liquid paraffin system; right: a micrograph of a suspension of $CaCO_3$ (treated with KR TTS)/liquid paraffin system demonstrating deagglomeration.

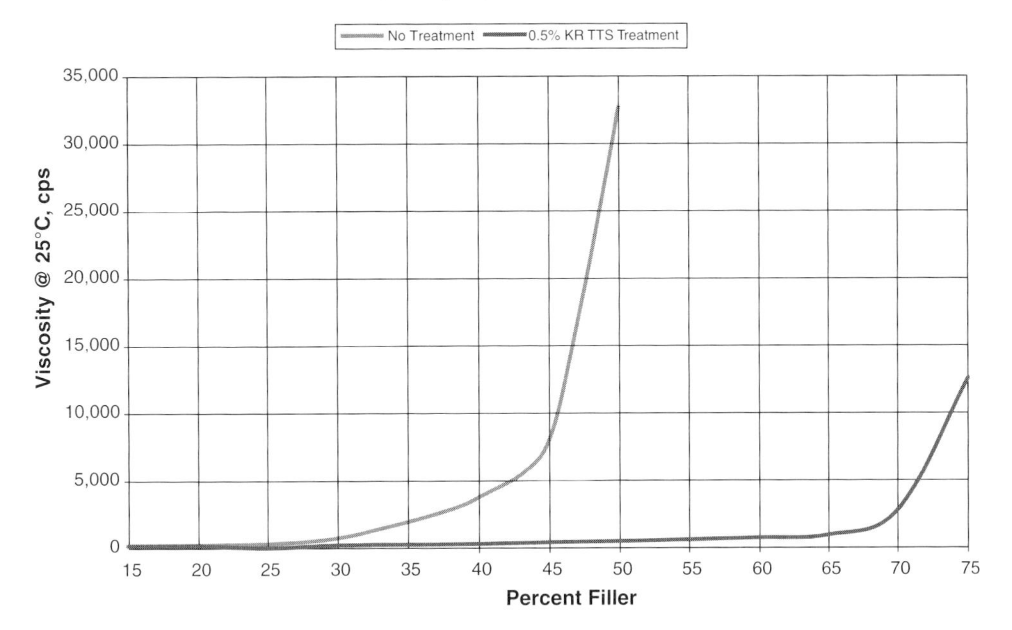

Figure 5.4 Pigment volume concentration curves comparing the critical pigment volume concentration point of untreated $CaCO_3$-filled mineral oil (left) with 0.5% KR TTS-treated $CaCO_3$-filled mineral oil (right) predicting the ability to fill plastics with higher loadings of filler without detracting from mechanical properties.

demand, and improved relative mechanical properties at any filler loading below the CPVC [1]. Figure 5.5 shows transitions from the model fluid to polypropylene (PP) as the organic phase and shows the flexibility imparted to a sample containing 70 wt% $CaCO_3$ (3 μm average particle size) in a PP homopolymer using 0.5 wt% KR TTS with respect to the filler. A PP composite containing 70% untreated $CaCO_3$ will normally break in a brittle manner, while the titanate treatment allows 180° bending with no white stress cracking.

Recently, Voelkel and Grzeskowiak [23] reported on the use of solubility parameters in the characterization of silica surface modified with titanates by inverse gas chromatography and compared their findings with earlier work on silanes.

5.2.1.3 Adhesion

One of the reasons why the dispersion of inorganics in plastics and adhesion of a plastic to an inorganic substrate is so difficult is that many thermoplastics and rubbers, such as olefin-based polymers, are nonpolar. Titanates and zirconates are well-established adhesion promoters, as discussed in Ref. [1]. A recent example of bonding olefins to metals appears in Ref. [24]. Another example of the adhesion of polyolefins to foil electrodes using KR TTS has been provided by Kataoka [25].

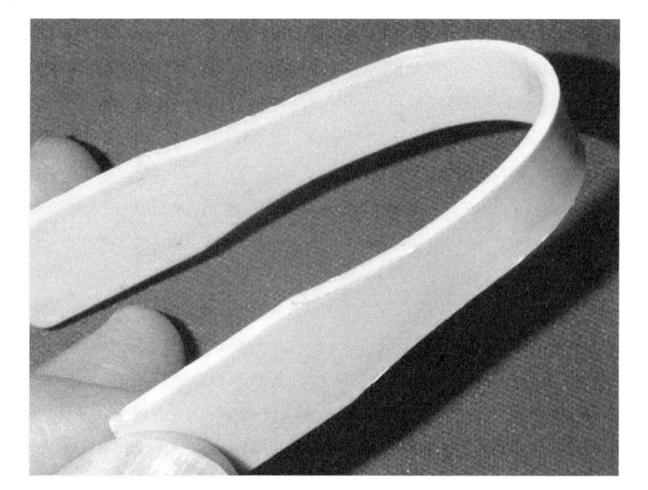

Figure 5.5 Demonstration of the flexibility of 70% $CaCO_3$ (2.5 μm), 0.5% KR TTS-treated filled PP homopolymer. Untreated 70% $CaCO_3$-filled PP would snap and break at the slightest deformation.

5.2.1.4 Hydrophobicity

Hydrophobicity is a desirable property to impart to fillers, functional particulates, and fibers in order to provide long-term protection to their composites against corrosion and aging. Increasing amounts of titanate on substrates such as $CaCO_3$ result in significant changes in hydrophobicity, as shown from contact angle measurements with water droplets. In a publication by Krysztafkiewicz *et al.* [26], hydrophobic modification of silica surface by silane and coupling agents was determined on the basis of the heat of immersion and infrared spectroscopy. The highest degree of hydrophobicity was observed for silicas modified with 1 wt% KR TTS and KR 33CS titanates (see Table 5.1), while it was a little lower after modification with 3 wt% aminosilane and methacryloxysilane. The authors also observed that water was necessary for the silane to couple to the silica whereas it was not needed in the case of the titanate.

5.2.1.5 Titanate Application Considerations

The correct usage of titanate coupling agents for optimum performance needs to take into account the following considerations:

- *In situ* coupling in the melt phase without the need for water of condensation is possible through monoalkoxy and neoalkoxy titanate or zirconate groups at temperatures above 200 °C. However, *in situ* coupling requires careful consideration of good compounding principles to avoid localization, inconsistent and incomplete coupling because of inadequate specific energy input (low shear) caused by reduced polymer viscosity induced by the coupling agent.
- Localization and physical absorption of the coupling agent on the filler or fiber that results in whole segments of uncoupled particulate surfaces can be largely overcome by using masterbatches of the coupling agent (see Table 5.2).

- In order to effect monomolecular level coupling, the titanate or zirconate must be solubilized in the organic phase (solvent, plasticizer, and polymer) or finely emulsified into water prior to the addition of the filler. If the organic phase has a high molecular weight, then sufficient shear and high mixing torque is needed to assure titanate distribution.
- Uniform distribution of the titanate in the dry powder ingredients or accurately dosing in the melt dictates the matching of the titanate form to that of the polymer or filler by using appropriate liquid, powder, or pelletized titanates (see Table 5.2).
- High specific energy input during melt compounding for maximum shear/work energy for dispersion and complete coupling. Titanates reduce the process temperatures of most thermoplastics by approximately 10% and by much more for certain thermoplastic polyesters, acting as transesterification catalysts [1, 10]. Therefore, when evaluating a titanate in an unfilled or filled thermoplastic, it is imperative to compare both compounds processed at the same specific energy input. The importance of specific energy input to the dispersion of fillers during single-screw extrusion compounding of particle-filled thermoplastics is discussed in Ref. [27].

An ideal amount of coupling agent to use is the amount that will form a monolayer on the surface of the filler to produce optimal filler dispersion effects, plus the amount that will have an optimal "repolymerization" catalytic effect on the polymer. Again, filler dispersion is defined as a complete deagglomeration of the filler as attrited or precipitated so as to allow all moisture and air in the interstices of the agglomerates to be replaced with a continuous organic polymer phase. Optimal "repolymerization" is defined as being the best balance of mechanical properties and system rheology. Various experimental methods for measuring viscosity changes induced by titanate in mineral oil/plasticizers are useful for determining the required amounts based on the organic polymer, on the filler, or on the combined organic and inorganic phases. One effective method for dry filler pretreatment is to apply the neat titanate, either by airless spraying or by adding it dropwise, over a period of 1 min, to a fluidized bed of the filler as created by a Henschel-type mixer operating at low speed (1800 rpms). The treated filler thus obtained can then be compared with an *in situ* treated control to test the effectiveness of the dry treatment method. This is necessary because for certain fillers, such as $Mg(OH)_2$, dilution or very slow dropwise addition is needed to avoid localization and uneven distribution. For example, in compounding linear low-density polyethylene (LLDPE) containing 60 wt% pretreated $Mg(OH)_2$, the required 0.7% LICA 38 first has to be diluted with 2.1% of an alkyl phosphate plasticizer, or the time for dropwise addition of the undiluted titanate has to be increased from 1 to 5 min.

5.2.2 Effects of Function (2)

"Repolymerization" is a patented [28] concept by the author to explain new and novel rheology and stress/strain effects in thermoplastics and thermosets obtained with titanates and zirconates that are independent of cross-linking and curative effects. The aromatic (e.g., phenyl, naphthyl, and styrenic) or aliphatic (e.g., ethyl, propyl, and butyl) groups that are typically present in the thermoplastic macromolecule, liquid chemical compounds, or thermoplastic elastomers are reactive with titanate (or zirconate) (functions 2–4), independent of any curative reaction mechanisms (function 5). Thus, the monolayered, organometallic-coupled particulate and/or fiber may be considered as a catalyst support bed for single-site, *in situ* metallocene-like "repolymerization" [28] of the surrounding polymer.

At present, published efforts in metallocene (titanocene and zirconocene) chemistry by major polymer producers appear to be centered on olefin polymers and copolymers. Metallocene-derived HDPE and engineering plastics seemingly remain a future goal, while titanate and zirconate esters appear to be efficacious to some degree in virtually all polymers synthesized by various routes [1]. Moreover, the titanocene or zirconocene catalysts used in synthesizing metallocene-derived polymers do not remain in the polymer. With "repolymerization," thermoplastics may be regenerated to virgin or recycled more efficiently since the thermally stable titanate or zirconate ester forms of the relevant organometallics "anneal" or "reconnect" polymer chain lengths that normally undergo scission during processing and remain in the polymer for subsequent repeat thermal cycles.

Table 5.3, extracted from the European Patent application "repolymerization" [28], shows the beneficial effects of titanates on most properties of ABS, PC, PP, and PS thermoplastics. As an additional example, Figure 5.6 shows the "repolymerization" effect of 0.2% LICA 12 on a 50/50 blend of LDPE and PP after six thermal cycles through a twin-screw extruder. The melt index of the control blend without titanate climbs from 17 to 38, while the value for the blend with titanate is only 24; this indicates a significant decrease in chain-scission due to the titanate.

"Repolymerization" appears to affect the isothermal recrystallization time, chain branching, and morphology of the polymer chains surrounding the particulate or

Table 5.3 "Repolymerization" effect of various neoalkoxy titanates and zirconates on the mechanical properties of unfilled ABS, PC, PP, and PS thermoplastics.

Resin	Titanate/ zirconate (%)	Tensile yield strength (MPa)	Elongation at break (%)	Flexural strength (MPa)	Flexural modulus (MPa)	Notched Izod impact strength (J/m)	% H_2O absorption 24 h
ABS	Control	48.9	18	82.7	2826	160	0.30
	L 44/H, 0.5	64.8	17	289	4757	256	0.08
PC	Control	66.9	65	89.6	2275	320	0.20
	NZ 12/H, 0.5	70.3	73	96.5	2344	421	0.14
PP	Control	33.8	120	—	1447	37.4	—
	L 12/H, 0.5	38.6	148	—	1516	74.8	—
PS	Control	35.2	10	65.5	2551	133	—
	L 12/H, 0.3	40.7	51	68.3	2551	197	—

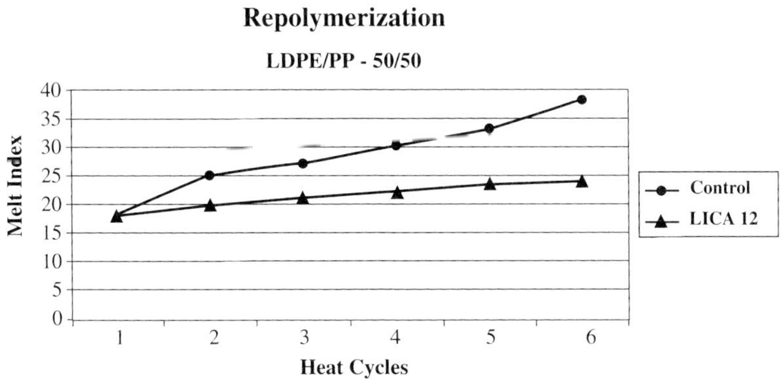

Figure 5.6 "Repolymerization" effect of 0.2% LICA 12 after six thermal cycles through a twin-screw extruder on a 50/50 blend of LDPE and PP. The control melt index increases from 17 to 38. The blend with titanate increases only to 24 indicating a significant regeneration of scissored polymer chains.

fiber. Table 5.4 gives examples of filled systems for which easier processing and better mechanical properties may be attributed to this effect.

5.2.3 Effects of Function (3)

Pyrophosphato and phosphato titanates (e.g., LICA 38 and LICA 12) may render $CaCO_3$-filled LLDPE flame retardant and provide synergistic effects with conventional flame retardants. Figure 5.7 depicts a theoretical monolayer of phosphato titanate (function 3) coupled to a substrate. Pyrophosphato titanates are efficacious on metal oxides such as antimony oxide, which is used in halogenated flame retardant systems. Because of their nontoxic smoke generation, recent efforts have been directed toward the use of ATH and $Mg(OH)_2$. ATH must be loaded to a level of 64 wt% of the total compound to generate enough steam to achieve a UL 94V-0 rating, while $Mg(OH)_2$ is usually loaded into thermoplastics in the 40–60 wt% range. The use of titanates allows highly filled polyolefins to be processed at approximately 10% lower temperatures, thus allowing water-releasing flame retardants to be processed more readily. Examples of the synergistic benefits of titanate in function (1) coupling, function (2) catalysis, and function (3) phosphato and pyrophosphato heteroatom intumescence are presented in Table 5.5.

Highly $CaCO_3$-filled polyolefins treated with phosphate titanate may be converted into fire extinguishing compositions. Data by the author on LLDPE filled with 44 wt% $CaCO_3$ and treated with 3 wt% LICA 38 demonstrated the flame retardant intumescent effect of the pyrophosphate titanate-containing composition as compared to a titanate-free control. This system has been used as a model for preparing powerful, yet safe, energetic composites containing high loadings of nitramine explosive (up to 85%) in a cellulose acetate butyrate matrix. In a related patent [43], the phosphate titanate is claimed to permit increased nitramine loadings, improve

Table 5.4 Examples of systems with function (2) effects.

System	Titanate coupling agent	Coupling agent effects	References
Talc-filled PP	LICA 12	Increase in PP MWD Increased dispersion, decreased melt viscosity Increased impact strength, reduced T_g	[29]
Nanoclay-filled SBS	KR TTS	Retention of SBS properties after compounding with modified filler Lower temperature of mixing and lower torque	[30]
$BaSO_4$-filled peroxide-cured polybutadiene	KR TTS	Increased elongation at break Increased tensile strength	[31]
Fiberglass-filled PPS	Lica 09, CAPOW NZ 97/H	Increase resin crystallization temperature, decrease isothermal crystallization Increase composite elongation at break, eliminate embrittlement	[1, 32]
Carbon black and organic pigments in printing inks	KR 46B	Higher solids contents	[33]
Cellulose fibers in mixed plastics waste	CAPS L 12/L	Strength increase despite little evidence of bonding; possible reactive compatibilization	[34]

flow characteristics, and control the spread of burning propellant spalls. Similarly, in solid rocket fuel containing aluminum and ammonium perchlorate in a hydroxy-terminated polybutadiene/polyurethane binder, the use of pyrophosphato titanate was shown to control the burning rate, in addition to providing improved dispersion of the fillers and improved mechanical properties of the propellant.

5.2.4 Effects of Function (4)

$CaCO_3$ and CB are the two largest volume fillers consumed in thermoplastics and elastomers. Before functions (5) and (6) are discussed, further discussion on the functional effects of titanates on these two fillers and a host of other inorganics/organics in thermoplastics, as noted by the author and other investigators, is instructional.

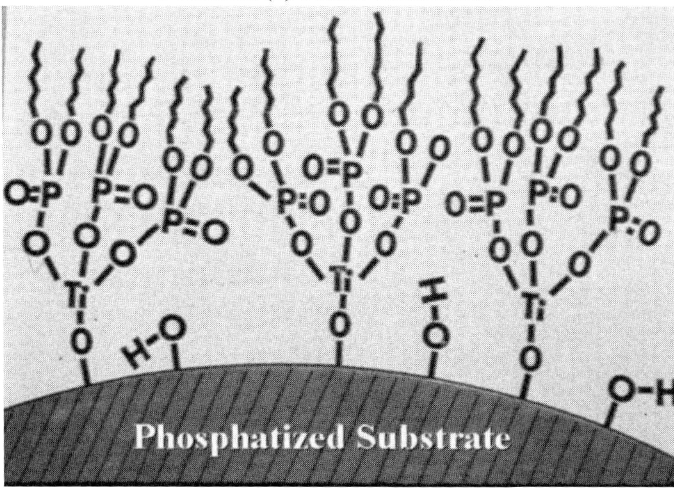

Figure 5.7 Graphic presentation of the function (3) monomolecular deposition of a phosphato titanate to create a phosphatized flame retardant or anticorrosive substrate.

5.2.4.1 $CaCO_3$-Filled Thermoplastics

The reactivity of titanates with $CaCO_3$ has been discussed herein and is well established [1]. For example, transparent polyolefin films containing titanate and 40 wt% 0.9 μm $CaCO_3$ and many other efficacious $CaCO_3$-filled thermoplastic compositions have been produced by the author [1], and significant commercial applications exist. Examples of work by others in the area of $CaCO_3$-filled polyolefins are shown in Table 5.6. Specific property data adapted from Ref. [45] are shown in Table 5.7. In general, if overall polyolefin composite strength is desired, and not just filler loading, then 20–30% fine particle $CaCO_3$ loading using 0.5 wt% titanate, all compounded at approximately 10% lower temperatures than would be used without titanates, is recommended by the author.

5.2.4.2 Carbon Black-Filled Polymers

Carbon black is the most extensively used filler in terms of volume in thermoplastic/thermoset elastomers. One method of quantifying carbon black dispersion is through the reduction in resistivity of insulating polymer matrices that occurs by virtue of the CB's ability to create a conductive three-dimensional particulate network. As the amount of carbon black is increased, the resistivity is decreased. In addition, the efficiency of a fixed amount of carbon black is increased as a result of the increased dispersion offered by the use of titanates. As an example, Table 5.8 shows the effect of increasing amounts of LICA 09 on the resistivity of a CB-filled styrene/butadiene block copolymer.

Yu *et al.* [47] investigated carbon black-filled polyolefins as positive temperature coefficient (PTC) materials by studying the effect of coupling agent treatment, composition, and processing conditions. Their data show that an 18 wt% CB

Table 5.5 Examples of systems with function (5) effects.

Filler	Polymer	Titanate	Comments	References
ATH, $Mg(OH)_2$	PP	Phosphato and pyrophosphate functionality	High loadings (64 wt%) to achieve flame retardancy without loss of mechanical properties	[1, 35]
ATH, $Mg(OH)_2$	Miscellaneous thermoplastics and elastomers	Titanates, zirconates	Nontoxic flame retardant	[36]
ATH	Dicumyl peroxide cross-linked LDPE, LLDPE	Isostearoyl titanate, isopropoxy tris(dioctyl pyrophosphoryl) titanate	Increased mechanical properties through synergism of titanate and a vinyltriethoxysilane	[37–39]
ATH	Polyolefin mixture	Phosphate functionality	Flame proof compositions	[40]
$Mg(OH)_2$	Polyolefins, EPDM, polyamide, ABS	Isopropyl-triisostearoyl, pyrophosphato	Fire-resistant wire and cable and moldings	[41, 42]
$CaCO_3$	LLDPE	Pyrophosphato	Self-extinguishing at 44 wt% filler	Author's data

Table 5.6 Titanate effects in $CaCO_3$-filled polyolefins.

Polymer	Titanate	Comments	References
PP/HDPE	LICA12	Optimum mechanical properties at 0.7% titanate	[44]
PP	LICA 12 (0.2, 0.3, and 0.4%)	Optimum elongation at break in the 20–40 wt% loading with impact strengths 35–65% more than the unfilled control	a)
PP	TTS	Improved dispersion, higher melt index, higher tensile elongation, improved optical properties	[45]
PP	TTS, calcium stearate	Titanates proved more effective in improving flow properties and impact strength	[46]

a) Doufnoune, R. and Haddaoui, M. (1999) Université FERHAT-Abbas, Institut de Chimie-Industrielle, LPCHP, Sétif-Algérie, private communication.

loading in LDPE treated with a pyrophosphate titanate (KR 38S) creates a composite with a stable PTC intensity, which is much lower than that achieved in the absence of titanate, thus allowing the production of novel "smart" polyolefin composites with more uniform conductivity control. From PTC intensity versus CB content plots, this 24-fold increase in PTC performance indicates that it would take about 7% additional untreated CB to reach the titanate-treated CB performance level.

It is obvious that any thermally conductive composition can be made more efficient by increasing the loading of the conductive material. Thus, it is necessary to load the polymer without losing the ability to process the compound and then to form a part that has suitable mechanical properties. A shift in CPVC induced by titanates is usually predictive of an increase in a functional particulate's performance. Refs [48, 49] provide examples of thermally conductive graphite elastomer (chloroprene and EPDM) compositions incorporating titanates. The effects of pretreating CB with KR TTS in terms of improving the low temperature flexibility of butyl rubber and the overall performance of a conductive isobutylene compound are described in Refs [50, 51]. It should be noted that KR TTS has been commercially sold in the United States since 1974 for masterbatching carbon black for polyolefin elastomers.

5.2.4.3 Other Functional Inorganics/Organics Used in Thermoplastics and Thermosets

Table 5.9 lists examples of polymer systems containing other widely used inorganic fillers that are treated with titanates/zirconates. The principal effects of the coupling agents in relation to the given application are also shown. Modifications of TiO_2 with pyrophosphate- and phosphito-coordinate titanates to produce a highly functional metal oxide [53] and an acrylic colorant [54] have been described. Titanates

Table 5.7 Comparison of properties of compression and injection-molded unfilled, and 40 and 50% $CaCO_3$-filled PP; untreated, 0.6 and 1.0% KR TTS-treated fillers.

Serial no.	Sample	MFI (g/10 min)	Tensile yield strength (MPa)	Elongation at break (%)	Brittle point (°C)	Notched Izod impact strength (J/m)
Compression molded						
1	PP	5.61	34.0	22	32	42
2	PP : $CaCO_3$ (50 : 50)	3.87	18.9	14	−7	39
3	PP : $CaCO_3$ (50 : 50) + 0.6% KR TTS	6.00	16.3	43	0	49
4	PP : $CaCO_3$ (60 : 40) + 0.6% KR TTS	6.11	20.4	50	22.5	51
5	PP : $CaCO_3$ (60 : 40) + 1% KR TTS	7.20	22.6	30	21.5	51
Injection molded						
1	PP	—	35.9	99	—	50
2	PP : $CaCO_3$ (50 : 50)	—	20.9	49	—	40
3	PP : $CaCO_3$ (50 : 50) + 0.6% KR TTS	—	20.6	65	—	45
4	PP : $CaCO_3$ (60 : 40) + 0.6% KR TTS	—	23.2	69	—	47
5	PP : $CaCO_3$ (60 : 40) + 1% KR TTS	—	23.8	61	—	52

Table 5.8 Resistivity of 3.75% XC-72R conductive black in styrene-butadiene block copolymer/PS – 10 mm thick test slab.

Wt% LICA 09 of 3.75% XC-72R	Resistivity	
	Surface (Ω/sq)	Volume (Ω cm)
Control	$>10^{16}$	7.8×10^{14}
0.67	1.7×10^{12}	3.0×10^{12}
1.00	2.1×10^{8}	4.3×10^{7}
2.00	5.7×10^{7}	3.7×10^{7}

have also been used as surface treatments of metal oxides for enhanced environmental stability [65] and in organic electroluminescent devices [68] for increased brightness.

5.2.5
Effects of Function (5)

Organofunctionality can be imparted to any inorganic particulate or substrate using suitable titanates or zirconates. For example, a pyrophosphato titanate can be used to treat silica to convert it to an anticorrosive pigment; an acrylic functional zirconate can convert TiO_2 to a UV-reactive pigment; or a water-insoluble pyrophosphato titanate or zirconate, or blend thereof, can be reacted with a methacrylamide functional amine to make a water-soluble, anticorrosive, acrylic functional organometallic additive for pigments, fillers, or surfaces. Such substrates would be compatible with water-based acrylics of high solids content.

As an example of surface functionalization, a recent patent [69] entitled "Surface functionalization of pigments and/or dyes for radiation-curable printing inks and coatings and their preparation" describes a UV-curable powder-coating composition containing TiO_2 treated with a zirconate containing a radiation-curable functional group [NZ 39(neopentyl(diallyl)oxytriacryl zirconate)]. The cured resin showed good optical properties and scratch resistance.

5.2.6
Effects of Function (6)

In the early stages of the introduction of the monoalkoxy titanates, hybrid titanates were introduced by the author, in which functions (3), (4), and (5) were intermixed by transesterification reactions at the Ti center of the molecules using various organic ligands. For example, the seventh in a series, titanate synthesized by the author was a hybrid titanate (KR 7) consisting of two moles of methacrylic acid and one mole of isostearic acid (see Table 5.1). The theory, borne out of commercial practice, was that the isostearoyl ligand would stabilize and protect the methacrylic ligands from autooxidation.

Table 5.9 Effects of titanate/zirconate in miscellaneous functional inorganics in thermoplastics and thermosets.

Filler	Coupling agent	Polymer	Comments	References
TiO_2	Pyrophosphato	Epoxy	Improved mechanical properties	[52]
$CaSO_4$	TTS	Acrylic acid-styrene copolymer	Use in cosmetic pastes	[55]
$CaCO_3$, talc, and $BaSO_4$	Miscellaneous	PP and EPR	Good processability at high loadings (70 wt%)	[56]
Fe_2O_3	KR 38S	Styrene/*n*-butyl methacrylate	Improved performance versus epoxysilane in copier toner	[57]
Fe_3O_4	KR TTS	Butyl acrylate-styrene copolymer	Use in magnetic toner	[58]
Hydroxyapatite	LICA 12, NZ 12	HDPE or starch/EVOH	Promote adhesion, catalysis, and improve mechanical properties in composites for prosthetic devices	[59, 60]
Tungsten carbide, titanium carbide	Miscellaneous	HDPE, polyolefin adhesive	Improved conductive positive temperature coefficient devices	[61, 62]
Saponite	NDZ-201	PP	Thermal-resistant automotive application	[63]
Barium ferrite	LICA 38	Thermoplastic natural rubber	Improved electrical and magnetic properties	[64]
Cadmium sulfide	KR 138S	Epoxy	Resistance to ozone attack in photosensitive plates	[1] (Table 34)
Clay	Miscellaneous	Rubber	Increased ozone resistance	[66]
Rare earth–iron–nitrogen powders	Miscellaneous	Miscellaneous	Improved coupling efficiency in bonded magnets	[67]

5.2.7
Dispersion of Nanofillers or "Making Nanofillers Work"

Making nanofillers work using 1.5 nm monolayer thickness of titanate and zirconate coupling agents requires a combination of many factors properly applied. The importance of selecting the right coupling agent chemistry has already been discussed. For example, LICA 38 has proven to be an effective dispersing agent of calcined clay (see platelet-type morphology in Figure 13.5) producing at 40 wt% loadings low viscosity, flowable dispersions in oil as shown in Figure 5.8.

Guidelines for applying titanates are given in Section 5.2.1.5. Some additional factors, important for the *in situ* exfoliation of nanoparticles, are as follows:

- **Type of interface:** The as-received nanoparticulates, such as montmorillonite nanoclays, should not have previously been exfoliated through quaternary ammonium salts (see Chapter 9) or have other surfactants on their surfaces; their presence would interfere with the interfacial reaction or coupling of the titanate to the nanoclay and the polymer. For example, when using titanates or zirconates, Southern Clay's unmodified Cloisite Na^+ grade would be preferred over an organomodified Cloisite 15A (see Chapter 9). Tables 5.10 and 5.11 indicate a continuous decrease in suspension viscosity with increasing dosage of titanate for the unmodified mineral, whereas for Cloisite 15A viscosity starts increasing after a certain additive concentration. Most water slurry-exfoliated particulates by quaternary ammonium salts contain up to 35% residual modifier after exfoliation and drying; this may provide compatibility issues with other additives in the formulation, including certain nonpolar polymers.

Figure 5.8 (Left) 40 wt% calcined clay in mineral oil – 2 000 000 cps viscosity. (Right) 0.7 wt% LICA 38 of calcined clay added *in situ* to mineral oil followed by 40 wt% clay – 9200 cps viscosity.

Table 5.10 Viscosity effects of varying dosages of LICA 38 added *in situ* to a slurry of 45 wt% unmodified Cloisite Na^+ nanoclay in mineral oil.

LICA 38 wt% based on clay	Brookfield (HBT) viscosity (cps at 25 °C)
0	720 000
0.5	400 000
1.5	144 000
3.0	57 600
6.0	51 200

- **Nanoparticle loading:** The amount of nanoparticulate that should be loaded into a polymer should be up to 10 wt% for metal oxides and 5 wt% for minerals such as nanoclays.
- **Titanate dosage based on surface area coupling:** The approximate amount of titanate needed for a 1.5 nm monolayer on a nanoparticulate is 2 wt% for metal oxides and 3 wt% for minerals such as nanoclays. This dosage is based on function (1) coupling.
- **Titanate dosage based on catalytic effect on polymer:** The amount of titanate on the nanoparticulate should also be at a level so that, in addition to the amount required for coupling, an amount of 0.2–0.4 wt% should be available to the polymer. This dosage is based on function (2) catalysis effects. For example, a nanoclay such as Cloisite Na^+ should be loaded at no more than 5% by total weight of the composite. A 3 wt% titanate forms a monolayer for coupling, but at this dosage, the polymer is exposed to only 0.15% for function (2) catalysis. A 6% titanate by weight on clay at 5 wt% in the composite would dose catalytically a more optimal 0.3% titanate by weight of the polymer.
- **In situ exfoliation:** Nanoparticulates can be exfoliated in the melt phase during extrusion, if the titanate has been first fully dissolved into the macromolecular network followed by addition of the nanoparticulate. Alternately, the soluble

Table 5.11 Viscosity effects of varying dosages of LICA 38 added *in situ* to a slurry of 39 wt% organomodified Cloisite 15A nanoclay in mineral oil.

LICA 38 wt% based on clay	Brookfield (HBT) viscosity (cps at 25 °C)
0	400 000
0.25	160 000
0.5	112 000
0.75	112 000
1.00	128 000
1.5	160 000
3.0	320 000
6.0	400 000

titanate may be sprayed on a fluidized bed of the nanoparticulate before it is added to the polymer melt. If the polymer system is water based such as a latex, the titanate must be made water compatible by techniques such as cosolvation, emulsification, or quaternization.

- **Initial boiling point:** The titanate or zirconate selected is a liquid ester with an initial boiling point (IBP). For *in situ* exfoliation, the initial contact of the organometallic ester with the polymer should be below the IBP of the titanate, which must be solvated into the heat sink of the polymer before the temperatures are raised to process temperatures. This usually means adding the titanate to the extruder hopper at room temperature and then allowing a transition up in temperature over the next two or three zones where the filler would also be present. Otherwise, the functionality of the titanate will be significantly diminished.

5.3 Summary and Conclusions

There is a wealth of other functional filler work that could be discussed, such as wood fiber/starch composites [70], the production of dyeable polypropylene through nanotechnology [71], and a host of fiberglass-, graphite-, and aramid-reinforced thermosets and thermoplastics. There is also much characterization work to be done to understand the function (2) catalysis, "repolymerization," copolymerization, and cross-linking effects reported in Refs [72–78], which can potentially make any filler with such catalytic properties more functional in plastics. The concluding discussion on thermally conductive thermoplastics highly filled with boron nitride and suitable for electronic packaging refers to a recent patent [79] with potentially significant commercial value. Of more than 2000 patents and references in the literature on titanates, this may serve as a representative case history since it covers so many of the issues raised in this chapter, specifically

- reactivity with a non-silane reactive inorganic, in this case boron nitride;
- *in situ* application of pellet or powder masterbatches of the coupling agent at temperatures in excess of 200 °C for reactive compounding in the melt;
- dispersion and adhesion effects, or stated another way, coupling and catalysis effects in relation to improved process rheology of filled polymers, increased flow of the polymer itself, shift in the CPVC of the filler to polymer ratio, and increased filler functionality such as thermal conductivity, enhanced mechanical properties, and enhanced end product performance;
- applicability to a host of polymers, processes, and equipment used in thermoplastics and thermoset manufacturing, particularly the ability to withstand the harsh temperature and shear conditions of commonly used high-volume melt processing such as extrusion, compounding, and injection molding.

The invention [79] relates to a thermally conductive moldable polymer blend comprising a thermoplastic polymer such as polyethylene terephthalate (PET), polybutylene terephthalate (PBT), polyphenylene sulfide (PPS), or polycarbonate (PC) with processing temperatures in the range of 200–300 °C and having a tensile at yield of at least 70 MPa, at least 60 wt% of a mixture of boron nitride powders with an average particle size of at least 50 μm, and a coupling agent such as a neoalkoxy or monoalkoxy titanate in a masterbatch form at a level of 0.1–5 wt%. Such a composition displays a thermal conductivity of at least about 15 W/m K and is capable of being molded using high-speed molding techniques such as injection molding. The coupling and/or dispersing agent serves to facilitate better wetting of the boron nitride fillers. It also helps reduce the melt viscosity of the composition and allows higher filler loading. In addition, the coupling and dispersing agent may also improve the polymer–filler interfacial adhesion and thus provide better physical and mechanical properties.

References

1 Monte, S.J. (1995) *Ken-React® Reference Manual: Titanate, Zirconate and Aluminate Coupling Agents*, 3rd edn, Kenrich Petrochemicals, Inc., Bayonne, NJ.

2 Monte, S.J. *et al.* (1988) 33rd International SAMPE Symposium, Anaheim, CA, USA, March 1988.

3 Monte, S.J. and Sugerman, G. (1988) 33rd International SAMPE Symposium (SAMPE II), Anaheim, CA, USA, March 1988.

4 Monte, S.J. and Sugerman, G. (1990) Corrosion '90, Las Vegas, NV, USA, April 1990, Paper No. 432.

5 Monte, S.J. and Sugerman, G. (1988) Water-Borne & Higher Solids Coatings Symposium, New Orleans, LA USA, February 1988.

6 Monte, S.J. and Sugerman, G. (1988) Corrosion/88, NACE, St. Louis, MO, USA, March 21–25, 1988.

7 Monte, S.J. and Sugerman, G. (1988) 2nd International Conference on Composite Interfaces (ICCI-II), Case Western Reserve University, Cleveland, Ohio, USA, June 1988.

8 Monte, S.J. *et al.* (1981) SPI Urethane Division 26th Annual Technical Conference, November 1981.

9 Monte, S.J. and Sugerman, G. (1985) SPI Urethane Division 29th Annual Technical/Marketing Conference, October 1985.

10 Monte, S.J. (1995) Proceedings of the SPE RETEC, White Haven, PA, USA, October 1995.

11 Monte, S.J. (1996) Proceedings of ACS Rubber Division Conference, Louisville, KY, USA, October 1996, Paper No. 57.

12 Monte, S.J. (1996) Rubber Technology International '96, UK & Int'l Press, a Division of Auto Intermediates Ltd.

13 Monte, S.J. (1997) Polyblends '97 SPE Div./Sect. Conference, NRCC, Montreal, Canada, October 1997.

14 Monte, S.J. (1989) Plastics Compounding, 59–65.

15 Glaysher, W.A. *et al.* (1990) High performance blow molding. Proceedings of the SPE RETEC, Conference, Itasca, IL, USA, pp. 311–335.

16 Monte, S.J. (1989) Proceedings of SPE Recycle RETEC, Charlotte, NC USA, October 30–31, 1989, Paper 9.8(13).

17 Monte, S.J. and Sugerman, G. (1990) Compalloy '90, New Orleans, LA, USA, March 1990.

18 Monte, S.J. (2001) RAPRA Addcon World 2001 Conference, Berlin, Germany, October 2001.

19 Monte, S.J. (2006) RAPRA Addcon 2006 Conference, Cologne, Germany, October 17, 2006.

20 Monte, S.J. (2007) FSCT: Future Coat – ICE 2007, Toronto, Canada, October 3, 2007.

21 Monte, S.J. (2008) World Adhesives Conference & Expo 2008, Miami, FL, USA, April 12, 2008.

22 Plueddemann, E.P. (1982) *Silane Coupling Agents*, Plenum Press, New York, p. 114.

23 Voelkel, A. and Grzeskowiak, T. (2000) *Chromatographia*, **51** (9/10), 606–614.
24 Yamazaki, A. (2001) JP 2001288440, Kyoritsu Chemical Industry Co., Ltd.
25 Kataoka, M. (2002) JP 2002025806 A2 20020125, Tokin Corp.
26 Krysztafkiewicz, A. *et al.* (1997) *J. Mater. Sci.* **32**, 1333–1339.
27 Wang, Y. and Huang, J.-S. (1996) *J. Appl. Polym. Sci.*, **60**, 1779–1791.
28 Monte, S.J. and Sugarman, G. (1998) US Patent 4,657,988, Kenrich Petrochemicals, Inc., 1987; EP Appl. 87301634.9-2109, Pub. No. 0 240 137 filed 25.02.87, issued.
29 Wah, C.A. *et al.* (2000) *Eur. Polym. J.*, **36**, 789–801.
30 Galanti, A. *et al.* (1999) *Kautschuk Gummi Kunststoffe*, **52** (1), 21–25.
31 Simonutti, F.M. (1998) JP 10108925, Wilson Sporting Goods Co.
32 Chen, C.-H. *et al.* (1994) US 5,340,861, Industrial Technology Research Institute, Hsinchu, Taiwan.
33 Fukae, K. and Yoshida, I. (2000) US Patent 6,132,922, Advance Color Technology, Inc.
34 Miller, N.A. *et al.* (1998) *Polym. Polym. Compos.*, **6** (2), 97–102.
35 Kato, H. *et al.* (1988) US Patent 4,769,179, Mitsubishi Cable Ind.
36 Eichler, H.-J. *et al.* (1999) WO 00015710, Alusuisse Martinswerk GmbH.
37 Jiang, P. *et al.* (2001) *Hecheng Shuzhi Ji Suliao*, **18** (5), 35–38.
38 Wang, G. *et al.* (2002) *J. Appl. Polym. Sci.*, **85** (12), 2485–2490.
39 Wang, G.L. *et al.* (2002) *Chin. J. Polym. Sci.*, **20** (3), 253–259.
40 Braga, V. *et al.* (2001) WO 2001048075, EP 1155080, Basell Technology Company BV, Netherlands.
41 Imahashi, T. and Kazuki, K. (2001) JP 2001312925, Kuowa Kagaku Kogyo K.K.
42 Chiang, W.-Y. and Hu, C.-H. (2001) *Composites, Part A*, **32A** (3–4), 517–524.
43 Monte, S.J. and Sugerman, G. (2001) US Patent 6,197,135, Kenrich Petrochemicals, Inc.
44 Ichazo, M.N. *et al.* (1999) Proceedings of the 57th SPE ANTEC, vol. 45, No. 3, pp. 3900–3902.
45 Sharma, Y.N. *et al.* (1982) *J. Appl. Polym. Sci.*, **27**, 97–104.
46 Szijártó, K. and Kiss, P. (1986) Polymer composites, *Filling of Polymers with the Aid of Coupling Agents*, Walter de Gruyter & Co., Berlin.
47 Yu, G. *et al.* (1998) *J. Appl. Polym. Sci.*, **70**, 559–566.
48 Ogino, M. (1991) JP 03070754, Bando Chemical Industries, Ltd.
49 Hatanaka, T. *et al.* (2002) JP 2002003670, Uchiyama Manufacturing Corp.
50 Kudo, M. *et al.* (2001) WO 2001092411, Denso Corporation.
51 Manabe, T. *et al.* (2001) JP 2001247732, Kanegafuchi Chemical Industry Co., Ltd.
52 Kuwabara, M. (1996) *Mater. Lett.*, **2** (6), 299–303.
53 Murakata, T. *et al.* (1998) *J. Chem. Eng. Jpn.*, **31** (1), 21–28.
54 Koike, Y. and Mano, S. (1998) JP 10315247, Tsuyakku K.K.
55 Bodelin-Lecomte, S. and Le Gars, G.Fr. (1998) EP 864322, L'OREAL.
56 Ren, Z. *et al.* (2002) *Polym. Polym. Compos.*, **10** (2), 173–181.
57 Young, Eugene F. and O'Keefe, D.J. (1996) US Patent 5,489,497, Xerox Corp.
58 Kozawa, M. *et al.* (1997) EP 794154, Toda Kogyo Corporation.
59 Sousa, R.A. *et al.* (2001) Proceedings of the 59th SPE ANTEC, vol. 47, pp. 2550–2554.
60 Vaz, C.M. *et al.* (2002) *Biomaterials*, **23** (2), 629–635.
61 Horibe, H. *et al.* (2002) US Patent Appl. 2002137831.
62 Kataoka, M. (2002) JP 2002226601, NEC Tokin Corp.
63 Yang, H. (2001) *Feijinshukuang Bianjibu J.*, **24** (2), 24–36.
64 Ahmad, S. *et al.* (1998) *Sci. Int.*, **10** (4), 375–377.
65 Uchida, N. *et al.* (2002) Eur. Patent Appl., EP 1225600, Toda Kogyo Corporation.
66 Chen, M. *et al.* (2001) *J. Appl. Polym. Sci.*, **82** (2), 338–342.
67 Imaoka, N. *et al.* (2001) JP 3217057, Asahi Chemical Industry Co., Ltd.
68 Kijima, Y. (2001) US Patent 6,312,837, Sony Corp.
69 Wang, Z. and Wu, B. (2002) WO 2002048272, UCB, S.A., Belgium.
70 Liu, Z.Q. *et al.* (2000) Wood fibre/starch composites: effects of processing and compounding. Proceedings of the 16th

Annual Meeting of the Polymer Processing Society, Shanghai, China, June 18–21, 2000.

71 Fan, Q. *et al.* (2002) Dyeable Polypropylene via Nanotechnology. NTC Project: C01-MD20 (formerly C01-D20), National Textile Center Ann. Report, University of Massachusetts, Dartmouth.

72 Kim, C.Y. *et al.* (1993) US Patent 5,237,042, Korean Institute of Science and Technology.

73 Kim, C.Y. *et al.* (1998) US Patent 5,714,570.

74 Cho, H.N. *et al.* (2000) US Patent 6,040,417.

75 Lee, S.-S. *et al.* (2001) *J. Polym. Sci., Part B: Polym. Phys.*, **39** (21), 2589–2597.

76 Kitani, I. *et al.* (2001) JP 2001232210, Lion Corp.

77 Kelley, D.W. (1989) US Patent 4,837,272.

78 Schut, J. (1996) Organometallic esters enhance recycling of PET/PC blends, *Plastics Formulat. Compound.*

79 Zhuo, Q. *et al.* (2000) WO 00/42098, Ferro Corporation.

6
Functional Polymers and Other Modifiers

Roger N. Rothon

6.1
Introduction

The interaction between the surface of particulate and fibrous fillers and a polymer matrix plays a key role in determining the processability and properties of filled composites. Unmodified filler surfaces often give poor interactions and this has led to the growth of an industry based on the use of additives to modify filler surfaces and improve their interactions with polymers. Several types of additives have been evolved for this purpose. Two of these, organofunctional silanes and titanates, have been described in Chapters 4 and 5. This chapter covers other approaches that have been studied, although only a few of these, notably fatty and unsaturated carboxylic acids and functionalized polymers, have achieved much commercial importance.

For the purposes of the present discussion, these additives are divided into noncoupling and coupling types. Both types have advantages and limitations, which are discussed below. Treatment of the noncoupling types is more extensive here than in many other works. This is for two reasons. First, they are commercially very important. Second, they provide considerable insight into basic issues such as the mechanism of reaction/interaction of important groups such as carboxylic acids and acid anhydrides, anchoring with the filler surfaces. These insights are equally applicable to the coupling types.

While division into coupling and noncoupling types is often based on chemical structures, this can be misleading. As will be shown later, some modifiers can act as either noncoupling or coupling types, depending on the formulation. The concept of reinforcement promoters, first introduced by Ancker *et al.* [1], and discussed in Section 6.2.3, is a potentially useful alternative way of classifying modifiers in some cases.

The coverage has largely been restricted to additives that can be considered to have some form of chemical attachment to the filler surface. This eliminates species such as glycols and some surfactants, which are only physically adsorbed. There is still a gray area, however, as some additives cannot only function as

Functional Fillers for Plastics: Second, updated and enlarged edition. Edited by Marino Xanthos

ISBN: 978-3-527-32361-6

polymer modifiers in their own right but may also react with a filler surface. Where such filler attachment is thought to be likely, and important to the effects observed, then such materials (e.g., bismaleimides (BM)) have been included in this chapter.

6.2 General Types of Modifiers and Their Principal Effects

6.2.1 Noncoupling Modifiers

These additives form a strong, essentially permanent, attachment to the filler surface, but only weakly interact with the polymer phase. They generally contain a filler reactive (anchor) group at the end of a linear hydrocarbon chain, and the main effects usually observed are as follows:

- Reduced filler surface polarity.
- Reduced filler water adsorption.
- Improved processing characteristics (faster incorporation, lower viscosity, reduced energy consumption, etc.; this can lead to less polymer degradation in some compounding operations).
- Reduced tendency of the filler to adsorb important compound additives, such as antioxidants and curatives, thus improving their efficiency.
- Improved filler dispersion.
- Improved surface finish of the final product.
- Low polymer/filler interaction, often resulting in an increase in stress whitening and a decrease in strength.
- Increased elongation at break and improved impact strength in thermoplastics.
- Reduced water adsorption by the composite, reducing swelling and stabilizing electrical properties.
- Changes in the nucleating effect of filler surfaces in some semicrystalline thermoplastics, notably polypropylene. This can affect the polymer microstructure and, hence, the composite properties.

6.2.2 Coupling Modifiers

These form strong bonds with both the filler and the polymer matrix, thus tying, or coupling, the two together. The coupling effects are superimposed on the noncoupling ones as described above.

The main effects of coupling are on compound properties, notably,

- strong filler/polymer adhesion;
- reduced stress whitening and increased strength;
- reduced elongation at break;

- impact strength that is often improved, but can be unaffected, or even reduced in some instances;
- reduced water adsorption and increased property retention under humid conditions.

The effects of coupling modifiers on surface polarity vary considerably; this is largely due to the nature of the polymer reactive groups, some of which, such as amino, can be quite polar themselves. The effects on processing are less clear-cut than with the noncoupling types. They usually depend on structural features other than coupling itself and can also be affected by at what point in the process the coupling is established (i.e., during compounding or extrusion/injection molding). Strong coupling on its own would be expected to increase melt viscosity and adversely affect processing. While this can be observed, it is often masked by other effects. There is also the potential for changes in filler nucleating effects similar to those mentioned for the noncoupling modifiers.

6.2.3 Reinforcement Promoters

This is an alternative way of describing the effects of surface modification, especially in thermoplastic systems. It was developed by Ancker *et al.* [1] and refers to modifiers that increase both strength and toughness. While most classical coupling agents increase strength, not all increase toughness. Although it is little used today, the Ancker approach is useful as it is independent of modifier structure or mechanism of action. It will be discussed further in Section 6.3.1.3.

6.3 Modifiers by Chemical Type

6.3.1 Carboxylic Acids and Related Compounds

6.3.1.1 General

The carboxylic acid group is widely used to attach organic species to filler surfaces. Anchoring to the filler is thought to proceed by salt formation with mineral fillers or esterification with organic fillers such as cellulosic products. The first type of attachment is effective on fillers with basic and amphoteric surfaces, such as carbonates and hydroxides, but not so useful on acidic surfaces such as found on silicas and silicates.

6.3.1.2 Saturated Monocarboxylic Acids (Fatty Acids) and Their Salts

Surface modification based on the use of fatty acids is the classic noncoupling approach and is widely used commercially. Fatty acids are saturated monocarboxylic acids with the general formula $C_nH_{2n+1}COOH$. The carbon chain can be linear or branched. Their name derives from the occurrence of some of the higher members, notably stearic acid, in natural fats. They are mainly obtained from natural product sources, with the products containing an even number of carbon atoms being much more abundant than those with odd numbers.

Before starting a detailed discussion, some words have to be said about the approach taken here, especially with respect to the fatty acid salts. There is considerable scope for confusion and it has to be admitted that the situation is quite complex and by no means clearly resolved at present. The problems stem from the fact that salts such as calcium stearate are frequently used as additives in their own right and can influence compound properties without having any filler surface effects. They are also often attracted to filler surfaces and may be formed when fatty acids react with filler surfaces. It is thus almost impossible to separate out the effects of surface and polymer modification, especially as filler surface treatments based on fatty acids may split off salts into the polymer phase, while salts initially in the polymer phase may become attached to the filler during processing. For consistency, the approach taken here is to discuss these additives in terms of filler surface attachment, but it is by no means clear that this is necessary for good effects are to be obtained with fatty (and other carboxylic) acids and their salts [2].

Methods of Application The method of application of the fatty acid coatings can have a strong influence on their structure and distribution. There are two methods in commercial use, namely, dry and wet coating.

Dry Coating In this method, the fatty acid is added to the filler while it is maintained in dispersed state, usually by high shear mixing. With the higher MW acids, it is essential for the mix to reach the melting point of the fatty acid, if true coating, as opposed to admixture, is to occur. Heat is also often necessary to drive off water formed by the reaction and to ensure all acid is converted to a salt form. In some equipment, the mixing procedure will produce sufficient heat, while in others external heating needs to be applied. In some cases, the fatty acid is dissolved in a small amount of solvent to aid the process. The conditions have to be carefully controlled if a tightly bonded surface layer is to be produced and free salt and residual acid are to be minimized. Free salt formation is favored by high local concentrations of acid, high additive acidity, and high reactivity of the filler surface. Thus, it is most prevalent when trying to coat with low molecular weight acids and with fillers such as magnesium hydroxide. In some instances, a fatty acid salt is used as the coating agent, instead of the free acid, although it is by no means clear as to whether this can give rise to strongly bound surface species.

Wet Coating Commercially, wet coating is usually carried out using an aqueous solution of a suitable salt of the acid. The sodium salt is most frequently employed, although this leads to some residual sodium in the product, which can be a problem if water adsorption and electrical stability are important. Ammonium salts have the advantage of not leaving any cationic residue, the ammonia being driven off either at the coating stage or during drying. The release of ammonia can, however, present handling problems. Again, care has to be taken if free salt formation is to be avoided.

Types of Acids and their Effects The fatty acids most commonly used for filler treatment are those with linear chains and with at least 14 carbon atoms, with stearic acid, $C_{17}H_{35}COOH$ being the most common. For cost purposes, blends rather than single acids are usually employed. These blends can contain significant amounts of unsaturated acids as well as nonacid material and have to be chosen with care and attention to the requirements of the final application. Unsaturated components can adversely affect the color of the final compound, with yellowing often being observed. Odor can also be an important consideration and is again affected by minor components. Significant levels of hydroperoxides can also be present in fatty acids, or in the coatings derived from them, and can adversely affect compound stability. Organic acids have a strong tendency to dimerize in solution, through hydrogen bonding between carboxyl groups, and this can influence the adsorption and reaction of fatty acids with filler surfaces.

Properly applied fatty acids provide the filler with a hydrocarbon-like surface, which is much less polar than the filler itself. For example, the treatment of a precipitated calcium carbonate with a fatty acid coating was found to reduce the dispersive component of surface energy from 54 to 23 mJ/m^2 [3]. As a result, the filler is made more compatible with many polymer types, resulting in benefits such as faster incorporation and mixing, better dispersion, less energy consumption, lower viscosity, and easier extrusion. The filler generally also has lower adsorbed water content.

The effects of fatty acid treatments on composite properties are generally those set out above for noncoupling modifiers. The low degree of interaction between the coated filler and polymer generally results in voiding at relatively low strains. As a result, tensile strength is usually decreased, while elongation and, sometimes, impact strength can be increased. Indeed, one of the reasons for the widespread use of fatty acid-coated calcium carbonate in homopolymer polypropylene is the high impact strength that can be achieved. The magnitude of the fatty acid effects can considerably vary and is minimized in the presence of significant amounts of similar surface-active species that are naturally present in some polymers. The most extreme case is probably found in elastomers such as SBR, where much larger effects are observed in solution-derived polymers than in emulsion ones, as the latter contain high levels of surfactants [4].

Structure of Surface Layers The adsorption of fatty acids onto polar surfaces has been widely studied, and it usually results in a layer with the carboxyl group at the surface and the hydrocarbon chains oriented vertically to the surface. If the surface is microscopically smooth enough, and the chains sufficiently long, then the layer can be considered as semicrystalline. When packed in this way, the area occupied at the surface by one molecule of a saturated, linear, fatty acid is about 0.21 nm^2. Fatty acids with branched chains, and those containing unsaturation, do not allow such close packing and hence occupy larger areas. Similar layers are believed to form on

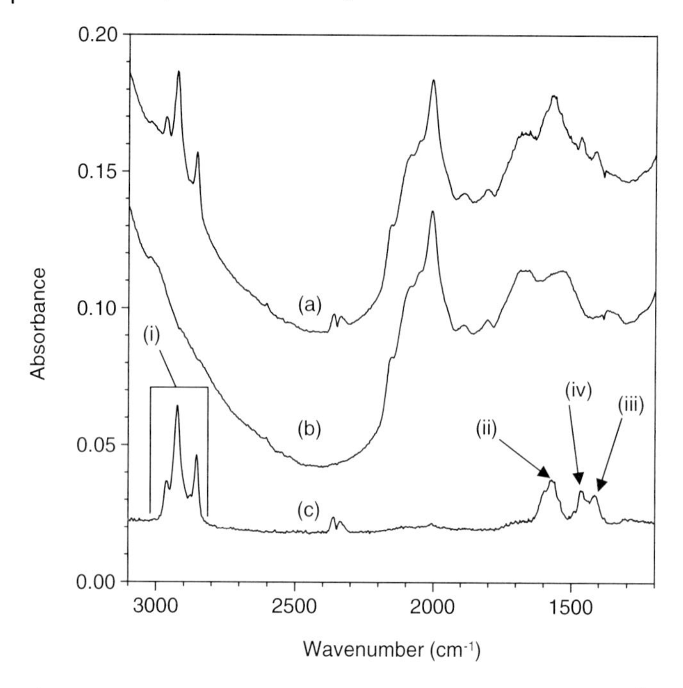

Figure 6.1 DRIFTS spectra (samples diluted to 5% w/w in finely ground KBr) showing the interaction of isostearic acid with aluminum hydroxide; (a) aluminum hydroxide (Alcan SF11-E) treated with about 0.7 monolayers of isostearic acid, (b) untreated aluminum hydroxide reference substrate, and (c) substrate subtracted spectrum (i.e., (a–b) multiplied by 5 to enhance peaks) showing (i) C—H stretching bands (2960–2850 cm^{-1}), (ii) asymmetric carboxylate carbonyl stretching band (1570 cm^{-1}), (iii) symmetric carboxylate carbonyl stretching band (1416 cm^{-1}), and (iv) methylenic C—H deformation band (1460 cm^{-1}). (Courtesy of Dr. C.M. Liauw, Manchester Metropolitan University, Manchester, UK.)

mineral fillers, but with the acid group converted to carboxylate. The conversion of isostearic acid to its carboxylate is illustrated by the FTIR data presented in Figure 6.1. Monolayer adsorption levels often correlate well with those predicted based on the known surface area of the filler and the area occupied by one molecule of surfactant.

When coatings are properly applied, changes in composite properties often peak at about the monolayer level. Thus, impact strength in calcium carbonate-filled PP homopolymer appears to be maximum at the theoretical monolayer level [5]. Rothon *et al.* have provided details for the effect of stearic acid-derived coatings on magnesium hydroxide in an ethylene vinyl acetate copolymer (18% vinyl acetate) [6]. Elongation at break was found to increase and tensile strength and secant modulus to decrease until the monolayer level was reached. Interestingly, excess coating above a monolayer was found to increase the aging rate of the composite (loss of properties at room or elevated temperature), with a commercial fatty acid blend being worse than a pure acid.

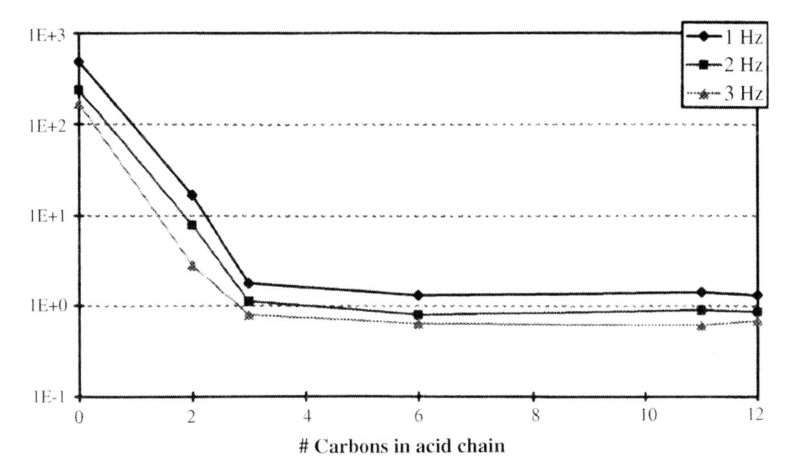

Figure 6.2 The effect of organic acid carbon chain length on the viscosity of calcium carbonate slurries in squalane (a model hydrocarbon fluid) measured at different frequencies. (Reproduced with permission from Ref. [7].)

Despite an obvious interest, little has been published regarding the effect of fatty acid chain length on filler performance. As mentioned above, blends approximating to stearic acid (C18) are most common and one would intuitively expect that quite long chains would be needed to increase surface hydrophobicity. This is considerably longer than found in most coupling agents. It is not clear if such long chains are essential or are merely used due to the relative availability and cost of the stearic acid type compositions. While it has been reported that chain lengths of about 14 carbons are necessary before crystallization can occur, the data in Figure 6.2 from work by DeArmitt and Breese show that good dispersion and reduced viscosity can be achieved with very short chains (two to three carbons) [7]. Hawarth and coworkers have reported chain length effects in polypropylene that seem to vary with the type of filler [8, 9]. Thus, they found no difference between C10 and C22 fatty acid treatment on a calcium carbonate but reported significant differences related to chain length in the same range when working with magnesium hydroxide. In the latter case, strength and stiffness were found to increase but toughness to decrease with increasing chain length.

The work of Tabtiang and Venables, although using unsaturated acids, appears to be relevant [10]. Under conditions where coupling did not appear to be a factor, their results indicated that a chain length of C10 was sufficient to lead to significant improvements in ductility of a calcium carbonate/polypropylene composite, as determined by elongation and impact properties. They found no marked additional effect up to chain lengths of C20.

Although thermal stability of the coating is little discussed, it is of potential interest when the filler is processed at high temperatures. Fatty acid surface treatments would be expected to start to oxidize below 200 °C. They are more stable in the absence of air, when decomposition probably starts in the range 250–300 °C and is quite complex, involving decarboxylation reactions. Stability will vary with the purity of the fatty acid source and the nature of the filler surface, with strongly basic surfaces tending to promote decarboxylation. It has been reported that magnesium stearate is less stable than calcium stearate. It has also been reported that coatings derived from fatty acids can destabilize some fillers; thus, stearate coating of magnesium hydroxide can lower its decomposition temperature [11].

The structure of the surface layers of fatty acid-treated fillers can be quite complex. In the ideal case, one would expect a single layer with all of the acid groups in the carboxylate form and interacting with a basic surface site (i.e., the coating to be in the form of a partial or half-salt, such as $M^{II}(OH)COOR$). In reality, a variety of other species can be present, including unreacted acid and fatty acid salt not attached to a surface site. In addition, there may be a separate fatty acid/salt phase in the bulk polymer, not associated with the filler surface. The relative importance of these will vary with the application. When multilayers are present, they will be only weakly attached to the surface monolayer.

Several studies have been made using organic solutions of fatty acid to treat the filler surface. Although this approach is generally not commercially viable, useful information can be obtained, provided care is taken in the interpretation. Of interest are the data for the adsorption of stearic and isostearic acids from heptane onto magnesium hydroxide reported by Liauw *et al.* [12]. Isotherms indicated a monolayer coverage close to the theoretical for the isostearic acid but significantly above theoretical for the stearic acid. Analysis of the supernatant liquors showed that considerable salt and no free acid was present in the case of the isostearic acid, while the nonadsorbed material from stearic acid was all in the form of unreacted acid. These results were interpreted as showing that both surface bonded half-salt Mg(OH)COOR and surface nonbonded full-salt $Mg(RCOO)_2$ were formed. In the case of isostearic acid, the full-salt is soluble enough to be removed, while it remains in the surface layers when stearic acid is used, thus contributing to the apparently anomalous monolayer level.

The mechanism of coating from aqueous solution has not been reported in any detail. Suess has made some measurements using C^{14}-labeled acid [13]. As with adsorption from organic solvent, a Langmuir-type isotherm showing monolayer adsorption was observed, but at a much lower level than for solvent adsorption. This was postulated to be due to adsorption as a hydrated complex salt. Presumably, this is converted to the stable coating layer during drying. In many instances, the aqueous phase will contain dissolved ions from the manufacturing process and, in such cases, the author has observed that the first stage can be precipitation of some form of fatty acid salt, which subsequently rearranges to form the surface coating.

The only nonlinear saturated fatty acid to receive any attention has been isostearic acid. Some patent literature describes the use of such acids, especially for treating aluminum hydroxide [14]. The molecular footprint area for isostearic acid has been determined to be about 0.45 nm^2, considerably higher than for stearic acid. This is because of its branched nature.

It should be noted that Chapter 16 contains an extensive discussion on the mechanism and type of stearate surface coating on calcium carbonate.

6.3.1.3 Unsaturated Carboxylic Acids and Related Compounds

The introduction of a reactive functionality into the polymer, such as unsaturation, offers the potential to use organic acids as coupling agents, but mixed results have been reported. A number of potentially suitable unsaturated products exist, notably, maleic acid and anhydride, acrylic and methacrylic acids, and unsaturated versions of fatty acids such as oleic acid. The acidities and double-bond reactivities of these compounds widely vary depending on their structures, and this probably accounts for the marked differences in their performance. Compounds such as acrylic acid have both high acidity and high double-bond reactivity due to the proximity of the carbonyl group to the double bond. The only commercial products specifically developed for use with filled polymer systems are from Lubrizol Advanced Materials (SOLPLUS). The exact composition of these additives is not disclosed, but they are described as unsaturated oligomeric carboxylic acids.

It has been demonstrated that effects consistent with coupling can be achieved with at least some of these additives, although not all structures are effective. It seems that the strength of the acid group and the reactivity and accessibility of the unsaturation are important factors.

Although not widely studied, the simplest case is that of peroxide cured or cross-linked systems, where unsaturation can take part in the curing reaction. Mori has reported very good results with some unsaturated carboxylic acids (e.g., $CH_2{=}C(CH_3)COOC_2H_4COOH$) in a calcium carbonate-filled unsaturated polyester resin [15]. Some of the results are presented in Table 6.1.

Most interest with unsaturated additives is in filled polyolefins such as polyethylene, polypropylene, and copolymers, where there is usually no added free radical source. In these circumstances, the simplest approach is to rely on mechanochemical processes to bring about the desired reaction. In this type of processing, free radicals are generated by shearing the polymer during compounding and can react with suitable sites on the filler surface modifier. These shear processes are greatest in the region around the filler particles, which aids the process.

Some of the basic effects obtained with aluminum hydroxide-filled EVA in the absence of peroxide and additional stabilizer are illustrated in Table 6.2, which contains a comparison between various unsaturated acids and three common organofunctional silane coupling agents. The silanes are all seen to significantly

Table 6.1 Effects of an unsaturated acid coupling agent[a)] on the mechanical properties of calcium carbonate-filled unsaturated polyester resin [15] (filler level not given).

Filler treatment	Tensile strength (MPa)	Tensile modulus (GPa)	Flexural strength (MPa)	Flexural modulus (GPa)	Izod impact strength (J/cm)
None	18.6	7.9	44.1	7.3	0.8
Unsaturated acid	36.3	10.3	96.0	9.7	1.6

a) $CH_2{=}C(CH_3)COOC_2H_4COOH$.

Table 6.2 A comparison[a] of various unsaturated acids and three organofunctional silane coupling agents as surface modifiers in 60 wt% ATH[b]-filled EVA[c] [16].

Additive and wt% level on filler	Final torque (Nm)	Tensile strength (MPa)	Elongation at break (%)
None	35.2	10.7	59
Methacrylic acid (0.5)	24.3	9.2	100
SOLPLUS C800[d] (0.8)	23.8	13.7	179
4-Allyloxy benzoic acid (1.0)	28.3	8.8	49
Sorbic acid (0.6)	25.8	10.0	48
A-172 vinylsilane[e] (0.6)	26.5	10.7	137
A-174 methacryloxysilane[e] (0.8)	24.9	10.2	150
A-1100 aminosilane[e] (0.8)	27.7	13.4	168

a) Brabender Plasticorder compounding, 180 °C initial temperature, followed by compression molding.
b) ATH is Superfine 7 (formerly Alcan Chemicals).
c) EVA is Evatane 1020 VN5 (formerly Elf Atochem).
d) SOLPLUS C800 from Lubrizol Advanced Materials is an unsaturated oligomeric carboxylic acid.
e) Silanes are from formerly GE Specialties.

reduce processing torque and increase elongation while maintaining or improving tensile strength. The unsaturated acids vary in performance, although they all significantly decrease torque. The best performance is from the commercial product, SOLPLUS C800, which significantly improves both tensile strength and elongation, giving results similar to those of the best silane coupling agent.

While mechanochemical grafting can be effective, it strongly depends on processing conditions and is often inhibited by the levels of antioxidant/stabilizer used in many commercial formulations. Antioxidant interference can be overcome by the addition of a very small amount of peroxide. Table 6.3 illustrates these effects.

When formulations of the type described above are processed in a twin-screw compounder, the reduction of torque may lead to both faster processing and improvement in mechanical properties. This is shown in Table 6.4, which compares a basic aluminum hydroxide/EVA system with one containing an unsaturated oligomeric carboxylic acid and a small amount of peroxide.

Selected unsaturated carboxylic acids are particularly useful as coupling agents for calcium carbonate-type fillers. They interact strongly with the basic surface of the carbonate, unlike silanes that only show a weak interaction. In addition, unlike simple saturated carboxylic acids, which just have a beneficial effect on impact strength, unsaturated carboxylic acids may increase tensile strength. Table 6.5 illustrates this effect in a 60% filled $CaCO_3$/PP homopolymer system. This system is heavily stabilized with both hindered phenol and phosphite-type antioxidants, and a small amount of peroxide was therefore added.

As already referred to, the approach of using unsaturated monomers, with or without peroxides, has a long history. In one of the earliest efforts, Bixler [17] used a coating based on a doubly unsaturated molecule such as butylene glycol

Table 6.3 The effect of antioxidant[a)] and peroxide on the performance of an unsaturated acid[b)] in a 60% w/w ATH-filled EVA[c)] [16].

Additives and wt% level on filler	Final torque[d)] (Nm)	Tensile strength[d)] (MPa)	Elongation at break[d)] (%)
None	35.2	10.7	59
Unsaturated acid (0.8)	24.6	13.2	104
Unsaturated acid (0.8) + antioxidant (0.3)	24.3	10.6	82
Unsaturated acid (0.8) + antioxidant (0.3) + *t*-butyl perbenzoate (0.01)	30.1	14.0	206

a) Irganox 1010, Ciba.
b) SOLPLUS C800, from Lubrizol Advanced Materials.
c) Source of ATH, EVA as in Table 6.2.
d) Brabender Plasticorder compounding, followed by compression molding. Initial temperature 180 °C except when peroxide present, then 160 °C.

dimethacrylate, an unsaturated anchoring molecule, such as maleic acid, and peroxide (in effect forming a polymeric coating with unsaturation and acid groups, at the surface). He reported good improvements with fillers such as clay and calcium carbonate in polyethylene.

Gaylord evaluated various clays in polyethylene using peroxide/maleic anhydride treatments [18]. He found very variable results, largely influenced by the nature of the clay surface, and concluded that some filler surfaces could actually inhibit the free radical grafting processes.

Aishima *et al.* used a variety of unsaturated acids as pretreatments for silicate fillers in PE and PP [19] and incorporated a polymerization inhibitor to prevent loss of reactivity during the filler coating process along with a small amount of peroxide. Improvements in both strength and toughness were reported (Table 6.6). They also found improvements in dynamic fatigue resistance. Fatigue is a property that is not often mentioned but is of considerable practical importance in some applications.

Table 6.4 Influence of an unsaturated oligomeric acid on processing speed and mechanical properties in a twin-screw extruder compounded[a)] 60% w/w ATH[b)]-filled EVA[c)] system.

Additives and wt % on filler[d)]	Throughput (kg/h)	Increase in throughput (%)	Tensile strength (MPa)	Elongation (%)
None	0.89	—	8.5	33
SOLPLUS C800 (0.8%) + dicumyl peroxide (0.04% active)	1.13	27	13.1	236

a) Processing in a 16 mm screw diameter, 400 mm barrel, Thermo twin-screw laboratory extruder.
b) EVA is Evatane 1020 VN5.
c) ATH is Martinal OL 104E from Albemarle.
d) Formulations also contain an antioxidant – 0.18% Irganox 1010 from Ciba Specialties.

Table 6.5 The effect[a)] of an unsaturated carboxylic acid in a highly filled (60% w/w) $CaCO_3$[b)]/PP homopolymer[c)] system.

Additives[d)] and wt % on filler	Tensile strength (MPa)	Unnotched impact strength (KJ/m^2)
None	22.7	34.1
SOLPLUS C800 (0.8%) + dicumyl peroxide (0.04% active)	10.7	20.9

a) Twin-screw compounding, barrel temperature 205 °C, followed by injection molding at 200 °C.
b) Carbital 110, from Imerys.
c) Eltex HV001 PF formerly from BP Solvay.
d) Formulations also contain antioxidants – 0.04% Irganox 1010 from Ciba Specialties and 0.04% Alkanox 240 from Chemtura.

In some cases, additives with multiple unsaturation may be employed. These can cross-link and strengthen the polymer, especially in the region around the filler particles where most polymer free radicals are generated. If cross-linking is concentrated around the particles, then processability should not be adversely affected. The concept of Ancker *et al.* of reinforcement promoters for filled thermoplastics, as opposed to coupling agents [1], involves the use of such agents in peroxide- and antioxidant-free systems. They argued that improvement in both strength and toughness was not an inevitable consequence of simple coupling of filler to polymer and that modifiers that actually achieved this should be called reinforcement promoters. Ancker *et al.* speculated that such reinforcement promoters function by forming a graded interface between the filler and the polymer. A way of quantifying the promoting ability of additives was developed, based on the number and reactivity of the double bonds present and on the affinity of the additive for the mineral filler surface. They found good correlation with experiment. Some of their results are given in Table 6.7. It should be stressed that all of this work is based on peroxide- and antioxidant-free systems and the authors admit that antioxidants can have very adverse effects, even on the best additives. Most of the

Table 6.6 The effect of various unsaturated acids on the properties of 50% w/w nepheline syenite (an anhydrous alkali aluminosilicate) in HDPE [19].

Filler treatment	Tensile strength (MPa)	Elongation at break (%)	Flexural strength (MPa)	Notched Izod impact strength (J/cm)
None	21.6	5	24.5	0.3
Acrylic acid	35.3	40	45.1	4.5
Methacrylic acid	30.4	11	34.3	1.5
Crotonic acid	26.5	9	33.3	1.1
Sorbic acid	29.4	16	36.3	1.6
Maleic acid	26.5	8	33.3	0.9
Itaconic acid	25.5	9	32.3	1.0

Table 6.7 Correlation of properties of filler surface-modified ATH in HDPE (50% w/w) with Anckers' promotion index [1].

Agent	Promotion index	Tensile strength (MPa)	Elongation at break (%)	Notched Izod impact strength (J/cm)
None	—	23.5	4.4	0.9
Trimethylolpropane triacrylate	3.0	38.6	16.3	1.2
Pentaerythritol triacrylate	4.0	35.0	35.0	3.1
Diethylene glycol diacrylate	−0.1	25.6	12.6	0.8
Glycerol monoacrylate	−2.5	24.9	10.0	0.8

additives (e.g., the triacrylates) identified as reinforcement promoters appear to have no obvious means of chemically attaching to a filler surface. They may just be strongly physically adsorbed or there may be some ester hydrolysis forming an unsaturated acid.

Tabtiang and Venables [20] studied the reaction of acrylic acid with the surface of calcium carbonate. The coating was carried out by means of a dry method and their results again demonstrate the complexities that can be encountered. IR spectrometric analysis of the coated fillers showed that all the acid was converted to the salt form, even when amounts well in excess of that required for a monolayer were used. However, the authors found no evidence of monolayer adsorption, with about half the added material becoming insoluble in xylene, a good solvent for calcium acrylate, at all levels of addition. They concluded that, with the method of coating employed, small droplets react with the filler to produce calcium acrylate, some of which is bound to the filler, or insolubilized in some way, while the rest remains as the free salt.

The same workers also examined the effects of the acrylic acid-coated filler in polypropylene. In order to detect grafting of the filler to the polymer, they extracted the polymer in hot xylene, followed by aqueous acid treatment. This method showed little or no grafting for coatings produced from acrylic acid alone, but extensive grafting was found when peroxide was incorporated into the coating.

The situation with unsaturated long-chain acids such as oleic acid is far from clear. Good results, especially for impact strength, have been reported for such coatings on magnesium hydroxide in polypropylene [21], although the observed effects are not necessarily those of classical coupling. In support of this, commercial products with such coatings appear to be available, although the present author and others have been unable to substantiate these claims. According to Ferrigno [22], it is necessary to stabilize unsaturated acids with antioxidants if they are to be successfully employed.

Tabtiang and Venables also examined several long-chain unsaturated acids, with and without peroxide [10]. The position of the double bond, and hence its accessibility and reactivity, varied significantly, complicating interpretation. With long-chain acids (>C10), there was little evidence for grafting to polypropylene, even with peroxide present. This may have been due to steric factors, however, as the double bonds were not at the end of the chain. With 10-undecanoic acid, containing a terminal double bond, there was some evidence of grafting, although nowhere close to that observed for the acrylic acid. The double bond in acrylic acid is much more reactive due to its proximity to the carboxyl group. All of the acids studied, except acrylic, significantly increased the impact strength, compared to uncoated filler, when no peroxide was used in the coating. In line with its effect on grafting, peroxide had no effect on the longer chain acids but did markedly reduce the impact strength obtained with the 10-undecanoic acid.

Acids derived from rosin are sometimes used. These are mostly a mixture of abietic and pimaric acids, which are three-ring, cyclic, aliphatic compounds carrying an acid group and some unsaturation in the rings themselves or pendant to them. They have been found by the author to give better results than stearic acid when used to treat precipitated calcium carbonates for use in EPDM compounds and there has reputedly been some commercialization of this application in Japan. The author has, however, not found any benefits from their use with fillers in polypropylene.

6.3.1.4 **Carboxylic Acid Anhydrides**

Cyclic anhydrides, such as maleic and succinic anhydride, appear to readily react with filler surfaces. This reaction has not been studied in detail but appears to be acid formation through hydrolysis, followed by salt formation. Most interest has been in anhydrides carrying some unsaturation, such as maleic anhydride, that have the potential to act as coupling agents. Some of the work with maleic anhydride has already been mentioned [17, 18]. Modeling shows that the area occupied at the filler surface is about $0.32\,nm^2$ for the anhydride and $0.45\,nm^2$ for the diacid resulting from hydrolysis [23]. It is not clear as to whether the reaction with the filler surface involves one or both acid groups.

Although there has been some interest in maleic anhydride itself, most interest has been in alkyl and especially in polymer-substituted succinic anhydrides, which are discussed in succeeding sections. Certain long-chain derivatives of succinic anhydride have been examined as possible filler surface modifiers, notably dodecyl and dodecenyl succinic anhydrides. The first is a possible noncoupling modifier, while the second, having some unsaturation, is a potential coupling agent. Because of the relatively large area occupied at the surface by the diacid, the organic tails cannot pack as tightly as with monocarboxylic acids.

Although significant effects have been obtained in the work carried out to date, both types have only given results typical of noncoupling additives, such as stearic acid. There has been little evidence of coupling ability in the case of the unsaturated version, even when used in conjunction with peroxide [10]. Presumably, this is due to steric factors.

Liauw *et al.* [23] found that when dodecenyl succinic anhydride was adsorbed onto aluminum hydroxide at 50 °C, most of the interaction was through acid groups formed by hydrolysis and there was little salt formation, with carboxylate salts appearing only at higher temperatures. Monolayer coverage levels were determined by a number of methods. Those based on effects such as slurry viscosity were found to correspond reasonably well with that predicted from the footprint area, while that determined by extraction was somewhat higher, presumably due to some multilayer formation.

Tabtiang and Venables studied the reaction of dodecenyl succinic anhydride with calcium carbonate, using dry blending [10]. They found that the anhydride was largely converted to the carboxylate salt, with little or no evidence of any remaining anhydride or free acid. The effects in polypropylene were those associated with a noncoupling additive, even when peroxide was present in the coating.

6.3.1.5 Bismaleimides

There has been some interest in this type of product, which was mentioned in patents way back in 1973 [24] and identified by Ancker as having a high reinforcement index [1]. They can be thought of as derivatives of maleic anhydride, produced by its reaction with aromatic or aliphatic diamines, and containing two very reactive double bonds. The product, which has attracted most interest, has been *m*-phenylenebismaleimide (BMI). Commercial usefulness has been limited by toxicity concerns and by color formation in some compounds. These products do seem to form strong attachment to many filler surfaces, but the mechanism is not clear, although it may involve acid groups resulting from hydrolysis of some of the imide linkages.

BMI has been shown to be an effective cross-linking agent for some elastomers [25]. The cross-linking is believed to proceed through free radicals generated by a charge transfer mechanism. BMI has also been found to give rise to significant property improvements in filled polymers such as polyethylene and polypropylene. Grafting to the polymer is believed to occur through the free radicals produced in a similar way as in elastomer cross-linking. As with the other unsaturated systems discussed earlier, it seems that antioxidants may interfere with the grafting process and that addition of peroxide can assist performance.

Khunova and coworkers have demonstrated beneficial effects from BMI with many types of mineral fillers in polyolefin compounds and also made a detailed study of the reactions occurring [26–29]. They found that BMI had little effect in unfilled compounds, but it significantly increased strength and elongation in filled systems. BMI also appeared to increase further the nucleating effect of fillers but reduced the total crystallinity. They showed, using IR methods, that BMI reacted with the polymer via the maleimide alkene, resulting in chain extension and cross-linking reactions. In polypropylene, these can offset some of the chain scission, which accompanies processing, especially at high filler loadings. The effects of BMI were thought to be predominantly in the regions around the filler particles. As with other unsaturated systems, stabilizers were found to inhibit the effects and peroxides to promote them.

Xanthos [30] has reported on the effects of BMI in wood fiber/PP composites. The temperatures needed for activation of the BMI system led to degradation of the

Table 6.8 The effect of bismaleimide structure on the mechanical properties of a 60% w/w magnesium hydroxide-filled polypropylene block copolymer[a] [31].

Modifier and level added to the formulation	Tensile stress at yield (MPa)	Elongation at break (%)	Impact strength (notched Charpy) (kJ/m^2)
Unfilled control	27	7 (yield)	18
Untreated	16.1	0.7	6.4
1,3-PDM (meta) 5 phr	26.5	11.1	10.9
1,2-PDM (ortho) 5 phr	25.7	3.3	7.0
1,4-PDM (para) 5 phr	26.9	9.2	9.0
C10DM 6.9 phr	25.5	16.0	8.6

a) $Mg(OH)_2$ Magnifin H5 (Martinswerk), in PP RS002P copolymer (Solvay); bismaleimides not precoated and used at equal molar amounts.

wood, but pregrafting of BMI to the polymer, using peroxide, was found to overcome this and gave a significant increase in tensile strength.

The effect of varying the position of the imide groups in phenylene bismaleimides (PDM) and of using aliphatic bismaleimides has been studied [31]. Using a magnesium hydroxide-filled PP copolymer, the *m*- and *p*-isomers (1,3- and 1,4-PDM, respectively) were found to be equally effective, while the *o*-isomer (1,2-PDM) was considerably less effective (see Table 6.8). An aliphatic bismaleimide was also included in the study and gave broadly comparable results.

BMI has quite a high melting point (190 °C), which may restrict its application in some polymers, such as EVA. It also has a tendency for color formation in the presence of many fillers and there is some concern over its possible toxicity. Aliphatic bismaleimides are of interest as they may be less toxic and less prone to color formation. They also have lower melting points, which may be useful in some polymers, allowing their use at lower processing temperatures. The results for some aliphatic bismaleimides with varying carbon chain length in aluminum hydroxide-filled thermoplastic EVA are presented in Table 6.9.

Table 6.9 Effect of γ-aminopropyltriethoxysilane and aliphatic bismaleimides with variable chain length on the properties of ATH/EVA-based composites[a] [31].

Modifier and level added to the formulation (phr)	Tensile strength (MPa)	Elongation at break (%)	Color
Untreated	10.7	59	Cream
Silane 5 phr	13.5	144	Cream
C4BM 5.0 phr	13.9	114	Bright pink
C6BM 5.6 phr	13.5	126	Dark pink
C12BM 7.5 phr	11.9	163	Cream

a) 60% w/w ATH (SF7E, Alcan Chemicals) in EVA (1020 VN5, 18% VA, Evatane, AtoFina); BM not precoated and used at equal molar amounts.

6.3.1.6 Chlorinated Paraffins

These additives are included here as they are another manifestation of the unsaturated additive approach. This coupling agent technology was developed by researchers at the Ford Motor Company, initially for use with phlogopite mica filled PP and later extended to other micas and glass-filled systems [32]. Although it is quite a complicated technology, it is able to provide the features of coupling at lower material costs than organosilanes. It is based on highly chlorinated paraffin waxes (e.g., Chlorez™ 700, a 70% w/w chlorinated paraffin in the C20–C30 range). These waxes are attracted to the filler surface during melt processing and, given enough residence time, undergo dehydrochlorination, resulting in very reactive, conjugated, unsaturation that can then mechanochemically graft with polypropylene radicals produced as part of the compounding/molding process. Ishida and Haung have demonstrated that grafting to PP takes place and that peroxide helps the process [33]. The mica surface is thought to promote the dehydrochlorination reactions, which seem to proceed more readily with this filler than with glass fiber.

An advantage of the system is that coupling can be delayed until the injection molding process, which can improve processability and effectiveness. The dehydrochlorination can be accelerated by the presence of alkaline species such as magnesium oxide, which also neutralizes the corrosive hydrogen chloride evolved. Other additives, such as maleic anhydride, peroxides, and organosilanes, can be incorporated.

Although this approach can impressively increase strength, it has several limitations. It requires careful control of processing conditions, long cycle times (which increase processing cost), to complete the surface reaction, and can cause corrosion problems during processing. Like all mechanochemical grafting approaches, it is also sensitive to the type and amount of antioxidants used. The acid-functionalized polymers described in Section 6.3.5 have ultimately proved to be a more cost-effective approach.

6.3.1.7 Chrome Complexes

Strange as it may seem now, the first commercial coupling agent was probably Volan A from DuPont, which was available in 1961 [34]. This was produced by condensing a methacrylate compound with trivalent chromic chloride and was successfully used with glass fibers in unsaturated polyester resins. It is another example of the use of unsaturated compounds. Today, this technology has been largely displaced by the organofunctional silanes, discussed in Chapter 4.

6.3.1.8 Other Carboxylic Acid Derivatives

There has been little success with derivatives other than the unsaturated ones referred to above. The main exception is with amino acids, which, as discussed later, are very successful coupling agents for nanolayer silicates into some polyamides. There has also been some interest in hydroxy acids, such as hydroxy stearic acid and its derivatives. The main use for these products currently seems to be as high-efficiency dispersants, especially for inorganic pigments.

6.3.2
Alkyl Organophosphates

Acid phosphates provide an alternative anchoring functionality to the carboxyl group and have been investigated for use on fillers such as calcium carbonate [35–37]. Attachment to the filler surface is again assumed to be through salt formation. With alkyl substituents, the properties achieved are generally similar to the less expensive fatty acids. Functional dihydrogen phosphates appear to be able to act as potential coupling agents, especially for calcium carbonate. Most work has been directed to elastomer systems, where improvements consistent with coupling have been reported for a number of treatments. Unsaturation was found to be effective in peroxide-cured elastomers and a mercapto function to give the best results in a sulfur-cured system. Somewhat surprisingly, little has been published on their performance in other polymer types, although some promising results have been reported for calcium carbonate in polypropylene and polyethylene [37].

6.3.3
Alkyl Borates

The use of these is described in the patent literature, but again they seem to offer no advantage over the less expensive fatty acids [38].

6.3.4
Alkyl Sulfonates

The use of alkyl sulfonic acids has not received much attention, but DeArmitt and Breeze have reported that they make good dispersants for dolomite in organic fluids [7].

6.3.5
Functionalized Polymers

This approach has ultimately proved to be the most successful of the non-silane coupling technologies. Here, the potential coupling agent is, in effect, prereacted with the polymer under controlled conditions, introducing filler reactive groups into the polymer. The most common groups used are carboxylic acids, anhydrides, and alkoxysilanes. Much of the basic information discussed above concerning the reaction of carboxylic acid and anhydride groups with filler surfaces is relevant to these additives.

There are two main types of functionalized polymers in use, polyolefins and polybutadienes, and these are discussed separately below.

6.3.5.1 Functionalized Polyolefins

These are the logical conclusion of the work with unsaturated acids and anhydrides discussed above, and are gaining considerable importance in thermoplastic applications. Instead of using a one-step process to achieve reaction with the filler and the polymer, the additive is pregrafted onto the polymer. While introducing an additional step, this overcomes many of the difficulties mentioned above, especially those related to the presence of high levels of antioxidant during the grafting stage. Although initially relatively expensive products, these grafted polymers are now very competitive in price and their use is rapidly expanding.

A variety of commercial products specifically aimed at being used in filled polymers are available, notably from Arkema (Lotader™ and Orevac™), DuPont (Fusabond™), Westlake Chemicals (Epolene™), ExxonMobil (Exxelor™), and Crompton (Polybond™). The terminology used in this discussion is as follows: acrylic acid grafted PE or PP = AA-g-PE or AA-g-PP; maleic anhydride grafted = MA-g-PE or MA-g-PP.

The principal commercial products are produced by peroxide grafting of maleic anhydride or acrylic acid onto polyethylene or polypropylene. Unfortunately, the grafting reaction is difficult to complete and, unless carefully controlled, can leave significant amounts of unreacted monomer and, possibly, peroxide in the composition; this would adversely affect the heat stability of the resulting composites. When applied to polypropylene, the peroxide grafting also leads to significant MW reduction resulting from chain scission. In some cases, the polyolefin is first thermally degraded to a wax-like material that has terminal unsaturation, which can then react with maleic anhydride either thermally by the -ene reaction or assisted by peroxide [39].

Maleic anhydride does not readily undergo homopolymerization and its grafting leads to the attachment of succinic anhydride groups. When acrylic acid is used, homopolymerization can occur, leading to longer side chains. These two effects are illustrated in Figure 6.3.

In some cases, copolymerization is used. Typical of the products made in this way are terpolymers of ethylene, *n*-butyl acrylate, and maleic anhydride, such as the Lotader products from Arkema, which have proved to be very good coupling agents in filled EVA copolymers.

Today, most commercial products for filler applications appear to be based on maleic anhydride, rather than acrylic acid, and what comparative data are available suggest that the maleated forms are the more effective and can be used at lower levels. The acrylic acid forms have higher functionality levels than the maleated ones.

A useful description of the grafting of acrylic acid onto polypropylene can be found in Ref. [40]. Good improvements in both strength and toughness (notched and unnotched Izod impact strength) of a calcium carbonate-filled polypropylene were obtained, although very high addition levels of the AA-g-PP were required. Improved bonding of filler to polymer was confirmed by electron microscopy.

Mai *et al.* reported on the use of AA-g-PP in aluminum hydroxide-filled PP homopolymer [41] and found the graft polymer to increase filler to polymer wetting and adhesion, as shown by electron microscopy of fracture surfaces. In addition, they found not only significant increases in flexural strength but also a loss in notched impact strength. The loss in impact strength was most marked at low to moderate filler levels and was least at the 60% level, typical for fire-retardant compounds. It was

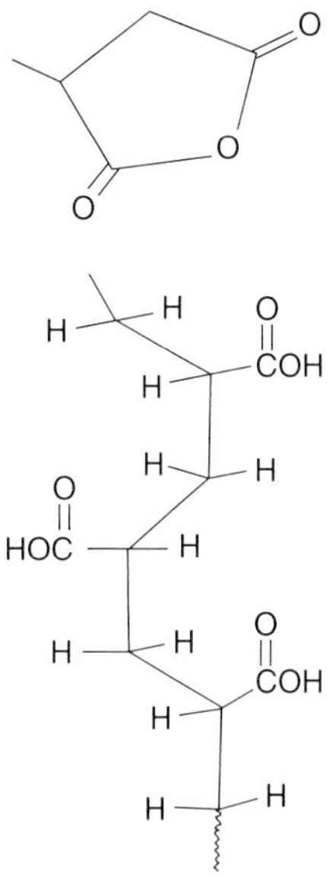

Figure 6.3 Filler reactive groups introduced into polymer backbones by maleic anhydride and acrylic acid grafting.

also least at the highest acid grafting level. The difference in impact results between this work and that of Ref. [40] may have stemmed from the nature of the filler. The calcium carbonate used was quite coarse, and probably reduced the impact strength of the polymer (data not given in the paper), while the aluminum hydroxide used was much finer and actually increased the impact strength of the polymer.

Adur [42] has also provided data on the effect of AA-g-PP on PP with a number of fillers, including glass fiber, and wollastonite. His results show significant increases in most properties (including notched and unnotched Izod impact strength and heat distortion temperature, HDT) when used with glass fibers. With wollastonite, there was a significant increase in tensile strength but no real effect on HDT or impact properties. Again, high levels of additive were needed.

Chun and Woodhams [39] reported significant increases in tensile strength but little effect on impact strength by using a maleated polypropylene wax with mica in a PP homopolymer. Lower levels of additive were needed than in the work with AA-g-PP described above. The authors also found a synergistic effect with some silane treatments.

Results obtained by this author and coworkers for a calcium carbonate-filled PP are presented in Table 6.10.

Sigworth has recently described the use of MA-g-PP and MA-g-PE in natural fiber-filled polyolefin compounds [43]. The additives were found to be effective on a wide range of natural products, including wood flour, and he reported significant improvements in the following properties with additive levels of about 5% based on the natural fiber: increased tensile, flexural and impact strengths; reduced room

Table 6.10 Effect of maleated PP on the properties of calcium carbonate filled PP.

Additive	Secant modulus (GPa)	Tensile strength (MPa)	Elongation at yield (%)	Impact strength (unnotched Charpy) (kJ/m^2)
None	2.4	24.7	1.5	5.6
MA-g-PP 7% w/w on filler	2.9	33.7	2.3	11.1

temperature creep; higher heat distortion temperature; reduced water adsorption; and improved property retention on moisture aging.

Functionalized polymers are mainly used as compounding ingredients rather than being precoated onto the filler. Some forms can be converted into aqueous emulsions, however, and it is believed that these are sometimes used to coat fillers. Their use as filler coatings can be useful in reducing dust problems in some cases.

As with the other acid and anhydride products already mentioned, bonding to a mineral filler surface is believed to be through salt formation. Bonding to the host polymer is believed to be a combination of chain entanglement, chain pinning, and even cocrystallization. It appears to be important to match the structure of the additive and the host polymer, at least to some degree. Thus, acid-functionalized homo- and copolypropylenes are available. It is also necessary to carefully choose the melt flow rate. High melt flow rates will favor processing but detract from stiffness and heat distortion temperature. While the level of acid groups is also clearly important, there is no clear advice on this from the manufacturers. Most grafted polypropylene products have quite low acid levels due in part to the MW reduction, which accompanies grafting.

The MW would seem to be important in determining the effectiveness of the interaction with the matrix polymer. There has been little useful data published, but Felix and Gatenholm [44] have shown that property improvements in a cellulose-filled polypropylene increase up to a limit of about 10 000 MW. As will be seen in the next section, a similar MW limit has been found for maleated polybutadienes (MPBD) in calcium carbonate-filled elastomers.

The current main uses for these functionalized polymers are as follows:

1) In glass fiber-reinforced polyolefins where the glass fiber is usually precoated with an amino functional silane that can react with the acid or acid anhydride through formation of an amide linkage (see Chapter 7).
2) In compounds based on fire-retardant fillers such as aluminum and magnesium hydroxides (see Chapter 17).
3) In wood- or natural fiber-filled polyolefins, where the acid or anhydride is thought to react with cellulosic hydroxyl groups (see Chapter 15).

The maleated polypropylenes also play an important role in the preparation of thermoplastic aluminosilicate nanocomposites [45]. This could become another important outlet if this technology is more widely adopted.

6.3.5.2 Functionalized Polybutadienes

Polybutadienes contain a high level of unsaturation suitable for taking part in polymerization and cross-linking reactions. They can also be readily further functionalized with the introduction of filler reactive groups. Two approaches are used to prepare coupling agents from them. They can be grafted with maleic anhydride or alkoxysilane groups can be introduced by hydrosilylation procedures, thereby generating products that can be used on fillers not responsive to acidic additives. Only the maleic anhydride approach is appropriate for this section.

As with the polyolefins, reaction with maleic anhydride results in the attachment of succinic anhydride groups, but in this case, no peroxide is necessary, the reaction proceeding as an -ene addition. High levels of anhydride can be grafted in this way, with the resulting products being water-soluble (in salt form) and, thus, suitable for filler coating. They can also be used as compounding additives. Liquid polybutadienes with MW in the region of 10–50 000 are most commonly used.

These additives are mainly used in elastomer applications, where they are suitable for both sulfur and peroxide cures. The effects of both acid content and MW have been reported in Ref. [46]. Property improvements were found to plateau at an average MW of about 10 000 and with about 20% w/w of grafted maleic anhydride.

While of most commercial interest in elastomers, the maleated polybutadienes can also provide good coupling for fillers in other polymers, such as cross-linked ethylene vinyl acetates, and thermosets such as polymethylmethacrylate (PMMA). Typical results in cross-linked EVA are presented in Table 6.11, where the MPBD additive is seen to give effects similar to a conventional vinylsilane. The effect on fire retardancy, as measured by limiting oxygen index, is included to illustrate that modifiers can affect such properties. A decrease in oxygen index is often seen when coupling agents are used in cross-linked formulations. Results for a variety of fillers in PMMA can be found in the patent literature [47].

Acid-functionalized polybutadienes are most effective with fillers such as calcium carbonate and aluminum and magnesium hydroxides but are not as effective with silicas and other silicates. As with the succinic anhydride derivatives discussed above, it is assumed that the anhydride functional polymers react with the filler surface by

Table 6.11 Effect of MPBD[a)] in a magnesium hydroxide-filled, peroxide cross-linked EVA composite[b)].

	Property		
Additive	**Tensile strength (MPa)**	**Elongation at break (%)**	**Limiting oxygen Index (%)**
None	10	325	42
MPBD 2% on filler	14	210	31
Vinylsilane (A-172) 2% on filler	13	190	32

a) Atlas G-3965, ICI.
b) 125 phr $Mg(OH)_2$ (DP 390, Premier Periclase, Ireland).

ring opening and salt formation, although there have been few confirmatory studies. Some useful work has been carried out with maleated polybutadiene containing about 20 wt% of grafted maleic anhydride. FTIR analysis showed that adsorption onto magnesium hydroxide from solution at room temperature gave mainly a mixture of salt and unreacted anhydride peaks, with only a small amount of free acid [48]. With aluminum hydroxide, the situation was somewhat different. As in the studies using dodecenyl succinic anhydride discussed earlier, unreacted anhydride and free acid predominated at room temperature, and temperatures of about 100 °C were necessary before salt formation was observed. This difference is probably due to the greater reactivity of the magnesium hydroxide surface.

Maleated polybutadienes have a number of uses, with filler modification being only a minor aspect. Synthomer Corporation is one of the main general producers (maleated Lithenes), although Evonik markets grades specially aimed at the use with calcium carbonate fillers (Polyvest).

6.3.6
Organic Amines and Their Derivatives

Alkyl and aryl amines have long had a minor role in treating certain types of filler surfaces, notably some clays. They are normally added in a water-soluble ammonium salt form, which can exchange ions with cations such as sodium on the clay surfaces. Their importance has recently received a significant boost, as a result of the growing interest in nanoclays as high-value effect fillers (see Chapter 9). This area has been the subject of an excellent review by Alexandre and Dubois [49]. Certain clays, notably montmorillonite, have an aluminosilicate layer structure with exchangeable cations such as sodium trapped between the layers. When these clays are swollen in water, the inorganic cations can be exchanged for long-chain cations from the ammonium salts. As a result, the layers can be separated into very thin (about 1 nm) high aspect ratio sheets and dispersed into polymers.

The ammonium salts can be derived from primary, secondary, and tertiary amines. The exact structure and degree of exchange are chosen to suit the application, especially the polarity of the polymer that the layers are to be dispersed in. Tertiary amines containing a mixture of methyl and long-chain alkyl groups are common. In some cases, where some polarity is desired, hydroxyl groups can be incorporated into the structures. A particularly important system is the use of amino acids, such as 12-aminododecanoic acid. These can provide chains bearing terminal acid groups, which can couple to the very thin plates to be incorporated into polymers such as nylon.

As briefly mentioned earlier, acid or anhydride functional polymers are used to ensure dispersion and exfoliation of the organoclays in polymers such as polypropylene. It is popularly believed that this is due to interaction of the functional polymer and the onium ion treatment on the clay, but this appears to be a misconception. It seems that the key feature, attracting the functional polymer into the clay galleries, is its interaction with free sites on the clay surface. Thus, complete coating of the surface is undesirable.

References

1 Ancker, F.H., Ashcraft, A.C., Leung, M.S., and Ku, A.Y. (1983) Reinforcement promoters for filled thermoplastic polymers, US Patent 4,385,136, Union Carbide Corp.

2 Fulmer, M., van der Kooi, J., and Koss, E.K. (2000) Improved processing of highly filled calcium carbonate compounds. Proceedings of the 58th Society of Plastics Engineers SPE ANTEC, vol. 46, p. 552.

3 Papirer, E., Schultz, J., and Turchi, C. (1984) Surface properties of calcium carbonate filler treated with stearic acid. *Eur. Polym. J.*, **20** (12), 1155–1158.

4 Rothon, R.N. (1984) The calcium carbonate challenge. *Eur. Rubber J.*, **166** (10), 37–42.

5 Taylor, D.A. and Paynter, C.D. (1994) Proceedings of Polymat'94 Conference, London, UK, September 1994, pp. 628–638.

6 Rothon, R.N., Liauw, C.M., Lees, G.C., and Schofield, W.C.E. (2002) Magnesium hydroxide filled ethylene vinyl acetate. *J. Adhesion*, **78**, 603–628.

7 DeArmitt, C. and Breese, K.D. (2003) Systematic dispersion optimisation method for mineral fillers and pigments in non-aqueous media. Proceedings of the Functional Fillers for Plastics 2003, Intertech Corp., Atlanta, GA, October 2003, Paper 16.

8 Haworth, B. and Raymond, C.L. (1997) Processing and fracture characterisation of polypropylene filled with surface modified calcium carbonate. Proceedings of the Eurofillers 97 Conference, Manchester, UK, September 1997, pp. 251–254.

9 Haworth, B. and Birchenough, C.L. (1995) Mechanical behaviour of polyolefins containing fillers modified by acid coatings of variable chain length. Proceedings of the Eurofillers 95 Conference, Mulhouse, France, September 1995, pp. 365–368.

10 Tabtiang, A. and Venables, R. (2000) The performance of selected unsaturated coatings for calcium carbonate filler in polypropylene. *Eur. Polym. J.*, **36**, 137–148.

11 Hancock, A. and Rothon, R.N. (2003) *Particulate Filled Polymer Composites*, 2nd edn (ed. R.N. Rothon), RAPRA Technology Ltd., Shawberry, Shrewsbury, Shropshire, UK, p. 90.

12 Liauw, C.M., Rothon, R.N., Lees, G.C., and Iqbal, Z. (2001) Flow micro-calorimetry and FTIR studies on the adsorption of saturated and unsaturated carboxylic acids onto metal hydroxides. *J. Adhesion Sci. Technol.*, **15** (8), 889–912.

13 Suess, E. (1968) Calcium Carbonate Interaction with Organic Compounds, PhD Dissertation, Lehigh University, USA.

14 Bonsignore, P.V. (1981) Surface modification of alumina hydrate with liquid fatty acids, US Patent 4,283,316, Aluminum Company of America.

15 Mori, A., Aizawa, M., and Kataoka, Y. (1987) Surface treatment agent, Japan Patent Appl. No. 62-190857 filed on 1987-07-30, Nippon Soda Co., Ltd.

16 Rothon, R.N., Schofield, J., Thetford, D., Sunderland, P., Lees, G., Liauw, C.M., and Wild, F. (2002) Functionalized acids as filler surface treatments. Proceedings of the Functional Fillers for Plastics 2002, Intertech Corp., Toronto, Canada, September 2002, Paper 5.

17 Bixler, H.J. and Fallick, G.J. (1969) Reinforcing filler, US Patent 3,471,439, Amicon Corp.

18 Gaylord, N.G., Ender, H., Davis, L., Jr., and Takahashi, A. (1979) *Polymer-Filler Composites Thru In Situ Graft Copolymerization: Polyethylene – Clay Composites*, ACS Symposium Series 121 (eds C.E. CarraherJr. and M. Tsuda), American Chemical Society, Washington, DC, pp. 469–474.

19 Aishima, I., Seki, J., Matsumoto, K., Furusawa, Y., Tsukisaka, R., and Takahashi, Y. (1980) Composition comprising a thermoplastic resin and mineral filler particles coated with an ethylenically unsaturated organic acid, US Patent 4,242,251, Asahi, Kasei Kogyo Kabushiki Kaisha.

20 Taibtiang, A. and Venables, R. (1999) Reactive surface treatment for calcium carbonate filler in polypropylene. *Compos. Interface*, **6** (1), 65–79.

21 Miyata, S., Imahashi, T., and Anubiki, H. (1980) Fire retarding polypropylene with

magnesium hydroxide. *J. Appl. Polym. Sci.*, **25** (3), 415–425.
22 Ferrigno, T.H. (1983) Stabilized modified fillers, US Patent 4,420,341.
23 Liauw, C.M., Hurst, S.J., Lees, G.C., Rothon, R.N., and Dobson, D.C. (1995) Filler surface treatments for particulate mineral/thermoplastic composites. *Plast. Rubber Comp. Proc. Appl.*, **24** (2), 211–219.
24 (a) Kishikawa, H. *et al.* (1973) Japan Kokai 73 45540; (b) (1974) *Chem. Abstr.*, **80**, 60620.
25 Hill, R.K. and Rabinovitz, M. (1964) Stereochemistry of "No-mechanism" reactions: Transfer Asymmetry in the reactions of olefins with dieneophiles. *J. Am. Chem. Soc.*, **86**, 965.
26 Khunova, V. and Sain, M.M. (1995) Optimization of mechanical strength of reinforced composites. 1. Study on the effect of reactive bismaleimide on the mechanical performance of a calcium carbonate filled polypropylene. *Angew. Makromol. Chem.*, **224**, 9.
27 Khunova, V. and Sain, M.M. (1995) Optimization of mechanical strength of reinforced composites. 2. Reactive bis-maleimide modified polypropylene composites filled with talc and zeolite. *Angew. Makromol. Chem.*, **225**, 11.
28 Khunova, V. and Liauw, C.M. (2000) Advances in the reactive processing of polymer composites based on magnesium hydroxide. *Chem. Papers*, **54** (3), 177–182.
29 Liauw, C.M., Khunova, V., Lees, G.C., and Rothon, R.N. (2000) The role of *m*-phenylenebismaleimide (BMI) in reactive processing of polypropylene/magnesium hydroxide composites. *Macromol. Mater. Eng.*, **279**, 34–41.
30 Xanthos, M. (1983) Processing conditions and coupling agent effects in polypropylene/wood flour composites. *Plast. Rubber Proc. Appl.*, **3** (3), 223–228.
31 Liauw, C.M., Lees, G.C., Rothon, R.N., Wild, F., Schofield, J.D., Thetford, D., and Sunderland, P. (2003) The effect of molecular structure of maleimide based interfacial modifiers on the mechanical properties of polyolefin/flame retardant filler composites. Proceedings of the Eurofillers03, Alicante, Spain, September 2003, pp. 145–147.
32 Meyer, F.J. and Newman, S. (1979) Proceedings of the 34th Annual Technical Conference, Reinforced Plastics/ Composites Division, SPI, Section 14-G.
33 Ishida, H. and Huang, Z.-H. (1984) Processing induced graft polymerization of chlorinated paraffin to polypropylene extruded with MgO and dicumyl peroxide. Proceedings of the 42nd Society of Plastics Engineers SPE ANTEC, vol. 30, pp. 205–208.
34 Yates, P.C. and Trebilcock, J.W. (1961) Proceedings of the 16th Annual Technical Conference, Reinforced Plastics/ Composites Division, SPI, Section 8.
35 Nakatsuka, T., Kawasaki, H., Itadani, K., and Yamashita, S. (1981) Topochemical reaction of calcium carbonate and alkyl dihydrogenphosphate. *J. Colloid Interface Sci.*, **82** (2), 298–306.
36 Nakatsuka, T., Kawasaki, H., Itadani, K., and Yamashita, S. (1982) Phosphate coupling agents for calcium carbonate. *J. Appl. Polym. Sci.*, **27**, 259–269.
37 Luders, W., Herwig, W., van Spankeren, U., and Burg, K. (1979) Plastic molding composition containing a filler, US Patent 4,174,340, Hoechst AG.
38 Fein, M.M. and Patnaik, B.K. (1978) Organic borate coupling agents, US Patent 4,073,766, Dart Industries.
39 Chun, I. and Woodhams, R.T. (1984) Proceedings of the 42nd Society of Plastics Engineers SPE ANTEC, vol. 30, pp. 132–135.
40 Rahma, F. and Fellahi, S. (2000) Performance evaluation of synthesized acrylic acid grafted polypropylene within $CaCO_3$/polypropylene composites. *Polym. Compos.*, **21** (2), 175–186.
41 Mai, K., Li, Z., Qui, Y., and Zeng, H. (2001) Mechanical properties and fracture morphology of $Al(OH)_3$/polypropylene composites modified by PP grafting with acrylic acid. *J. Appl. Polym. Sci.*, **80** (13), 2617–2623.
42 Adur, A.M. (1989) Seminar papers, in *Reactive Processing: Practice and Possibilities*, RAPRA Technology Ltd., Shawberry, Shrewsbury, Shropshire, UK, Paper 5.8.
43 Sigworth, W. (2002) Coupling agents: performance additives for natural fibre-

filled polyolefin composites. Proceedings of the Functional Fillers for Plastics 2002, Intertech Corp., Toronto, Canada, September 2002, Paper 19.

44 Felix, J.M. and Gatenholm, P. (1991) Formation of entanglements. *J. Appl. Polym. Sci.*, **50**, 699.

45 Manias, E., Touny, A., Wu, L., Strawhecker, K., Lu, B., and Chung, T.C. (2001) Polypropylene/montmorillonite nanocomposites. Review of the synthetic routes and materials properties. *Chem. Mater.*, **13**, 3516–3523.

46 Rothon, R.N. (1990) Unsaturated bifunctional polymeric coupling agents, in *Controlled Interphases in Composite Materials, Proceedings of the 3rd International Conference on Composite Interfaces* (ed. H. Ishida), Elsevier, New York, pp. 401–406.

47 Rothon, R.N. (1988) British Patent 0,295,005, filed, ICI PLC.

48 Rothon, R.N. (1991) Improving the performance of mineral filled composites with additives based on bi-functional polymers. Proceedings of the High Performance Additives Conference, PRI/BPF, London, UK, May 1991, Paper 12.

49 Alexandre, M. and Dubois, P. (2000) Polymer-layered silicate nanocomposites: preparation, properties and uses of a new class of materials. *Mater. Sci. Eng.*, **28**, 1–63.

Part Three
Fillers and Their Functions

A High Aspect Ratio Fillers

Functional Fillers for Plastics: Second, updated and enlarged edition. Edited by Marino Xanthos

ISBN: 978-3-527-32361-6

7
Glass Fibers

Subir K. Dey and Marino Xanthos

7.1
Background

The term glass covers a wide range of inorganic materials containing more than 50% silica (SiO_2) and having random structures. They are often considered as supercooled liquids in a state referred to as the vitreous state. According to the Roman historian Pliny, Phoenician sailors made glass when they attempted to cook a meal on a beach, over some blocks of natron (a mineral form of sodium carbonate) they were carrying as cargo; in doing so, they melted the sand beneath the fire and the mixture later cooled and hardened into glass. The first true glass was probably made in western Asia, perhaps Mesopotamia, at least 40 centuries ago as a by-product from copper smelter. The science of glass making was developed over a long period of time from experiments with a mixture of silica sand (ground quartz pebbles) and an alkali binder fused on the surface. The basic raw materials for the manufacture of ordinary soda glass used for windows and bottles are sand, sodium carbonate (soda), and calcium carbonate (limestone). Typical compositions of these soda-lime-silica glasses, expressed in percent oxides, are about 72–74% SiO_2, 14–16% Na_2O, 5–10% CaO, 2.5–4% MgO, and minor amounts of Al_2O_3 and K_2O [1, 2].

It was found that by varying the chemical composition, the mechanical, electrical, chemical, optical, and thermal properties of the glasses, and the ease with which they could be drawn into fibers, could be modified. Pyrex® glass, which contains about 80% SiO_2 and a relatively large amount of B_2O_3 (typically 13%), Na_2O (4%), and minor quantities of Al_2O_3 and K_2O, is stronger than the soda-lime glass, has better chemical resistance, and has a lower coefficient of thermal expansion but is not so easily drawn into fibers [1]. Various compositions that are more readily drawn into fibers have been designed for plastics' reinforcement purposes. In general, the most commonly used reinforced plastic is E-glass, a lime-borosilicate glass derived from a Pyrex composition.

The commercialization of glass fibers in the mid-1930s and the development of polyester resins during the same period were instrumental in the introduction and establishment of reinforced plastics/composites (RP/C) based on glass fibers as new

Functional Fillers for Plastics: Second, updated and enlarged edition. Edited by Marino Xanthos

ISBN: 978-3-527-32361-6

construction materials. Significant advances in the processing and applications of molded and laminated products were made in the United States of America during World War II, and these were followed in the postwar era by the penetration of RP/C in many markets such as automotive, marine, aircraft, appliance, recreational, and corrosion-resistant equipment [3]. During the manufacture of glass fibers, after being collimated into strands, the continuous fibers were further processed into various forms suitable for use with thermosetting or thermoplastic matrices. Initial application of glass fibers were as continuous (long) reinforcements in thermosetting resins. Discontinuous (short) glass fiber reinforced thermoplastics (RTP), which constitutes the focus of this chapter, were first developed in the late 1940s to early 1950s as glass reinforced nylon and polystyrene [4], and since then there has been a significant growth in the use of glass fibers in a variety of commodity and engineering thermoplastics.

7.2 Production Methods

Liquid glass is formed by blending the appropriate ingredients for a given composition (e.g., sand, metal salts, boric acid for E-glass) in a high-temperature furnace at temperatures up to 1600 °C. The liquid is passed through electrically heated platinum alloy bushings that may contain up to 4000 holes, through which the glass is metered into filaments. The diameter of the filaments that can, in principle, be varied from 2.6 to 27.3 μm is controlled by composition, viscosity, tip diameter, drawing temperature, cooling rate, and rate of attenuation. The filaments are drawn together into a strand (closely associated) or roving (loosely associated), cooled and coated with proprietary formulations of organic chemicals, the "sizing," to provide filament cohesion through film formation, lubrication, protection of the glass from abrasion, and compatibility with the polymer matrix. The collimated filaments from the fiber-forming stage can be wound on a tube by means of a high-speed winder to give continuous strands that can be characterized according to the diameter of their filaments, the linear weight or tex count (g/km), the direction of twist, and the number of turns per meter. Strands used in reinforced plastics contain filaments with diameters in the 9.1–23.5 μm range; those with diameters ranging from 9.1 to 15.9 μm are specifically used for thermoplastics [3]. Figure 7.1 provides a schematic representation of glass fiber production [5]. The continuous strands may be converted into various other forms suitable for open-mold and other thermoset applications, such as rovings, woven rovings, fabrics, and mats. They may also be cut to specific lengths to produce chopped strands or milled to finer sizes for a variety of thermoplastic and thermoset applications. By varying the composition of the initial mixture of raw materials, different types of glasses can be produced.

As discussed in Chapter 2, the surface free energy of glass varies depending on whether the glass is bare or sized. The high surface energy of bare (pristine) glass filaments, which permits their easy wetting, will rapidly decrease due to adsorption of atmospheric water. In addition, as shown in Figures 7.2 and 7.3, the strength of virgin

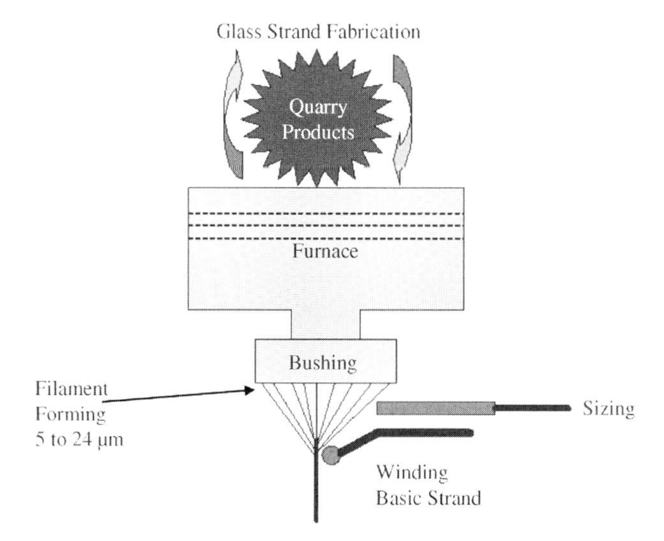

Figure 7.1 Flow diagram of glass fiber production. Adapted from Ref. [5].

filaments decreases rapidly in air or in water as well as at elevated temperatures [6]. Data from glass manufacturers show differences between a standard borosilicate glass (E-glass) and a higher strength, higher modulus glass (R-glass) of different compositions. Sizings applied on the fiber surface immediately after the fibers leave

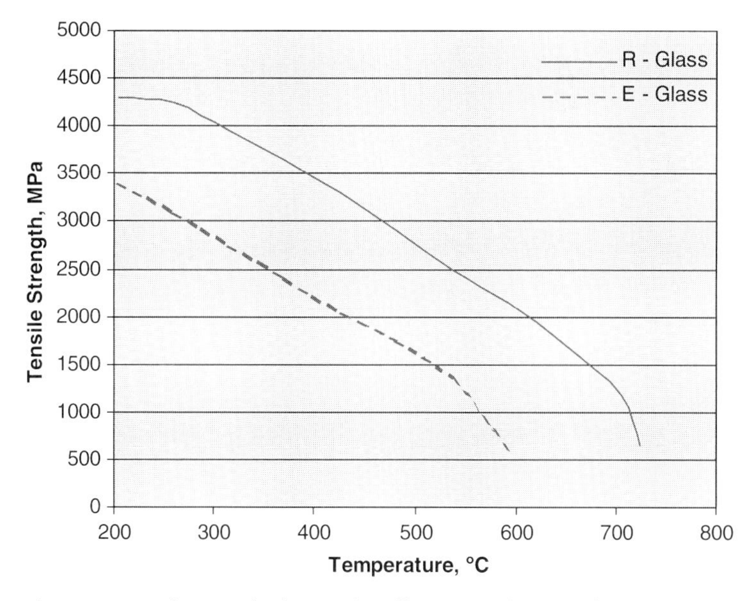

Figure 7.2 Tensile strength of virgin glass filaments as function of temperature. Adapted from Ref. [6].

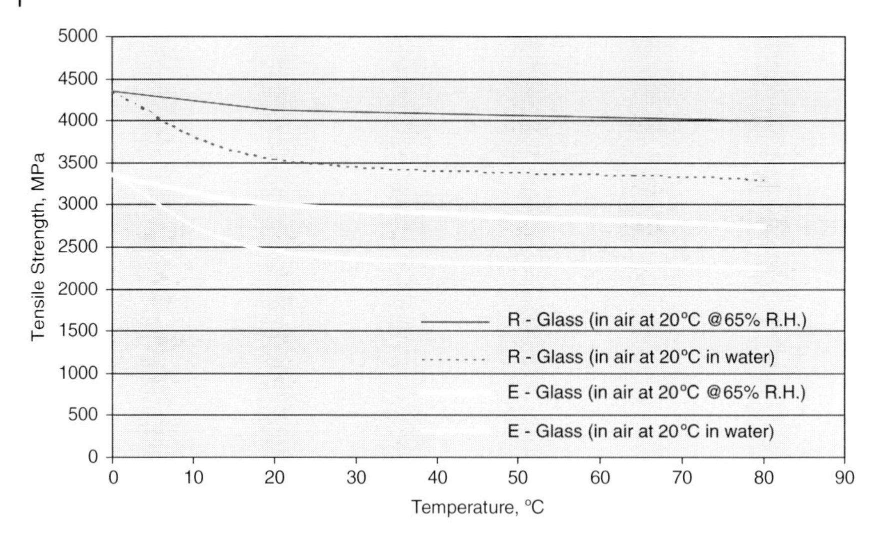

Figure 7.3 Tensile strength of virgin glass filaments as function of temperature from manufacturing in different environments. Adapted from Ref. [6].

the bushing minimize the strength loss during subsequent processing. Each sizing is specially designed for a given molding or compounding process and for a different matrix type. In addition to ingredients added to improve the handling characteristics of the fibers, the use of adhesion promoters such as silane coupling agents contributes to enhancing the mechanical properties of composites and particularly their resistance to aging (see also Chapter 4). The applied coatings tend to reduce surface free energy and, in general, wetting or spreading is not as favorable with these sizings as it is with truly bare fibers.

The importance of sizing composition in the development of glass fibers grades suitable for a specific application cannot be underestimated. Sizing composition and its solubility in the polymer are different for room temperature cure of thermosets (HSB, high solubility), high-temperature press moldings (LSB, low solubility) or, for use in reinforced thermoplastics [3]. Sizing compositions for chopped strands or milled fibers that are commonly used with thermoplastics and certain thermosets include both LSB and RTP types. As an example, compositions for glass fibers used for polypropylene (PP) reinforcement are typically based on a hydrolyzed aminoalkyltrialkoxysilane as the coupling agent, and an aqueous, colloidal dispersion of the alkali metal salt of a low MW maleated PP containing an appropriate emulsifying agent. The resulting coating provides fiber surfaces with an acceptable degree of wettability by molten PP and a reasonable compatibility with crystallized PP, thus affording composites with a good, overall spectrum of mechanical properties. The complexity of commercial sizing compositions is illustrated with three examples of water-based formulations suitable for three different polymers in Table 7.1 [7].

Table 7.1 Composition of glass fiber sizings [7].

Components	Polyester compatible (US Patent 4 752 527)	Polyvinyl acetate compatible (US Patent 4 027 071)	Polyurethane compatible (US Patent 3 803 069)
Solvent	Water	Water	Water
Coupling agent	γ-Methacryloxy-propyltrimethoxysilane	γ-Ethylenediaminepropyltrimethoxysilane or methacrylic acid complex of chromic chloride	γ-Aminopropyltriethoxysilane
Film former	Unsaturated bisphenolic glycol-maleic polyester	Polyvinyl acetate	Curable blocked polyurethane resin emulsion
Antistatic agent	Cationic organic quaternary ammonium salt		
Lubricant	Polyethyleneimine polyamide	Cationic fatty acid amide or tetraethylene pentamine	
Strand-hardening agent	Aqueous methylated melamine-formaldehyde resin		
pH control	Acetic acid	Acetic acid	
Emulsifying agent			Condensate of polypropylene oxide with propylene glycol

7.3
Structure and Properties

The conventional concept of an all silica vitreous structure is that of silicon–oxygen tetrahedral building blocks linked at their corners and randomly organized in a network. In glasses, the structure is broken or distorted by the addition of monovalent fluxes and modified by substitution of such ions as aluminum and boron. Thus, the melting point and the viscosity of the vitreous silica are reduced by fluxing with sodium oxide, which is added to the melt as sodium carbonate. Calcium and magnesium, which are other constituents of glass, enter the network structure as network modifiers. These modifiers make the structures more complex to hinder crystallization of the molten mass during the cooling process. This is why the glass is often referred to as a supercooled liquid having no crystallization or melting point and no latent heat of crystallization or fusion.

Glass fibers are characterized by a relatively high strength and a reasonable cost, but lower modulus as compared to other fibrous reinforcements such as carbon and aramid. The fibers are incombustible, have excellent high-temperature resistance and chemical resistance, and offer chemical affinity to a variety of resins or coupling agents through surface silanol groups. Table 7.2 compares properties of glass yarns with those of aramid and carbon [8].

The fibers can be further characterized by their physical and chemical properties, which are governed primarily by the composition of the glass. There are several glass fiber types, with different chemical compositions for different applications. They include the following:

A-glass: The most common type of glass for use in windows, bottles, and so on, but not often used in composites due to its poor moisture resistance.

C-glass: High chemical resistance glass used for applications requiring corrosion resistance.

D-glass: The glass with improved dielectric strength and lower density.

E-glass: A multipurpose borosilicate type and the most commonly used glass for fiber reinforcement.

Table 7.2 Comparative properties of continuous fiber reinforcements for plastics [8].

Property	E-glass	Carbon[a]	Aramid (Kevlar 49)
Tensile strength (MPa)	3450	3800–6530	3600–4100
Elasticity modulus (GPa)	73	230–400	131
Elongation to break (%)	3–4	1.40	2.5
Density (g/cm^3)	2.58	1.78–1.81	1.44
Relative cost	3.7	52–285	44

a) Values depend on the type of fiber (standard, intermediate, or high modulus).

S-glass: A magnesia–alumina–silicate composition with an extra high strength-to-weight ratio, more expensive than E-glass and used primarily for military and aerospace applications.

The chemical composition and physical properties of these glasses and their fibers are shown in Table 7.3.

Other specialty glasses include R-glass, which has higher strength and modulus than S-glass, AR-glass that has a high zirconia content and is much more resistant to alkali attack than E-glass; and AF-glass, a general purpose, alkali borosilicate with improved durability compared to container and sheet glasses.

7.4 Suppliers

Table 7.4 contains information on the major global glass fiber suppliers and their products.

7.5 Cost/Availability

Total glass fiber demand in the United States was forecasted to reach 3.5 million tons in 2007 with a significant portion of the envisaged increases being based on opportunities in reinforced plastics [10]. Continuous forms normally used in thermosets are strands (compactly associated bundles of filaments), rovings (loosely associated bundles of untwisted filaments or strands) and woven roving fabrics, continuous filament mats (felts of continuous filaments distributed in uniform layers held together by a binder), and chopped strand mats (felts or mats consisting of glass strands chopped to lengths of mostly 50 mm and held together by a binder). Discontinuous forms for use in thermoplastics' processes (extrusion, injection, molding) and certain thermoset fabrication methods (injection and compression molding and reinforced resin injection molding, RIM) include chopped strands and milled fibers. The market price of glass fibers depends on their type with A and E glasses being the lowest in price. For E-glass, the most common reinforcement grade, prices are in the range of 1.70–3.00 US$/kg depending on the form and quantities.

Chopped glass strands are produced from continuous rovings and are available at different lengths varying from 3 to 4.5 mm for thermoplastics and from 4 up to 13 mm for thermosets. Nominal aspect ratios with a typical 10 μm diameter fiber (before compounding) thus vary from 300 to 1300. Suitably sized grades are available depending on the particular thermoplastic (PA, PET, PBT, PP, PC, PPS, PPO, styrenics) or thermoset application (mostly unsaturated polyesters (UP) as sheet and bulk molding compounds (SMC, BMC)).

Milled fibers are produced from continuous E-glass roving by hammermills and are used in both thermoplastics and thermosets. Unlike chopped strands, which are

Table 7.3 Chemical composition and physical properties of various glass fibers.

	A-glass	C-glass	D-glass	E-glass	S-glass
Composition					
SiO_2	72–73.6	60–65	74	52–56	65
Al_2O_3	0.6–1.0	2–6	0.3	12–16	25
B_2O_3		2–7	22	5–10	—
K_2O	0–0.6	—	1.5	0–2	—
Na_2O	14–16	8–10	1.0	0–2.0	—
MgO	2.5–3.6	1–3	—	0–5	10
CaO	5.2–10	14	0.5	16–25	—
TiO_2	—	—	—	0–1.5	—
Fe_2O_3	—	0–0.2	Trace	0–0.8	—
Li_2O	0–1.3	—	0.5	—	—
SO_3	0–0.7	0–0.1	—	—	—
F_2	—	—	—	0–1.0	—
Property					
Density (g/cm^3)	2.50	2.49–2.53	2.14–2.16	2.52–2.65	2.5
Softening point (°C)	700	689–750	775	835–860	970
Modulus of elasticity (GPa)	70–75	69	55	70–75	85
Tensile strength (MPa)	2450	2750	2500	3400	4600
Elongation at break (%)	4.3	—	4.5	4.5	—
Poisson's ratio	0.23	—	—	0.22	0.23
Hardness (Mohs)	6.5	—	—	6.5	—
Refractive index	1.51–1.52	1.54	1.47	1.55–1.57	1.52
Dielectric constant at 1 MHz	6.9	6.24	3.56–3.85	5.8–6.7	4.9–5.3
Dielectric strength (kV/mm)	—	—	—	8–12	—
Volume resistivity (Ω cm)	—	—	—	10^{13}–10^{14}	—
Coefficient of linear expansion (10^{-6} °C^{-1})	5–8	9.4	3.5	5	5.9
Thermal conductivity (W/(m K))				1.0–1.3	
Specific heat (J/(kg K))				810–1130	
Refractive index	1.51–1.52	1.54	1.47	1.55–1.57	1.52

Note: Data compiled from various manufacturers web sites and Refs [2, 3, 9].

Table 7.4 List of major glass fiber suppliers.

Name, web site	Location, product
AGY Holding Corp. – http://www.agy.com	United States, manufacturers of E-glass yarns and chopped strand mat
Camely AF – http://www.camelyaf.com.tr	Turkey, E-glass for thermoset and thermoplastic
Central Glass Co., Ltd – http://www.cgco.co.jp/	Japan, E-glass yarn and direct sized roving, chopped strand mat, milled glass
Composite Reinforcements, Ltd – http://www.composite-reinforcements.co.uk/	United Kingdom, distributor of glass fiber rovings, fabrics, and chopped strand mats
Fiberex, Ltd – http://www.fiberex.com/	Canada, boron- and fluoride-free E-CR glass fiber rovings for composite reinforcement applications
Glasseiden GmbH – http://www.glasseide-oschatz.de/	Germany, glass fiber strands, mats, and rovings
JINWU Glass Fibre – http://www.jwfg.com	China, E- and C-glass fiber fabrics, yarns, and rovings
Johns Manville – http://www.jm.com	E-glass assembled and direct rovings yarns, chopped strands
KCC Glass Fiber – http://www.kccworld.co.kr/eng/product/kk/introduction/enggf.htm	Korea, E-glass direct and assembled rovings, milled glass, woven roving, and veil
Lauscha Fiber International GmbH – http://www.lfifiber.com/	Germany, chopped strands and microglass fiber
Lintex Co., Ltd – http://www.glasstextile.com/	China, chopped glass fiber strands and rovings
Millennium Dragon Fibreglass Trading Co., Ltd – http://www.fiberglassking.com/	China, manufacturers of fiberglass yarns, rovings, and chopped strands
Nanjing Fiberglass Research Institute – http://www.fiberglasschina.com/ÿnglish/	China, specialty fibers and fabrics
Nippon Electric Glass Co., Ltd – http://www.neg.co.jp/	Japan, E-glass chopped strands, rovings, and yarns for composite reinforcement applications
Nitto Boseki Co., Ltd – http://www.nittobo.co.jp/	Japan, yarns, rovings, chopped strand mat, filament mat
Owens Corning – http://www.owenscorning.com/	United States, glass fiber and composites for industrial end uses
PPG Industries – http://corporateportal.ppg.com/	United States, glass fiber for nonwoven and textile applications. Fiberglass fabrics, products, and composite structures for automotive, construction, and manufacturing industries

(Continued)

Table 7.4 (*Continued*)

Name, web site	Location, product
Saint-Gobain Group – http://www.saint-gobain.fr/	France, glass fiber filaments, chopped strands, fabrics, and tapes for the automotive, construction, electronics, and household appliance industries
Saint-Gobain Vetrotex, Inc. – http://www.sgva.com/	United States, glass fiber chopped strands and rovings for composite reinforcement applications, continuous strand mats
Texas Fiberglass Group – http://www.fiberglass.to	Hong Kong, China, manufacturer of AR-glass and E-glass roving, chopped strand, yarn, tape, mesh, fabric, cloth, scrim, gun roving
Tianma Group – http://www.tm253.com/	China, diversified group of companies, active in glass fiber products, plastics, and chemicals. E-glass yarns and rovings for composite reinforcement applications
Twiga Fiberglass Ltd – http://www.twigafiber.com/	India, E-glass-chopped strand, chopped strand mat
Vertex AS – http://www.vertex.cz/	Czech Republic, glass fiber filaments and chopped strands for composite reinforcement applications. Part of the Saint-Gobain Group
Vesta Intracon BV – http://www.vesta-intracon.nl/	The Netherlands, distributors of glass fiber rovings, yarns, and chopped strands for composite reinforcement applications

reduced in size to specific lengths, the milled glass fiber is reduced in size to an average volume. Their length (50–350 μm) is substantially smaller than that of chopped strands. Milled glass is produced in powder and floccular forms with the powdery form having in general shorter fiber length and higher bulk density than the floccular form. Various grades available for different applications (thermoplastics, thermosets, PTFE) differ in bulk density and type of sizing. More information on the types and characteristics of available glass fiber products can be found in the suppliers' web sites and in Refs [3, 9, 11]. Silver-coated conductive glass fibers with powder resistivity from 1.6 to 3.5 mΩ cm and mean lengths varying from 175 to 50 μm are also available for applications requiring electromagnetic interference (EMI) shielding properties [12].

7.6 Environmental/Toxicity Considerations

Commercial glass fibers consist of the basic glass component plus organic surface sizings at concentrations $<5\%$ of the overall weight. Short-term exposure to glass fibers may cause irritation of the skin and possibly irritation of the eyes and upper respiratory tract (nose and throat). Fiberglass is a nonburning material although the organic binders may burn. It is generally considered to be an inert solid waste not requiring hazardous disposal procedures.

Health effects after long-term exposures to glass fibers have been the topic of numerous studies. The potential for inhaled glass fibers to cause any health hazard depends on its "respirability," that is, their potential to enter the lower respiratory tract. Only fibers of less than approximately 3.5 μm in diameter are considered respirable and hence potentially carcinogenic. Other criteria for respirable fibers [13] are length/diameter ratio larger than 3 and length larger than 5 μm.

According to a study conducted at the National Cancer Institute in 1970, glass fibers less then 3 μm in diameter and greater than 20 μm in length were found to be "potent carcinogens" in the pleura of rats. In another laboratory study [14], animals exposed to very high concentrations of respirable microfibers with mean diameter of 0.5 μm on a long-term basis developed lung tumors, fibrosis, and mesotheliomas. Since then, studies have continued to appear, showing that fibers of this size not only cause cancer in laboratory animals, but also cause changes in the activity and chemical composition of cells, leading to changes in the genetic structure in the cellular immune system. Although these cell changes may be more common (and possibly more important) than cancer, it is the cancer-causing potential of glass fibers that has attracted most attention.

According to various reports, the concentrations of glass fibers to which the workers in fiberglass manufacturing plants are exposed to are far lower than the concentrations to which asbestos workers were exposed. However, statistically significant increased levels of lung cancer among the workers handling glass fibers have been reported [15–18] in several industry-sponsored epidemiological studies conducted in the late 1980s in the United States, Canada, and Europe. Recently, it has been shown that an 8 h exposure to 0.043 glass fibers per cubic centimeter of air is sufficient to cause lung cancer in one-in-every-thousand exposed workers during a 45-year working lifetime [19]. Other major epidemiological studies in the United States, Europe, and Canada involving 21 500 workers in the fiberglass manufacturing showed no increased incidence in lung cancer or nonmalignant respiratory disease [13]. Other epidemiological studies published in 1997 and listed in Ref. [13] did not provide evidence of increased incidence of cancer in populations working in the plants for a long time. For more current information, please refer to the appropriate MSDS from glass fiber suppliers.

OSHA considers glass fibers as a "nuisance" dust with a permissible occupational exposure limit (8 h time-weighted average, TWA) of 15 mg/m^3 as total dust and 5 mg/m^3 as respirable dust. ACGIH's TWA limit for inhalable glass fiber dust is 5 mg/m^3 and 1 fiber/cm^3 for the respirable fraction. The International Agency for Research on Cancer (IARC) and the US National Toxicology Program (NTP) and OSHA do not list continuous filament glass fibers as a carcinogen. The diameters of most fibers used for plastic reinforcement are above 6 μm, and therefore, they do not conform to the "respirable" criteria, although chopped, crushed, or severely mechanically processed fibers may contain a very small amount of respirable fibers. Repeated or prolonged exposure to respirable fibers may cause fibrosis, lung cancer, and mesothelioma. However, the measured airborne concentration of these respirable fibers in environments in which fiberglass has been extensively processed has been shown to be extremely low and well below the threshold limiting value (see, e.g., Ref. [20]).

7.7 Applications

As a result of their inherent physical and mechanical properties and availability in high aspect ratios, the primary function of short glass fibers is the enhancement of mechanical properties of a variety of thermoplastics and thermosets. The characteristics of glass fiber thermoplastic composites, such as high strength to weight ratio, good dimensional stability, good environmental resistance, good electrical insulation properties, ease of fabrication, and relatively low cost, make them particularly suitable in a variety of applications such as automotive, appliances, business equipment, electronics, sports, and recreational.

Typical glass fiber concentrations in commodity thermoplastics, engineering thermoplastics, and thermosets may range from 10 to 50 vol%. For several resins, 50% by volume translates to about 70% by weight glass fiber, values achievable only by orienting the fibers, as for example in pultrusion. Such high loadings are not common in commercial compounding techniques, where excessively high fiber loadings attribute to severe attrition, fiber breakage, and, hence, reduced composite mechanical properties. The effects on the mechanical properties of a specific polymer, as discussed in Chapters 2 and 3, largely depend on the type of polymer, filler concentration, fiber aspect ratio retention in the final molded product, dispersion and fiber orientation, and the type of surface treatment (coupling agent, sizing). Figures 7.4–7.8 contain representative data on the mechanical properties of miscel-

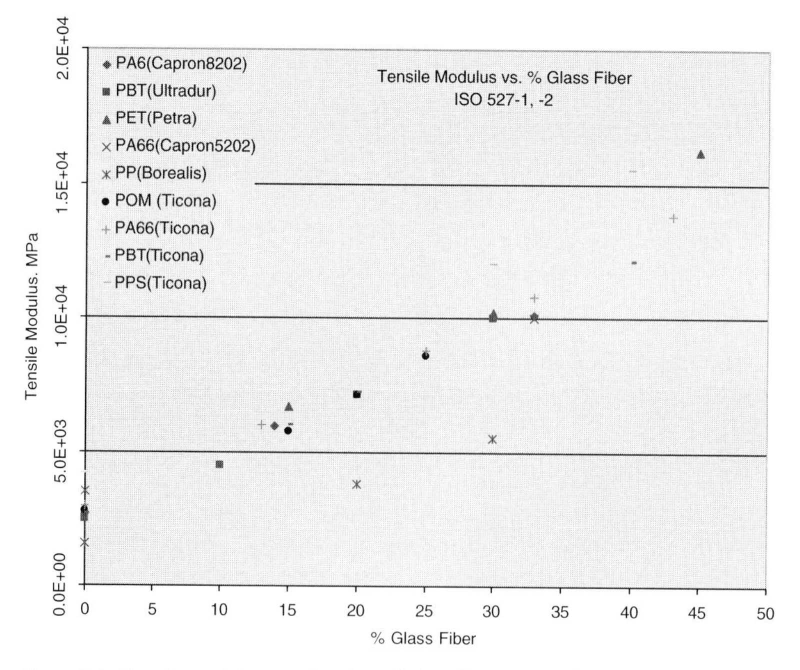

Figure 7.4 Tensile modulus as a function of glass fiber content for several engineering thermoplastics. PP data are included for comparison.

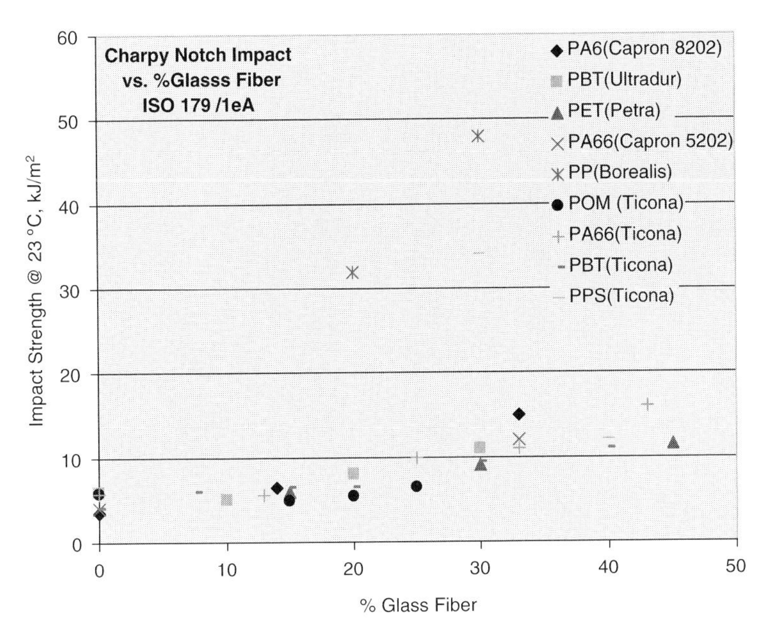

Figure 7.5 Charpy impact strength as a function of glass fiber content for several engineering thermoplastics. PP data are included for comparison.

laneous glass fiber reinforced thermoplastics. The figures have been constructed from replotted data generated by materials suppliers and included in Ref. [21]. In all cases, the glass used was chopped strands of E-glass and was assumed to bear appropriate surface treatments.

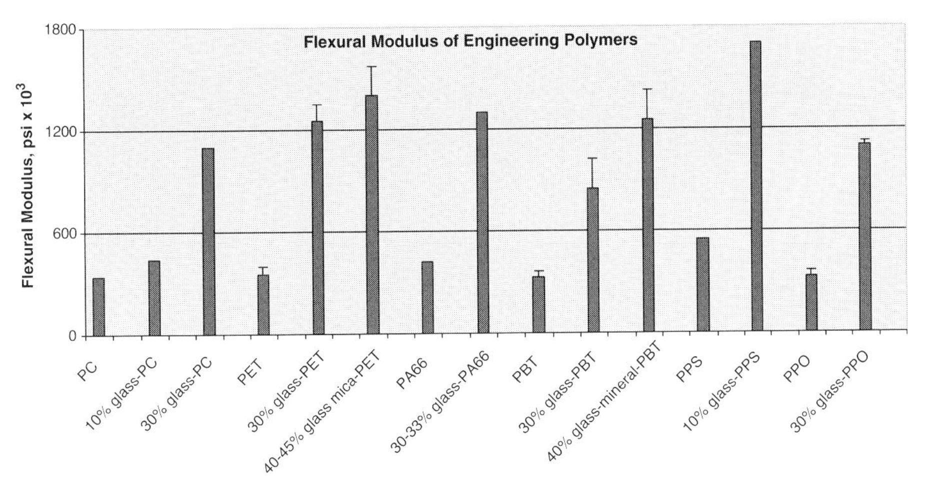

Figure 7.6 Comparison of flexural modulus of glass fiber reinforced engineering thermoplastics. Data on the unfilled resins are included for comparison. (To convert from psi to Pa multiply by 6895.)

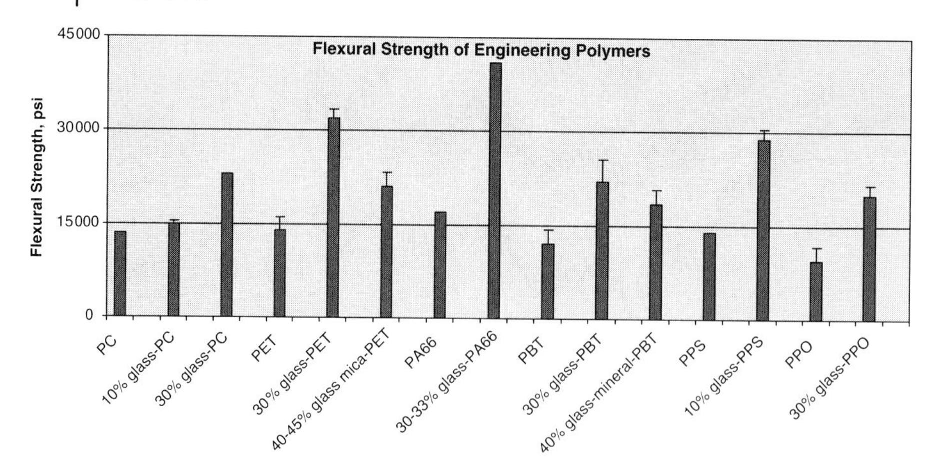

Figure 7.7 Comparison of flexural strength of glass fiber reinforced engineering thermoplastics. Data on the unfilled resins are included for comparison. (To convert from psi to Pa multiply by 6895.)

Figures 7.4 and 7.5 show the effects of increasing glass concentration on the tensile moduli and Charpy impact strength values of several injection-molded engineering thermoplastics; data on polypropylene, a much lower modulus commodity resin commonly reinforced with glass fibers, are also included for comparison purposes.

Figures 7.6–7.8 contain data on the flexural properties and Izod impact strength of six engineering thermoplastics. In general, modulus and strength increase versus the unfilled resin with the addition of glass. Impact strength also shows increases with the exception of the inherently tough PC and PPO resins. It is of interest to note that for certain ductile resins such as polyphenylene ether sulfone (PPSU) and PC-ABS, and PC-PBT blends, it has been shown that surface coatings with low adhesion to the matrix can give improved impact strength values at the expense of flexural strength, but without a modulus compromise [22]. For comparison purposes, the data in

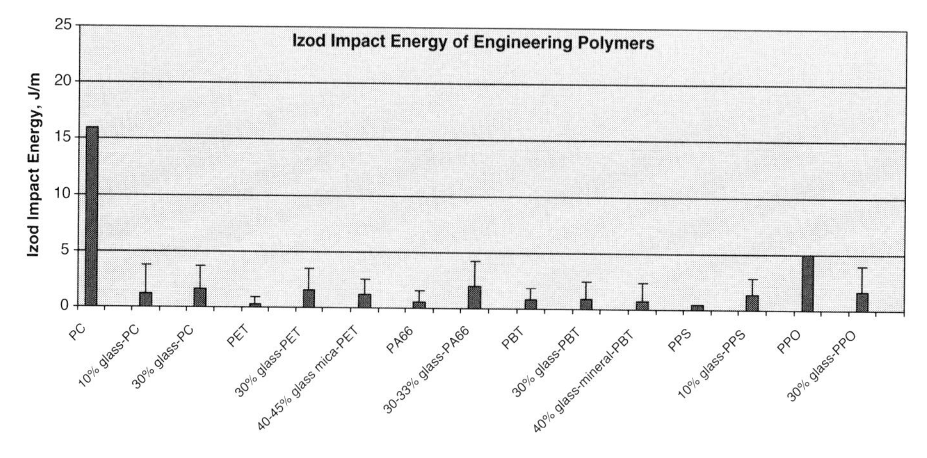

Figure 7.8 Comparison of impact strength of glass fiber reinforced engineering thermoplastics. Data on the unfilled resins are included for comparison.

Table 7.5 Properties of injection-molded glass fiber reinforced polyolefin before and after maleation of the matrix [23].

Property	Unfilled unmodified matrix	Unmodified matrix + 20% glass	Maleated matrix 20% glass
Tensile yield strength (MPa)	24.8	—	—
Tensile break strength (MPa)	10.8	36.5	42.1
Tensile yield elongation (%)	9.2	—	—
Tensile break elongation (%)	23	3.5	4.8
Flexural modulus (MPa)	1030	2620	2340
Flexural strength (MPa)	30.3	49.4	51.3
Izod impact notched (J/m)	48.0	53.4	80.1
Izod impact unnotched (J/m)	1014	208	272
HDT at 1.82 MPa (°C)	47	79	84

Note: Fiberglass OCF457AA, Owens Corning; Matrix, polyolefin based on a model recyclable stream containing at least 80% HDPE, modified with peroxide/maleic anhydride in twin-screw extruder.

Figures 7.6–7.8 include combinations of glass fibers with mica or mineral fillers that are normally used to control warpage and improve dimensional stability. Additional extensive information on the properties of injection-molded glass reinforced thermoplastics can be accessed via the web sites of glass suppliers, resin producers, and compounders, and found in Ref. [3].

Adhesion in polyolefin-glass fiber composites can be increased by suitable reactive modification of the nonpolar matrix. The effect of glass fibers on the properties of a predominantly polyethylene matrix, before and after grafting with maleic anhydride, is shown in Table 7.5 [23]. Maleation of the matrix further enhances its tensile properties, impact strength, and heat distortion temperature (HDT) by virtue of improved adhesion as a result of reaction of the aminosilane groups present on the glass surface with the pendant polymer anhydride groups (see also Chapters 4 and 6). Figures 7.9 and 7.10 show a comparison of the fracture surfaces of unmodified and maleated composites; the microphotographs clearly show pullout regions and bare fiber surfaces in the case of the unmodified, nonpolar matrix.

As an example of the importance of aspect ratio, Table 7.6 contains data on injection-molded 30 wt% glass fiber polyamide composites containing fibers of different diameters [24]. Assuming that the average fiber length is approximately the same in all molded samples, it is clearly evident that increasing fiber diameter (decreasing aspect ratio) results in lower mechanical properties (with the exception of notched impact strength). As a result of the compounding and molding processes, the initial fiber length was reduced from 3–5 mm to less than 0.45 mm. Similar trends were reported at higher glass fiber concentrations (50 and 63 wt%). The observed aspect ratio effects are more pronounced in long fiber reinforced thermoplastics (LFRTP) where pellets with long initial fiber length are produced by pultrusion-type operations rather than extrusion compounding. Longer initial fiber lengths are expected to yield longer final lengths and, hence higher aspect ratios, in the injection-molded composites.

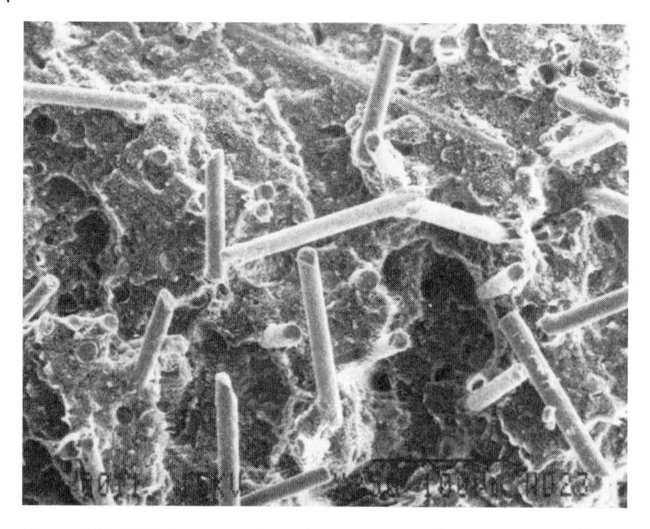

Figure 7.9 SEM fracture surface showing poor adhesion between the unmodified matrix and glass fibers [23].

7.8
Environmental Impact

Manufacturers, users, and nongovernmental organizations (NGO) are constantly demanding environmental impact data for the materials used in the production of various items. Environmental impact data, normally include items such as total energy used, green house gas emissions (GHG), toxic elements released, biological

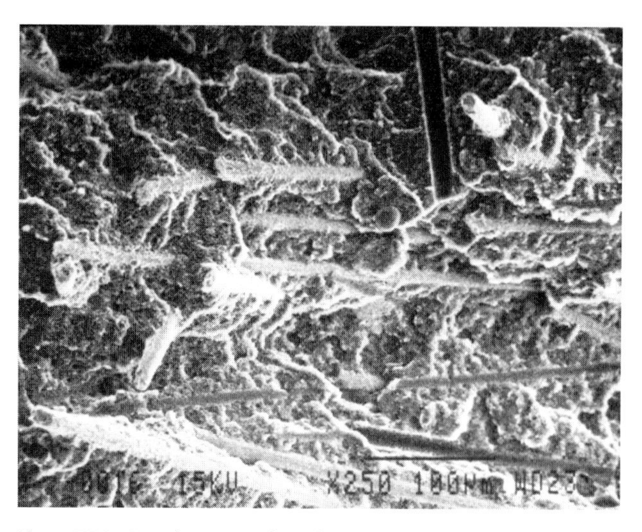

Figure 7.10 SEM fracture surface showing improved adhesion between the maleated matrix and glass fibers [23].

Table 7.6 Effect of glass fiber diameter on properties of 30% E-glass fiber containing Nylon 66 [24].

Property	Glass fiber diameter (μm)			
	10	11	14	17
Tensile strength (MPa)	184	183	174	164
Tensile modulus (GPa)	9.73	9.76	9.55	9.5
Tensile strain (%)	2.83	2.8	2.7	2.49
Flexural strength (MPa)	287	285	270	249
Flexural modulus (GPa)	9.2	9.21	9.29	9.01
Unnotched impact strength (J/m)	979	894	775	568
Notched impact strength (J/m)	130	130	135	139

oxygen demand (BOD), and chemical oxygen demand (COD). Lifecycle assessment tools such as the European Eco-indicators 95 and 99 are science-based impact assessment methods for life cycle analysis and pragmatic ecodesign methods. In the United States, EPA has been actively working with industry and academe in collecting such data. An increasing number of buyers demand from their suppliers to provide sustainability numbers. Companies such as Sonoco [25] have developed their own software to evaluate packaging options using sustainability numbers.

Environmental impact data for some materials of different densities including glass fibers, natural fibers, polymers, and composites are compiled from Refs [26–28] and are listed in Table 7.7. It is clear that the type of the polymer matrix would be the controlling factor for the environmental impact of composites based on either natural fibers or glass fibers. Data from Ref. [26] suggest that pallets made of natural fiber/PP are environmentally superior to similar pallets made from glass fiber/PP. However,

Table 7.7 Environmental impact data for various fibers and polymers [26–28].

	Glass fiber	China reed fiber	PP	ABS	Epoxy resin	Nylon 66	Nylon 66 + 30% GF
Density (g/cm^3)	2.5–2.6	1.4	0.9	1.1	1.2	1.14	1.37
Energy used (MJ/kg)	48.3	3.6	77	95	141	139	114
Green House gas emissions (kg/kg)	2	0.65	2	3	6	1.78	1.67
BOD to water (mg/kg)	1.75	0.36	34	34	1 200	3.6	0.033
COD to water (mg/kg)	19	2.3	179	2 200	51 000	15	5.7

Table 7.8 Environmental impact of glass and flax fibers.

Materials/ processes	Material	Eco-indicator (mPt/kg)	Remarks
Fibers	Flax	0.34	Hackled long fibers, value might be too low, since production of used pesticides is not taken into account
	Glass	2.31	Including extraction of raw materials, transport, and production
Matrix materials	EP resin	10.2	Including extraction of raw materials, transport, and production, mean European data
	UP resin	9.45[a)]	Value might be too low because production of energy carriers is not taken into account
	PP	2.99	Including extraction of raw materials, transport and production, mean European data

a) This value is for hand layup, for closed mold processing the Eco-indicator would be 3.08 mPt/kg.

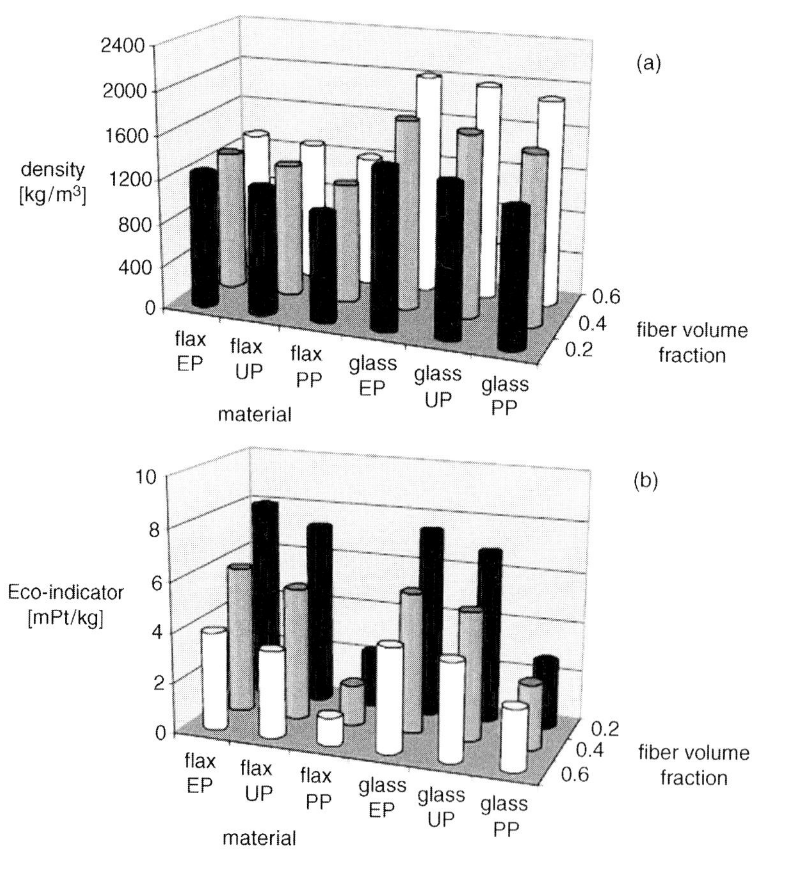

Figure 7.11 Density and Eco-indicator of glass fiber and flax thermoset and thermoplastic composites [29].

the environmental impact of the natural fiber/PP pallet may be worse if its expected life is lower than that of the glass fiber/PP pallet (3 years versus 5 years).

Table 7.8 lists the Eco-indicator of two fibers and three polymer matrices in milliPoints/kg material (mPt/kg) [29]. As shown before, the environmental impact of glass fibers is higher than that of flax, but again for a given composite, controlling factors are the type of matrix and the corresponding volume fractions. Figure 7.11 shows a comparison of the density and Eco-indicator of glass fibers in epoxy (EP), UP, and PP composites with those of equivalent flax composites at different volume fractions.

References

1 Piggott, M.R. (1980) Chapter 3, in *Load Bearing Fibre Composites*, Pergamon Press Inc., Oxford, England.

2 Callister, W.D., Jr. (2003) Chapter 13, in *Materials Science and Engineering, An Introduction*, 6th edn, John Wiley & Sons, Inc., Hoboken, NJ.

3 Milewski, J.V. and Katz, H.S. (eds) (1987) Chapter 14, in *Handbook of Reinforcement for Plastics*, Van Nostrand Reinhold, New York.

4 Sheldon, R.P. (1982) Chapter 1, in *Composite Polymeric Materials*, Applied Science Publishers, Ltd, Barking, Essex, England.

5 Saint-Gobain Vetrotex technical information; http://www.vetrotexna.com/business_info/gstrand.html#11.

6 Saint-Gobain Vetrotex technical information; http://www.vetrotextextiles.com/pdf/RGlass%20DS2000.pdf.

7 Michigan State University, Intelligent Systems Laboratory; http://islnotes.cps.msu.edu/trp/rtm/siz_basc.html.

8 See Ref. [2], Chapter 16 and Appendix B.

9 Hohenberger, W. (2001) Chapter 17, in *Plastics Additives Handbook* (ed. H. Zweifel), Hanser Publishers, Munich.

10 Freedonia Group (2003) Reinforced Plastics to 2007, Market report; summary accessed http://freedonia.ecnext.com.

11 Wypych, G. (2000) *Handbook of Filllers*, ChemTec Publishing, Toronto, Ont., Canada, pp. 187–188.

12 http://www.pottersbeads.com/markets/sgfibers.asp, Potters Industries, Inc. *Technical Information on Conduct-O-Fil® Silver Coated Glass Fibers*, Potters Industries Inc., Valley Forge, PA, 2001.

13 European Glass Fiber Producer Association (2002) *Continuous Filament Glass Fibre and Human Health*, APFE Publication, Brussels.

14 Cullen, R.T. *et al.* (2000) Pathology of a Special Purpose Glass Microfiber (E Glass) relative to Another Glass Microfiber and Amosite Asbestos, *Inhal. Toxicol.*, **12** (10), 959–977.

15 Shannon, H.S. *et al.* (1987) Mortality Experience of Ontario Glass Fibre Workers–Extended Follow-up, *Ann. Occup. Hyg.*, **31** (4B), 657–662.

16 Goldsmith, J.R. (1986) Comparative Epidemiology of Men Exposed to Asbestos and Man-made Mineral Fibers, *Am. J. Ind. Med.*, **10** (5–6), 543–552.

17 Marsh, G.M. *et al.* (1990) Mortality among a Cohort of US Man-made Mineral Fibers: 1985 follow-up. *J. Occup. Med.*, **32**, 594–604.

18 Boffetta, P. *et al.* (1992) Lung cancer mortality among workers in the European production of Man-made Mineral Fibers–A Poisson regression analysis. *Scand. J. Work Environ. Health*, **18**, 279–286.

19 Infante, P.F. *et al.* (1994) Fibrous Glass and Cancer. *Am. J. Ind. Med.*, **26**, 559–584.

20 PPG Industries, Inc., Material Safety Data Sheet for "Fiber Glass Continuous Filament", revised February 2004.

21 CAMPUS® Computer Aided Material Preselection by Uniform Standard; http://www.campusplastics.com/.

22 Galluci, R.F. (2004) Proceedings of the 62nd SPE ANTEC, vol. 50, p. 2718.

23 Xanthos, M., Grenci, J., Patel, S.H., Patel, A., Jacob, C., Dey, S.K., and Dagli, S.S. (1995) Thermoplastic composites from maleic anhydride modified post-consumer plastics. *Polym. Compos.*, **16** (3), 204.
24 BASF Corp, An Advanced High Modulus (HMG) Short Glass Fiber Reinforced Nylo6: Part I, Technical article available at http://www.basf.com/PLASTICSWEB/displayanyfile?id=0901a5e180086583.
25 Design for Sustainability (DFS) Software, Sonoco, Hartsville, SC.
26 Corbiere-Nicollier, T., Laban, B.G., Lundquist, L., Leterrier, Y., Manson, J.A.E., and Jolliet, O. (2001) Lifecycle assessment of biofibers replacing glass fibers as reinforcement in plastics. *Resour. Conservat. Recycl.*, **33**, 267–287.
27 Josihi, S.V., Drzal, L.T., Mohanty, A.K., and Arora, S. (2004) Are natural fiber composites environmentally superior to glass fiber composites? *Composites: Part A*, **35**, 371–376.
28 Boustead, I. (2002) *Ecoprofiles of Plastics and Related Intermediates*, Association of Plastic Manufacturers of Europe (APME), Brussels, Belgium (downloadable http://www. apme.org).
29 van Dam, J.E.G. and Bos, H.L. (2009) The environmental impact of fiber crops in industrial applications, www.fao.org/es/esc/common/ecg/343/en/Environment_Background.pdf.

8
Mica Flakes

Marino Xanthos

8.1
Background

Mica is a term for a group of more than 35 phyllosilicate minerals with a layered texture and perfect basal cleavage. This perfect cleavage, due to weak bonding between the layers, results in splitting or delamination of the mica layers into thin sheets. Micas compose roughly 4% of the earth's crystal minerals and are common in all three major rock varieties: igneous, sedimentary, and metamorphic [1]. Micas as a group are variable in chemical composition and in physical and optical properties. They are basically complex potassium aluminosilicates with some aluminum atoms replaced by magnesium and iron and may contain minor amounts of a variety of other elements. Muscovite and phlogopite, the most important commercial types, have unique characteristics such as chemical inertness, superior electrical and thermal insulating properties, high thermal stability, and excellent mechanical properties.

Micas are used in sheet and ground forms. High quality sheet mica is used principally in the electronic and electrical industries. Built-up mica produced by mechanized or hand setting of overlapping splittings and alternate layers of binders and splittings, and reconstituted mica (mica paper) are primarily used as electrical insulation materials. Commercial micas are divided into "wet ground" and "dry ground" depending on the method of production. In addition to its more recent widespread uses as functional filler for plastics, dry-ground mica has several other applications. It is used in tape-joint cement compounds for gypsum dry wall, in the paint industry as a pigment extender, in the well-drilling industry as an additive to drilling muds, in the rubber industry as a mold release compound, and in the production of rolled roofing and asphalt shingles. Wet-ground mica, which retains the brilliancy of its cleavage faces, is mostly used in pearlescent paints and in the cosmetics industry [1].

Since the early days of the development of phenolic molding compounds for electrical applications, mica has been extensively used as a filler of choice. It was not until the late 1960s/early 1970s that the realization of the importance of the aspect

Functional Fillers for Plastics: Second, updated and enlarged edition. Edited by Marino Xanthos

ISBN: 978-3-527-32361-6

ratio and interfacial adhesion in platelet/flake-containing polymers initiated R&D efforts at Canadian Universities and industrial laboratories that led to the commercialization of grades suitable as reinforcements for plastics. Early work was carried out in the University of Toronto, Toronto, Ontario [2–5], Fiberglass Canada, Sarnia, Ontario [6], and Marietta Resources International, Boucherville, Quebec; in 1975 the latter opened a plant that now has a capacity of 30 000 tons per annum, producing high aspect ratio (HAR) (surface treated and untreated) phlogopite grades, in addition to traditional mica products [7]. During the same period, fundamental work on the mechanics of flake reinforcement resulted in the development of predictive equations for modulus, strength, and toughness of flake-reinforced composites (see Chapter 2), which, in most cases, were confirmed by experiments. Research efforts from that period were reviewed by Woodhams and Xanthos in 1978 [8].

In the mid-1970s/early 1980s, the potential of high aspect ratio mica as a reinforcement in a variety of thermoplastics and thermosets, and the parameters affecting its performance, was described in a series of technical bulletins from Marietta Resources International and presentations at Society of Plastics Engineers and Society of Plastics Industry conferences (see, e.g., Ref. [9]). This early work complemented by publications from the Canadian and the US University researchers and Canadian Government laboratories provided significant information on the understanding of

- the coupling differences between muscovite and phlogopite, particularly in polypropylene;
- the rheological characteristics of flake-containing melts;
- the importance of flow-induced flake orientation on properties and weld-line strength;
- the differences between flake reinforcement and reinforcement with fibers and irregular fillers.

During that period, it also became clear that, as for fragile glass fibers, retention of flake aspect ratio and flake orientation (prerequisites for effective reinforcement) strongly depend on the type of polymer (thermoplastics versus thermosets) and processing/shaping method. It also became obvious that retention of high aspect ratio for large diameter flakes would only be favored by low-shear processing methods, usually applicable to thermosets; these would ensure flake planar orientation without having to consider the complications arising from the flow of filled thermoplastic melts in circular channels and irregularly shaped dies. It is unfortunate that, even at the present time, high aspect ratio, smaller diameter, thin flakes that would be less susceptible to mechanical breakdown are not, in general, available. The status of the developments in mica-reinforced plastics during the aforementioned period was reviewed by Hawley in 1987 [7]. It should be noted that the principles of flake reinforcement developed on mica more than 25 years ago are directly applicable to the much higher aspect ratio montmorillonite-based nanoclays introduced in the last 15 years as plastic reinforcement (see also Chapter 9).

The multiple functions of mica have been outlined in Chapter 1 of this book, along with an example of its role in the search of multifunctional fillers for polypropylene compounds for automotive applications. Mica-reinforced thermoplastics such as polypropylene, polyethylene, nylon, and polyesters are now established in a variety of automotive applications and consumer products where mica supplements or replaces glass fibers and other mineral fillers. The wider use of mica in many applications has been limited by low impact strength and low weld-line strength in certain plastics. These issues are the focus of continuing R&D efforts by materials suppliers and compounders/molders.

In addition to its primary function as a high aspect ratio mechanical property enhancer, mica is also used as a modifier of electrical properties and as an important component of sound-deadening formulations; it is also used for reducing permeability, improving dimensional stability, and as a modifier of optical properties. The multiple functions of mica are compared to those of other fillers in Table 1.4.

8.2 Production Methods

Mica occurs worldwide, with large deposits in the Unites States, Canada, France, Korea, Malaysia, Mexico, Russia, Finland, Madagascar, and India with smaller deposits in some European countries. Muscovite mica, the most common form, is found in acidic igneous rocks such as granites and also forms very large "books" in pegmatites. In metamorphic rocks, muscovite occurs in lower grades of purity. Phlogopite, the second most common form of mica, is found in ultrabasic igneous rocks, which reflects its high magnesium content. Biotite, the lesser of the micas in terms of commercial importance, contains more iron than magnesium, is brown to black in color, and is found in granites and intermediate igneous rocks [1].

Mining mica is typically accomplished through quarrying, although it is occasionally feasible to use underground mining methods. In producing large mica sheets for electronic and high temperature applications, the fragility of mica limits the feasible methods that can be used for its extraction. Mica blocks are split into thin splittings and individually assessed, trimmed, and sorted by size, color, and quality.

Mica is relatively easy to cleave while in coarse flake form, but as grinding proceeds breakage perpendicular to the cleavage, plane supersedes delamination. Thus, the production of high aspect ratio, well-delaminated, small-sized flakes is a challenge. The production of ground mica generally involves steps that are determined by the purity of the host ore. Preliminary size reduction and purification from other minerals may involve various flotation steps, magnetic separation, flotation cells and hydrocyclones, or air-table separation. The actual dry grinding may involve impact processors such as rotor mills, high speed hammer and cage mills, and pin mills. In the Suzorite™ process [7] using ore from a high purity, that is, with up to 90% phlogopite deposit, impurities such as feldspar and pyroxene are removed after primary milling. Further milling takes place in closed systems containing air classifiers that remove well-delaminated flakes and return thicker flakes for reprocessing. Finer flakes can be made by feeding these materials into fluid energy mills. A variety of other methods have been proposed for separating high aspect ratio flakes [3, 7].

Wet grinding may also be employed, although at higher cost and with lower capacity than in dry grinding. Wet grinding is still performed batchwise in chaser mills, where rollers revolve on the surface of the mica with a smearing action for several hours. The heavier thick flakes settle out, whereas the well-delaminated mica overflows into settling tanks. Other wet-grinding methods (log mills, vibro-energy milling, high-pressure water jets, and ultrasonics) and techniques for separating high aspect ratio flakes (e.g., sedimentation, water elutriation) are reviewed in Refs [7, 10].

8.3 Structure and Properties

Commercial micas are only available in the muscovite and phlogopite forms. Their compositions are subject to variations due to isomorphous substitutions. The basic mica structure is a sandwich, where the outer layers are silica tetrahedra in which some of the silicon atoms have been substituted by aluminum and the middle layer consists of aluminum, magnesium, iron, and fluorine plus hydroxyl groups arranged in an octahedral fashion. Muscovite, $KAl_2(AlSi_3O_{10})(OH)_2$, has a gibbsite, $Al(OH)_3$, structure sandwiched between the silica sheets (Figure 8.1). It is water-white with a

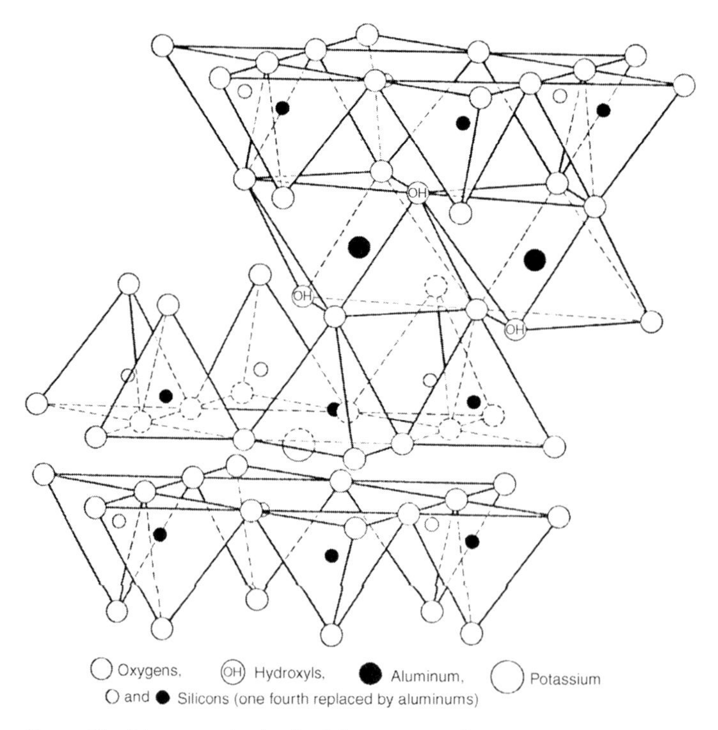

Figure 8.1 Diagrammatic sketch of the structure of muscovite [11].

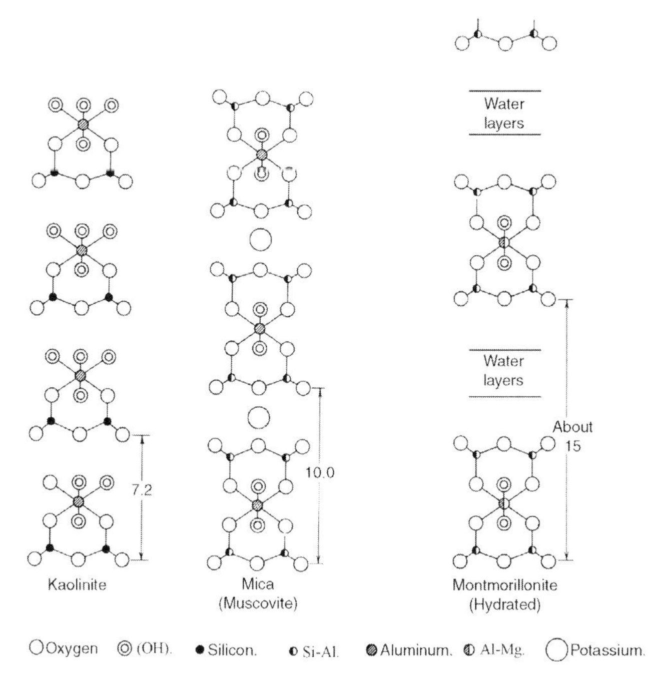

Figure 8.2 Comparison of the structure of mica with kaolinite and montmorillonite. Adapted from Ref. [12].

pinkish or greenish hue. Commercial phlogopite mica, $K(Mg, Fe)_3(AlSi_3O_{10})(OH, F)_2$, is dark brown to black in color depending on the iron content. The middle layer of the sandwich is brucite, $Mg(OH)_2$, with iron substitution in both Fe(II) and Fe(III) forms. These three-layer units are about 100 nm thick, are held loosely by potassium atoms in 12-fold coordination with the oxygen atoms, so that the interlayer forces are rather weak permitting cleavage. In Figure 8.2, the structure of mica is compared with those of two other phyllosilicate fillers, namely kaolinite and montmorillonite.

Chemical analyses of typical commercial muscovites and phlogopites are given in Tables 8.1 and 8.2, which summarizes typical physical and mechanical properties. As expected, data vary depending on mineral source, sample type and size, and method of property measurement. Certain micas expand upon heating at temperatures above 600 °C. Muscovites generally lose their combined water at lower temperatures. Characteristic properties of both muscovite and phlogopite are high modulus (more than double the modulus of glass fibers; 172 GPa versus 70 GPa) and high strength, low coefficient of thermal expansion, high thermal conductivity and temperature resistance, good dielectric properties, low hardness, low coefficient of friction, and good chemical resistance.

A key characteristic of mica that is not apparent from Table 8.2 is its planar isotropy due to its plate-like nature. The listed properties are similar in both the "x" and the "y" directions in the plane (but not always in the "z" direction), giving rise to the isotropic

Table 8.1 Chemical analysis of commercial phlogopite and muscovite micas [8, 13, 14].

	Phlogopite[a)]	Muscovite[b)]	Muscovite[c)]
SiO_2	40.7	47.9	45.6
Al_2O_3	15.8	33.1	33.1
MgO	20.6	0.69	0.38
FeO	7.83	—	—
Fe_2O_3	1.21	2.04	2.48
K_2O	10.0	9.8	9.9
Na_2O	0.1	0.8	0.6
BaO	0.5	—	—
CaO	Trace	0.5	0.2
TiO_2	0.1	0.6	Trace
Cr_2O_3		—	—
MnO	Trace	Trace	Trace
F	2.16	—	—
P	Trace	0.03	Trace
S	Trace	0.01	Trace
H_2O combined	1.0	4.3	2.7
H_2O free	0.01	0.1	0.25

a) Suzorite Mica Products.
b) The English Mica Co.
c) Inderchand Rajgarchia & Sons (P) Ltd.

properties characteristic of oriented mica composites (see Chapter 2). Thus, unlike fibers, mica reinforces equally in the two directions in the plane.

The morphological features of ground micas vary depending on the method used for their delamination. Dry-ground mica flakes are not of even thickness but are stepped due to uneven delamination, which also results in feathered edges. Figure 8.3 shows sharp edges, broad size distribution, and incomplete delamination for a particular grade of a commercial dry-ground phlogopite; irregular surfaces are also shown in Figure 1.2 for ultrasonically delaminated flakes. In contrast, the wet-grinding process polishes the flakes to an even thickness and rounds off the edges (see Figure 1.5), usually producing a high aspect ratio product. Thus, wet-ground mica has more lubricity and has a higher "sparkle" than the dry-ground product.

Natural micas have a very low ion-exchange capacity as compared to the bentonite clays described in Chapter 9 and those sites where exchange ions are all on the outer surfaces. However, the reactive surface groups of micas, some appearing on the faces after delamination, are amenable to treatment with a variety of additives that may improve dispersion/adhesion in a variety of polar or nonpolar polymeric matrices. Details on additives and methods of surface treatments are included in Refs [7, 17]. For polar polymers (nylons, thermoplastic polyesters, and polyurethanes), amino-silanes and aminostyrylsilanes have been found efficient. With nonpolar polymers such as polyolefins, appropriate coupling is needed to increase strength values above those of talc or calcium carbonate filled systems and closer to those of glass fibers filled systems. Azidofunctional and aminostyrylsilanes have been found particularly

Table 8.2 Properties of muscovite and phlogopite micas [8, 13–16].

Property	Muscovite	Phlogopite
Color	White, off-white, ruby, green	Amber, yellow, light brown
Crystal structure	Monoclinic	Monoclinic
Hardness (Mohs)	3–4	2.5–3.0
Density (g/cm^3)	2.7–3.2	2.75–2.9
pH, aqueous slurry	6.5–8.5	7.5–8.5
Water solubility	Trace	Trace
Refractive index	1.55–1.61	1.54–1.69
Tensile modulus (GPa)	172	172
Tensile strength (MPa)	255–296[a)] 3100[b)] 690–900[c)]	255–296[a)] 690–900[c)]
Linear coefficient of thermal expansion (per °C)		
Perpendicular to cleavage	$15–25 \times 10^{-6}$	—
Parallel to cleavage	$8–9 \times 10^{-6}$	$13–15 \times 10^{-6}$
Chemical resistance	Very good	Good
Dielectric constant at 10^4 Hz	2.0–2.6	5.0–6.0
Maximum temperature with little or no decomposition (°C)	500–530	850–1000
Thermal conductivity (W/(m K))	2.5×10^{-5}	2.5×10^{-5}

a) Measured on sheets with stressed edges [8].
b) Measured on sheets with edges stressed [8].
c) Calculated for the effective strength of high aspect ratio flakes in plastics [8].

Figure 8.3 SEM microphotograph of commercial dry-ground phlogopite mica. Courtesy of Dr S. Kim, Polymer Processing Institute, Newark, NJ.

effective for PP homopolymers. Maleic anhydride or acrylic acid functionalized polyolefins, often in combination with aminosilanes, are lower cost, effective coupling agents [18] and are discussed in more details in Chapter 6. The effect of acid number of anhydride-based coupling agents on mica-filled PP has been reviewed in Ref. 19. Low MW chlorinated polyolefins are also a low-cost alternative to silanes and are discussed in Chapter 6 and in Ref. [7]. Other examples of less conventional, but highly effective surface treatments, also discussed in Chapter 6, include bismaleimides that were shown to increase the tensile strength of PP/40% mica injection-molded samples by 25% [20]. The response of micas to a variety of adhesion promoters including silanes, titanates, functionalized polyolefins, and others has been reviewed in Refs [17, 21]. More recent research work confirmed the beneficial effect of introducing maleated PP or maleated SEBS on the interfacial adhesion of mica PP composites [22, 23] and titanates on the properties of mica nylon 6 composites [24, 25].

8.4 Suppliers

More than 25 suppliers have been reported worldwide [26] producing different mica products. According to the US Geological Survey [1, 27] in January 2009, there were 10 domestic mica producers of dry- and wet-ground mica. North Carolina was the major producing state with 51% of domestic production, and the remainder was produced in Alabama, Georgia, South Carolina, and South Dakota. Current major US producers include Imerys that recently acquired the Suzorite Mica Products phlogopite business and the King's Mountain Mines muscovite business. Other US producers include Azco Mining, Inc., BASF Corp., and Pacer Corp. European producers include Acim Jouanin S.A. (France), Cogebi NV (Belgium), Kemira Pigments Oy (Finland), Microfine Minerals Ltd (UK), Sigmund Lindner GmbH (Germany), Ziegler & Co., GmbH (Germany), CMMP (France), and Kärtner Montanindustrie (Austria). Several Indian producers are mostly specializing in mica sheets rather than ground products.

Total mica production in the US was about 99 000 tons in 2008, which is about one-fourth of the worldwide mine production estimated to be about 390 000 tons [27]. Countries with the highest levels of mine production, and correspondingly large estimated reserves, are the United States and Russia, followed by Finland, the Republic of Korea, France, Canada, Brazil, and India, among others.

8.5 Cost/Availability

In 2008, average prices for the US mica ranged from $200 to $400/ton for dry ground and from $700 to $1000/ton for wet-ground mica [27, 28]. Micronized mica was quoted at $700–1000/ton [28]. Prices for surface-treated grades vary and are usually more than twofold higher than those for their untreated counterparts; however, the profit margin for such grades decreases as more compounders are using non-silane reactive compounds or functionalized polymers as adhesion promoters with untreated grades. Grades suitable for plastics compounding can be phlogopite or muscovite, dry ground or wet ground, untreated or surface treated for use with polar or nonpolar polymers. Typical sizes for grades suitable for plastic applications are −40 + 100 mesh, −60 + 120 mesh, −100 mesh, and −325 mesh (below 45 μm). Flake thickness depends on the degree of delamination. Average aspect ratios, although not always specified, may range from 10 : 1 up to 100 : 1 depending on the aspect ratio definition (usually equivalent diameter over thickness) and method of measurement (see e.g., Refs [2, 3, 29]). Grades for paints and pearlescent pigments are usually 85–95% −325 mesh wet ground or micronized.

8.6 Environmental/Toxicity Considerations

Finely divided mica is generally considered as a nuisance dust with an applicable OSHA PEL of 3 mg/m^3 and a respirable ACGIH TLV of 3 mg/m^3 (TWA, 8 h period). This is valid for mica containing less than 1% crystalline silica [30, 31]. Mica itself is not listed as a carcinogen by OSHA, NTP, or IARC. However, crystalline silica that may be present as a contaminant is classified as carcinogenic to humans with an ACGIH TLV (respirable) of 0.05 mg/m^3 and an OSHA PEL (respirable) of 0.1 mg/m^3. Crystalline silica levels in Suzorite phlogopite mica (CAS No: 12001-26-2) may vary in the range 0.1–1% [30]. Mica with less than 1% silica is considered an uncontrolled product according to Canadian WHMIS. Mica is unreactive and nonflammable and is not classified as hazardous waste. It meets FDA criteria covering its safe use in articles intended for food contact use and is listed in the US Code of Federal Regulations Title 21 part 175 and 177 under “indirect food additives” [30].

8.7 Applications

8.7.1 General

Processing methods for mica filled thermoplastics include extrusion, injection molding, thermoforming, structural foam molding, blow molding, and rotomolding. Mica may be incorporated in thermoplastics by twin-screw extruder melt compounding or, with certain polymer types, particularly in powder form, by direct injection molding of dry blends. In general, the free flowing mica flakes disperse in molten resins more easily that fillers containing aggregates that need to be broken or glass fibers that need to be separated into filaments. The relatively low Mohs hardness of 2.5–3.0 results in lower abrasion and wear to metallic equipment than with glass fibers. Polypropylene/mica compounds have found significant uses in the automotive industry for such products as under-the-hood components, trim, dashboard, and grille opening panels. Uses in other thermoplastics include packaging HDPE films, rotomolded HDPE and LLDPE tanks, polyolefin structural foam for automotive parts, speaker cabinets, and thermoplastic polyesters for automotive applications such as distributor systems.

In thermosets, traditional methods of processing were casting (as for epoxies) and compression/transfer molding (as for phenolic compounds) or compression molding/lamination as in the case of impregnated mica paper. In new applications, mica has been incorporated in certain hybrid glass fiber/unsaturated polyester composites for marine, automotive, and household applications produced by sprayup, layup, and combinations thereof. It is also finding use in automotive applications in RTM and reinforced RIM polyurethanes, where it is added in the polyol component replacing part of the milled glass. Applications of micas in thermoplastics and thermosets have been reviewed in Ref. [7].

8.7.2
Primary Function

The primary function of mica that has led to significant applications in automotive and other industries is modification and improvement of mechanical properties. General effects are significant increases in modulus, which in most cases is independent of degree of interfacial adhesion but is still strongly dependent on orientation; usually an increase in tensile and flexural strength normally, the effects being strongly dependent on the degree of adhesion and extent of orientation. Elongation usually decreases. Often, mica's outstanding performance in improving stiffness is unsatisfactorily offset by reduced impact strength, the latter depending on the type of test (notched versus unnotched, falling dart), type of polymer, flake orientation, size, loading level, and interfacial adhesion. Thermomechanical properties such as heat distortion temperature and creep resistance generally improve upon the addition of mica, the effects being strongly dependent on adhesion [7].

As discussed in Chapter 2, for most processing methods for thermoplastics, the flow-induced orientation of flakes is predominantly parallel to the flow direction with a region of misalignment in the core. In injection molding, such morphologies can be modified through the application of shear to the melt as it cools (e.g., SCORIM™), which has a marked effect on orientation and physical properties. In mica-reinforced thermosets, processes such as compression molding and lamination ensure mostly planar orientation (Figure 1.2) and isotropic properties as discussed in Chapter 2. The issue of reduced weld-line strength due to the resulting unfavorable flake orientation when two flow fronts meet may be mitigated by proper selection of the process conditions (injection speed, melt temperature), mica concentration, and mica size [32, 33].

Comparison of the effects of mica, glass fibers, talc, calcium carbonate on the properties of a polyolefin matrix (unmodified and chemically modified by maleation)

Table 8.3 Comparison of properties of untreated and azidosilane-treated 40% mica/PP injection-molding composites with commercial 30% glass fiber/PP and unfilled PP [7, 9, 17].

Property	Unfilled PP	40% Untreated mica	40% Azidosilane-treated mica	30% Glass fibers
Tensile strength (MPa)	32.9	28.7	43.4	44.1
Flexural strength (MPa)	31.5	45.5	66.5	70.7
Flexural modulus (GPa)	1.26	6.51	7.70	6.51
Izod impact strength notched (J/m)	24.0	32.0	34.7	74.8
Izod impact strength unnotched (J/m)	No break	203	235	502
HDT at 1.85 MPa (°C)	56	89	108	125

is shown in Tables 16.2 and 16.3 [34]. Even after normalizing for equal loadings, the effects on mechanical properties follow in general the pattern glass fibers > mica flakes > talc > calcium carbonate. Table 8.3 shows the effect of pretreatment with a sulfonylazidosilane (now discontinued) on the properties of injection-molded 40% filled mica compounds and a comparison with commercially available 30% glass fiber compounds. The possibility of approaching the mechanical properties of glass fiber compounds through selection of the appropriate surface treatment is clear.

8.7.3 Other Functions

The mechanical properties of ternary PET composites containing 40–45% total glass fibers and mica are shown in Figures 7.6–7.8 and compared with those of composites containing 30% glass. Flexural modulus is higher for the glass/mica composites than for the all-glass composites; however, the addition of mica decreases flexural strength and impact strength. In such composites, mica acts as multifunctional filler, controlling warpage and improving dimensional stability. Mica reduces thermal expansion and shrinkage, the effects becoming more pronounced with increasing filler concentration and increased interfacial interactions. Such effects have been discussed earlier in relation to injection-molded PP/mica containing PP-g-MA and more recently in relation to rotomolded MDPE/mica and LLDPE/mica containing PE-g-MA [35, 36]. Warpage resulting from residual stresses and differential shrinkage can be high in fiber-reinforced thermoplastics, but is reduced by the planar orientation of mica flakes. Combinations of glass (10–15%) and mica (20–30%) are often used as a best compromise between shrinkage and warpage and mechanical properties. For example, a commercial mica/glass fiber PET compound has notched Izod impact strength of 74.7 J/m and a relative warpage measured on an annealed disk of 5; the corresponding values for an all glass fiber compound are 128 J/m and 125 respectively [7].

Mica flakes embedded in a polymer and properly oriented in a plane can provide a tortuous path to vapors and liquids, similarly to the natural composites shown in Figure 1.1. Barrier properties can be imparted in blow-molded containers, packaging films, and corrosion-resistant coatings not only by mica but also by other impermeable lamellar fillers, including glass flakes, talc, and nanoclays. In blown LDPE film, the addition of 10% mica was found to reduce the oxygen permeability from 4.16 to 3.03 Barrer [37]. Assuming an impermeable, fully oriented lamellar filler, Eq. (8.1), [38] may be used to predict the composite permeability, P_c, perpendicularly to the filler plane as a function of the matrix permeability, P_m, filler volume fraction, V_f, matrix volume fraction, V_m, and filler aspect ratio α

$$P_c/P_m = V_m/t_f, \tag{8.1}$$

where the factor

$$t_f = 1 + (\alpha/2)V_f \tag{8.2}$$

represents tortuosity (actual path of solvent or gas over film thickness).

In addition to particle aspect ratio and concentration, permeability is affected by several other factors including polymer crystallinity, particle orientation, and adhesion. Modified equations to accommodate misalignment effects and the presence of the (assumed) impermeable to oxygen crystalline phase of HDPE in mica/HDPE films have been proposed [37].

Mica can also modify optical properties in semicrystalline polymers by acting as a nucleating agent and as a substrate for oxide deposition in pearlescent pigments produced by platelet core–shell technologies as discussed in Chapter 1. Figure 1.5 shows a cross section of an anatase/mica pigment particle produced by this technology.

An additional function of mica may be manifested in damping applications. Fillers may introduce a broadening in the damping ($\tan \delta = G''/G'$) transition region of a polymer, shifting it to longer times or higher temperature. The broadening of the transition region may be useful in vibration-damping and sound-deadening materials. Flake-filled elastomers and plastics often have high mechanical damping and for this reason vibration-damping materials may contain flakes that facilitate the conversion of the energy of vibration into heat rather than emitting it to air. Part of this damping may result from one layer of a flake such as mica or graphite sliding over another layer when the material is deformed [39]. For a specific polymer system, it has been shown that mica flakes are more effective than talc, $CaCO_3$ or TiO_2, with damping increasing with filler concentration [40].

References

1 Hedrick, J.B. (2009) Mica, *Minerals Yearbook 2006*, United States Geological Survey, accessed at www.minerals.usgs.gov.

2 Lusis, J., Woodhams, R.T., and Xanthos, M. (1973) The effect of flake aspect ratio on the flexural properties of mica reinforced plastics. *Polym. Eng. Sci.*, **13** (2), 139.

3 Kauffman, S.H., Leidner, J., Woodhams, R.T., and Xanthos, M. (1974) The preparation and classification of high aspect ratio mica flakes for use in polymer reinforcement. *Powder Technol.*, **9**, 125.

4 Woodhams, R.T. (1974) US Patent 3,799,799.

5 Woodhams, R.T. and Xanthos, M. (1978) US Patent 4,112,036.

6 Maine, F.W. and Shepherd, P.D. (1974) Mica reinforced plastics: a review. *Composites*, **5** (5), 193.

7 Hawley, G.C. (1987) Flakes, Chapter 4, in *Handbook of Reinforcement for Plastics* (eds H.S. Katz and J.V. Milewski), Van Nostrand Reinhold, New York.

8 Woodhams, R.T. and Xanthos, M. (1978) Chapter 20, in *Handbook of Fillers and Reinforcements for Plastics* (eds H.S. Katz and J.V. Milewski), Van Nostrand Reinhold Co., NY.

9 Xanthos, M., Hawley, C.G., and Antonacci, J. (1977) Parameters affecting the engineering properties of mica reinforced thermoplastics. Proceedings of the 35th SPE ANTEC, vol. 23, p. 352.

10 Hawley, G.C. (1988) Mica, *Pigment Handbook: Vol. 1: Properties and Economics* (ed. P.A. Lewis), John Wiley & Sons, Inc., New York, pp. 227–256.

11 Grimm, R.E. (1962) *Applied Clay Mineralogy*, McGraw-Hill, New York, p. 23.

12 Kingery, W.D. (1967) *Introduction to Ceramics*, John Wiley & Sons, Inc., New York, pp. 130–131.

13 ZEMEX Industrial Minerals, Boucherville, Que, Canada, Technical information on "Suzorite mica – products and applications" (ZEMEX is acquired by Imerys).

14 *Physical Properties of Mica*, Inderchand Rajgarchia & Sons (P) Ltd; accessed at http://www.icrmica.com.

15 Engelhard Corp., Hartwell, GA, USA, Technical information on mica (Engelhard is acquired by BASF).

16 Hohenberger, W. (2001) Chapter 17, in *Plastics Additives Handbook* (ed. H. Zweifel), Hanser Publishers, Munich.

17 Hawley, G.C. (1999) Proceedings of the Coupling Agents and Surface Modifiers '99, Intertech Corp., Atlanta, GA, September 22–24, 1999.

18 Xanthos, M. (1988) Interfacial agents for multiphase polymer systems: recent advances. *Polym. Eng. Sci.*, **28**, 1392.

19 Olsen, D.J. and Hyche, K. (2002) Proceedings of the 47th SPE ANTEC, vol. 35, p. 1375.

20 Xanthos, M. (1983) Processing conditions and coupling agent effects in polypropylene/wood flour composites. *Plast. Rubber Proc. Appl.*, **3** (3), 223.

21 Canova, L.A. (2000) Proceedings of the 58th SPE ANTEC, vol. 46, p. 2211.

22 Yazdani, H., Morshedian, J., and Khonakdar, H.A. (2006) Effects of silane coupling agent and maleic anhydride-grafted PP on the morphology and viscoelastic properties of PP-mica composites. *Polym. Compos.*, **27**, 491–496.

23 Yazdani, H., Morshedian, J., and Khonakdar, H.A. (2006) Effect of maleated pp and impact modifiers on the morphology and mechanical properties of PP/mica composites. *Polym. Compos.*, **27**, 614–620.

24 Bose, S. and Mahanwar, P.A. (2005) Effects of titanate coupling agent on the properties of mica reinforced Nylon-6 composites. *Polym. Eng. Sci.*, **18**, 1480–1486.

25 Bose, S., Raghu, H., and Mahanwar, P.A. (2006) Mica reinforced Nylon-6 – effect of coupling agent on mechanical, thermal and dielectric properties. *J. Appl. Polym. Sci.*, **100**, 4074–4081.

26 Global Industry Analysts Inc., Mica, Market research report, 06/2004; Summary accessed at http://www.the-infoshop.com.

27 Hedrick, J.B. (1990) Mica (natural), scrap and flake, mineral commodity summaries, US Geological survey, January 2009; accessed at www.minerals.usgs.gov.

28 Mineral PriceWatch, Issue 165, 2008, www.indmin.com.

29 Canova, L.A. (1990) Proceedings of the 45th Ann. Conf., Compos. Instit., SPI, February 12–15, 1990, vol. 17-F, pp.1–5.

30 Suzorite Mica Products, Inc., Boucherville, Que., Canada, MSDS Suzorite Mica, September 2003.

31 NIOSH, The Registry of Toxic Effects of Chemical Substances, "Silicate, mica", accessed at http://www.cdc.gov/niosh/rtecs.

32 Ferro, J.P. (2003) Proceedings of the Functional Fillers for Plastics 1995, Intertech Corp. Houston, TX, USA, December 4–6, 2003.

33 Dharia, A. and Rud, J.O. (2002) Proceedings of the 60th SPE ANTEC, p. 48.

34 Xanthos, M., Grenci, J., Patel, S.H., Patel, A., Jacob, C., Dey, S.K., and Dagli, S.S. (1995) Thermoplastic composites from maleic anhydride modified post-consumer plastics. *Polym. Compos.*, **16** (3), 204.

35 Robert, A. *et al.* (2000) Proceedings of the 58th SPE ANTEC, vol. 46, p. 1399.

36 Robert, A. and Crawford, R.J. (1999) Proc. 57th SPE ANTEC, 45, p. 1478.

37 Xanthos, M., Faridi, N., and Li, Y. (1998) Processing/structure relationships of mica-filled polyethylene films with low oxygen permeability. *Intern. Polym. Process.*, **13** (1), 58.

38 Nielsen, L.E. (1967) Models for the permeability of filled polymer systems. *J. Macromol. Sci.*, **A1**, 926.

39 Nielsen, L.E. and Landel, R.F. (1994) Chapter 8, in *Mechanical Properties of Polymers and Composites,* 2nd edn, Marcel Dekker Inc., New York.

40 Wypych, G. (2000) *Handbok of Filllers,* ChemTec Publishing, Toronto, Ont., Canada, pp. 112–115, 807–808.

9
Nanoclays and Their Emerging Markets

Karl Kamena

9.1
Introduction

9.1.1
Clays, Nanoclays, and Nanocomposites

"Nanoclay" is the term generally used when referring to a clay mineral with a phyllosilicate or sheet structure with dimensions of the order of 1 nm thick and surfaces of perhaps 50–150 nm. The mineral base can be natural or synthetic and is hydrophilic. The clay surfaces can be modified with specific chemistries to render them organophilic and therefore compatible with organic polymers. Surface areas of nanoclays are very large, about 750 m^2/g. When small quantities are added to a host polymer, the resulting product is called a nanocomposite.

Nanoclays and nanocomposites have generated a tremendous amount of research interest and curiosity, and it is estimated that hundreds of millions of dollars have been invested globally in order to investigate relevant technologies and products. Commercialization has not been rapid to date, but realistically the understanding and development of this concept to and through product stages is detailed and time-consuming. It is likely that nanoclays and nanocomposites will continue to satisfy niche applications and markets and will begin to grow in substantial volume increments as producers and users grow more comfortable with this developing technology.

9.1.2
Concept and Technology

The nanocomposite concept appears to have its origin in pioneering research conducted in Japan by Unitika, Ltd. in the 1970s [1] and separately by Toyota Central Research and Development Laboratories in the late 1980s [2]. The theory was that if nanoclays could be fully dispersed or exfoliated to high aspect ratio platelets into polymers at relatively low levels (2–5 wt%), a number of mechanical and barrier properties would be enhanced. The original work at both Unitika and Toyota CRDL

Functional Fillers for Plastics: Second, updated and enlarged edition. Edited by Marino Xanthos

ISBN: 978-3-527-32361-6

Table 9.1 Tensile properties and impact strength of nylon 6 nanoclay materials.

Specimen type (montmorillonite, wt%)	Tensile strength (MPa)	Tensile modulus (GPa)	Charpy impact strength (kJ/m^2)
NCH-5 (4.2)	107	2.1	2.8
NCC-5 (5.0)	61	1.0	2.2
Nylon 6 (0)	69	1.1	2.3

was based on an *in situ* process for the preparation on nylon 6 nanocomposites. According to this method, a nanoclay is introduced into the caprolactam monomer stage of the process, and the caprolactam is intercalated into the clay galleries. Under appropriate reactor conditions, the caprolactam polymerizes and the platelets are further expanded and become exfoliated to become an integral part of the bulk polymer [3, 4]. Toyota reported that NCH (nanocomposite nylon 6/clay hybrid) materials provided significant improvements in mechanical, thermal, and gas barrier properties at 2–5 wt% loadings of montmorillonite. Toyota CRDL has also prepared nylon 6/clay nanocomposites (NCC) by melt-compounding techniques. Typical mechanical properties are listed in Table 9.1 [2].

Other methods to prepare nanocomposites include a solvent-assisted process, whereby a cosolvent is employed to help carry the monomer into the galleries and is subsequently removed from the polymer system, and direct polymer melt intercalation methods, which involve the direct addition of nanoclays to a polymer melt under shear conditions at elevated temperatures, allowing their direct exfoliation into the polymer [5].

Following on the Unitika and Toyota CRDL work, there has been a large amount of investigation in many industrial and academic environments, and much of the effort has been targeted at achieving exfoliation in a technologically and economically feasible manner [6]. As an example, during the period from November 2000 to September 2002, there were 11 international conferences devoted specifically to developments in nanocomposites, with over 350 papers presented by academic, government, and industry researchers [7]. Numbers of conferences, published papers, and patents grew exponentially thereafter.

9.2 Production Methods

9.2.1 Raw and Intermediate Materials

Many nanoclays are based on the smectite clay, montmorillonite, a hydrated sodium calcium aluminum magnesium silicate hydroxide, $(Na,Ca)(Al, Mg)_6(Si_4O_{10})_3 (OH)_6 \cdot nH_2O$. Montmorillonite is found throughout the world in small quantities in its natural geological state. In large deposits, where the mineral is found in greater than 50% concentrations admixed with a variety of other minerals, it is known as bentonite.

Commercially attractive deposits of bentonite are located in many geographical areas ranging from the United States (particularly in Wyoming) to Western Europe, the Mideast, China, and so on.

Natural montmorillonite clays are most commonly formed by the *in situ* alteration of volcanic ash resulting from volcanic eruptions in the Pacific and Western United States during the Cretaceous period (85–125 million years ago). Opinions differ concerning the process and time of alteration of the ash to clay. Certainly the change began with contact with water. The instability of the ash made for ease of dissolution and reaction with the available marine chemistry. Probably the most important single factor in the formation of the clay was the availability of sufficient magnesium in the marine sediment environment. Ensuing chemical and structural changes took place throughout the deposits' entire geologic history. It is estimated that the resulting deposits in Wyoming, for example, alone contain over 1 billion tons of available clay.

Geological maps are available from areas around the globe where clay deposits have been detected. Such maps and associated area photographs and topographical maps assist in guiding the decision for exploration drilling. Such activity is termed "borehole" drilling and its first part is to drill borehole samples on centers covering areas of 50–300 ft (15–90 m). The use of global positioning satellite (GPS) techniques ensures accurate surveying to within a few centimeters.

Generally speaking, in characterizing a deposit the prospector/producer will look for the following information:

- Purity
- Crystallography
- Chemistry
- Particle size
- Morphology
- Charge
- Dispersion characteristics
- Response to processing parameters

If the decision is to mine, the results from the borehole tests guide the miners and heavy equipment operators in removing the clay from the earth and depositing it in stockpiles. Depending on the ash fall volume, the depth of a deposit can be from a few centimeters to several meters and the length can be up to hundreds of meters. The borehole data give the profile of the deposit, and the mining techniques follow accordingly.

After the overburden is removed, layers of clay are formed into disks and allowed to sundry before removal. The clay is removed from the pit in layers, and the ensuing stockpile is constructed layer by layer. This construction is done in an exacting manner to maximize crude clay homogenization.

9.2.2
Purification and Surface Treatment

The diagram in Figure 9.1 summarizes the process for separating the montmorillonite clay from other nonclay minerals, such as quartz, gravel, limestone followed by

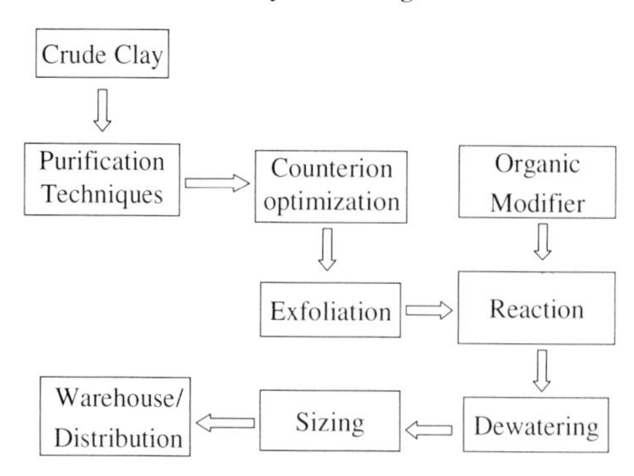

Figure 9.1 Flow chart of clay processing.

surface treatment. Copious quantities of water are used to ensure that the montmorillonite clay is in an exfoliated condition so that larger particles can be removed through various separation techniques. Strategic sampling points exist throughout the process. Statistical process control tools assist the production operators in maintaining the process within natural variation limits. Important control factors in the process are as follows:

- Solids/water ratio
- Counter ion optimization
- Purity
- Preorganic reaction particle size
- Organic/inorganic ratio
- Postorganic reaction dispersive characteristics
- Postmilling solids/moisture ratio
- Postmilling dispersive characteristics
- Postmilling particle size
- Packaging aesthetics

The organic modifier used in the surface treatment is generally a quaternary ammonium compound, although other onium ions, that is, phosphonium, can be considered. The reaction that occurs is an ion-exchange reaction, wherein the positively charged quaternary salt replaces the sodium cations on the clay surface. During the reaction, as the clay is being converted to an organoclay, it changes from hydrophilic in nature to oleophilic.

9.2.3 Synthetic Clays

Synthetic clays may be prepared using a variety of chemical sources providing the necessary elements, namely silicon, oxygen, aluminum, and magnesium among others. Synthetic clays are the subject of current research, but there is little public knowledge concerning the various technologies being investigated. Natural clays would appear to have an inherent raw material cost advantage, but the ability to control purity, charge density, and particle size is an appealing objective.

One synthetic clay that has been in the market for several years is a synthetic mica prepared from a natural raw material, talc, which is treated in a high temperature electric furnace with alkali silicofluoride. The chemical structure is $NaMg_{2.5}SiO_4O_{10}(F_{\alpha}OH_{1-\alpha})_2$, $0.8 \leq \alpha \leq 1.0$. The producer of this material claims lower impurity levels and higher aspect ratios than with natural montmorillonite species.

9.3 Structure and Properties

Silicon and oxygen are common to all clay minerals, and the combination with other elements, such as aluminum, magnesium, iron, sodium, calcium, and potassium, and the numerous ways in which the elements can be linked together make for a large number of configurations. An important distinction in clay mineral properties is the capacity of certain clays to change volume by absorbing water molecules from other polar ions into their structure. This is called the swelling property. Clays are divided into swelling and nonswelling-type materials, and swelling types are called smectites. Of the many smectite varieties, montmorillonite appears to be the most suitable as the basis for a nanoclay.

Silica is the dominant constituent of montmorillonite clays, with alumina being essential, as well. Clays have a sheet structure consisting of two types of layers, the silica tetrahedral and alumina octahedral layers. The silica tetrahedral layer consists of SiO_4 groups linked together to form a hexagonal network of repeating units of composition Si_4O_{10}. The alumina layer consists of two sheets of close-packed oxygens or hydroxyls, between which octahedrally coordinated aluminum atoms are embedded in such a position that they are equidistant from six oxygens or hydroxyls. The two tetrahedral layers sandwich the octahedral, sharing their apex oxygens with the latter. These three layers form one clay sheet. Figure 9.2 shows the structure of a dioctahedral smectite consisting of two-dimensional arrays of silicon-oxygen tetrahedra and two-dimensional arrays of aluminum- or magnesium-oxygen-hydroxyl octahedra.

If the octahedral positions were occupied by alumina, the structure would not correspond to montmorillonite but to that of the inert mineral pyrophyllite. So, extremely important to the structure of clays is the phenomenon of isomorphous substitution. Replacement of trivalent aluminum by divalent magnesium or iron(II) results in a negative crystal charge. The excess negative charge is compensated on the clays' surface by cations that are too large to be accommodated in the interior of the crystal. Further, in low pH environments, the edges of the clay crystal are positive, and compensated by anions. The structure of montmorillonite is compared with the structures of kaolinite and muscovite mica in Figure 8.2.

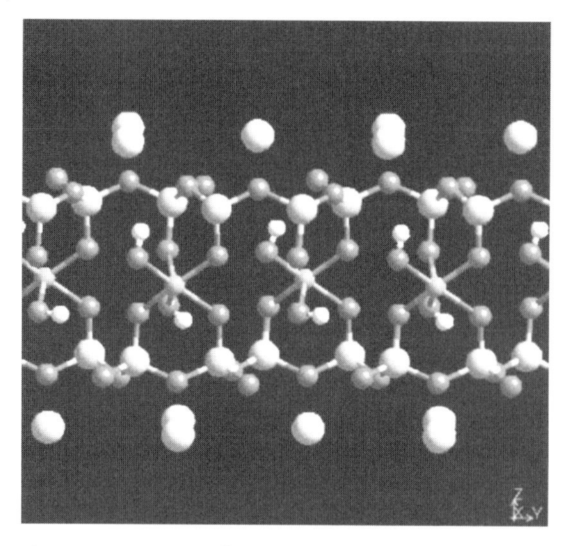

Figure 9.2 Structure of smectite clay.

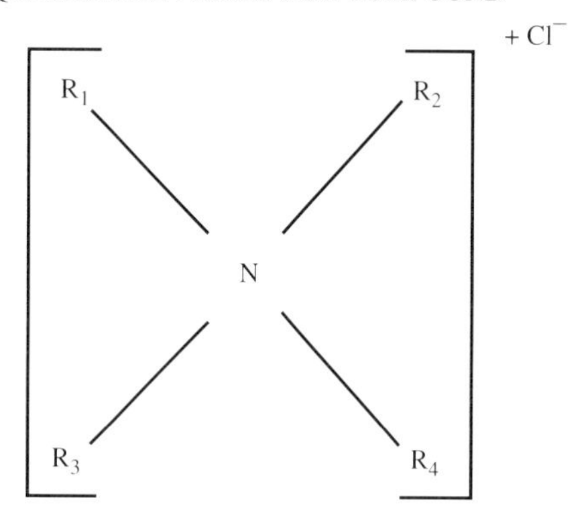

Figure 9.3 Structure of quaternary ammonium compound.

In order to convert the montmorillonite clay into a nanoclay compatible with organic polymers, an ion exchange process is performed to treat the clay surfaces. Generally, an organic cation, such as from a quaternary ammonium chloride, is used to change the hydrophilic/hydrophobic characteristics of the clay (Figure 9.3).

Typical characteristics of montmorillonite clays are as follows:

- Shape: platelet.
- Size: 1nm thick, 75–150 nm across.

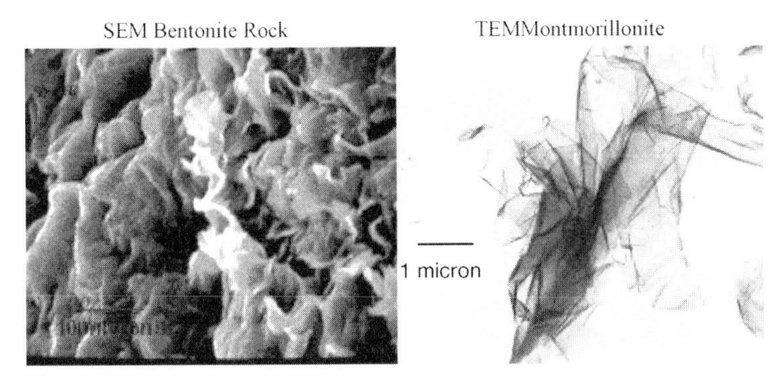

Figure 9.4 Morphologies of bentonite rock and exfoliated montmorillonite.

- Charge: unit cell 0.5–0.75 charge 92 meq/100 g clay.
- Surface area: >750 m^2/g.
- Specific gravity 2.5 (lower for alkyl quaternary ammonium bentonites).
- Modulus: ~170 GPa.
- Particle: robust under shear, not abrasive (Mohs hardness 1–2).

Figure 9.4 shows nanoplatelets that can be produced from their precursor, a bentonite rock.

9.4 Suppliers

The following are among the major nanoclay suppliers worldwide:

1) Southern Clay Products, Inc., Gonzales, TX (USA)
2) Nanocor (Division of AMCOL Int'l.), Arlington Heights, IL (USA)
3) Laviosa Chimica Mineraria S.p.A., Livorno, Italy
4) Kunimine Industries, Tokyo, Japan
5) Elementis Specialties, Inc., Heightstown, NJ (USA)
6) CO-OP Chemical, Ltd., Tokyo, Japan

Typical products are untreated and surface-treated grades. Surface-treated grades differ in their degree of hydrophobicity and type of cation introduced through ion exchange.

9.5 Cost/Availability

Nanoclays are relatively new commercial products and, as such, the cost/price structure is immature. Although a variety of products appear to be available from the suppliers listed, quantities being sold are small and reflect specialty and developmental pricing policies. As the market grows and matures, it is expected that prices for materials will be in the range US $2.50–4 per pound (US $5.5–9 per kg).

9.6 Environmental/Toxicity Considerations

The health and environmental issues for nanoclays specifically are minimal and manageable. Sister organoclay products have been used for many years in a host of industrial and consumer products. The perception that nanoclays are somehow different because of the prefix "nano" may be the problem. Nanoclays only become "nano" when they are placed in a host–polymer matrix, whereupon they cannot be separated or distinguished form the bulk polymer and other constituents. Crystalline silica is a naturally occurring component that may be present in commercial alkyl quaternary ammonium bentonite (CAS No. 68953-58-2) at concentrations <0.5% [8]. Crystalline silica dust (see also Chapter 19) when inhaled is a health hazard in humans and is regulated to very low permissible exposure limits.

9.7 Applications

Nanoclays, in addition to their primary function as high aspect ratio reinforcement, also have important additional functions such as thermal and barrier properties and synergistic flame retardancy. Some of the factors responsible for good performance in nanocomposites are as follows:

- Intercalation (surfactant and polymer)
- Interfacial adhesion or wetting
- Exfoliation (dispersion and delamination)

Under appropriate conditions, the gallery spaces can be filled with monomer, oligomer, or polymer. This increases the distance between platelets, swelling the clay. Clay platelets swollen with polymer are said to be intercalated. If the clay swells so much that it is no longer organized into stacks, it is said to be exfoliated as shown in Figure 9.5.

Figure 9.6 is a schematic of the various dispersion mechanisms operative in producing nanoplatelets of very high aspect ratio. The nominal size of a dry nanoclay particle is about 8–20 μm. Comprising the particle are approximately 1–3 million clay platelets, consisting of bundles of platelets called tactoids. Through a combination of chemistry and processing/shear techniques, the particle is separated into tactoids and the platelets are peeled from the tactoid to become fully dispersed or exfoliated.

The primary appeal of a clay/polymer nanocomposite is that much smaller quantities of the nanoclay can be used to enhance polymer performance without detracting from other key characteristics. A comparison of properties achieved with talc and with nanoclay in TPO (thermoplastic elastomer) is shown in Figure 9.7.

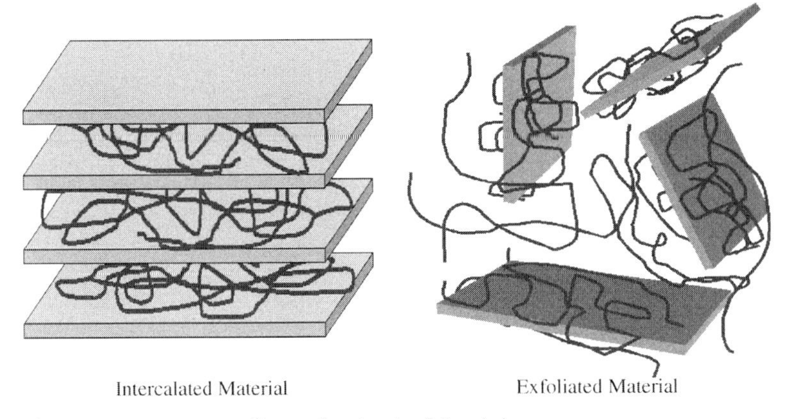

Figure 9.5 Comparison of intercalated and exfoliated clays.

Indeed, one of the major challenges has been to develop fully exfoliated products to obtain the maximum benefit of nanoclays. During the dispersion process, particles are sheared into tactoids and platelets peel from the tactoids to become fully dispersed or exfoliated in the host matrix (see Figure 9.8). During compounding, important process parameters are clay feed position, type of twin-screw extruder, and screw design/speed. There are numerous publications discussing the effects of process conditions on degree of exfoliation [9, 10].

A variety of potential host systems including polyamides, polyolefins, PVC, TPU, PLA, EVA, ionomers, rubber, recycled streams, and polymer blends have been evaluated for nanoclays. Although exfoliation has been achieved in many polymers, it has not led to significantly improved mechanical properties other than modulus. The high degree of interest in the nanocomposite concept has not yet resulted in a

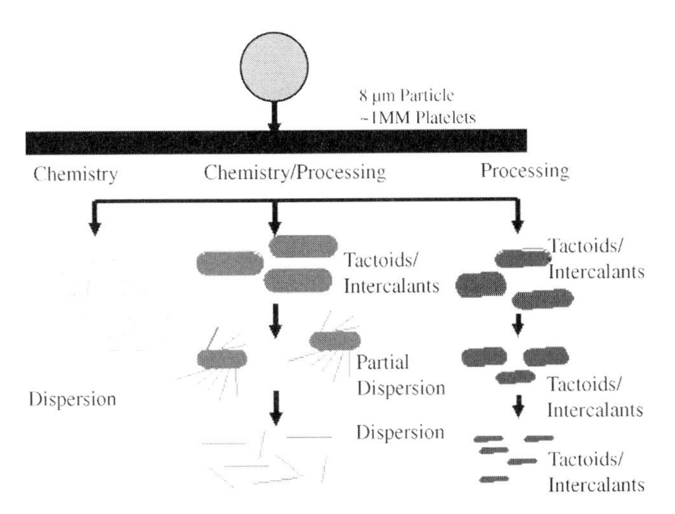

Figure 9.6 Schematic of the different clay dispersion mechanisms.

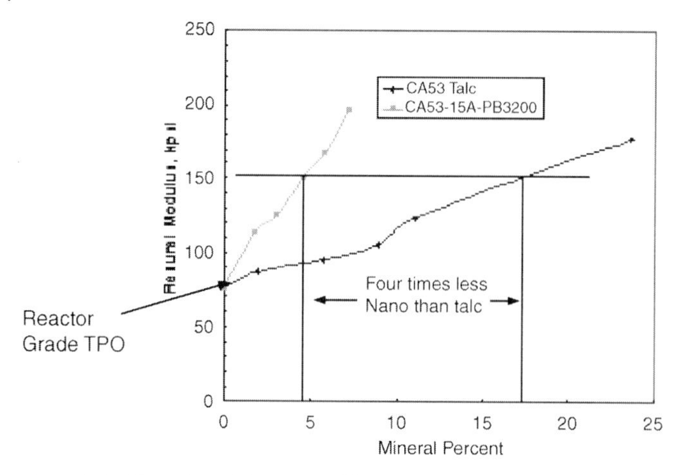

Figure 9.7 Flexural modulus versus filler concentration in a thermoplastic elastomer, TPO; comparison of talc (CA53) with nanoclay (CA53-15A) in the presence of maleated PP (PB3200).

plethora of commercial products. However, products are emerging with increasing frequency as producers, processors, and users gain more experience with the products and envisage potential commercial applications. In the case of the nanocomposites used in the 2004 Chevrolet Impala side moldings, General Motors reported a weight savings of 7% as well as a better overall surface quality because the filler is so fine it does not disrupt the surface of the part. The fine fillers are also said to improve mar resistance. Usually, when a stone hits a rocker panel, the white color that appears is due to the eye being able to see the filler. In the case of the nanocomposites, this effect should be lessened due to the inherently smaller filler size. As with any new technology and product, challenges had to be overcome. For example, the tooling required some design changes, shrinkage rates were different, and color recipes needed to be changed. Manufacturing the nanocomposite was also a

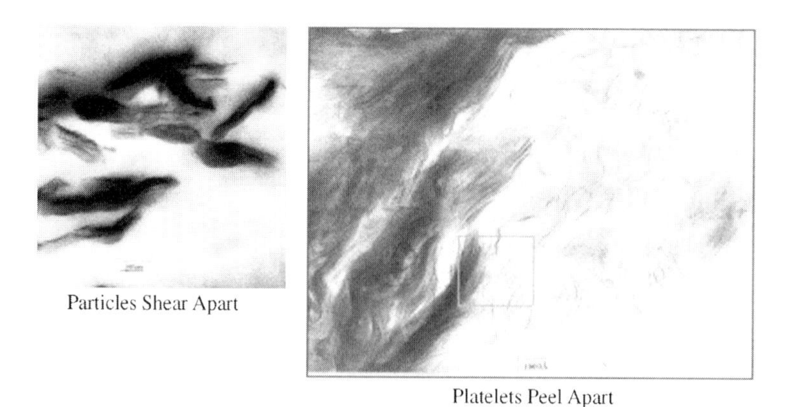

Figure 9.8 Monitoring the dispersion of nanoclays by transmission electron microscopy.

challenge: introducing a nanoclay at small levels into a TPO requires good distribution and dispersion of the dry product into the polymer melt so that the nanoclay can be substantially exfoliated.

The following are additional examples of the multifunctional character of nanoclays:

Thermal: A nanoclay - nylon 6 has been commercialized by Unitika for an engine cover that required substantially higher heat distortion temperatures than achieved with nylon 6. At a 4 wt% loading of synthetic mica, the DTUL (at 1.8 MPa) increased from 70 °C for neat nylon 6 to 152 °C. Also, flexural strength increased from 108 to 158 MPa and flexural modulus from 3.0 to 4.5 GPa [11].

Barrier: Several companies offer commercial nanonylon products with improved barrier properties and maintaining clarity. Targeted packaging applications include multilayer PET bottles for high oxygen barrier demands for beer bottles and flexible multilayer films for meats and cheeses. The permeability of nylon barrier resins is generally reduced by a factor of 2 to 4 with less than 5% nanoclay addition [12]. Nylon nanocomposites are also being considered for automotive applications such as fuel tanks and lines [13].

Synergistic flame retardancy: Nanocomposites have been demonstrated to reduce flammability, particularly through lowering peak heat release in cone calorimeter experiments. In combination with conventional flame retardants such as magnesium hydroxide or aluminum trihydrate, several polyolefin-based wire and cable products have been developed that incorporate 5% nanoclay to reduce the use of conventional fire retardant agents and to improve physical properties [14, 15].

Low-density SMC: In 2007, Yamaha announced the development of NanoXcel, a low-density sheet molding compound to be used in their WaveRunner line of personal watercraft. Nanoclays enable the reduction of fiberglass and other bulking (clay) components' content, thus lowering the SMC specific gravity by about 25% (from about 1.9 to 1.5) without sacrificing mechanical properties [16].

References

1 Fujiwara, S. and Sakamoto, T. (1976) Japanese Patent No. JPA51-109998.

2 Kato, M. and Usuki, A. (2000) Chapter 5, in *Polymer-Clay Nanocomposites*, John Wiley & Sons, Inc., New York.

3 Usuki, A. *et al.* (1993) *J. Mater. Res.*, 8, 1179–1184.

4 Usuki, A. *et al.* (1993) *J. Mater. Res.*, 8, 1174–1178.

5 Vaia, R. (2000) Chapter 12, in *Polymer-Clay Nanocomposites* (eds T.J. Pinnavaia and J.W. Beal), John Wiley & Sons, Inc., New York.

6 Kamena, K. (2001) Nanocomposites: the path to commercialisation. Conference Proceedings, Principia Partners, Baltimore, June 4–5, 2001.

7 Kamena, K. (2002) Functional Fillers for Plastics 2002, Conference Intertech Toronto Corp., Canada, September 18–20 2002.

8 Southern Clay Products, Inc . (2003) Materials Safety Data Sheet, "Cloisite 20A", Revised September 2003.

9 Pinnavaia, T.J. and Beall, G.W. (eds) (2000) *Polymer-Clay Nanocomposites, multiple*

chapters, John Wiley & Sons, Inc., New York.

10 Utracki, L.A. and Cole, K.C. (2004) Proceedings of the 2nd International Symposium on "Polymer Nanocomposites 2003", National Research Council Canada, also Polym. Eng. Sci., **44**, 6.

11 Yasue, K., Katahira, S., Yoshikawa, M., and Fujimoto, K. (2000) Chapter 6, in *Polymer-Clay Nanocomposites* (eds T.J. Pinnavaia and J.W. Beal), John Wiley & Sons, Inc., New York.

12 Defendini, B. (2002) High barrier polyamide-6 nanocomposite and oxygen scavenger. Proceedings of the Nanocomposites 2002, Conference, European Plastics News, Amsterdam, The Netherlands, January 28–29, 2002.

13 Nakamura, K. (2004) Examining progress towards developing polyamide nanocomposites for automotive applications. Proceedings of the Nanocomposites 2004, Conference, European Plastics News, Brussels, Belgium, March 17–18, 2004.

14 Gilman, J. and Kashiwagi, T. (2000) Chapter 10, in *Polymer-Clay Nanocomposites* (eds T.J. Pinnavaia and J.W. Beal), John Wiley & Sons, Inc., New York.

15 Beyer, G. (2001) *Polym. News*, **26**, 370–378.

16 Plueddeman, C. (2007) Lighter, Stronger, Faster, Boating Magazine, November 2007, pp. 84–86.

10
Carbon Nanotubes/Nanofibers and Carbon Fibers

Zafar Iqbal and Amit Goyal

10.1
Introduction

Carbon nanotube (CNT)- and nanofiber-reinforced polymer nanocomposites and micron-sized carbon fiber-based polymer composites look set to have significant impact on emerging advanced products ranging from aerospace, automotive, and PEM (proton exchange membrane) fuel cell parts to surgical implants and to components for nanoelectronics. The area of micron-scale carbon fiber filled composites, unlike that of the emerging field of carbon nanotube and nanofiber-based nanocomposites, is relatively mature. Although both areas are discussed in this chapter, we focus more on the rapid advances being made in the field of carbon nanotube-based nanocomposites, discuss some new developments in conventional micron-scale and submicron scale carbon fiber composites, and point out possible synergies.

10.2
Materials

10.2.1
Types of Carbon Nanotubes/Nanofibers and Their Synthesis

There has been intense interest in carbon nanotubes since their discovery by Iijima in 1991 [1], in large part because they possess unique structural and electronic properties. Single-wall carbon nanotubes (SWNTs) are the fundamental form of carbon nanotubes, with unique electronic properties that emerge due to their one dimensionality; a SWNT is a single hexagonal layer of carbon atoms (a graphene sheet) that has been rolled up to form a seamless cylinder. Three types of SWNTs with differing chirality can be formed, as depicted in Figure 10.1. The one-dimensional unit cell shown has a circumference given by the chiral vector $C = na + mb$, where n and m are integers equivalent to the roll-up vectors and a and b are unit vectors of the

Functional Fillers for Plastics: Second, updated and enlarged edition. Edited by Marino Xanthos

ISBN: 978-3-527-32361-6

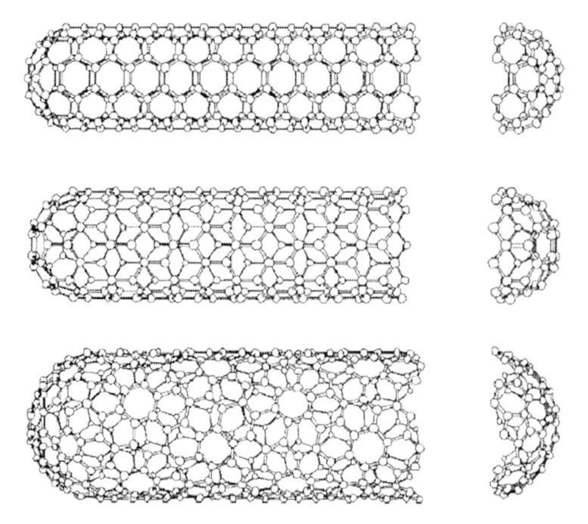

Figure 10.1 Armchair ($n,m = 5,5$) (top); zigzag (9,0) (middle); and chiral (10,5) single-wall nanotubes. All armchair tubes are metallic, whereas only 1/3 of the chiral tubes have metallic character. (n,m), the roll-up vectors are proportional to the tube diameter. The dangling bonds at the tube ends are saturated by hemispherical fullerene caps.

hexagonal lattice. A multiple-wall carbon nanotube (MWNT) is a stack of graphene sheets rolled up into concentric cylinders. This stacking results in a loss of some of the unique one-dimensional properties present in the single (SWNT) and double (DWNT) tube structures. The walls of each MWNT layer are parallel to the central axis. A stacked cone or herringbone arrangement is also formed by catalytic chemical vapor deposition (CVD), which can grow with a hollow, tubular center. These structures have relatively large diameters (typically $\geq$50 nm) compared with the near molecular-scale dimensions of SWNTs and the low nanoscale dimensions of most MWNTs, and are therefore referred to as carbon nanofibers (CNFs).

MWNTs were first synthesized using a noncatalytic carbon arc-discharge method by Iijima [1]. SWNTs were initially synthesized in 1–2% yields in soot generated in an arc struck between graphite electrodes containing a few percent Fe, Co, or Ni by Bethune *et al.* [2] and Iijima and Ichihashi [3]. Smalley and coworkers [4] then scaled up SWNT synthesis using a dual laser ablation technique with transition metal particles incorporated in the graphite target. This method could produce SWNTs with yields of up to 70%. The carbon nanotubes are formed catalytically in the extremely high temperature of the ablation plume with a narrow distribution of diameters around 1.3 nm and, due to van der Waals forces, generally assemble into bundles or ropes of parallel SWNTs. Soon afterwards Journet *et al.* [5] showed that about 50% yields of SWNT bundles, similar in size to those produced by laser ablation, can be obtained using the arc-discharge method when catalyst particles of rare earths such as Y are incorporated together with transition metals in the graphite rods. CVD methods involving the decomposition of hydrocarbon precursor gases, typically ethylene and acetylene, in the presence of transition metal (iron, cobalt, or nickel) catalysts on a support material such as alumina or silica, have been used to make CNFs [6, 7] and

MWNTs [8] at temperatures in the 550–1000 °C range. MWNTs grown by the CVD technique, however, have high defect densities in their structures. Arc-grown MWNTs, on the other hand, are largely defect-free because growth occurs at plasma-generated temperatures in excess of 2000 °C. Recently, plasma-enhanced CVD (PE-CVD) growth of MWNTs has emerged as a technique for the growth of vertically aligned MWNTs and CNFs [9, 10].

Since the late 1990s, largely defect-free SWNTs have been grown at near or above 90% purity by CVD techniques involving the catalytic decomposition of methane at temperatures near 1000 °C [11, 12], by the catalytic disproportionation of carbon monoxide (CO) under high pressures (the so-called high-pressure carbon monoxide or HIPCO process) and temperatures above 1000 °C [13], and at 1 atm and temperatures below 1000 °C [14–17], using catalyst supported on silica and MgO. Cheng *et al.* [18] produced SWNTs at 1200 °C and undetermined purity levels by heating flowing benzene together with ferrocene and thiophene precursors to form floating catalytic particles, whereas Maruyama *et al.* [19] generated SWNTs at temperatures down to 550 °C, using ethanol under low-pressure conditions. Because of the etching effect of OH radicals produced on decomposition of alcohol, non-SWNT phases, such as amorphous and nontubular nanocarbons and MWNTs, were not formed. Maruyama's group has also been able to grow vertically aligned SWNTs on catalyst-coated quartz substrates using ethanol as precursor [20] in a thermal CVD process. Low-pressure conditions using either ethylene or propylene as the carbon source were employed by Sharma and Iqbal [21] to grow and observe in real time both SWNTs and MWNTs *in situ* in an environmental transmission electron microscope.

As-synthesized SWNTs are typically bundled and consist of a range of tube diameters and chiralities. A method to grow single-diameter, individual SWNTs is to form them inside zeolites with selected pore sizes. Catalyst-free SWNTs with a diameter of 0.42 nm corresponding to that of the smallest fullerene, C_{20}, were grown by Wang *et al.* [22] by this method. Another catalyst-free method [23] that provides thin bundles of SWNTs, with a narrow diameter distribution in the 1.2–1.6 nm range, involves horizontal templated growth of the tubes on the Si face of hexagonal silicon carbide (6H-SiC), although growth occurs only at temperatures above 1500 °C. Recently, PE-CVD with methane as carbon source has been used for the first time to grow SWNTs in the 550–900 °C temperature range. In the first study reported, SWNTs were grown bridging the pores of a zeolite positioned on a Ni plate [24]. In the second study, largely semiconducting SWNTs were grown on ferritin (a precursor for nanoscale Fe catalyst particles) on silica [25], and in the third study, SWNTs were formed on sol–gel produced bimetallic Co-Mo catalysts on MgO [26]. It remains to be seen as to whether more controlled alignment of SWNTs can be obtained by the PE-CVD technique.

10.2.2 Types of Carbon Fibers and Their Synthesis

The presently used micron-sized carbon fibers contain at least 90% carbon and are produced by heat treatment or controlled pyrolysis of different precursor fibers. The

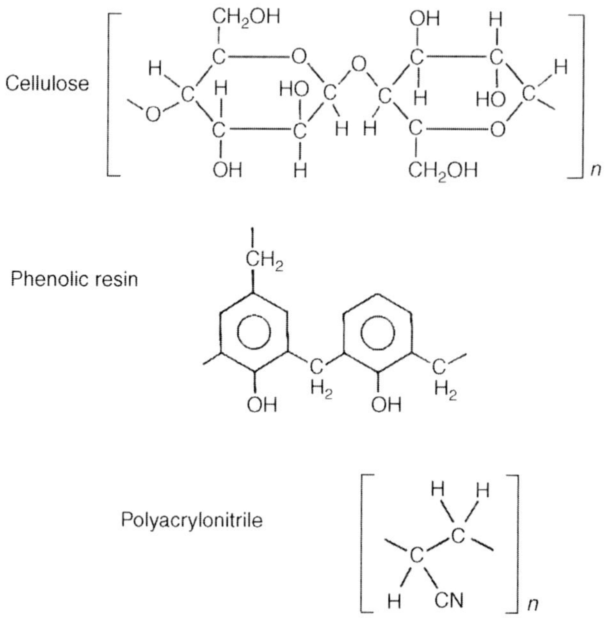

Cellulose

Phenolic resin

Polyacrylonitrile

Figure 10.2 Molecular structures of polymeric precursors for micron-size carbon fibers.

most prevalent precursors are polyacrylonitrile (PAN), cellulose fibers (such as viscose rayon and cotton), petroleum or coal tar pitch, and certain phenolic fibers. Pitch is a tar-like mixture of hundreds of branched organic compounds with differing molecular weights formed by heating petroleum or coal. The so-called mesophase of pitch is in a liquid-crystalline state. The structures of the PAN, cellulose, and phenolic resin are depicted in Figure 10.2.

Micron-sized carbon fibers can be classified in terms of the precursor fiber materials as PAN-based, mesophase or isotropic pitch-based, rayon-based, and phenolic-based. The synthesis process involves a heat treatment of the precursor fibers to remove oxygen, nitrogen, and hydrogen to form the carbon fibers. It is well established in the literature that the mechanical properties of the carbon fibers are improved by the increasing crystallinity and orientation and by reducing defects in the fiber. The best way to achieve this is to start with a highly oriented precursor and then maintain the high orientation during the process of stabilization and carbonization through tension.

10.2.2.1 **PAN-Based Carbon Fibers**

There are three successive stages in the conversion of a PAN precursor into high-performance carbon fibers:

(a) **Oxidative stabilization:** The PAN precursor is first stretched and simultaneously oxidized in the 200–300 °C temperature range. This treatment converts thermoplastic PAN to a nonplastic cyclic or ladder compound.

(b) **Carbonization:** After oxidation, the fibers are carbonized at about 1000 °C without tension in an inert atmosphere (normally nitrogen) for a few hours. During this process, the noncarbon elements are removed as volatiles to give carbon fibers with a yield of about 50% of the mass of the original PAN.
(c) **Graphitization.** Depending on the type of fiber required, the fibers are treated at temperatures in the range 1500–3000 °C; this step improves the ordering and orientation of the crystallites in the direction of the fiber axis.

10.2.2.2 **Carbon Fibers from Pitch:**
Carbon fiber fabrication from pitch generally involves the following four steps:

(a) **Pitch preparation:** Essentially it is an adjustment in the molecular weight, viscosity, and crystallite orientation for spinning and further heating.
(b) **Spinning and drawing:** In this step, the pitch is converted into filaments, with some alignment in the crystallites to achieve directional characteristics.
(c) **Stabilization:** In this step, cross-linking is introduced to maintain the filament shape during pyrolysis. The stabilization temperature is typically between 250 and 400 °C.
(d) **Carbonization:** The carbonization temperature is typically in the range 1000–1500 °C.

Carbon fibers made from the spinning of molten pitches are of interest because of the carbon yield approaching 99% and the relative low cost of the starting materials. The formation of melt-blown pitch webs is followed by stabilization in air and carbonization in nitrogen. Processes have been developed with isotropic pitches and with anisotropic mesophase pitches. The mesophase pitch-based and melt-blown discontinuous carbon fibers have a structure comprised of a large number of small domains, each domain having an average equivalent diameter from 0.03 to 1 mm, and a nearly unidirectional orientation of folded carbon layers assembled to form a mosaic structure on the cross section of the carbon fibers. The folded carbon layers of each domain are oriented at an angle to the direction of the folded carbon layers of the neighboring domains on the boundary.

Carbon fibers from isotropic pitch Isotropic pitch or a pitch-like material, such as molten polyvinyl chloride, is melt spun at high strain rates to align the molecules parallel to the fiber axis. The thermoplastic fiber is then rapidly cooled and carefully oxidized at a low temperature (<100 °C). The oxidation process is rather slow, so as to ensure stabilization of the fiber by cross-linking to make it infusible. However, upon carbonization, relaxation of the molecules takes place, producing fibers with no significant preferred orientation. This process is not industrially attractive due to the lengthy oxidation step, and because only low-quality carbon fibers with no graphitization are produced. These fibers are used as fillers in various plastics to form thermal insulation materials.

Carbon fibers from anisotropic mesophase pitch High molecular weight aromatic pitches that are mainly anisotropic in nature are referred to as mesophase pitches.

The pitch precursor is thermally treated above 350 °C to convert it to mesophase pitch, which contains both isotropic and anisotropic phases. Due to shear stresses occurring during spinning, the mesophase molecules orient parallel to the fiber axis. After spinning, the isotropic part of the pitch is made infusible by cross-linking in air at a temperature below its softening point. The fiber is then carbonized at temperatures up to 1000 °C. The main advantage of this process is that tension is not required during stabilization or graphitization, unlike in the case of rayon or PAN precursors.

10.2.2.3 Carbon Fibers from Rayon

The conversion of rayon fibers into carbon fibers is a three-stage process:

1) **Stabilization:** Stabilization is basically an oxidative process that involves different steps. In the first step, from 25 to 150 °C, there is physical desorption of water. The next step is dehydration of the cellulose unit between 150 and 240 °C. Finally, thermal cleavage of the cyclosidic linkage and scission of ether bonds and some C–C bonds occurs via free radical reactions (240–400 °C) followed by aromatization.
2) **Carbonization:** Heat treatment between 400 and 700 °C converts the carbonaceous residue into graphite-like layers.
3) **Graphitization:** Graphitization is carried out under strain at 700–2700 °C to obtain high modulus fibers through a longitudinal orientation of the planes.

10.2.2.4 Carbon Fibers from Phenolic Resins

Micron-sized carbon fibers are synthesized from phenolic resin fibers such as Kynol [27]. The carbon fibers prepared are typically in an activated form, which produces well-developed mesopores for use in applications as high-surface area adsorbents.

10.2.2.5 Vapor-Grown Carbon Fibers

Vapor-grown carbon fibers (VGCFs) comprise a large family of filamentous nanocarbons. They can be distinguished in terms of the arrangement of the graphene layers in their molecular scale structure: they can be "plate-like," with near-parallel graphene layers that are approximately perpendicular to the fiber axis, or they can have the "fish-bone" microstructure with stacked cones of graphene planes. Submicron (50–200 nm diameter) VGCFs of the "fish-bone" structure approach the dimensions of MWNTs and are referred to as CNFs (see above) in this chapter. VGCFs and CNFs are generally grown by depositing carbon by the high-temperature (typically in the 900–1200 °C range) decomposition of a hydrocarbon (usually methane) catalyzed by finely divided transition metal catalyst particles. Depending on the catalyst, different growth forms are found: one-directional growth (the fiber grows with the catalyst at the tip: "tip-mode" or at the rear: "rear-mode"), bidirectional growth (simultaneous growth in two opposite directions with the catalyst particle in the middle), multidirectional growth (more than two fibers grow out of one catalyst particle: "octopus fiber") as well as branched growth (a larger catalyst particle explodes during the growth resulting in a branched growth of a number of smaller fibers).

10.2.3
Chemical Modification/Derivatization Methods

The development of carbon nanotube-based nanocomposites was initially impeded by the inability to uniformly disperse the nanotubes in the polymer matrix due to a lack of compatibility between the chemical structures of the two components. Compatibility has now been achieved in many cases by chemical modification or derivatization of the nanotube sidewalls. Some degree of derivatization or functionalization is achieved following the nanotube chemical vapor deposition synthesis by the adsorption of electron-withdrawing oxygen on the tube walls and the net formation of acidic –COOH groups as a result of acid purification procedures to remove the catalyst and support as well as amorphous/microcrystalline carbon produced as impurity during synthesis. Derivatization also allows for solubility of the nanotubes in specific organic solvents and in water and enables covalent interaction between the nanotube sidewalls and the polymer side groups, leading to better adhesion at the nanotube-polymer interface and the formation of nanocomposites with exceptionally high mechanical strength.

Two approaches have been utilized to achieve derivatization. The first has involved chemical modification of the nanotube surface, while the second has involved chemical interaction with various defects on the graphitic walls of the tubes and at the tube ends. The surface modifications reported in the literature for nanotubes have been somewhat similar to that achieved on the C_{60} fullerene, although closer examination has revealed sizable differences in reaction type, location, and symmetry of the chemistry involved. On the other hand, defect site functionalization involves chemistry that is not applicable to the fullerenes because they are free from similar defects.

A large amount of literature exists on the chemical modification of carbon nanotubes, but detailed understanding is still lacking because of the paucity of theoretical calculations and simulations. Several research groups have reported the successful functionalization of both SWNTs and MWNTs [28–33]. These modifications have involved the direct attachment of functional groups such as fluorine or hydrogen to the graphitic walls, reactions with nitrenes and carbenes, or the use of carboxylic acid groups bonded to the nanotube walls produced on oxidation of shortened and unbundled tubes. Chen *et al.* [28] first reported the use of acid groups for attaching long alkyl chains to SWNTs via amide linkages. There is now ample evidence that nanotube-bound carboxylic acid groups are the sites at which a variety of functional groups for the solubilization of both shortened and full-length carbon nanotubes are attached. For example, it has been shown that esterification of the carboxylic groups can be used to functionalize and solubilize nanotubes of any length [34–36]. Mono-, di-, and tri-nitroanilines have been recently attached to SWNTs via carboxylic groups and reaction with thionyl chloride [37]. Multiple sulfonate, $-OSO_3H$, groups have been chemically introduced on MWNTs [38] and compounded with emeraldine base polyaniline to form composites with enhanced electrical conductivity and thermal properties due to concomitant doping of the polymer by the sulfonated nanotubes in the course of composite fabrication. Solubilization in water has been achieved by wrapping with polymers such as polyvinyl pyrrolidone (PVP) [39] and polyethylene imine (PEI) [40][1)], and by reaction with glucosamine [41]. SWNTs have also been effectively dispersed/solubilized in water by their sonication in the presence of the single-stranded version of the central polymeric molecule in biology, DNA [42] and enzymes suitable for use in biosensing and biofuel cells [43, 44]. In the case of DNA, molecular modeling suggests that single-stranded DNA binds to SWNTs through π-stacking interactions that result in helical wrapping to the nanotube sidewalls [42].

For carbon nanofibers and conventional microfibers, the key to the formation of high-strength polymer composites is the adhesion of the fibers to the polymeric matrices. The adhesion forces are still not fully understood, primarily because the surfaces of the carbon fibers are complex with respect to their structure and chemistry. The forces result from different interactions across the interface, which include dispersive interactions of the van der Waals type involving London forces, nondispersive interactions involving acid–base processes, and covalent chemical bonds. Typical surface treatment involves oxidation treatment in air or ozone to form oxygen containing functional groups. Alternative approaches involve the use of plasma-induced surface modification [45] or electrochemical anodization in an acidic electrolyte such as phosphoric acid [46]. Surface groups produced consist of basic pyrone-like structures, neutral quinines, and acidic carboxylic groups. The strength of the composites formed has been correlated with the surface roughness observed by means of detailed scanning electron and tunneling microscopies [47]. Recently, microwave and ultraviolet irradiation techniques have been used to functionalize SWNT sidewalls with acid ($-SO_3^- H^+$) and hydroxyl (OH^-) groups [48, 49], respectively. Interestingly, hydroxyl group functionalization of SWNTs dispersed with surfactants in water using 254 nm UV-radiation is diameter and nanotube-type sensitive, and can, therefore, provide a method for the separation of metallic from semiconducting SWNTs.

10.2.4
Polymer Matrices

As discussed in Section 10.2.3, PVP and DNA have been used to wrap and water-solublize SWNTs. For specific actuator, electrical and electro-optic applications, SWNTs have been wrapped by piezoelectric polyvinylidene fluoride and trifluoroethylene copolymer [50] or with conjugated polymers [51, 52]. The conjugated polymer used to form a composite with MWNTs and an electron-transport layer in light-emitting diodes is poly(*m*-phenylene-vinylene-*co*-2,5-dioctyloxy-*p*-phenylene-vinylene) (PmPV) [53]. Wrapping coupled with electron doping has been achieved with polyethylene imine to form p–n junction devices ([40], see footnote 1).

1) Zhang, Y., Grebel, H., and Iqbal, Z. Polyethylene-imine functionalization of single wall carbon nanotubes for air-stable n-type doped coatings, unpublished data.

Thermosetting epoxy resins are widely used in the fabrication of carbon fiber-based composites for aerospace applications. High-temperature amorphous thermoplastics with high impact strength, which include polycarbonate, polysulfones, polyetherimide, polyethersulfones, and partially crystalline polyetheretherketone, are alternative polymers bearing functional groups that can undergo selective interactions with the functional groups formed on the carbon fiber surface. For the fabrication of electrically conductive bipolar plates for proton exchange membrane fuel cells, chemically passive polymers such as polypropylene (PP) are preferred [54], whereas poly(acrylonitrile-butadiene-styrene) (ABS), polystyrene (PS), and high impact polystyrene (HIPS) are used in the fabrication of composites for applications where high impact strength is required.

10.3 Polymer Matrix Composites

10.3.1 Fabrication

In contrast to short carbon fiber reinforced thermoplastics, which are processed by conventional melt processing techniques, the limited availability, the high cost, and the difficulties encountered in achieving a high degree of dispersion continue to present challenges in the manufacture of carbon nanotube composites. Currently, most carbon nanotube-reinforced composites are prepared in the laboratory using the so-called solution–evaporation method [55–58]. The solution and curing agent may vary with different polymer matrices. The general procedure involves dissolving the polymer to form a first solution, dispersing/dissolving SWNTs or MWNTs to form a second solution, mixing the two solutions with the aid of ultrasonication, and finally casting films or solid parts from the mixed solution and subjecting them to a curing process.

To achieve more uniform nanotube dispersion in composites, Haggenmueller *et al.* [59] developed an alternative melt mixing method consisting of a combined solution–evaporation technique to prepare a thin SWNT-polymer film followed by repeated compression molding of the latter. The resulting product was reported to yield compositionally uniform films. Using a small batch mixer, adequately dispersed nanotube composites from polypropylene, poly(acrylonitrile-butadiene-styrene), polystyrene, and high impact polystyrene have been prepared [60].

Another technique, known as the dry powder mixing method, has been employed by Cooper *et al.* to produce nanotube-reinforced polymethylmethacrylate (PMMA) composites [61]. Like most of the currently used fabrication methods for nanotube-based polymer composites, this technique is a combination of several protocols including solution–evaporation, sonication, kneading, and extrusion. More specifically, these workers used ultrasonic techniques to blend carbon nanotubes with PMMA particles, and the blend was later extruded to orient the nanotubes. Yang *et al.* [62] prepared small-scale batches of ABS nanocomposites without the use of solvents or ultrasonic techniques with good dispersion of the nanotubes.

Another method used known as extrusion freeform fabrication (EFF) belongs to a family of manufacturing processes in which different parts are built in layers. It is a solid free-form fabrication (SFF) technique, where the feed is in the form of a solid. A 3D computer model is generated and transferred to a computer supported by the SFF software. The model is sliced into layers and the geometrical information is fed for each layer of the part, which is then built layer by layer. Carbon nanotubes and carbon fibers are very suitable for this technique because they do not clog the nozzles. In EFF, the solid feed material is placed in a heated head and is forced through a nozzle by a piston into a specified shape. Once a layer is complete, the support base is lowered in the z-direction. The EFF process helps in tailoring the alignment of fibers in composites since the extrusion path can be changed in different parts. A study of SWNTs and VGCFs mixed with ABS polymer using Banbury mixing and EFF was conducted by Shofner *et al.* [63]. A high degree of dispersion of the nanotubes and fibers was achieved without porosity. For VGCF and SWNTs, sizable tensile strength and modulus improvements were observed.

A recently reported method for producing novel SWNT-polymer nanocomposites involved the use of self-assembled SWNT nanopaper films produced by vacuum filtration of SWNTs dispersed with surfactants in aqueous solution. The free-standing nanopaper films were soaked in a polymer resin followed by drying and hot pressing to form a composite in which the polymer is intercalated into the free volume between the SWNT bundles [64].

10.3.2
Mechanical and Electrical Property Modification

Carbon fibers have been used in both thermosetting and thermoplastic polymer composites for a long time, imparting higher modulus and strength and lighter weight than glass fibers (see comparison of properties in Table 2.1), electrical and thermal conductivity, chemical resistance, and reduced wear. Specific examples of their effects on thermoplastics and thermosets may be found in the handbooks and general references listed in Chapters 1 and 2. However, with the discovery of the near-molecular scale carbon nanotubes and advances in understanding their mechanical and electrical properties over the past decade, new nanocomposites based on these novel materials are now possible. Experimental estimates of SWNT strength are in the range of 13–52 GPa and tensile modulus is of the order of 1 TPa [65–67], values that are much higher than those for carbon fibers (also see Table 2.1). Electrical resistivity and thermal conductivity measurements along the length of a bundle of SWNTs indicate values of approximately 10^{-4} Ω cm and 200 W/(m K), respectively [68]. Two main issues to be addressed for effective use of nanotubes in composites are alignment and uniform dispersion in the polymer matrix. This is because SWNTs form in bundles and tend to agglomerate with weak van der Waals forces. A great deal of work has been done aiming at overcoming this problem and several surfactant-based and organic solutions have been identified to disperse and chemically functionalize carbon nanotubes [69–74]. Another issue being addressed is that of interfacial bonding between the nanotubes and the polymer matrix, which affects the efficiency of load transfer across the nanotube-polymer interface.

Several studies on the characterization and fabrication of carbon nanotube-polymer nanocomposites have highlighted the important roles of the parameters discussed in Chapter 2 (such as, orientation, dispersion, and interfacial adhesion) in determining the properties of the composites. Jia *et al.* [75] used an *in situ* process for the fabrication of a PMMA/MWNT composite. An initiator was used to open up the π bonds of the MWNTs in order to increase the linkage with the PMMA. The formation of C–C bonds results in a strong interface between the nanotubes and the PMMA. For samples mixed with carbon nanotubes, smaller amounts of initiator are required and improved mechanical properties are obtained. In another example, simple sonication of SWNTs in solvents such as DMF was found not to yield good dispersion according to the method of Haggenmueller *et al.* [59] introduced above; therefore, repetitive film forming with sonication and drying followed by mixing at higher temperatures and pressure was required to obtain a uniform dispersion (Figure 10.3). The melt could be formed as a film or spun into fibers. With the introduction of SWNTs, the draw ratio is reduced and a roughened surface results for the fiber, as viewed under an optical microscope. Mechanical and electrical properties are improved with increasing SWNT concentration. Melt processing, therefore, appears

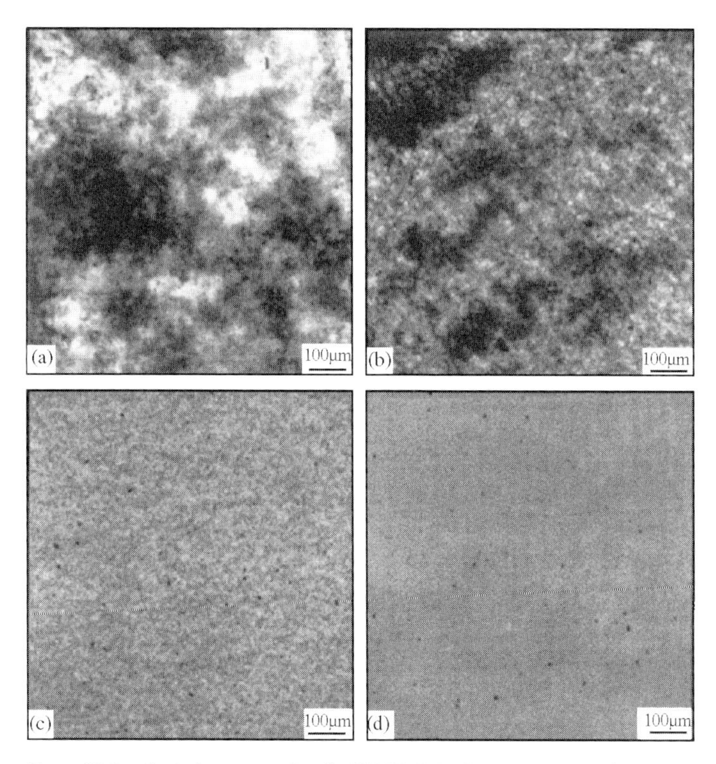

Figure 10.3 Optical micrographs of a SWNT–PMMA nanocomposite having 1 wt% purified soot: (a) only sonication and drying. The as-cast film is repeatedly subjected to hot pressing (180 °C, 3000 lb, 3 min) and is shown here after; (b) 1 cycle; (c) 5 cycles; (d) 20 cycles. (Reproduced with permission from Elsevier [59].)

to be a very effective method for realizing targeted mechanical and electrical properties in the bulk composites.

Nanotubes were found to be oriented in the extrusion flow direction, increasing the impact strength of the PMMA nanocomposites formed by the method of Cooper *et al.* [61]. Jin *et al.* [72] proposed a method of casting a suspension of carbon nanotubes in a solution of thermoplastic polymer polyhydroxyaminoether (PHAE) in chloroform. In this study, it was found that the resulting nanocomposites could be stretched up to five times their original length without breaking under varying mechanical loads in a temperature range of 90–100 °C. The nanotubes were aligned inside the polymer matrix, as indicated by X-ray diffraction and transmission electron microscopy (Figure 10.4). Highly aligned SWNTs in polystyrene and polyethylene have been obtained by Haggenmueller *et al.* [76] using a twin-screw extruder.

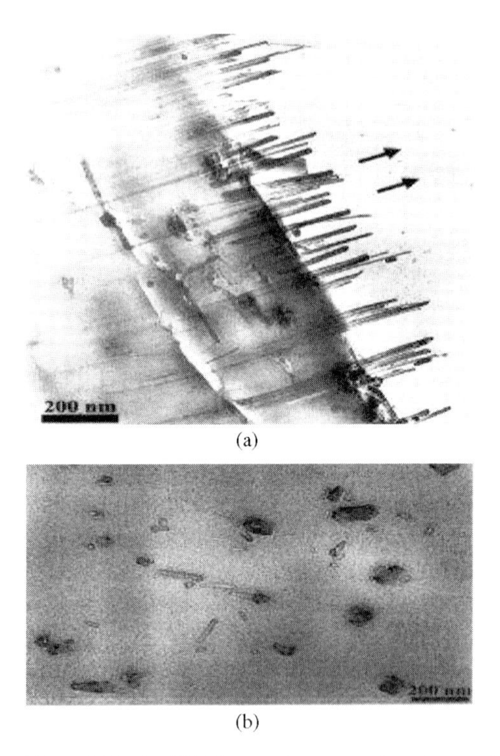

Figure 10.4 (a) TEM image of an internal fracture surface of a composite about 90 nm in thickness after being microtomed parallel to the stretching direction. The nanotubes are aligned parallel to the stretching direction and fiber pullout is observed. In some areas, nanotubes bridge the microvoids (or microcracks) in the matrix and presumably enhance the strength of the composite. (b) Cross-sectional view of the same composite microtomed perpendicular to the stretching direction. Cross sections of the nanotubes and nanoparticles are observed. (Reproduced with permission from the American Institute of Physics [72].)

Composite fibers obtained with 20% nanotube loading showed a 450% increase in elastic modulus relative to polyethylene fibers.

Carbon nanotube-polystyrene nanoporous membranes with aligned MWNTs traversing the membrane thickness have recently been fabricated by Hinds *et al.* [77]. The structures formed are depicted in Figure 10.5. These nanoporous membranes have the ability to gate molecular transport through the cores of the nanotubes, offering potential applications in chemical separations and sensing.

(a)

(b)

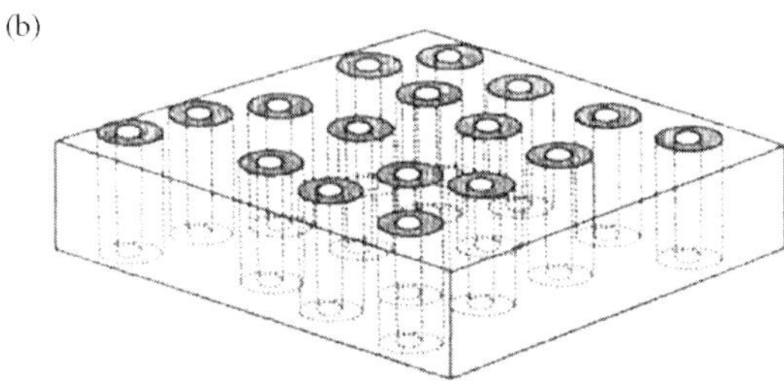

(c)

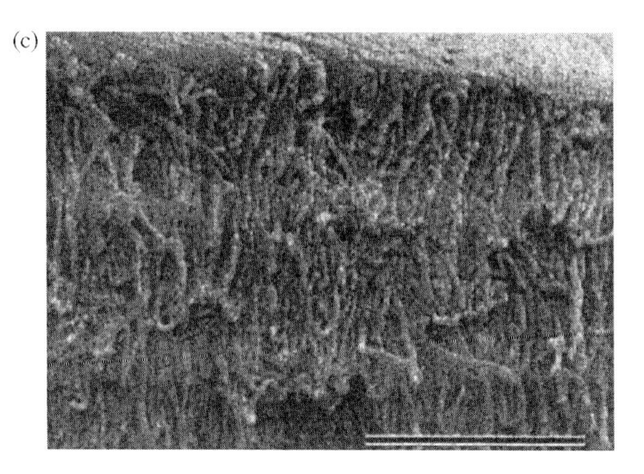

Figure 10.5 (a) As-grown aligned MWNTs produced by a Fe-catalyzed chemical vapor deposition process. (b) Schematic of a target membrane structure. (c) Scanning electron micrograph of MWNT-polystyrene composite membrane. Scale bar represents 2.5 μm. (Reproduced with permission from Science AAAS [77].)

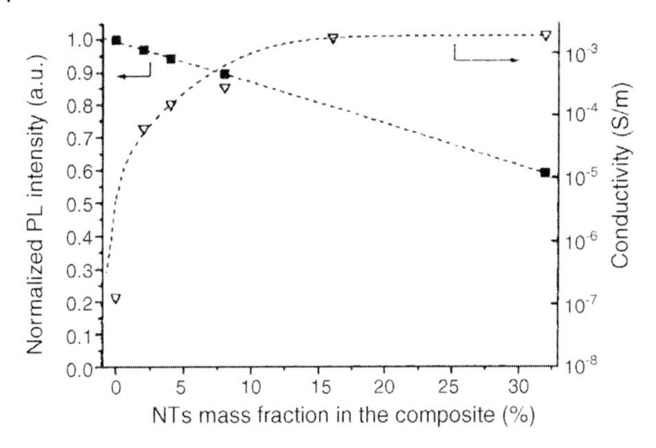

Figure 10.6 Normalized photoluminescence (PL) intensity and conductivity for conducting polymer-nanotube composite films as a function of nanotube to polymer mass ratio. (Reproduced with permission from the American Institute of Physics [51].)

Carbon nanotubes have been introduced into conducting polymers, such as poly (*m*-phenylene-vinylene-*co*-2,5-dioctyloxy-*p*-phenylene-vinylene) (PmPV), as an electron transport layer in organic light-emitting diodes. Their introduction led to a significant increase in efficiency and an increase in electrical conductivity by four orders of magnitude [51]. The normalized photoluminescence intensity and electrical conductivity as a function of MWNT loading for these composites are shown in Figure 10.6. Helical wrapping of the conducting polymer around the nanotubes has been modeled by Lordi and Yao [78]; such a model is depicted in Figure 10.7.

10.4
Cost/Availability

Tables 10.1 and 10.2 list the US and international companies that supply nanotubes, related nanocarbon materials, and carbon fibers. The prices at the time of writing are also given where available. Note that the cost of pure SWNTs still remains very high.

10.5
Environmental/Toxicity Considerations

Fullerene soot with a high SWNT content was tested to assess its biochemical activity [79]. The dermatological trial results did not show any signs of health hazards related to skin irritation and allergic risks. To determine whether carbon nanotubes and, in particular, SWNTs can induce any significant health hazards, Huczko *et al.* [80] performed tests routinely used in the pathophysiological testing of asbestos-induced disease. No abnormalities of pulmonary function or measurable inflammation were detected in guinea pigs. However, more recent studies by Poland *et al.* [81] showed that

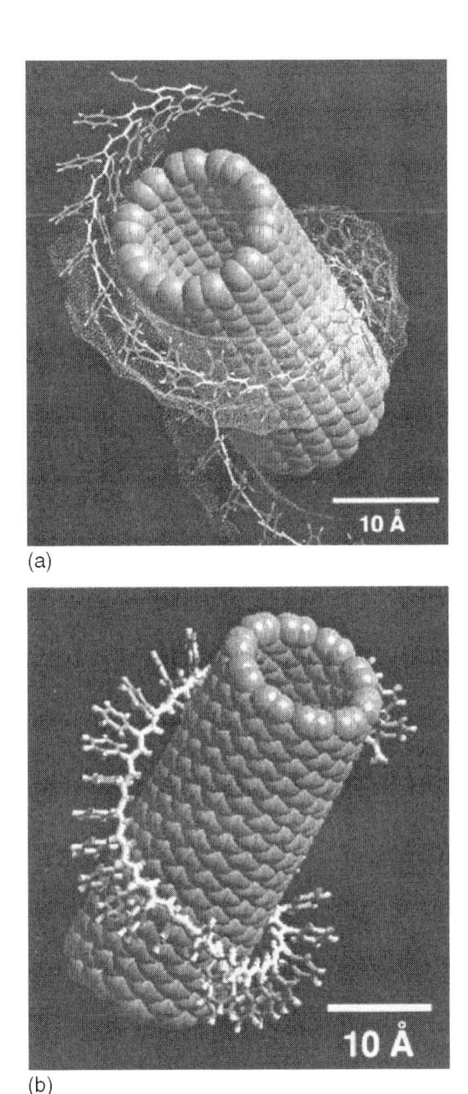

Figure 10.7 (a) Model of *cis*-poly(phenylacetylene) wrapped perfectly around a (10,10) SWNT. (b) Model of *trans*-poly(phenylacetylene), which has a slightly smaller diameter, distorts around the SWNT and wraps more tightly. (Reproduced with permission from Ref. [78].)

on exposure of the mesothelial lining of the chest cavity of mice to long multiwall carbon nanotubes resulted in asbestos-like pathogenic behavior, but the study does not reveal whether the nanotubes are able to persist long enough to reach the lung tissue after inhalation. In another recent experiment, mice breathed air containing 40 μm long multiwall carbon nanotubes with very little inflammatory or fibrogenic effects [82], most likely because they do not persist in the body as long as asbestos fibers do. Therefore, as also discussed in Chapter 1, care should be exercised with the use of nanofillers including carbon nanotubes since their toxicology has not yet been fully explored.

Table 10.1 List of carbon nanotube and related material suppliers.

Company	Products	Price[a)]	Location
Applied Science Inc.	MWNTs and CNFs (nanofibers)	CNFs: $65/lb	Cedarville, OH, USA
Bucky USA	Fullerenes, MWNTs and SWNTs	MWNT: $60–150/g SWNT: $150/g	Houston, TX, USA
Carbolex	SWNTs	$100/g	Broomall, PA, USA
Carbon Solutions	SWNTs	$50–400/g	Riverside, CA, USA
Catalytic Materials Ltd	Suppliers of nanofibers and MWNTs	MWNT: $ 55–40/g	Pittsboro, NC, USA
Hyperion Catalysis International	MWNTs	N/A	Cambridge, MA, USA
Hanwha Chemical Corp.	MWNTs and SWNTs	N/A	Jung-Ku, Seoul, Korea
MER Corporation	Various (from fullerenes to SWNTs)	MWNT: $10–25/g SWNT: $60/g	Tucson, AZ, USA
Mitsui XNRI	MWNTs	N/A	Tokyo, Japan
NanoCarbLab	SWNTs	$60–380/g	Moscow, Russia
Nanocs International	MWNTs	MWNT: $80/g SWNT: $ 350/g	New York, USA
Nanocyl	MWNTs and SWNTs	N/A	Namur, Belgium
Nanolab	MWNTs	MWNT: $125–165/g	Newton, MA, USA
NanoMaterials	Inorganic nanotubes and nanospheres	N/A	Longmont, CO, USA
Nanomirae	Nanofibers – Herringbone and spiral MWNTs	N/A	Guro-gu, Seoul Korea
Rosseter Holdings Ltd	MWNT, SWNTs and Nanohorns	$20/g MWNT	Limassol, Cyprus
SouthWest Nano Technologies, Inc.	SWNTs	$500/g SWNT	Norman, OK, USA
Sun Nanotech	MWNTs	N/A	Jiangxi, PR China
Tsinghua-Nafine Nano-Powder	MWNTs	N/A	Beijing, PR China
Unidym	HIPCO SWNTs		Menlo Park, CA, USA
Xintek Inc.	Nanotube materials (MWNTs/SWNTs)	N/A	Chapel Hill, NC, USA

a) Prices when available are given at the time of writing. Price ranges reflect purity and quantities.

Carbon microfibers easily form dust during handling and get dispersed in the atmosphere. The fibers also tend to stick to the human skin or mucous membranes causing pain and itching. Protective gear for skin, eyes, and throat therefore need to be worn to prevent these hazards. Local air exhausts and ventilators can help in removing the dust. Protective cream or gloves need to be used during handling of the fibers. Since the fibers are electrically conductive, care should be taken around exposed electrical circuits and outlets. Some general purpose grades of carbon fiber

Table 10.2 List of carbon fiber suppliers.

Company	Products	Trademark	Location
Amoco Fabrics and Fibers	Pitch-type fibers	THORNEL	USA
Asahi Kasei Corporation		HI CARBOLON	USA
Toho Rayon Co.	PAN-type fibers	BESFIGHT	Japan
Toray	PAN-type fibers	TORACA	Japan
Mitsubishi Rayon	Pitch-type fibers	Dialead	Japan
NGF	Pitch-type fibers	N/A	
BASF	PAN-type fibers	CELION	USA
Hercules	PAN-type fibers	MAGNAMITE	USA
SGL Carbon		SIGARTEX	Germany/USA
Zoltex		PANEX	USA
RK Carbon Fibers Ltd		CURLON	UK
Courtauld Ltd		COURTELLE	UK
Ashland		CARBOFLEX	USA

may ignite at temperatures lower than 150 °C in the presence of air or fuel. If heated to higher than 400 °C in the presence of air or fuel, the fibers slowly burn but stop burning as soon as the burning fuel is removed.

Carbon fiber waste should be treated as industrial rather than household waste. Local governments may have their own local codes for disposing carbon fiber wastes. On the positive side, carbon is thought to have good compatibility with human tissue. Carbon fibers and fiber composites have therefore been used extensively as components for artificial body parts and devices.

10.6 Applications

Carbon fiber composites are widely used in the aerospace industry, and with the decreasing price of the fibers they are increasingly being used in automobile, marine, sports, and construction industries. In aerospace, epoxy/carbon composites are used in the space shuttle payload door, its manipulator arm, and its booster tail and fins. There is extensive use of carbon fiber/epoxy composites in helicopter structures as well as in commercial aircraft. Carbon fiber composites have started to be used in automobiles, mainly for saving weight. Here, carbon nanotubes, in particular cost-effective MWNTs and CNFs, are being used in car bumpers and gasoline tanks. In the future, with decreasing prices, nanotubes could be used in these composites to obtain much higher strength and at much lower loading levels.

In addition to their primary function as mechanical property modifier, the high electrical conductivity of carbon fibers provides carbon-based composites with static dissipation and radio frequency shielding characteristics. This opens up a whole range of applications and with carbon nanotubes this can be achieved at enormously low loading levels. One application with future potential is the use of carbon nanotubes to fabricate bipolar interconnecting flow-field plates for fuel cells [54]. A whole range of futuristic applications in nanoelectronics is also emerging with SWNTs and MWNTs. The high thermal conductivity of carbon fibers and the diamond-like thermal conductivity of SWNTs make their composites highly attractive for heat sinks in electronics. The low density of the composites compared to copper makes them even more attractive for aerospace electronics.

References

1 Iijima, S. (1991) Helical microtubules of graphitic carbon. *Nature*, **354**, 56–58.

2 Bethune, D.S., Kiang, C.H., DeVires, M., Gorman, G., Savoy, R., Vazquez, J., and Beyers, R. (1993) Cobalt-catalyzed growth of carbon nanotubes with single-atomic-layer walls. *Nature*, **363**, 605–607.

3 Iijima, S. and Ichihashi, T. (1993) Single-shell carbon nanotubes of 1-nm diameter. *Nature*, **363**, 603–605.

4 Thess, A., Lee, R., Nikolaev, P., Dai, H., Petit, P., Robert, J., Xu, C., Lee, Y.H., and Smalley, R.E. (1996) Crystalline ropes of metallic carbon nanotubes. *Science*, **273**, 483–487.

5 Journet, C.W., Matser, K., Bernier, P., Laiseau, L., Lefrant, S., Deniard, P., Lee, R., and Fischer, J.E. (1997) Large-scale production of single-walled carbon nanotubes by the electric-arc technique. *Nature*, **388**, 756–758.

6 Tibbetts, G.G. (1984) Why are carbon filaments tubular? *J. Cryst. Growth*, **66**, 632–638.

7 Tennent, H.G. (1987) Carbon fibrils, method for producing same and compositions containing same, US Patent No. 4,663,230.

8 Amelinckx, S., Zhang, X.B., Bernaerts, D., Zhang, X.F., Ivanov, V., and Nagy, J.B. (1994) A formation mechanism for catalytically grown helix-shaped graphite nanotubes. *Science*, **265**, 635–639.

9 Ren, Z.F., Huang, Z.P., Xu, J.W., Wang, J.H., Bush, P., Siegel, M.P., and Provencio, P.N. (1998) Synthesis of large arrays of well-aligned carbon nanotubes on glass. *Science*, **282**, 1105–1107.

10 Meyyappan, M., Delzeit, L., Cassell, A., and Hash, D. (2003) Carbon nanotube growth by PECVD. A review. *Plasma Sources Sci. Technol.*, **12**, 205–216.

11 Kong, J., Soh, T.H., Cassell, A.M., Quate, C.F., and Dai, H. (1998) Synthesis of individual single-walled carbon nanotubes on patterned silicon wafers. *Nature*, **395**, 878–879.

12 Kong, J., Cassell, A., and Dai, H. (1998) Chemical vapor deposition of methane for single-walled carbon nanotubes. *Chem. Phys. Lett.*, **292**, 567–574.

13 Nikolaev, P., Bronikowski, M.J., Bradley, R.K., Rohmund, F., Colbert, D.T., Smith, K.A., and Smalley, R.E. (1999) Gas-phase catalytic growth of single-walled carbon nanotubes from carbon monoxide. *Chem. Phys. Lett.*, **313**, 91–97.

14 Kitiyanan, B., Alvarez, W.E., Harwell, J.H., and Resasco, D.E. (2000) Controlled production of single-wall carbon nanotubes by catalytic decomposition of CO on bimetallic Co-Mo catalysts. *Chem. Phys. Lett.*, **317**, 497–503.

15 Lan, A., Iqbal, Z., Aitouchen, A., Libera, M., and Grebel, H. (2002) Growth of single-wall carbon nanotubes within an ordered array of nanosize silica spheres. *Appl. Phys. Lett.*, **81**, 433–435.

16 Lan, A., Zhang, Y., Iqbal, Z., and Grebel, H. (2003) Is molybdenum necessary for the growth of single-wall carbon nanotubes from CO? *Chem. Phys. Lett.*, **379**, 395–400.

17 Goyal, A. and Iqbal, Z. (2008) Scaleable production of single wall carbon nanotubes. *Chem. Phys. Lett.*, submitted for publication.

18 Cheng, H., Li, F., Su, G., Pan, H.Y., He, L.L., Sun, X., and Dresselhaus, M. (1998) Large-scale and low-cost synthesis of single-walled carbon nanotubes by the

catalytic pyrolysis of hydrocarbons. *Appl. Phys. Lett.*, **72**, 3282–3284.

19 Maruyama, S., Kojima, R., Miyauchi, Y., Chiashi, S., and Kohno, M. (2002) Low-temperature synthesis of high-purity single-walled carbon nanotubes from alcohol. *Chem. Phys. Lett.*, **360**, 229–234.

20 Murakami, Y., Chiashi, S., Miyauchi, Y., Hu, M., Ogura, M., Okubu, T., and Maruyama, S. (2004) Growth of vertically aligned single-walled carbon nanotube films on quartz substrates and their optical anisotropy. *Chem. Phys. Lett.*, **385**, 298–303.

21 Sharma, R. and Iqbal, Z. (2004) In situ observations of carbon nanotube formation using environmental transmission electron microscopy. *Appl. Phys. Lett.*, **84**, 990–992.

22 Wang, N., Li, G.D., and Tang, Z.K. (2001) Mono-sized and single-walled 4 Å carbon nanotubes. *Chem. Phys. Lett.*, **339**, 47–52.

23 Derycke, V., Martel, R., Radosavljevic, M., Ross, F.M., and Avouris, P. (2002) Catalyst-free growth of ordered single-walled carbon nanotube networks. *Nano Lett.*, **2**, 1043–1046.

24 Kato, T., Jeong, G.H., Hirata, T., Hatakeyama, R., Tohji, K., and Motogima, K. (2003) Single-walled carbon nanotubes produced by plasma-enhanced chemical vapor deposition. *Chem. Phys. Lett.*, **381**, 422–426.

25 Li, Y., Mann, D., Rolandi, M., Kim, W., Ural, A., Hung, S., Javey, A., Cao, J., Wang, D., Yenilmez, E., Wang, Q., Gibbons, J.F., Nishi, Y., and Dai, H. (2004) Preferential growth of semiconducting single-walled carbon nanotubes by a plasma enhanced CVD method. *Nano Lett.*, **4**, 317–321.

26 Maschmann, M.R., Amama, P.B., Goyal, A., Iqbal, Z., Gat, R., and Fisher, T.S. (2006) Parametric study of synthesis conditions in plasma-enhanced CVD of high quality single wall carbon nanotubes. *Carbon*, **44**, 10–18.

27 Lin, R.Y. and Economy, J. (1973) Preparation and properties of activated carbon fibers derived from phenolic precursor. *Appl. Polym. Symp.*, **21**, 143–152.

28 Chen, J., Harmon, M.A., Hu, H., Chen, Y., Rao, A.M., Eklund, P., and Haddon, R.C. (1998) Solution properties of single-walled carbon nanotubes. *Science*, **282**, 95–98.

29 Michelson, E.T., Huffman, C.B., Rinzler, A.G., Smalley, R.E., Hague, R.H., and Margrave, J.L. (1998) Fluorination of single-wall carbon nanotubes. *Chem. Phys. Lett.*, **296**, 188–194.

30 Bahr, J.L., Yang, J., Kosynik, D.V., Bronikowski, M.J., Smalley, R.E., and Tour, J.M. (2001) Functionalization of carbon nanotubes by electrochemical reduction of aryl diazonium salts: a bucky paper electrode. *J. Am. Chem. Soc.*, **123**, 6536–6542.

31 Pekker, S., Salvetat, J.P., Jakab, E., Bonard, J.M., and Forro, L. (2001) Hydrogenation of carbon nanotubes and graphite in liquid ammonia. *J. Phys. Chem.*, **B105**, 7938–7943.

32 Lin, Y., Taylor, S., Huang, W., and Sun, Y.P. (2003) Characterization of fractions from repeated functionalization reactions of carbon nanotubes. *J. Phys. Chem.*, **107**, 914.

33 Sun, Y.P., Fu, K., Lin, Y., and Huang, W. (2002) Functionalized carbon nanotubes: properties and applications. *Acc. Chem. Res.*, **35**, 1096–1104.

34 Sun, Y.P., Huang, W., Lin, Y., Fu, K., Kitaygorodskiy, A., Riddle, L.A., Yu, Y., and Carroll, D.L. (2001) Soluble dendron-functionalized carbon nanotubes: preparation, characterization, and properties. *Chem. Mater.*, **13**, 2864–2869.

35 Riggs, J.E., Walker, D.B., Carroll, D.L., and Sun, Y.P. (2000) Optical limiting properties of suspended and solubilized carbon nanotubes. *J. Phys. Chem.*, **B104**, 7071–7076.

36 Fu, K., Huang, W., Lin, Y., Riddle, L.A., Carroll, D.L., and Sun, Y.P. (2001) Defunctionalization of functionalized carbon nanotubes. *Nano Lett.*, **1**, 439–441.

37 Wang, Y., Malhotra, S., and Iqbal, Z. (2004) Nanoscale energetics with carbon nanotubes. *Mater. Res. Soc. Symp. Proc.*, **800**, AA9.1.1–AA9.1.9.

38 Dai, L. and Mau, A.W.H. (2001) Controlled synthesis and modification of carbon nanotubes and C60: carbon nanostructures for advanced polymeric composite materials. *Adv. Mater.*, **13**, 899–913.

39 O'Connell, M.J., Boul, P., Ericson, L.M., Huffman, C.B., Wang, Y., Haroz, E., Kuper, C., Tour, J.M., Ausman, K.D., and Smalley, R.E. (2001) Reversible water-solubilization of single-walled carbon nanotubes by polymer wrapping. *Chem. Phys. Lett.*, **342**, 265–271.

40 Shim, M., Javey, A., Kam, N.W.S., and Dai, H. (2001) Polymer functionalization for air-stable n-type carbon nanotube field-effect transistors. *J. Am. Chem. Soc.*, **123**, 11512–11513.

41 Balzano, L., Herrera, J.E., Alvarez, W.E., Pompeo, F., and Resasco, D.E. (2002) Characterization of single-walled carbon nanotubes (SWNTs) produced by CO disproportionation on Co-Mo catalysts. *Chem. Mater.*, **14**, 1853–1858.

42 Zheng, M., Jagota, A., Semke, E.D., Diner, B.A., Mclean, R.S., Laustig, S.R., Richardson, R.E., and Tassi, N.G. (2003) DNA-assisted dispersion and separation of carbon nanotubes. *Nat. Mater.*, **2**, 338–342.

43 Wang, Y., Iqbal, Z., and Malhotra, S. (2005) Functionalization of carbon nanotubes with enzymes and nitramines. *Chem. Phys. Lett.*, **402**, 96–101.

44 Wang, S.C., Yang, F., Silva, M., Zarow, A., Wang, Y., and Iqbal, Z. (2009) Membrane-less and mediator-free enzymatic biofuel cell using carbon nanotube/porous silicon electrodes. *Electrochem.Commun.*, **11**, 34–37.

45 Mittal, K.L. (1999) Proceedings of the 2nd International Symposium on Polymer Surface Modification, Newark, New Jersey, May 24–26, 1999.

46 Park, S.-J., Kim, M.H., Hong, Y.T., and Lee, J.-R. (2001) Surface characteristics of electrochemically modified carbon fibers in phosphoric acid solution: effect of surface treatment on interfacial mechanical behaviors of composites. *J. Chem. Eng., Jpn.*, **34**, 396–400.

47 Krekel, G., Hüttinger, K.J., Hoffman, W.P., and Silver, D.S. (1994) The relevance of the surface structure and surface chemistry of carbon fibers in their adhesion to high-temperature thermoplastics. Part I. Surface structure and morphology. *J. Mater. Sci.*, **29**, 2968–2980.

48 Wang, Y., Iqbal, Z., and Mitra, S. (2006) Rapidly functionalized, water-dispersed carbon nanotubes at high concentration. *J. Am. Chem. Soc.*, **128**, 95–99.

49 Alvarez, N.T., Kittrell, C., Schmidt, H.K., Hauge, R.H., Engel, P.S., and Tour, J.M. (2008) Selective photochemical functionalization of surfactant-dispersed single wall carbon nanotubes in water. *J. Am. Chem. Soc.*, **130**, 14227–14233.

50 El-Hami, K. and Matsushige, K. (2003) Covering single walled carbon nanotubes by the poly(VDF-co-TrFE) copolymer. *Chem. Phys. Lett.*, **368**, 168–171.

51 Fournet, P., Coleman, J.N., Lahr, B., Drury, A., Blau, W.J., O'Brien, D.F., and Hörhold, H.-H. (2001) Enhanced brightness in organic light-emitting diodes using a carbon nanotube composite as an electron-transport layer. *J. Appl. Phys.*, **90**, 969–975.

52 Coleman, J.N., Curran, S., Dalton, A.B., Davey, A.P., McCarthy, B., Blau, W., and Barklie, R.C. (1998) Percolation-dominated conductivity in a conjugated-polymer-carbon-nanotube composite. *Phys. Rev.*, **B58**, R7492–R7495.

53 Czerw, R., Woo, H.S., Carroll, D.L., Ballato, J.M., and Ajayan, P.M. (2001) Tailoring hole transport and color tunability in organic light-emitting devices using single-wall carbon nanotubes. *Proc. SPIE*, **4590**, 153–161.

54 Iqbal, Z., Pratt, J., Matrunich, J., Guiheen, J., Narasimhan, D., and Rehg, T. (2003) Nanocomposite for fuel cell bipolar plate, US Patent No. 6,572,997.

55 Mitchell, C.A., Bahr, J.L., Arepalli, S., Tour, J.M., and Krishnamoorti, R. (2002) Dispersion of functionalized carbon nanotubes in polystyrene. *Macromolecules*, **35**, 8825–8830.

56 Wood, J.R., Zhao, Q., and Wagner, H.D. (2001) Orientation of carbon nanotubes in polymers and its detection by Raman spectroscopy. *Composites: Part A*, **32**, 391–399.

57 Qian, D. and Dickey, E.C. (2001) In-situ transmission electron microscopy studies of polymer-carbon nanotube composite deformation. *J. Microsc.*, **204**, 39–45.

58 Sandler, J., Shaffer, M.S.P., Prasse, T., Bauhofer, W., Schutle, K., and Windle, A.H. (1999) Development of a dispersion process for carbon nanotubes in an epoxy

matrix and the resulting electrical properties. *Polymer*, **40**, 5967–5997.

59 Haggenmueller, R., Gommans, H.H., Rinzler, A.G., Fischer, J.E., and Winey, K.I. (2000) Aligned single-wall carbon nanotubes in composites by melt processing methods. *Chem. Phys. Lett.*, **330**, 219–225.

60 Qian, D., Dickey, E.C., Andrews, R., and Randell, T. (2000) Load transfer and deformation mechanisms in carbon nanotube-polystyrene composites. *Appl. Phys. Lett.*, **76**, 2868–2870.

61 Cooper, C.A., Ravich, D., Lips, D., Mayer, J., and Wagner, H.D. (2002) Distribution and alignment of carbon nanotubes and nanofibrils in a polymer matrix. *Comp. Sci. Technol.*, **62**, 1105–1112.

62 Yang, S., Castilleja, J.R., Barrera, E.V., and Lozano, K. (2004) Thermal analysis of an acrylonitrile–butadiene–styrene/SWNT composite. *Polym. Degrad. Stab.*, **83**, 383–388.

63 Shofner, M.L., Rodriguez-Macias, F.J., Vaidyanathan, R., and Barrera, E.V. (2003) Single wall nanotube and vapor grown carbon fiber reinforced polymers processed by extrusion freeform fabrication. *Composites: Part A*, **34**, 1207–1217.

64 Wang, Z., Liang, Z., Wang, B., Zhang, C., and Kramer, L. (2004) Processing and property investigation of single-walled carbon SWNTs (SWNT) buckypaper/epoxy resin matrix nanocomposites. *Composites: Part A*, **35**, 1225–1232.

65 Yu, M.F., Files, B.S., Arepalli, S., and Rouff, R.S. (2000) Tensile loading of ropes of single wall carbon nanotubes and their mechanical properties. *Phys. Rev. Lett.*, **84** (24), 5552–5555.

66 Walters, D.A., Ericson, L.M., Casavant, M.J., Liu, J., Colbert, D.T., Smith, K.A., and Smalley, R.E. (1999) Elastic strain of freely suspended single-wall carbon nanotube ropes. *Appl. Phys. Lett.*, **74** (25), 3803–3805.

67 Gere, J.M. and Timoshenko, S.P. (1990) *Mech. Mater.*, PWS Publishing Company, Boston.

68 Hone, J., Llaguno, M.C., Nemes, N.M., Johnson, A.T., Fischer, J.E., Walters, D.A., Casavant, M.J., Scmidt, J., and Smalley, R.E. (2000) Electrical and thermal transport properties of magnetically aligned single wall carbon nanotube films. *Appl. Phys. Lett.*, **77** (5), 666–668.

69 Dufresne, A., Paillet, M., Pautaux, J.L., Canet, R., Carmona, F., Delhaes, P., and Cui, S. (2002) Processing and characterization of carbon nanotube/poly (styrene-co-butyl acrylate) nanocomposites. *J. Mater. Sci.*, **37**, 3915–3923.

70 Barraza, H.J., Pompeo, F., O'Rear, E.A., and Resasco, D.E. (2002) SWNT-filled thermoplastic and elastomeric composites prepared by miniemulsion polymerization. *Nano Lett.*, **2**, 797–802.

71 Safadi, B., Andrews, R., and Grulke, E.A. (2002) Multiwalled carbon nanotube polymer composites: synthesis and characterization of thin films. *J. Appl. Polym. Sci.*, **84**, 2660–2669.

72 Jin, L., Bower, C., and Zhou, O. (1998) Alignment of carbon nanotubes in a polymer matrix by mechanical stretching. *Appl. Phys. Lett.*, **73**, 1197–1199.

73 Thostenson, E.T. and Chou, T.W. (2002) Aligned multi-walled carbon nanotube-reinforced composites: processing and mechanical characterization. *J. Phys. D: Appl. Phys.*, **35**, L77–L80.

74 Thostenson, E.T. and Chou, T.W. (2003) On the elastic properties of carbon nanotube-based composites: modelling and characterization. *J. Phys. D: Appl. Phys.*, **36**, 573–582.

75 Jia, Z., Wang, Z., Xu, C., Liang, J., Wei, B., Wu, D., and Zhu, S. (1999) Study on poly (methyl methacrylate)/carbon nanotube composites. *Mater. Sci. Eng.*, **A271**, 395–400.

76 Haggenmueller, R., Zhou, W., Fischer, J.E., and Winey, K.I. (2003) Production and characterization of polymer nanocomposites with highly aligned single-walled carbon nanotubes. *J. Nanosci. Nanotechnol.*, **3**, 105–110.

77 Hinds, B.J., Chopra, N., Rantell, T., Andrews, R., Gavalas, V., and Bachas, L.G. (2004) Aligned multiwalled carbon nanotube membranes. *Science*, **303**, 62–65.

78 Lordi, V. and Yao, N. (2000) Molecular mechanics of binding in carbon-nanotube-polymer composites. *J. Mater. Res.*, **15**, 2770–2779.

79 Huczko, A. and Lange, H. (2001) Carbon nanotubes: experimental evidence for a null risk of skin irritation and allergy. *Fullerene Sci. Technol.*, **9**, 247–250.

80 Huczko, A., Lange, H., Calko, E., Grubek-Jaworska, H., and Droszcz, P. (2001) Physiological testing of carbon nanotubes: are they asbestos-like? *Fullerene Sci. Technol.*, **9**, 251–254.

81 Poland, C.A., Duffin, R.D., Kinloch, I., Maynard, A., Wallace, W.H., Seaton, A., Stone, V., Brown, S., MacNee, W., and Donaldson, K. (2008) Carbon nanotubes introduced into the abdominal cavity of mice show asbestoslike pathogenicity in a pilot study. *Nat. Nanotechnol.*, **3**, 423–428.

82 Ryman-Rasmussen, J.P., Tewksbury, E.W., Moss, O.R., Cesta, M.F., Wong, B.A., and Bonner, J.C. (2008) Inhaled multiwalled carbon nanotubes potentiate airway fibrosis in a murine model of allergic asthma. *Am. J. Respir. Cell Mol. Biol.*, **40**, 349–358.

B Low Aspect Ratio Fillers

Functional Fillers for Plastics: Second, updated and enlarged edition. Edited by Marino Xanthos

ISBN: 978-3-527-32361-6

11
Natural Fibers

Craig M. Clemons

11.1
Introduction

The term "natural fibers" covers a broad range of vegetable, animal, and mineral fibers. However, in the composites industry, it usually refers to wood fiber and plant-based bast, leaf, seed, and stem fibers. These fibers often contribute greatly to the structural performance of the plant and, when used in plastic composites, can provide significant reinforcement. Below is a brief introduction to some of the natural fibers used in plastics. More detailed information can be found elsewhere [1–4].

Although natural fibers have been used in composites for many years, interest in these fibers has waned with the development of synthetic fibers such as glass and carbon fibers. However, recently there has been a resurgence of interest, largely because of ecological considerations, legislative directives, and technological advances. One of the largest areas of recent growth in natural fiber plastic composites is the automotive industry, particularly in Europe, where the low density of the natural fibers and increasing environmental pressures are giving natural fibers an advantage. Most of the composites currently made with natural fibers are press-molded although a wide range of processes have been investigated [1, 5].

Flax is the most used natural fiber (excluding wood) in the European automotive industry, most of which is obtained as a by-product of the textile industry [5]. However, other natural fibers such as jute, kenaf, sisal, coir, hemp, and abaca are also used. Natural fibers are typically combined with polypropylene, polyester, or polyurethane to produce such components as door and trunk liners, parcel shelves, seat backs, interior sunroof shields, and headrests [6].

European consumption of natural fibers in automotive composites was estimated at 26 000 tons in 2003 and is expected to grow by 10% per year [7]. Worldwide consumption in all applications by 2010 has been estimated at 110 000–120 000 tons per annum in a variety of applications including automotive, building, appliances and business equipment, and consumer products.

Functional Fillers for Plastics: Second, updated and enlarged edition. Edited by Marino Xanthos

ISBN: 978-3-527-32361-6

11.2
Structure and Production Methods

The major steps in producing natural fibers for use in plastics include harvesting of the fiber-bearing plants, extraction of the fibers, and further processing of the raw fiber to meet required purity and performance aspects for use in plastic composites.

Methods exist for harvesting most natural fibers since they are used in manufacturing of products other than composites. For example, fibers derived from wood are used in the paper and forest products industries, flax fiber is used to make linen and cigarette papers, and jute fiber is used in making rope and burlap [4]. Since many natural fibers are an annual crop, issues such as storage and variability in the growing season need to be considered. Europe is making large investments in new harvesting and fiber separation technologies for natural fibers such as flax [8].

Fiber extraction procedures will depend on the type and portion of plant the fibers are derived from (e.g., bast, leaves, wood) as well as the required fiber performance and economics. Fiber-bearing plants have very different anatomies (e.g., tree versus dicotyledonous plants) and often fibers are derived from agricultural residues or by-products from industry [8]. Consequently, the processing needs can differ greatly.

Wood is primarily composed of hollow, elongated, spindle-shaped cells (called tracheids or fibers) that are arranged parallel to each other along the trunk of the tree [9]. These fibers are firmly cemented together and form the structural component of wood tissue. Fibers are extracted from wood by mechanical or chemical means during the pulping process.

Bast fibers such as flax or kenaf have considerably different structure than wood and, consequently, are processed quite differently. They exist in the inner bark of the stems of dicotyledons, which are typically less than 30% of the stem [8]. Inside the inner bark is a woody core (called the "shive") with much shorter fibers [4]. Fiber strands are removed from the bast. These fiber strands are several meters long and are actually fiber bundles of overlapping single ultimate fibers. Bast fibers are processed by various means that may include retting, breaking, scutching, hackling, and combing [4, 8]. The exact process depends, in a large part, on the type of plant and fiber source. For example, flax fiber can be obtained from different flax plants or from by-products from linen or flax seed production [4]. Useful natural fibers have also been derived from other parts of the plant including leaves (e.g., sisal), seeds (e.g., cotton, coir), or grass stems [3]. The production of these fibers varies greatly depending on fiber type. Some natural fibers can be spun into continuous yarns or made into nonwoven mats that allow expanded processing options for composite production. For example, much of the natural fibers used in automotive composites are currently made into fiber mats that are often needled, thermally fixed with small amounts of polymeric fibers, or otherwise modified to improve handling, and then press molded [3]. However, this additional processing comes with increased cost. The use of short fibers in more conventional processes such as injection molding is projected to increase in the future opening up new markets as new technologies overcome processing hurdles [10].

11.3 Properties

11.3.1 Chemical Components

The structure and chemical makeup of natural fibers varies greatly and depends on the source and many processing variables. However, some generalizations are possible. Natural fibers are complex, three-dimensional, polymer composites made up primarily of cellulose, hemicellulose, pectins, and lignin [12]. These hydroxyl-containing polymers are distributed throughout the fiber wall. The major chemical components of selected natural fibers are listed in Table 11.1.

Cellulose varies the least in chemical structure of the three major components and can be considered the major framework component of the fiber. It is a highly crystalline, linear polymer of anhydroglucose molecules with a degree of polymerization (n) around 10 000. It is the main component providing the strength, stiffness, and structural stability. Hemicelluloses are branched polymers containing 5- and 6-carbon sugars of varied chemical structure and whose molecular weights are well below those of cellulose but which still contribute as a structural component of wood [13]. Portions of the hemicelluloses are polymers of 5-carbon sugars and are called pentosans [1].

Lignin is an amorphous, cross-linked polymer network consisting of an irregular array of variously bonded hydroxy- and methoxy-substituted phenylpropane units [13]. The chemical structure varies depending on its source. Lignin is less polar than cellulose and acts as a chemical adhesive within and between fibers.

Pectins are complex polysaccharides whose main chain is a modified polymer of glucuronic acid and residues of rhamnose [3]. Side chains are rich in rhamnose, galactose, and arabinose sugars. Chains are often cross-linked by calcium ions improving structural integrity in pectin-rich areas [3]. Pectins are important in nonwood fibers especially bast fibers. The lignin, hemicelluloses, and the pectins collectively function as matrix and adhesive, helping to hold together the cellulosic framework structure of the natural composite fiber.

Natural fibers also contain lesser amounts of additional extraneous components including low molecular weight organic components (extractives) and inorganic matter (ash). Though often small in quantity, extractives can have large influences on

Table 11.1 Chemical composition of selected natural fibers [3].

Species	Cellulose	Lignin	Pectin
Flax	65–85	1–4	5–12
Kenaf	45–57	8–13	3–5
Sisal	50–64	—	—
Jute	45–63	12–25	4–10
Hardwood	40–50	20–30	0–1
Softwood	40–45	36–34	0–1

properties such as color, odor, and decay resistance [13]. The high ash content of some natural materials, such as rice hulls, causes some concern about their abrasive nature.

11.3.2 Fiber Dimensions, Density, and Mechanical Performance

Due to different species, a natural variability within species, and differences in climates and growing seasons, natural fiber dimensions as well as physical and mechanical performance can be highly variable. Methods of producing fibers with more reproducible properties are a major research effort [14].

Most natural fibers have a maximum density of about 1.5 g/cm^3. Though some natural fibers, such as wood, are hollow and have low densities in their native state, they are often densified during processing. Nevertheless, even the maximum density of these fibers is considerably less than that of inorganic fibers such as glass fibers. As such, their low density makes them attractive as reinforcement in applications where weight is a consideration.

Table 11.2 summarizes the dimensions and Table 11.3 the mechanical properties of selected natural fibers. Though variable, high aspect ratios are found especially for flax and hemp. The mechanical performance of the fibers is good but not as good as synthetic fibers such as glass. Variability in mechanical properties can be large and is due to influences such as species effects, growing conditions, and fiber harvesting or processing methods. However, their densities are considerably lower. The balance of significant reinforcing potential, low cost, and low density is part of the reason that they are attractive to industries such as the automotive industry.

11.3.3 Moisture and Durability

The major chemical constituents of natural fibers contain hydroxyl and other oxygen containing groups that attract moisture through hydrogen bonding [16]. The

Table 11.2 Dimensions of selected natural fibers.

	Length (mm)		Width (μm)		
Fiber type	Average	Range	Average	Range	References
Flax	33	9–70	19	5–38	[23]
Hemp[a)]	25	5–55	25	10–51	[23]
Kenaf	5	2–6	21	14–33	[23]
Sisal	3	1–8	20	8–41	[23]
Jute	2	2–5	20	10–25	[23]
Hardwood	1	—	—	15–45	[9]
Softwood	—	3–8	—	15–45	[8]

a) Industrial hemp is listed as a controlled substance in the United States and cannot be used in commercial production of composites.

Table 11.3 Mechanical properties of selected organic and inorganic fibers.

Fiber/fiber bundles	Density (g/cm^3)	Stiffness (GPa)	Strength (MPa)	Elongation at break (%)	References
Glass	2.49	70	2700	—	[24]
Kevlar	1.44	124	2800	2.5	[25]
Nylon 6	1.14	1.8–2.3	503–690	17–45	[25]
Polypropylene	0.91	1.6–2.4	170–325	80–100	[25]
Polyester (staple)	1.38	1.5–2.1	270–730	12–55	[25]
Flax[a)]	1.4–1.5	50–70	500–900	1.5–4.0	[3]
Hemp[a),b)]	1.48	30–60	300–800	2–4	[3]
Jute[a)]	1.3–1.5	20–55	200–500	2–3	[3]
Softwood	1.4	10–50	100–170	—	[3]
Hardwood	1.4	10–70	90–180	—	[3]

a) Fiber bundles.
b) Industrial hemp is listed as a controlled substance in the United States and cannot be used in commercial production of composites.

moisture content of these fibers can vary greatly depending on fiber type. The processing of the fiber can also have a large effect on moisture sorption. Table 11.4 shows the wide range of moisture contents for different natural fibers at several relative humidities.

This hygroscopicity can create challenges both in composite fabrication and in the performance of the end product. If natural fibers are used, a process that is insensitive to moisture must be used or the fibers must be dried before or during processing. Natural fibers absorb less moisture in the final composites since they are at least partially encapsulated by the polymer matrix. However, even small quantities of absorbed moisture can affect performance. Moisture can plasticize the fiber, altering the composite's performance. Additionally, volume changes in the fiber associated with moisture sorption can reduce fiber–matrix adhesion and damage the matrix [15]. Methods of reducing moisture sorption include adequately dispersing the fibers in

Table 11.4 Equilibrium moisture content at 27 °C of selected natural fibers [23].

	Equilibrium moisture content (%)		
Fiber	30% Relative humidity	65% Relative humidity	90% Relative humidity
Bamboo	4.5	8.9	14.7
Bagasse	4.4	8.8	15.8
Jute	4.6	9.9	16.3
Aspen	4.9	11.1	21.5
Southern pine	5.8	12.0	21.7
Water hyacinth	6.2	16.7	36.2
Pennywort	6.6	18.3	56.8

the matrix, limiting fiber content, improving fiber–matrix bonding, chemically modifying the fiber, or simply protecting the composite from moisture exposure.

Natural fibers undergo photochemical degradation when exposed to UV radiation [16]. They are degraded biologically because organisms recognize the chemical constituents in the cell wall and can hydrolyze them into digestible units using specific enzyme systems [16]. Though the degradability of natural fibers can be a disadvantage in durable applications where composites are exposed to harsh environments, it can also be an advantage when degradability is desired.

Due to their low thermal stability, natural fibers are generally processed with plastics where high temperatures are not required (less than about 200 °C). Above these temperatures, many of the polymeric constituents in natural fibers begin to decompose. Since cellulose is more thermally stable than other chemical constituents, highly pulped fibers that are nearly all cellulose have been used to extend this processing window [11, 17].

The release of volatile gases can, before, during, and after processing, lead to odor issues in applications where the composite is in an enclosed environment such as in many automotive applications and especially when moisture is present [18].

11.4
Suppliers

Natural fibers are used to manufacture a variety of products – linen, geotextiles, packaging, and specialty papers, for example. Natural fibers can be obtained from growers, distributors, importers, and as by-products from other manufacturing processes. Additionally, some companies sell semifinished products made from natural fibers (e.g., nonwoven mats) that can be further processed into composites. Due to the huge variety and diverse nature of natural fibers, there are currently few good resources that list a wide number of manufacturers. However, with the growing use of natural fibers in plastics, they are beginning to be listed in plastics industry resources. For example, the following is a list of some of the major suppliers of natural fibers to the North American composite industry from one industry resource [19][1]:

- Danforth Technologies (Point Pleasant, NJ, USA)
- JRS Rettenmaier (Schoolcraft, MI, USA)
- Kenaf Industries of South Texas (Lasara, TX, USA)
- Creafill Fibers (Chestertown, MD, USA)
- Rice Hull Specialty Products (Stuttgart, AR, USA)
- Stemergy (Delaware, Ontario, Canada)

Some of the major European fiber suppliers listed by another online database [20] are

- AGRO-Dienst GmbH, Germany
- Badische Naturfaseraufbereitung GmbH, Germany

1) Private communications with Principia Partners, Exton, PA.

- K.E.F.I. – Kenaf Eco Fibers Italia S.p.A., Guastalla, Italy
- Holstein Flachs GmbH, Mielsdorf, Germany
- Procotex SA Corporation, Belgium
- SANECO, France

11.5 Cost/Availability

Cost and availability of various natural fibers depend greatly on locale, region, import markets, and competing applications. For example, jute is commonly grown in India and Bangladesh, flax is prevalent in Europe, and many nonwood, natural fibers have to be imported in the United States. Although nonwood agricultural fibers and agricultural fiber wastes are abundant worldwide, their source can be diffuse and infrastructure for collection, purification, and delivery is sometimes limiting. Although there is increasing interest in commercial uses of industrial hemp worldwide, it is listed as a controlled substance in the United States and cannot be used in the commercial production of composites.

11.6 Environmental/Toxicity Considerations

The environmental benefits of wood and other natural fibers have been an important influence on their use, particularly in Europe. Natural fibers are derived from a renewable resource, do not have a large energy requirement to process, and are biodegradable [21].

Generally speaking, natural fibers are not particularly hazardous. However, natural fibers have low thermal stability relative to other reinforcing fibers and can degrade, release volatile components, and burn. Some basic precautions include avoiding high processing temperatures, using well-ventilated equipment, eliminating ignition sources, and using good dust protection, prevention, and control measures. Due to the wide variety of fibers classified as natural fibers, it is difficult to make specific comments. For information on environmental and health risks, users should consult their suppliers.

11.7 Applications (Primary and Secondary Functions)

Recently, there has been a resurgence of interest in the use of natural fibers as reinforcements in plastics due to their good mechanical performance, increasing ecological considerations in selecting materials, and legislative direction. Considerable funds have been expended on research trying to introduce flax, hemp, kenaf, and other natural fibers especially in the automotive industry with the greatest success in the use of mat technologies in panel applications. Considerable research and development is being undertaken to overcome existing limitations and expand natural fiber use into other areas such as injection-molded products.

The mechanical performance of natural fiber reinforced plastics varies greatly depending on the type of natural fibers, fiber treatments, type of plastics, additives, and processing methods. Natural fibers are added to plastics to improve mechanical performance such as stiffness and strength without increasing the density or cost too much. Though lower in mechanical performance than glass, the balance of properties of natural fibers along with other advantages such as lower density, aesthetics, and low abrasiveness during processing offer advantages in some applications. However, the generally low impact performance of natural fiber composites tends to limit their use [1].

Natural fibers are hydrophilic and do not tend to be easily wetted or bond well with many matrix materials, particularly the commodity thermoplastics. Coupling agents, such as maleated polyolefins, silanes, and isocyanates, are often necessary for adequate performance. A wide variety of coupling agents, fiber surface modifications, and treatments has been investigated for use in natural fiber plastic composites and are reviewed elsewhere [22].

Table 11.5 shows the mechanical performance of polypropylene composites made with several different natural fibers. Not surprisingly, the fibers (i.e., pulp, kenaf) are more effective reinforcements than the particulate (i.e., wood flour – low aspect ratio wood fiber bundles). Wood fibers are an order of magnitude stronger than the wood from which they derive [12] and the higher aspect ratio improves stress transfer efficiency, particularly when a coupling agent is used. Adding a maleated polypropylene coupling agent improves performance, especially flexural and tensile strengths and unnotched impact strength. Fiber preparation methods have a large effect on reinforcing ability. The high-performance dissolving pulp fibers have nearly all noncellulose components removed and are more effective than the lower cost, thermomechanical pulp fibers.

Recently, new compounding methods have been investigated to produce long, natural fiber-reinforced thermoplastic pellets and improve composite mechanical properties [31, 32]. For example, pellets have been formed by melt impregnation of continuous natural fiber yarns by pultrusion followed by cooling and chopping. Another method involves commingling of continuous forms of natural and synthetic fibers that are then heated, consolidated, and chopped.

Many reasons such as lack of familiarity and the current, limited availability have so far prevented large-scale penetration of these materials into broader injection-molded markets. However, double-digit growth is still expected as large companies enter the market and technological advances are made [32].

Environmental considerations are also driving increased use of wood and other natural fibers since they are derived from renewable resources, do not have a large energy requirement to process, and are biodegradable [6]. They are lighter than inorganic reinforcements, which can lead to benefits such as fuel savings when their composites are used in transportation and packaging applications. A comparison of the environmental impact of natural fibers versus glass fibers is presented in

Table 11.5 Effect of selected natural fibers on the mechanical performance of several polypropylenes (all composites contain 40 wt% fiber).

Filler type	Coupling agent	Izod impact[a)] Notched (J/m)	Izod impact[a)] Unnotched (J/m)	Flexural properties[b)] Maximum strength (MPa)	Flexural properties[b)] Modulus (GPa)	Tensile properties[c)] Maximum strength (MPa)	Tensile properties[c)] Modulus (GPa)	Tensile properties[c)] Elongation at break (%)	References
PP-1[d)]									
None	No	20.9	656	38.3	1.19	28.5	1.53	5.9	[26]
Wood flour[e)]	No	22.2	73	44.2	3.03	25.4	3.87	1.9	[26]
Wood flour	Yes[f)]	21.2	78	53.1	3.08	32.3	4.10	1.9	[26]
Thermomechanical pulp (softwood)[g)]	No	22.2	90	48.9	3.10	29.7	3.68	2.1	[26]
Thermomechanical pulp (softwood)	Yes[f)]	21.3	150	76.5	3.50	50.2	3.89	3.2	[26]
PP-2[h)]									
None	No	24	—	41	1.4	33	1.7	$\gg$10	[27]
Dissolving pulp (softwood)[i)]	Yes[f)]	—	—	82.7	3.43	60.4	4.67	4.5	j)
Kenaf[k)]	Yes[l)]	28	160	82	5.9	56	6	1.9	[27]

a) ASTM D-256 [28].
b) ASTM D-790 [29].
c) ASTM D-638 [30].
d) Fortilene 3907, polypropylene homopolymer, melt flow index = 36.5 g/10 min, Solvay Polymers, Deer Park, TX, USA.
e) Grade 4020 (American Wood Fibers, Schofield, WI, USA).
f) Maleated polypropylene (MP880, Aristech, Pittsburgh, PA, USA).
g) Laboratory produced from a mixture of pines.
h) Fortilene 1602, polypropylene homopolymer, melt flow index = 12 g/10 min, Deer Park, TX, USA.
i) High purity cellulose pulp (Ultranier-J, Rayonier Inc., Jessup, GA, USA).
j) Clemons, C.M. Unpublished data.
k) Kenaf strands obtained from AgFibers, Inc., Bakersfield, CA, USA.
l) Maleated polypropylene, G-3002, Eastman Chemical Company, Longview, TX, USA.

Chapter 7. With changing consumer perceptions, some manufacturers use the natural look of these composites as a marketing tool. Others have added natural fibers to increase bio-based material content. Natural fibers are often a preferred choice when reinforcing biodegradable polymers since natural fibers themselves are biodegradable.

References

1. Bledski, A.K. *et al.* (2002) Natural and Wood Fibre Reinforcement in Polymers, *Rapra Review Reports*, Rapra Technology Ltd, Shawberry, Shrewsbury, Shropshire, UK, 13 (8), 144 pp.
2. Rials, T.G. and Wolcott, M.P. (1997) Physical and mechanical properties of agro-based fibers, Chapter 4, in *Paper and Composites from Agro-Based Resources* (eds R.M. Rowell, R.A. Young, and J.K. Rowell), CRC Press, Inc., Boca Raton, FL, pp. 63–81.
3. Lilholt, H. and Lawther, J.M. (2000) Natural organic fibers, Chapter 1.10, in *Comprehensive Composite Materials, Vol. 1: Fiber Reinforcements and General Theory of Composites* (ed. T.-W. Chou), Elsevier, New York, pp. 303–325.
4. McGovern, J.N. *et al.* (1987) Other fibers, Chapter 9, in *Pulp and Paper Manufacture, Vol. 3: Secondary Fibers and Non-Wood Pulping,* 3rd edn (eds F. Hamilton, B. Leopold, and M.J. Kocurek), TAPPI, Atlanta, GA, pp. 110–121.
5. Plackett, D. (2002) The natural fiber-polymer composite industry in Europe – technology and markets. Proceedings of the Progress on Woodfibre-Plastic Composites Conference 2002, University of Toronto and Materials and Manufacturing Ontario, Toronto, ON.
6. Suddell, B.C. and Evans, W.J. (2003) The increasing use and application of natural fibre composites materials within the automotive industry. Proceedings of the Seventh International Conference on Woodfiber-Plastic Composites, Forest Products Society, Madison, WI.
7. Müssig, J., Karus, M., and Franck, R.R. (2005) Bast and leaf fibre composite materials, Chapter 10, in *Bast and Other Plant Fibres* (ed. R.R. Franck), Woodhead Publishing Limited, Cambridge, UK, pp. 345–376.
8. Young, R.A. (1997) Processing of agro-based resources into pulp and paper, Chapter 6, in *Paper and Composites from Agro-Based Resources* (eds R.M. Rowell, R.A. Young, and J.K. Rowell), CRC Press, Inc., Boca Raton, FL, pp. 63–81.
9. Miller, R.B. (1999) Structure of Wood, Chapter 2. *Wood Handbook: Wood as an Engineering Material,* General Technical Report FPL-GTR-113, USDA Forest Service, Forest Products Laboratory, Madison, WI, 463 pp.
10. Kaup, M. *et al.* (2002) Evaluation of a market survey 2002: the use of natural fibre in the German and Austrian automotive industries: status, analysis and trends. Presented at EcoComp 2003 Conference, September 1–2, 2003, London, UK.
11. Jacobson, R. *et al.* (2001) Low temperature processing of ultra-pure cellulose fibers into nylon 6 and other thermoplastics. Proceedings of the 6th International Conference on Woodfiber-Plastic Composites, Forest Products Society, Madison, WI, pp. 127–133.
12. Rowell, R.M. (1992) Opportunities for lignocellulosic materials and composites, Chapter 2, in *Emerging Technologies for Materials and Chemicals from Biomass* (eds R.M. Rowell, T.P. Schultz, R. Narayan), American Chemical Society, Washington, DC.
13. Pettersen, R.C. (1984) The chemical composition of wood, Chapter 2, in *The Chemistry of Solid Wood* (ed. R.M. Rowell), American Chemical Society, Washington, DC, pp. 76–81.
14. Kenny, J.M. (2001) Natural fiber composites in the European automotive industry. Proceedings of the 6th International Conference on Woodfiber-

Plastic Composites, Forest Products Society, Madison, WI, pp. 9–12.
15 Peyer, S. and Wolcott, M. (2000) Engineered Wood Composites for Naval Waterfront Facilities, 2000 Yearly Report to Office of Naval Research, Wood Materials and Engineering Laboratory, Washington State University, Pullman, WA, 14 pp.
16 Rowell, R.M. (1984) Penetration and reactivity of cell wall components, Chapter 4, in *The Chemistry of Solid Wood* (ed. R.M. Rowell), American Chemical Society, Washington, DC, p. 176.
17 Sears, K.D. *et al.* (2001) Reinforcement of engineering thermoplastics with high purity cellulose fibers. Proceedings of the Sixth International Conference on Woodfiber-Plastic Composites, Forest Products Society, Madison, WI, pp. 27–34.
18 Bledzki, A.K. *et al.* (2003) Odor measurement of natural fiber filled composites used for automotive parts. Proceedings of the 9th Annual Global Plastics Environmental Conference: Plastics Impact on the Environment, Detroit, MI, Society of Plastics Engineers.
19 Anonymous (2003) Special report: who's who in WPC, *Natural Wood Fiber Compos.*, **2** (8), 4.
20 N-FibreBase (2002) M-Base Engineering + Software GmbH, Aachen, Germany, www.n-fibrebase.net.
21 Suddell, B.C. and Evans, W.J. (2003) The increasing use and application of natural fibre composite materials within the automotive industry. Proceedings of the 7th International Conference on Woodfiber-Plastic Composites, Forest Products Society, Madison, WI.
22 Lu, J.Z. *et al.* (2000) Chemical coupling in wood fiber and polymer composites: a review of coupling agents and treatments. *Wood Fiber Sci.*, **32** (1), 88–104.
23 Rowell, R.M. *et al.* (2000) Characterization and factors effecting fiber properties. Proceedings of the Natural Polymers and Agrofibers Based Composites: Preparation, Properties, and Applications, Embrapa Instrumentação Agropecuária, São Carlos, Brazil, pp. 115–134.
24 Chamis, C.C. (1983) Laminated and reinforced metals, in *Encyclopedia of Composite Materials and Components* (ed. M. Grayson), John Wiley and Sons, Inc., New York, p. 613.
25 Billmeyer, F.W. Jr (1984) *Textbook of Polymer Science*, 3rd edn, John Wiley and Sons, Inc., New York, pp. 502–503.
26 Stark, N.M. (1999) *Forest Prod. J.*, **49** (6), 39–46.
27 Sanadi, A.R. *et al.* (1995) *Ind. Eng. Chem. Res.*, **34** (5), 1889–1896.
28 ASTM International (2002) Test Method D256-02e1. *Standard Test Methods for Determining the Izod Pendulum Impact Resistance of Plastics, Annual Book of ASTM Standards*, ASTM International, West Conshohocken, PA
29 ASTM International (2002) Test Method D790-03. *Standard Test Methods for Flexural Properties of Unreinforced and Reinforced Plastics and Electrical Insulating Materials*, Annual Book of ASTM Standards, ASTM International, West Conshohocken, PA
30 ASTM International (2002) Test Method D638-02a. *Standard Test Methods for Flexural Properties of Unreinforced and Reinforced Plastics and Electrical Insulating Materials, Annual Book of ASTM Standards*, ASTM International, West Conshohocken, PA.
31 Bledski, A.K., Jaskiewicz, A., Murr, M., Sperber, V.E., Lützendorf, R., and Reußmann, T. (2008) Processing techniques for natural- and wood-fibre composites, Chapter 4, in *Properties and Performance of Natural-Fibre Composites* (ed. K.L. Pickering), Woodhead Publishing Ltd, Cambridge, UK, pp. 163–192.
32 Carus, M. and Gahle, C. (2008) Injection molding with natural fibres, *Reinforced Plastics*, April, pp. 18–25.

12
Talc

Vicki Flaris

12.1
Production Methods

Talc is a natural mineral found worldwide and is the major constituent of rocks known as soapstone or steatite [1, 2]. Montana ores have 85–95% talc, while New York, Vermont, and Canada ores contain 35–60% talc, with the remainder of the ore being magnesium carbonate. The purest of talcs are found in Montana, while the whitest come from California; however, the latter are more abrasive because of asbestos-related hard contaminants. Vermont talcs contain higher percentages of magnesium and iron. Talcs can be gray, green, blue, pink, and even black [1, 3].

Talc is mined in open-pit (for the majority of talc deposits) or underground operations [3, 5]. There are seven to eight steps to talc production. The first step involves overburden removal. This involves the removal of waste rock covering the talc vein by giant shovels (which can shift up to 1500 tons of rock an hour) [4]. Second, the exposed talc is extracted using shovels, and different ore types are sorted; this step is known as the talc extraction step. Third, the crude ore is crushed with rollers or jaw crushers to a size of 10–15 cm and sorted according to content and brightness with techniques such as hand sorting or state-of-the-art laser image analysis technology.

The type of further processing depends on the purity of the ore (dry versus wet processing). Pure Montana talc can be dry processed. The fourth and fifth steps in talc manufacturing involve grinding and classification. It is in these areas that most advances have been made in the past 25 years as consumers have realized the effectiveness of fine particles (large surface area) over coarser ones [6]. Only for more demanding applications such as in the pharmaceutical industry is further purification necessary.

The size of the crushed ore can be further reduced using roller mills or cone crushers. The requirements of the talc–plastic composite determine the fineness of the talc. A standard grinder roller produces 50 μm particles, finer grades are between 10 and 40 μm, and the finest grades are between 3 and 10 μm [2]. Roller-milled products are used in low impact strength polymer-filled parts, such as fans in automotive under-the-hood applications. Fine micronization (1–12 μm) is carried

Functional Fillers for Plastics: Second, updated and enlarged edition. Edited by Marino Xanthos

ISBN: 978-3-527-32361-6

out in hammer mills, tube mills, pebble mills, fluid-energy mills, and now by using a new delaminating process [7]. Ball or rod milling with steel media can discolor the talc, so ceramic grinding media are used instead. Acceptable grinding rates can be achieved [5]. Grind is measured in the plant by top size, loose bulk density, and/or median particle size. Most suppliers report the last measurement. The median particle size can be measured using laser light scattering technique or using Stokes law settling rates [3]. It has only been possible to obtain submicron products since the advent of improved classification methods [6]. Classification is usually carried out by air methods. To obtain the desired grades, the choice of process conditions and type of equipment are critical.

High-purity talc (97–98%) is obtained by means of wet methods. Manufacturers use techniques such as froth flotation, sedimentation, spray drying, magnetic separation, centrifugation, and hydrocycloning. After techniques such as flotation have been applied, the material is filtered, dried, and milled by jet mill micronization or in an impact mills. Bleaching agents are used where brightness is a major concern.

The sixth and seventh steps involve treatment of certain talcs. Some talc grades are silane surface-treated for the rubber industry. Others are treated with glycol stearate to improve dispersibility and processing. Amine-coated talcs are used for fertilizers and cationic talcs for pitch control in papermaking. The surface treatment also helps with compatibilization reactions of certain components of polymer blends [1, 4].

The last step in the production of talc involves delivery of the powdered mineral in bags, semibulk bags, or bulk. Talc can also be delivered in pellet form or as slurry.

A major talc producer in Montana has achieved a 26% reduction in greenhouse gas emissions at one of its mills by increasing operational efficiency and equipment upgrades using robotics. Decreases of 26 and 15% in natural gas and water, respectively, have also been achieved by using a dry compacting process for densification.

12.2 Structure and Properties

Pure talc is a hydrated magnesium silicate with the chemical formula $Mg_3Si_4O_{10}(OH)_2$ that belongs to the group of phyllosilicates [9]. The center brucite plane is chemically bonded by bridging oxygen atoms to two tetrahedral silica planes (see Figure 12.1). Talc, unless heated at above 800 °C, has a plate-like structure (see Figure 12.2). The planar surfaces of the individual platelets are held together by weak van der Waals forces, which means that talc can be delaminated at low shearing forces. This makes the mineral easily dispersible and accounts for its slippery feel [1, 5]. The greasy feel is also due to its very low hardness (Mohs hardness = 1) and density of 2.7–2.8 g/cm^3 [9]. In contrast, the layers in mica are held together by ionic forces, whereas in kaolin hydrogen bonding forces hold aluminosilicate layers together. The size of an individual talc platelet can vary from 1 to >100 μm, depending on the deposit. The platelet size determines the talc's lamellarity. Highly lamellar talc has large individual platelets, whereas compact

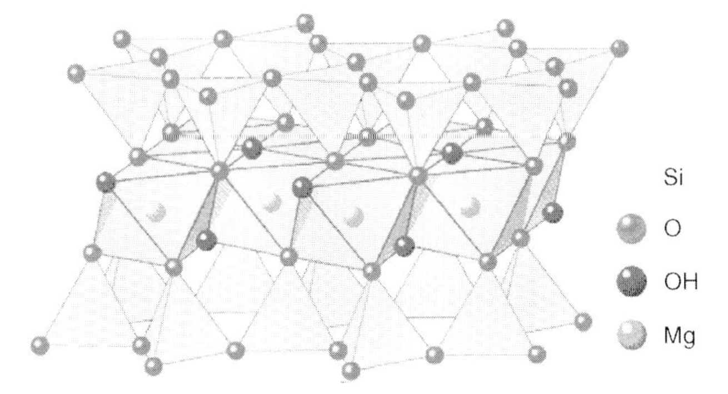

Figure 12.1 Molecular structure of talc. Magnesium has octahedral coordination. (Courtesy of Luzenac® Inc.)

Figure 12.2 Micrograph of Ultratalc 609. (Courtesy of Specialty Minerals, USA.)

(microcrystalline) talc has smaller platelets [4]. Talc is accompanied by the mineral chlorite in which Mg^{2+} ions have been substituted by Al^{3+} or Fe^{3+}. Other mineral contaminants are magnesium carbonate, mica, quartz, sericite, and often tremolite, a type of amphibole asbestos [10]. The composition of talc depends on the source and presence of tremolite. The US Montana talcs are considered to be free of asbestos and tremolite. California platy talcs contain minor amounts (<3%) of tremolite. Hard talcs contain 5–25% tremolite. Industrial talcs mined in New York contain 25–50% tremolite and other asbestiform minerals [1]. Certain talc compounds have odor issues depending on the source. The odor is believed to result from interaction of the talc with stabilizers added to the polymer-based formulation.

The theoretical chemical composition of pure talc by weight is 19.2% magnesium, 29.6% silicon, 50.7% oxygen, and 0.5% hydrogen. In terms of metal oxides, it is 31.7% MgO and 63.5% SiO_2 with the remaining 4.8% being H_2O. Other elements found in impure talcs in variable amounts are Ca, Al, and Fe. Trace elements include Pb, As, Zn, Ba, and Sb [1, 2].

Talc-filled composites have low gas permeability and high resistivity because of the plate-like nature of the impermeable talc particles and the resulting tortuous, complicated diffusion path. Talc is also unique in its ability to easily delaminate and can be used as a lubricant. Talc is the softest mineral on the Mohs hardness scale and is used as a standard. Commercial grades are, however, usually somewhat harder as a result of impurities [5]. In general, as the mineral is soft it is also less abrasive. This is advantageous as there is reduced wear on processing equipment (such as extruders). The surface of talc is hydrophobic, which has been explained in terms of the high ionic character of the central magnesium plane, uniform polarity, and symmetry of structure and neutrality of the layers. The hydrophobic nature of talc allows it to be more compatible with polymer resins. Hydrophobicity can be further increased through coating with zinc stearate. Physical and chemical properties of talc are summarized in Table 12.1 [1, 4, 5].

Particle size and shape are very critical with regard to the final mechanical properties of the composite. For a medium porosity particle material, the specific surface area is in the range 3–20 m^2/g. Typical particle size distributions of coarse and fine products are shown in Figure 12.3. Coarser products have an average size of about 10–20 μm and top size of up to 75 μm. For the coarser grades, particle size

Table 12.1 Characteristics of talc [1, 2, 35].

Property	Data
Crystal structure	Monoclinic
Typical chemical composition (wt%)	
MgO	24.33–31.9
SiO_2	46.4–63.5
CaO	0.4–13
Al_2O_3	0.3–0.8
Fe_2O_3	0.1–1.8
Platelet aspect ratio	5–20
Density (kg/m^3)	$2.58–2.83 \times 10^3$
pH	9.3–9.6
Oil absorption (ASTM D281)	20–57%
Refractive index	1.54–1.59
Mohs hardness	1–1.5
Brightness	78–93
Thermal conductivity (W/(K m))	0.02
Specific heat (J/(kg K))	8.7×10^2
Coefficient of thermal expansion (K^{-1})	8×10^{-6}

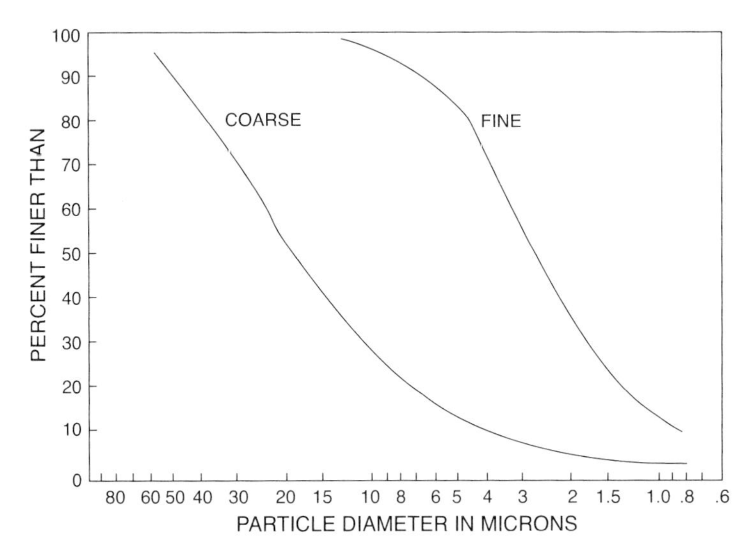

Figure 12.3 Typical particle size distribution of talc products [5]. (With permission from the publisher.)

information is indicated by top particle size sieve data. Particle size information for the finer grades is obtained by either sieve or sedigraphic light scattering water techniques. Fine talcs may be as low as 1 μm in size with an upper size limit of 12 μm. Particle thickness varies in the range 0.2–6 μm. Fine talcs, for example, the recent introduction of nanotalcs from Nanova LLC, are very important in plastics applications [6].

Talc is inert to most chemical reagents. It contains about 5 wt% water, which is chemically bound to the magnesium oxide or the brucite layer. The associated water is lost between 380 and 500 °C. Above 800 °C, talc progressively loses hydroxyl groups and above 1050 °C, recrystallizes forming enstatite (anhydrous magnesium silicate) through an endothermic reaction and liberation of water. The estimated melting point of talc is 1500 °C [1, 3, 5].

The surface chemistry of talc is not well understood. Reactive groups that form upon talc's fractured surface are (a) weakly acidic terminal hydroxyl groups ($HO{-}Si \rightarrow H^+ + [O{-}Si]^-$); (b) sites of proton release through polarization of water molecules ($Mg^{2+} + H_2O \rightarrow 2H^+ + MgO$); (c) Lewis acid sites (metal ions react with paired electrons on water molecules); (d) octahedral iron Ox/redox sites (Fe^{II}/Fe^{III} for any Fe present in talc ore); (e) strongly acidic Brønsted sites ($Mg^{2+} + HO{-}Si{-} \rightarrow H^+ [Si{-}O{-}Mg]^+$); and (f) weakly basic sites ($Mg(OH)_2$) [3]. Interaction with polymeric matrices may presumably be achieved through some of the above functionalities. Luzenac® R7 talc is a result of improvements in a proprietary surface modification technology. R7 talc synergistically interacts with the rubber domains of a thermoplastic olefin elastomer (TPO) matrix to improve flexural modulus and maintain impact strength. It also improves UV stability, color retention, and thermal

Table 12.2 Thermal stability of homopolymer PP (a comparison between unmodified and surface-modified talc of same particle size) [11].

	Days to failure			
	0 wt % talc	15 wt % talc	27 wt % talc	40 wt % talc
Luzenac® R7 talc	132	118	105	90
Unmodified talc	132	95	80	18

stability (see Table 12.2). R talc is effective with ethylene propylene diene monomer (EPDM) and ethylene propylene rubbers (EPR) but not as effective with metallocene elastomers containing TPOs. Recently, Clark [11] presented evidence of improvements in scratch and mar resistance when talc is grafted to TPO. The need for higher performance materials will drive the development of new surface modification technologies [3].

12.3 Suppliers

Rio Tinto Minerals is responsible for 25% of the world talc production and is the largest producer in North America with largest mine in Montana. Its total production increased from 1.26 million tons in 2000 to 1.33 million tons in 2002 following the acquisition of the Three Springs mine in Australia [12]. Production for 2007 was around 1.28 million tons affected by the slow housing market [13]. Major talc suppliers and grades recommended for use in plastics are listed in Table 12.3.

12.4 Cost/Availability

In mid-2008, most suppliers quoted typical grade prices in the range of US $0.15–0.88/kg ($0.07–0.40/lb), with some premium grades costing up to $1.54/kg ($0.70/lb). Availability of any of the talcs does not seem to be an issue. Talc production declined by 6%, and sales by 9%, from that of 2006 due to a slow housing market. US exports of talc increased by 23% and US imports decreased by 33% compared to 2006; these are attributed to the decline in the US dollar relative to other currencies. Canada remained the major destination for US talc, accounting for 11% of tonnage [8]. The major input sources to the United States are China (48%) and Canada (35%). The talc industry in Texas has been consolidating for the past 2 years. A European investment firm has purchased a major talc producer with mines in Finland, with its plants there, and also in the Netherlands,

Table 12.3 Major talc suppliers.

Supplier	Grades
ACC Resources Co., LP (distributors of Asada Japan) One Maynard Drive, Park Ridge, NJ 07656, USA Tel: (201) 307 1500 Fax: (201) 307 0540 Web site: www.accr.com	BHS-325, 400, 600, 800, 1000, 1250, 1500, 2000, 5000
American Talc Co. (bought Zemex Industrial Minerals) P.O. Box 1048, Van Horn, TX 79855, USA Tel: (432) 283 2330 Fax: (432) 283 2569 Web site: http://americantalc.com	
IMI Fabi USA (produces talc at Natural Bridge, NY, Benwood, WV sites) Benwood Plant, Second & Marshall Street, Benwood, WV 26031, USA Tel: (304) 233 0050 Fax: (304) 232 0793 Web site: www.hitalc.com; www.imifabi.com	Benwood 2202, 2203, 2204, 2207, 2210, 2213 Talc HTPultra 5, 5c, 10, 10c Talc HTP 05, 05c, 1c, 2, 3, 4 PP500, 1250
Non-Metals Inc. – affiliate of CNMIEC, China 1870 W. Prince Rd., Suite 67, Tucson, AZ 85705, USA Tel: (520) 690 0966, 1-800-320 0966 Fax: (520) 690 0396 Web site: www.nonmetals.com	
Rio Tinto Minerals America (made up of Luzenac®, Borax and Dampier Salt) 8051 E. Maplewood Avenue, Building 4, Greenwood Village, CO 80111, USA Tel: (303) 713 5000 Fax: (303) 713 5769 Web site: www.riotinto.com	Cimpact® 550, 610, 699, 710, 710R, 710HS, CB7 Jetfil® T290, P200, T390, P350, P500, 575C, 625C, 700C, 7C Nicron® 403, 674 Mistron® Vapor R, Vapor RE, ZSC, AB, NT, 400C, 554 Artic Mist® Stellar® 420, 510 Vertal® 92, 97, 7, 77, 503 Luzenac® 8230 Luzenac® R7 Silverline® 002, 403 Jetfine 3CA, 1CA, 0.7CA Steaplast® 9502 Luzenac® HAR® T-84, W-92 Mistron® HAR

(Continued)

Table 12.3 (*Continued*)

Supplier	Grades
Specialty Minerals Inc. (wholly owned subsidiary of Mineral Technologies Inc.) 35 Highland Ave., Bethlehem, PA 18017, USA Tel: 1-800-801 1031 Fax: (610) 882 8726 Web site: www.mineralstech.com	MICROTALC® MP 10-52, 12-50, 30-36, 44-26 MICROTALC® MPD 10-52, 12-50 ULTRATALC® 609, 609D ABT® 2500 CLEARBLOC® 80 MICROTUFF® AG 101, 111, 121, 191, 262, 445, 609 AGD 111, 191, 609 POLYBLOC® antiblock 9102, 9103, 9103S, 9107, 9110, 9310m 9405, 9410 OPTIBLOC® 10, 25 FLEX TALC® 122, 405, 610, 815 ULTRATALC® 609 & MICROTUFF® AG 609 – ultrafine grades less than 1 μm; new FLEX TALC® 405 (0.6 μ)
R.T. Vanderbilt Company Inc. 30 Winfield St., P.O. Box 5150, Norwalk, CT 06856-5150, USA Tel: (203) 853 1400 Fax: (203) 853 1452 Web site: www.rtvanderbilt.com	Nytal® 100, 100HR, 200, 3X, 300, 3300, 400, 5X, 6600, 7700 Vantalc® 6H, 6H-II, F2003, F2504, PC, R, 2000, 2500, 3000, 3100, 3500, 4000, 4500

becoming the second leading producer of talc in the world and leading supplier of talc in the European paper industry.

12.5 Environmental/Toxicity Considerations

Safety issues regarding talc have been controversial. The debate revolves around asbestos-type impurities, particularly tremolite. Most commercial talc grades contain no asbestos, and those that do, contain only trace amounts. The Occupational Safety and Health Administration (OSHA) and Mine Safety and Health Administration (MSHA) have an 8 h time-weighted exposure limit for non-asbestos containing talc dust of $3 \times 10^{-6}\,kg/m^3$. For personal product industries, more rigorous controls are instituted and set by the American Industrial Hygiene Association (AIHA) and the Cosmetic Toiletry and Fragrance Association (CTFA). Talc is approved by the US Food and Drug Administration (FDA) for use in polymeric compounds in contact with food (Listing – Title 21, Code of Federal Regulations Part 178.3297, "Colorants for Plastics"). There are minimal concerns with skin contact, other than dryness with continuous exposure. Eye contact causes a mild mechanical irritation, while ingestion is of no concern. Approved dust masks should be worn when the American Conference of Governmental Industrial Hygienists (ACGIH) threshold limit value (TLV) of 2×10^{-6} kg/m^3 is exceeded according to the National Institute of Occupational Safety and Health (NIOSH) and now also OSHA. Studies by Porro *et al.*, Siegal *et al.*, McLaughton *et al.*, and Kleinfeld *et al.* [14] have shown that long-term inhalation can lead to mild scarring of the lungs (pneumonoconiosis symptoms: wheezing, chronic cough, and shortness of breath). Pneumonoconiosis was found to be caused by fibrous varieties of talc, and the particle length, rather than the composition of the talc, seemed important. Other adverse health effects are associated with covering of the lungs (pleural thickening) [3, 5, 15].

12.6 Applications

12.6.1 General

Talc is an important reinforcing filler for plastics, in particular polypropylene (PP). The major benefits of incorporating talc into plastics are summarized in Table 12.4 according to its primary and secondary functions, with examples obtained from Refs [1–3, 8, 16–18]. Primary reasons for using talc include improvements in mechanical properties such as heat deflection temperature (HDT), rigidity, creep resistance and sometimes impact resistance, and lower shrinkage. Additional secondary benefits, because of the flaky nature of talc, include improvements in dimensional stability (as it orients along flow lines during molding) and lower permeability; other benefits include reduced coefficient of linear thermal expansion (CLTE), increased brightness, reduction of injection molding cycle due to nucleation and its use as an antiblock additive [10, 11, 16, 19, 20]. Adverse effects include reduction in toughness and elongation at break in certain polymers, reduced weld line strength, and for certain polymer/stabilizer package combinations reduction in long-term thermal aging and UV resistance. In the rubber industry, talc is used to increase stiffness and processability.

Talc-filled masterbatches are available up to 75 wt% loadings. Talcs densified through zero force compaction/densification technology (Rio Tinto) have shown better mechanical properties and higher throughput rates during processing than regular talcs using compressive forces of pelletization [3]. Dark talc is used where color is not important, such as in certain exterior and interior automotive applications, and when the reinforcing properties are not as critical. White talcs are used in applications involving washers and dryers and garden furniture. Lower flexural modulus composites are based on talcs containing more granular minerals such as quartz and magnesite. Higher flexural strength and modulus composites are based on talc grades that are more homogeneous in particle size distribution and contain higher percentage of coarser particles [10].

Table 12.4 Primary and secondary benefits of talc in plastics.

Functionality	Examples
Primary function	
Major improvements in HDT	• In homopolymer PP increase by 60 °C at 40% loading • In PP copolymer increase by 75 °C at 40% loading • In commingled postconsumer stream (80% PE, remaining PET, PS, PP, PVC) at 40% loading, a 6 °C increase; in maleated stream with 40% loading, a 13 °C increase
Major improvements in modulus	• In homopolymer PP double stiffness at 40% loading • In PP copolymer fourfold increase at 40% loading • In commingled stream at 40% loading, a 130% increase
Minor improvements in impact resistance for certain polymers	• In PS with an elastomer and talc (see Table 12.6) • Only maleated commingled stream with 40% loading showed an increase in Izod impact strength (22%)
Increase in tensile strength	• 2–10% concentration of talc in HDPE increases tensile strength by 15–80% • In commingled postconsumer stream at 40% loading an 8% increase, whereas in maleated stream with 40% loading, a 19% increase
Secondary function	
CLTE/mold shrinkage decrease	• At 30% loading, mold shrinkage decreased by 57% in homopolymer PP and 39% in PP copolymer
Lower permeability	• With 2–10% talc loading in HDPE, a decrease in permeability of 15–55%
Nucleation capacity	• Talc-filled PS in PS foam
Enhance moisture barrier	• 20% loading in PP increased barrier by 50%
Antitack agent	• Rubber blends

12.6.2 Applications by Polymer Matrix

12.6.2.1 Polyolefins

Owing to its platy nature, talc is used in linear low-density polyethylene (LLDPE) as an antiblocking agent preventing two or more contacting film layers from sticking together (see Chapter 20) and as a nucleating agent [10]. Recently, talc has found some new applications, such as reducing the necessary dosage of fluorocarbon elastomer polymer processing aids (PPAs) and enhancing the moisture barrier in packaging films [17, 21–24]. Because of its nonpolar nature, a major application of talc in high-density polyethylene (HDPE) (where a 2–10 wt% concentration leads to a 15–80% increase in tensile strength) is in wire and cable. Grades coated with zinc stearate are used in cross-linked low-density polyethylene (XLPE) wire coatings as flame retardancy aids. These talc grades increase char build up and also act as thixotropic agents and reduce dripping.

Table 12.5 Typical properties for talc-filled homopolymer and copolymer PP [5, 18].

Property	Homopolymer			Copolymer		
	Unfilled	20 (wt%)	40 (wt%)	Unfilled	20 (wt%)	40 (wt%)
Density (10^3 kg/m^3)	0.903	1.05	1.22	0.899	1.04	1.22
Flexural modulus (MPa)	1655	2482	3275	756	2206	2896
Yield tensile strength (MPa)	35.5	34.1	31.4	27.6	27.9	25.8
Rockwell R hardness	99	98	95	82	87	85
Heat deflection temperature (°C) (455 kPa)	97	123	131	85	117	12 7
Notched Izod impact strength, J/m (22 °C)	45.7	32.0	20.9	133.5	53.4	32.0

Talc is mostly used with PP (over 200 000 tons each year) with typical loadings in the range 10–40 wt% [25, 26]. It provides benefits such as stiffness, dimensional stability, enhanced thermoforming, opacity, whiteness, and high-temperature heat resistance for automotive (fan shrouds and blades) and appliance (washer tubs, pump housings, and spin baskets) applications [27]. Some small amounts of stabilizers are also needed for more demanding automotive applications, such as under-the-hood, bonnet, dashboard, bumper, and interior and exterior trim. Other applications include fascias, grills, front module panels, and kickplates [25]. The purity of talc affects its efficiency in improving thermal properties since even low levels of metal ions can catalyze polymer degradation. Disadvantages of talc in PP applications include low scratch resistance and low impact strength. Table 12.5 summarizes the effects of mineral loading on the properties of PP homopolymers and copolymers [5, 18].

Increasing talc concentration has a direct effect on stiffness, while particle size has a less pronounced effect. Macrocrystalline talc (talc with large aspect ratio, length to thickness) will impart greater stiffness than microcrystalline talc. The addition of talc generally reduces notched impact strength, the effect depending largely on particle size. For example, in a 30% loaded copolymer blend, fine talc can promote ductile fracture whereas coarse talc can promote a brittle fracture. The fineness of talc is even more critical for low-temperature impact strength. Figure 12.4 summarizes the wide range of stiffness/toughness properties achieved with different talc grades in polypropylene resins [28].

Studies have shown that the best balance of Izod impact strength and flexural modulus is achieved by blending PP containing more than 20 wt% of a metallocene α-olefin, such as ethylene/1-octene, with fine talc (median particle size 1 μm) at levels greater than 10% [29–31]. Alternatively, polypropylene-talc compounds (containing 30 wt% of high aspect ratio talc), offer excellent stiffness/impact balance without the elastomer phase. These compounds further enhance mechanical and thermal properties, as well as reduce shrinkage after molding. They also facilitate recycling possibilities due to the absence of the elastomer phase and allow selective sorting by flotation because of their low densities [32].

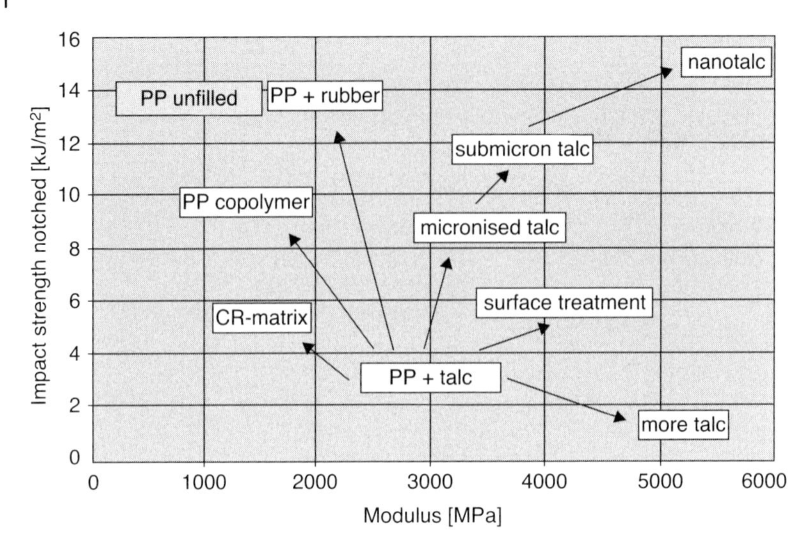

Figure 12.4 Polypropylene–talc systems covering a wide range of stiffness/toughness properties [28]. (With permission from the publisher.)

Among other properties, talc is beneficial in reducing mold shrinkage. Clark and Steen [3] have shown that at a 30 wt% loading, mold shrinkage is reduced by 57% in homopolymers and by 39% in impact copolymers. In 30% talc-filled PP compounds, the filler decreases the coefficient of thermal expansion by half in the temperature range 50–150 °C; thermal expansion does not seem to be affected by particle size. At a 15 wt% loading, talc increases the PP crystallization temperature from 115 to 123 °C. Talc can also be used to impart barrier properties in PP films. At a 20 wt% loading, the moisture vapor transmission rate (MVTR) in a 0.2–0.25 mm film is reduced by 50%. The thermal stability is affected by the crystallinity of the talc (high purity is thermally more stable), specific surface area, and its heavy metal content. Heavy metals linked to the silica lattice slightly affect thermal stability, while carbonate ions can decrease the thermal performance of talc-filled PP compounds [3, 18].

Surface treatment improves properties by increasing dispersion of the particles and providing adhesive sites [33]. It has been shown that maleic anhydride-based coupling agents can increase tensile and flexural properties [16]. For 20 and 40% talc-filled PP, levels of 20 wt% of an acrylic acid-modified polypropylene yield optimal physical properties, such as HDT, tensile, and flexural properties, but do not affect the impact strength [33]. The bonding mechanism is believed to involve (a) bonding between magnesium ions of the talc surface and acrylic acid of the modified PP; a magnesium salt is formed and complexed by oligomeric acrylate ions and held between platelets of the poly ($Si_4O_{11}{}^{6-}$) ions and (b) cocrystallization of the chains of the acid-modified PP with the PP matrix. Particle size recommendations for talc-filled polyolefins are summarized in Figure 12.5 [3].

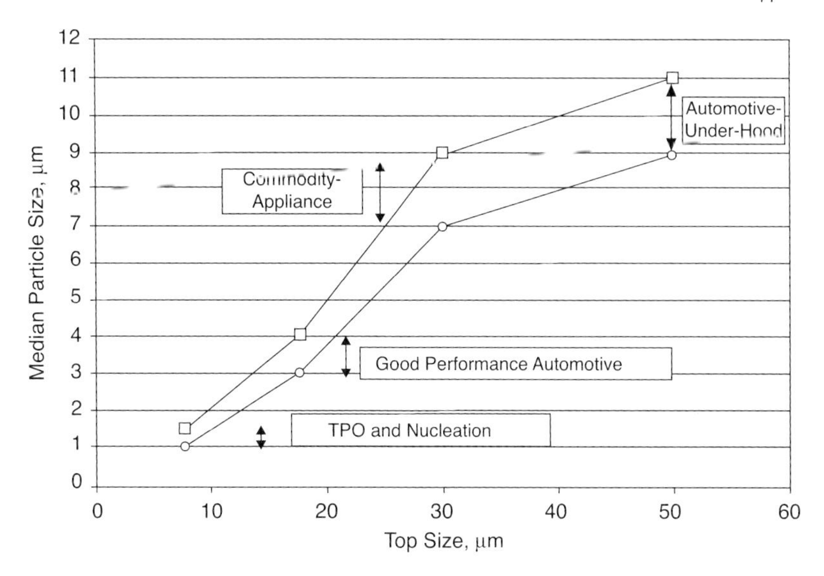

Figure 12.5 Range of particle sizes for talc-filled polyolefin applications [3]. (With permission from the publisher.)

12.6.2.2 **Polyvinyl Chloride**

A major application of talc in polyvinyl chloride (PVC) is in building products such as flooring where talc loadings can be as high as 50%. The talc has to be ultra dry in this application to prevent streaking (moisture vaporization during processing). Addition of 20% talc to PVC increases its flexural modulus by about 56%. This increase will be manifested at elevated temperatures, up to a level at which the resin begins to soften. Impact strength can be increased by the addition of relatively high levels (10 phr) of an acrylic impact modifier [34]. Another minor application of surface treated talcs is in the wire and cable industry, where requirements of tensile strength and dielectric properties must be met [5].

12.6.2.3 **Styrenics and Thermoplastic Elastomers**

In polystyrene (PS), talc is used in combination with an elastomer to overcome reduction in impact properties (note the improvement in Table 12.6 from Ref. [5]).

Table 12.6 Properties of elastomer-modified polystyrene-containing talc [5].

	High impact polystyrene	40% surface-treated talc, 10%TPE, 50% crystal PS	30% surface-treated talc, 10% TPE, 60% crystal PS
Falling weight impact strength (J)	4.3	3.6	3.6
Notched Izod impact strength (J/m)	69.4	74.8	101.5
Flexural modulus (GPa)	2.38	4.2	3.71
Tensile yield strength (MPa)	37.1	35.0	37.1

Another commercial application is in PS foams, where it is used as a cell-nucleating agent. In rubber blends, talc is used as an antitack agent to prevent newly formed goods from sticking together. In some specialty elastomers, it reduces air and fluid permeability [5].

Other market applications of talc involving polymers outside the filled thermoplastics industry include [3, 5]

- interior and exterior architectural coatings, where it is used to control gloss and sheen, improve opacity, tint strength, and weatherability, and enhance viscosity and sag resistance;
- industrial coatings to provide flatting, sandability, and enhance package stability and water and chemical resistance;
- paints for hiding power, matting effect, and satin finish;
- roofing.

References

1 Wypych, G. (1993) *Fillers*, ChemTec Publishing, Toronto, Ont., Canada, pp. 43–46.
2 Wypych, G. (2000) *Handbook of Fillers*, ChemTec Publishing, Toronto, Ont., Canada, pp. 150–153, 663–667.
3 Clark, R.J. and Steen, W.P. (2003) Chapter 8, in *Handbook of Polypropylene and Polypropylene Composites* (ed. H.G. Karian), Marcel Dekker Inc., New York, pp. 281–309.
4 www.riotinto.com.
5 Sekutowski, D. (1996) Section IV, in *Plastics Additives and Modifiers Handbook* (ed. J. Edenbaum), Chapman & Hall, London, UK, pp. 531–538.
6 Harris, P. (2003) Nanominerals, *Ind. Miner.*, **443**, 60–63.
7 Lamellar filler process for treatment of polymers, PCT WO 98/45374.
8 http://minerals.usgs.gov.
9 Kromminga, T. and Van Esche, G. (2001) Chapter 7, in *Plastics Additives Handbook* (ed. H. Zweifel), Hanser Publishers, Munich.
10 Cordera, M. (1998) Proceedings of the Functional Fillers and Fibers for Plastics 98, Intertech Corp. 5th International Conference, Beijing, PR China, June 1998.
11 Clark, R. (2003) Proceedings of the Functional Fillers for Plastics 2003, Intertech Corp., Atlanta, GA, October 2003.
12 www.roskill.com/reports/talc.
13 asfdssfd www.riotinto.com/what we produce/452=talc_4034asp.
14 www.cdc.gov/niosh/pe188/14807-96.html.
15 www.rtvanderbilt.com.
16 Xanthos, M., Grenci, J., Patel, S.H., Patel, A., Jacob, C., Dey, S.K., and Dagli, S.S. (1995) Thermoplastic composites from maleic anhydride modified post-consumer plastics. *Polym. Compos.*, **16** (3), 204.
17 Deutsch, D.R. and Radosta, J.A. (1999) Proceedings of the POLYOLEFINS XI, The SPE International Conference on Polyolefins, Houston, TX, pp. 657–677.
18 Radosta, J.A. (1995) Proceedings of the Functional Fillers for Plastics 95, Intertech Corp. Houston, TX.
19 Harris, T. (2003) Proceedings of the Functional Fillers for Plastics 2003, Intertech Corp. Atlanta, GA, October 2003.
20 Gill, T.S. and Xanthos, M. (1996) Effects of fillers on permeability and mechanical properties of HDPE blown films. *J. Vinyl Addit. Technol.*, **2** (3), 248.
21 Fazzari, A.M. (1997) *Modern Plastics Encyclopedia*, vol. **74**, No. 13, McGraw-Hill, New York, p. C3.
22 Graff, G. (1998) *Mod. Plast.*, **75** (5), 32–33.
23 Chapman, G.R. *et al.* (1998) The SPE International Conference on Additives for Polyolefins, Houston, TX, pp. 149–171.
24 Amos, S. and Deutsch, D.R. (1999) *Proceedings of the TAPPI Polymers,*

Laminations and Coatings Conference, TAPPI Press, Atlanta, GA, pp. 829–847.

25 Kochesfahani, S., Crépin-Leblond, J., and Jouffret, F. (2007) SPE Automotive TPO Global Conference, Sterling Heights, MI (CD).

26 Kochesfahani, S., Leblond, J., and Jouffret, F. (2007) SPE International Polyolefins Conference, Houston, TX.

27 Posch, W. (2003) Proceedings of the Functional Fillers for Plastics 2003, Intertech Corp., Atlanta, GA, October 2003.

28 Holzinger, T. and Hobenberger, W. (2003) *Ind. Miner.*, **443**, 85–88.

29 Wernett, P.C. *et al.* (2004) TPO Conference, Dearborn, MI.

30 Huneault, M.A. *et al.* (1998) Proceedings of the 56th Annual Technical Conference Society of Plastics Engineers, SPE, vol. 44.

31 Walton, K.L. and Clayfield, T. (2000) Proceedings of the 58th Annual Technical Conference Society of Plastics Engineers, SPE, vol. 46, pp. 2623–2627.

32 Shearer, G. and Kochesfahani, S. (2007) SPE Automotive TPO Global Conference, Sterling Heights, MI (CD).

33 Adur, A.M. and Flynn, S.R. (1987) Proceedings of the 45th Annual Technical Conference Society of Plastics Engineers, SPE, vol. 33, p. 508.

34 Wiebking, H.E. (2005) Proceedings of the 63rd Annual Technical Conference Society of Plastics Engineers, SPE, pp. 3874–3878.

35 Hohenberger, W. (2001) Chapter 17, in *Plastics Additives Handbook* (ed. H. Zweifel), Hanser Publishers, Munich.

13
Kaolin*

Joseph Duca

13.1
Introduction

The term kaolin encompasses a group of minerals, the dominant one being kaolinite. In industry, the term kaolin mainly refers to the mineral kaolinite, and this term will be used throughout in this chapter. The lesser minerals of the kaolin group comprise hydrated aluminosilicates such as dickite, nacrite, and halloysite [1]. Structurally, kaolinite consists of an alumina octahedral sheet bound on one side to a silica tetrahedral sheet, stacked alternately. The two sheets of kaolinite form a tight fit with the oxygen atoms forming the link between the two layers (Figure 13.1) [2]. The theoretical composition for the $Al_2O_3{\cdot}2SiO_2{\cdot}2H_2O$ mineral is 46.3% SiO_2, 39.8% Al_2O_3, and 13.9% H_2O [3].

Kaolin is considered to be a phyllosilicate mineral. Phyllosilicates are characterized by an indefinitely extended sheet of rings, in which three of the tetrahedral oxygens are shared whereas every fourth oxygen atom is apical and points upward. These phyllosilicates also often have a hydroxyl group centered between the apical oxygens. This occurs through bonding of the silica sheet to a continuous sheet of octahedra, with each octahedron tilted onto one of its triangular sides. In kaolin, these octahedra contain the trivalent aluminum cation. To balance the charge, only two of every three aluminum octahedral positions are occupied by aluminum cations to form a gibbsite structure. Hence, a layer of silica rings is joined to a layer of alumina octahedra through shared oxygens resulting in a plate-like morphology. The individual kaolin particle has an oxygen surface on one side and a hydroxyl surface on the other side as shown in Figure 13.1 [2]. This means that such layers can stack through hydrogen bonding to the lamella above and below. Consequently, kaolin is often shown to be in what are called "booklets," which are stacks of plates, one on top of the other and connected through hydrogen bonding (Figure 13.2). Interestingly, talc, on the other hand, has a characteristic soft, slippery feel, due to easy sliding and delamination of its platelets, which

* This chapter is essentially version that appears in 2005 final edition, with minor modifications by the Editor to reflect changes in the suppliers' names. Following the acquaintance of Engelhard Corporation by BASF Corp. Mr. J. Duca was unavailable to update/revise this chapter.

Functional Fillers for Plastics: Second, updated and enlarged edition. Edited by Marino Xanthos

ISBN: 978-3-527-32361-6

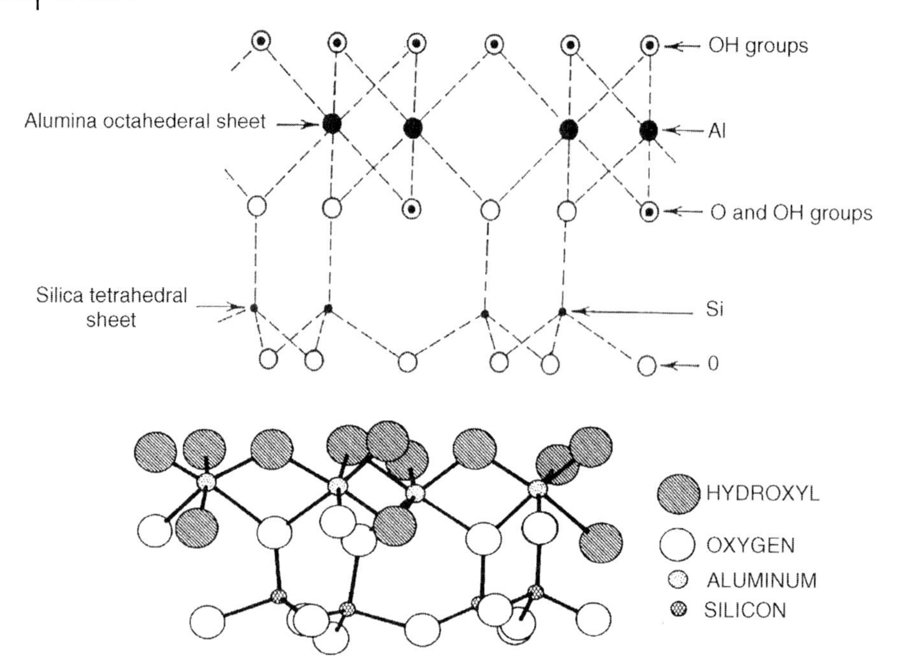

Figure 13.1 Structure of kaolin.

are held through relatively weak van der Waals forces (see also Chapter 12) [1]. In contrast, kaolin interlaminar bonding is due to stronger hydrogen bonding involving an oxygen face from the silica layer bonded to a hydroxyl face from the gibbsite layer. Unlike talc, kaolin booklets can be delaminated only by significant grinding [1].

Kaolin is an extremely versatile white functional filler, with applications in paper coating, paper filling, paints, plastics, rubber, and inks. The plastics and adhesive

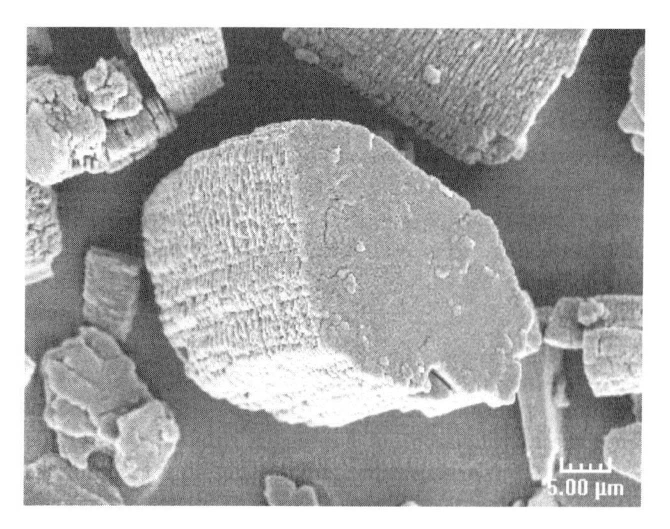

Figure 13.2 "Booklets" of kaolin.

industries consume some 65 000 tons of kaolin in the United States per year. This can be expected to increase, especially as the price of resins increases. Among other markets are pharmaceuticals, cracking catalysts, and ceramics.

13.2 Production Methods

Kaolins are classified as either primary or secondary. Primary kaolins are formed by the alterations of crystalline rocks such as granite. The source of this kaolin is found where it is formed. Conversely, secondary kaolin deposits are sedimentary and are formed by erosion of primary deposits. The secondary deposits contain much more kaolinite (about 85–95%) than the primary deposits, which contain only 15–30%. The balance of the ore consists of quartz, muscovite, and feldspar in the primary deposits and quartz, muscovite, smectite, anatase, pyrite, and graphite in the secondary deposits. Kaolin, also known by the common term clay, is usually open-pit mined in the United States from vast deposits in Georgia, South Carolina, and Texas. The ore is not processed in one singular way. There are also distinct methods of ore beneficiation, each adding value to the mineral.

13.2.1 Primary Processing

Commercially available kaolin grades can be produced by air floating or wet processing. Air-floated kaolin is the least processed, and therefore the least expensive. Here, the mineral is crushed, dried, and pulverized. Thereafter, it is floated in an air stream and classified using an air classifier. The finer particles are separated from the coarse ones and from the non-kaolin particles, which are referred to as grit.

The much more sophisticated wet process delivers products that are higher in purity, with consistent high quality and with a broader range of particle sizes and distributions. The wet process starts near the mining site with the formation of a clay–water suspension in a step referred to as "blunging." In this step, kaolin slurry is produced to contain 40–50% solids. This slurry is then pumped into the processing plant for degritting, manipulation of particle size distribution and/or morphology, leaching of color bodies, and other value-adding processing steps. Finally, apron drying and pulverization or spray drying into bead form takes place. A schematic of the wet process containing the various beneficiation steps discussed below is shown in Figure 13.3. The final product is packaged in 25 kg–1 ton bags. Twenty-ton containers are often employed for overseas shipments. In addition, the final product might arrive as a high solid content slurry, which is transported to the customer in rail cars.

13.2.2 Beneficiation

Particle size and shape may be manipulated by centrifugation using a combination of solid-bowl and disk-nozzle centrifuges to classify the particles according to size. The

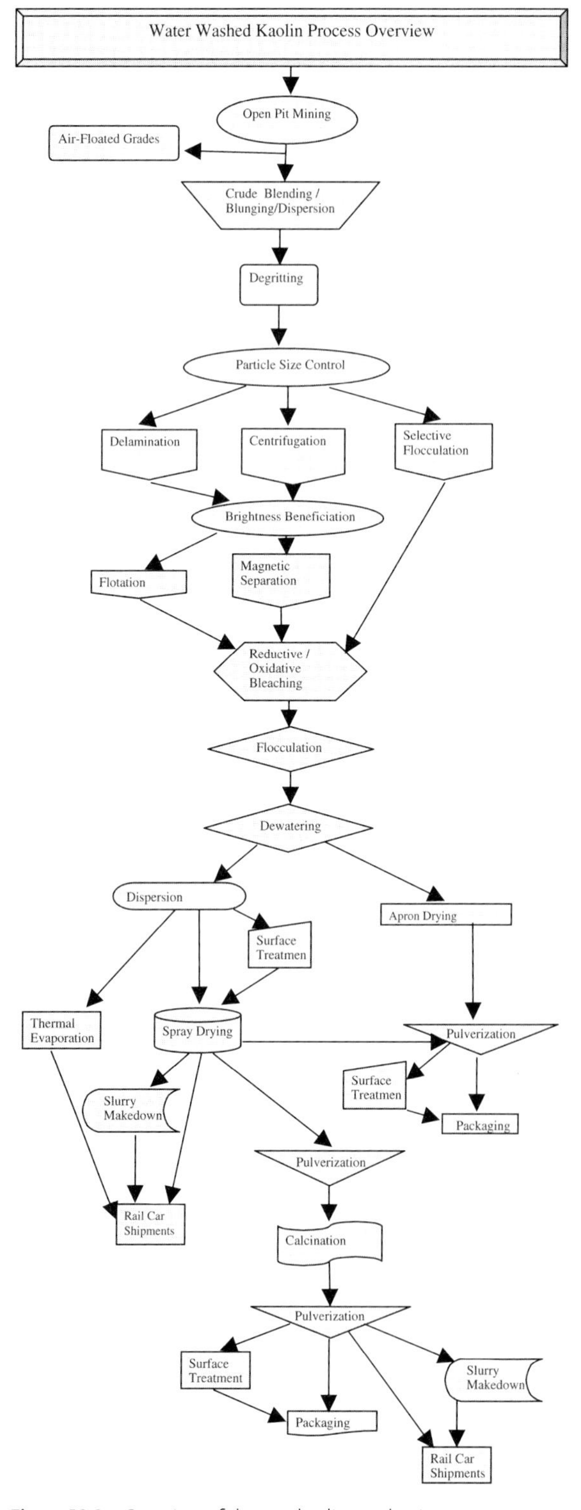

Figure 13.3 Overview of the wet kaolin production process.

interplay between the two types of centrifuges allows a wide array of particle size distributions. The efficiency of separation of coarse and fine particles depends on many process variables, including the initial particle size distribution, the viscosity of the feed stream, the retention time in the centrifuge, and the mechanical configuration of the centrifuge itself.

An additional manipulation is delamination, which changes both the apparent particle size distribution and the morphology of the particles. Coarse kaolin slurry is pumped into an agitation tank containing plastic or glass-bead medium. The objective is to cleave the stacks of kaolin without breaking the platelets and causing excessive fines. Delamination results in the formation of higher aspect ratio individual platelets by breaking the weak hydrogen bonds that hold the platelets stacked together as booklets (see Figure 13.4). The delaminated platelets originating from Georgia, USA, where many kaolin suppliers mine out, average about 0.15 μm in thickness and about 0.6 μm in diameter. Higher aspect ratios are possible depending on the kaolin feed into the delaminating process. As discussed in Chapter 2, high surface area and high aspect ratio are essential for efficient stress transfer in polymer composites, although they may result in higher viscosity. The particle size distribution also has an impact on melt viscosity, with a narrower distribution corresponding to a higher viscosity. This effect is more significant at higher kaolin loadings.

The brightness and color of the kaolin can be improved through various methods. High-intensity magnetic separation (HIMS) is a continuous, semibatch process that uses cryogenic, superconducting magnets cooled by liquid nitrogen and helium. Kaolin slurry is passed through a steel-wool matrix within a magnetic coil. The matrix retains paramagnetic impurities such as iron oxide and anatase. HIMS increases the GE brightness of the mineral by five points, the latter defined as the directional reflectance at 457 nm [4]. Periodically, the HIMS is shut down and the steel-wool matrix is flushed with water. The discharge is often a deep-red slurry of magnetic rejects consisting mainly of dark reddish-purple hematite and yellowish iron-enriched anatase. Another option for improved brightness is froth flotation, which

Figure 13.4 SEM micrograph showing individual kaolin platelets.

removes (although not entirely) anatase from the kaolin slurry. Flotation decreases the content of anatase color bodies from about 2% down to 0.7%. During the process, the iron-stained anatase-laden froth rises to the top of the vessel and is skimmed off and discarded. Another step is reductive bleaching that converts the iron's valence from +3 to +2, thereby dispelling the color and enhancing the brightness by about five points [5].

By lowering the pH of the kaolin slurry to 3 with sulfuric acid, some of the iron is solubilized. Adding a strong reducing agent, such as sodium hydrosulfite, reduces the iron and keeps it in a soluble ferrous state, such that it can be removed by rinsing during filtration [6]. Oxidative bleaching is used to render organic color bodies colorless since less than 1% organic carbon can discolor the kaolin to an unacceptable degree [7]. Two oxidative bleaching agents are often used: sodium hypochlorite, which is added to the slurry in aqueous form, and ozone, which is bubbled through a gas–liquid contact tower. The organic matter is oxidized to colorless carbon dioxide gas. Another way to improve brightness is selective flocculation, which involves the flocculation of discrete TiO_2 particles, followed by quiescent settling of a portion of the anatase at the bottom of a collection vessel and its subsequent removal. Also, kaolin can be flocculated with a high molecular weight anionic polymer, leaving the titaniferrous contaminants in the suspension and thereby discarded.

13.2.3 Kaolin Products

Kaolin products are shipped in both wet and dry forms. For the plastics industry, the filler, obviously, needs to be in dry form. Filtering takes place on vacuum filters to create a cake consisting of about 60% solids. The cake is rinsed with water to remove soluble salts, dried in rotary dryers, and finally pulverized. The finished products obtained in this way are referred to as acid clays since their final pH is 3.5–5. On the other hand, the filter cake can be redispersed by adding anionic dispersants and adjusting the pH to neutral. At this point, the dispersed slurry may be spray-dried to a moisture level of typically less than 4%, resulting in a predispersed product with high bulk density and good bulk flow properties. The product is in its final form at this stage, although it may be pulverized prior to packaging depending on the end-use application. It can be packaged in different bag sizes or into bulk hopper cars. For the production of high solid content slurries, the dispersed filter product may be passed through a vacuum evaporator, increasing the solid content from 60 up to 70%. Another method commonly used to achieve 70% solid content is to blend the spray-dried product with the filter product in a highly agitated vessel.

13.2.4 Calcination

Calcination is used to produce value-added kaolin products. Commonly, there are two families of calcined kaolin. One is "metakaolin," which is produced by heating spray-dried, pulverized kaolin at approximately 550–600 °C. At this temperature, the crystalline-bound water of hydration is released. With this 14% loss in weight and the concomitant changes in the crystalline structure, the mineral becomes highly reactive and is rich in soluble alumina. In PVC insulation, such as low-voltage wire applications, metakaolin significantly increases volume resistivity as a result of its high dielectric capacity and good thermal insulating properties. The second family of products produced by calcination is referred to as "fully calcined." These are formed as a result of increasing the calcination temperature to approximately 1000 °C, at which an exothermic reaction occurs. At this higher temperature, the kaolin is significantly improved in terms of its whiteness and brightness, provides better light scatter, and, hence, can extend TiO_2 pigments with concomitant significant cost savings. Calcined products are pulverized for residue control of free mica or abrasive silica, which would otherwise shorten the life of processing machine parts. Calcination of either type (meta or full) changes the kaolin structure from crystalline to amorphous. Sintering or fusing of individual clay particles takes place creating a porous material (Figure 13.5), which enhances light scatter. The calcined materials, due to their higher porosity, have considerably higher oil absorption and an increased refractive index. The mineral becomes somewhat harder, in addition to other key properties that alter as a result of calcination.

13.2.5 Surface Treatment

Surface treatment is another value-added step that can improve the performance of kaolin. Since the filler is naturally very hydrophilic due to its hydroxyl groups, a treatment can be applied to render its surface hydrophobic or organophilic. These surface-modified kaolins are useful especially in plastics and rubber industries, where they improve adhesion and dispersion and hence act more effectively as functional fillers. Silanes, titanates, and fatty acids as discussed in Chapters 4–6, respectively, may be used to modify the surface characteristics of either hydrous or calcined kaolins, promoting deagglomeration, often lower viscosities, and improved mechanical and electrical properties.

13.3 Properties

Typical chemical compositions of hydrous and calcined kaolins are shown in Table 13.1 [8]. The major effects of the removal of water of hydration are an increase in refractive index, a moderate increase in the otherwise low Mohs hardness, a decrease or increase in specific gravity depending on the extent of calcination, and a decrease in dielectric constant (Table 13.2) [8]. The calcined materials, due to their higher void structure, are capable of considerably higher oil absorption, as indicated in Table 13.3, which also includes comparisons of additional properties of various

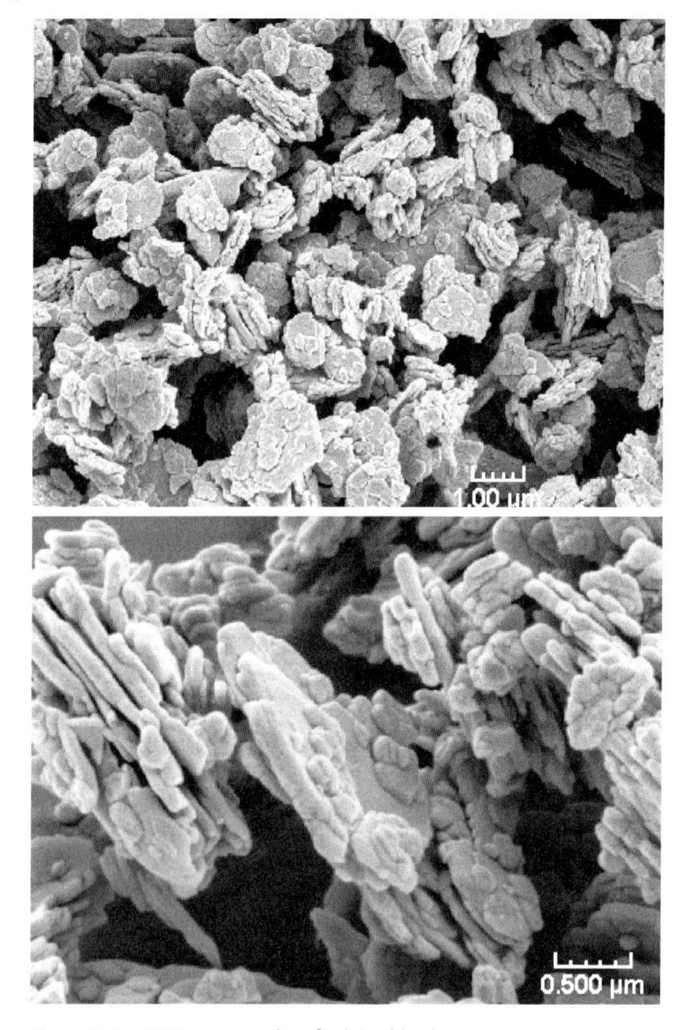

Figure 13.5 SEM micrographs of calcined kaolin.

kaolin grades. Tables 13.4 and 13.5 [9] summarize some of the features of hydrous and calcined kaolins and the corresponding benefits observed in various plastic and rubber formulations.

13.4 Suppliers

Major suppliers of kaolin with their corresponding capacities in million tons per annum (mtpa) are Imerys with >5.0 mtpa, Huber Engineered Materials (business sold in 2008 to IMin Partners, Fort Worth, TX) with ~2.0 mtpa, Engelhard (business

Table 13.1 Typical chemical composition (%) of water-washed hydrous and calcined kaolin[a)] [8].

Metal oxide	Hydrous grade	Calcined grade
Al_2O_3	38.8	44.7
SiO_2	45.2	52.5
Na_2O	0.05–0.3	0.2
TiO_2	0.6–1.7	0.6–1.8
CaO	0.02	0.1
Fe_2O_3	0.3–0.9	0.3–1.0
MgO	0.03	0.3
K_2O	0.05–0.2	0.2
Loss on ignition	13.6–14.2	< 0.5

a) Volatile-free basis.

Table 13.2 Typical physical properties of water-washed hydrous and calcined/silane-treated kaolin grades [8].

Property	Hydrous grade	Calcined and/or silane-treated grades
Melt temperature (°C)	~1800	~1800
Specific heat capacity (cal/(g °C))	0.20–0.22	0.20–0.22
Free moisture (%)	<1.0	<0.5
Mohs hardness	2–2.5	2.5–3
Refractive index	1.56	1.62
Dielectric constant	2.6	1.3
Specific gravity	2.58	2.5–2.63

Table 13.3 Miscellaneous properties of specific kaolin grades.

Property	Air-floated	Water-washed hydrous	Water-washed delaminated hydrous	Calcined and silane surface treated
Residue >44 μm	0.3–1.5	0.01	0.01	0.015
GE brightness	70–81	83–92	87–92	90–96
pH	4.5–6.5	3.5–8.0	6.0–8.0	5.0–6.0
Median particle size (ESD[a)], μm)	0.3–1.3	0.2–4.5	0.4–1.0	0.8–2.0
Oil absorption rubout (g/100 g)	28–36	31–46	38–46	50–95
Surface area BET (m^2/g)	10–22	12–22	11–15	7–12
Specific gravity	2.58	2.58	2.58	2.63
Cost ($/kg)	0.11–0.22	0.26–0.55	0.26–0.48	0.35–1.21

a) Equivalent spherical diameter.

Table 13.4 Features and benefits of hydrous kaolin [9].

Features	Benefits
Inert	Low soluble salts Corrosion and chemical resistance
Soft plate-like particles	Nonabrasive; low wear on equipment Barrier properties
Particle size distribution	Wide range – very fine to coarse; narrow distributions
Brightness	Good color in nonblack compounds
Low residual contaminants	Good dispersion and low abrasion High purity
Delamination	Higher aspect ratio; improved modulus and barrier properties

sold in 2008 to BASF Corp.) with ~2.4 mtpa, and CADAM/PPSA (suppliers to paper market only) with ~1.4 mtpa (2 mtpa by 2006/2007) [10].

Imerys leads the European market for polymer end use. The aforementioned four companies supply the majority of the world's water-washed kaolin and are major exporters of the mineral. Other key producers are Thiele with 1.25 mtpa and Quarzwerke owned by Amberger Kaolinwerke (AKW) with 0.9 mtpa. Other US producers of air-floated grades are KT Clay (owned by Imerys) with >500 000 tpa and Unimin with >500 000 tpa. It should be noted that European kaolin comes mostly from primary deposits and does not lend itself to the air-flotation process [10].

13.5 Cost/Availability

Kaolin is readily available in Middle Georgia, the United States, where the major US suppliers mine, process and package, and export to Europe, Asia, and throughout the world. Other sources of kaolin are likewise mined and sold outside the United Stated. The cost of the mineral fluctuates, especially with energy prices having a direct impact on the production of calcined kaolin. In general, the 2004 truckload price structure, FOB, was

Air-floated grades:	0.07–0.22 $/kg
Water-washed hydrous grades:	0.33–0.55 $/kg
Calcined grades:	0.51–1.10 $/kg
Surface-treated grades:	0.99–1.52 $/kg

US products are available in 50 lb (22.5 kg) bags, semibulk bags, and in bulk for certain products.

In North America, about 11 000 tons of hydrous kaolin was sold for $2.1 million to the plastics industry in 1999; forecast for 2004 was a 2.5% growth in tons per annum to 12 000 tons [11]. Of the calcined kaolin consumed, the use in plastics accounted for

Table 13.5 Features and benefits of calcined kaolin [9].

Features	Benefits
Inert	Low soluble salts
	Electrical, corrosion, and chemical resistance
Structured and porous particles	Abrasion resistance
	Higher oil absorption
	TiO_2 extension
Surface treatment	Improved dispersion and adhesion
	Higher modulus
	Improved tear and impact strength
	Hydrophobic
Brightness	Good color in nonblack compounds
Low residual contaminants	Good dispersion and low abrasion
	High purity
Low free moisture	Compatibility with resins; little effects on cure rate

10% of the total volume, behind that used in paper and paint. In North America, about 58 000 tons sold at $23.8 million in 1998; forecast for 2004 was a 5% growth in tons per annum to 74 000 tons. For air-floated kaolin in North America about 22 000 tons sold for $1.4 million to the plastics industry in 1999, with practically no growth forecast for 2004 [11].

13.6 Environmental/Toxicity Considerations

With the exception of possible dust issues, kaolin is a nontoxic mineral. A dust mask approved by NIOSH is highly recommended. The Food and Drug Administration gives kaolin a "Generally Regarded as Safe" status (GRAS). Kaolin is used in many consumer products in the cosmetic and pharmaceutical industries. It is employed in household products, such as sunscreen lotions, toothpastes, and even antidiarrhea treatments. The total crystalline silica in china clay is less than 0.1%. Kaolin is considered a naturally occurring chemical substance per TSCA, 40 CFR 710.4(b). In the event of slurry shipment, there will be low levels (ppm) of biocide added for long-term protection against microbial degradation.

13.7 Applications

13.7.1 Primary Function

Kaolin, being plate-like filler with a relatively low aspect ratio [6, 7, 12, 13], may impart certain mechanical property improvements in thermoplastics. Similar to other plate-like materials, kaolin can improve dimensional stability (warp resistance), isotropy, and surface smoothness, a common problem with high aspect ratio fibrous reinforcements. Calcined kaolin, particularly after surface treatment with aminosilanes, provides an array of benefits in nylon matrices. Surface-treated calcined kaolin results in maximum dispersion and has the highest effect on mechanical properties. Hydrous kaolin, due to its water of hydration, would not be a candidate for use in hydrolytically unstable polymers. The particle size of the mineral and the method of application and concentration of the silane are important. Finer grade kaolin will enhance mechanical properties of a composite more than coarse grades, assuming complete dispersion of the particles in the matrix. Data support the fact that silane pretreatment delivers superior performance compared to the *in situ* addition of the coupling agent (see also Chapter 4). Figures 13.6 and 13.7 [14] show graphical comparisons of results obtained with the two methods for injection-molded nylon 6,6 containing 40% calcined kaolin with a mean particle size of 1.4 μm. Both methods, that is, *in situ* treatment and pretreatment, lead to performance improvements; however, the pretreatment method provides significantly better results. This is

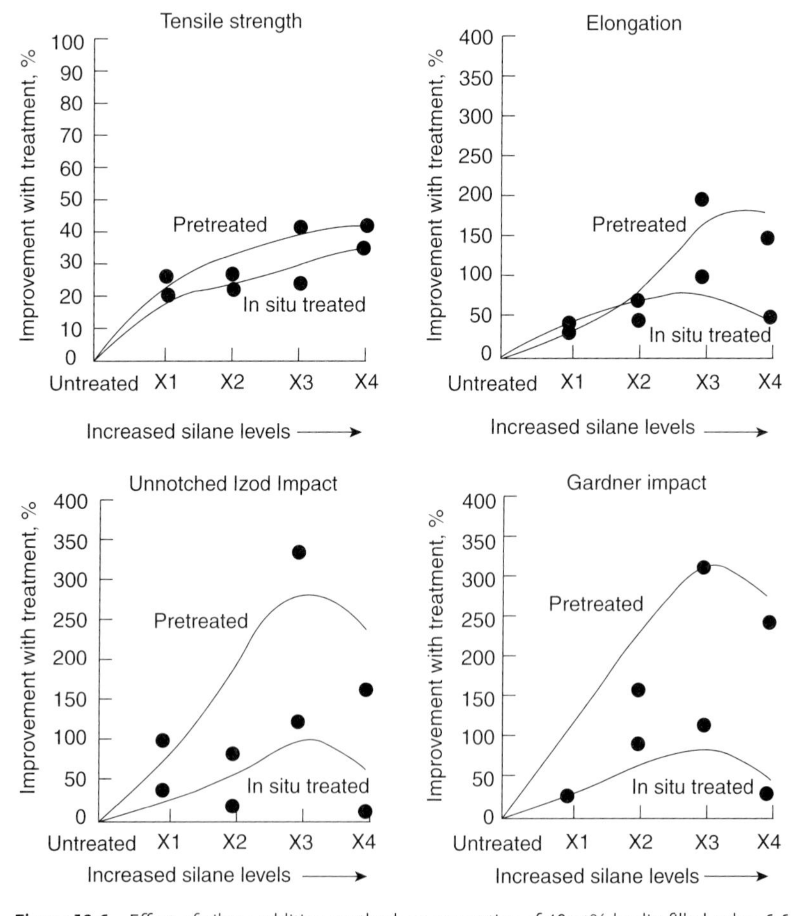

Figure 13.6 Effect of silane addition method on properties of 40 wt% kaolin-filled nylon 6,6.

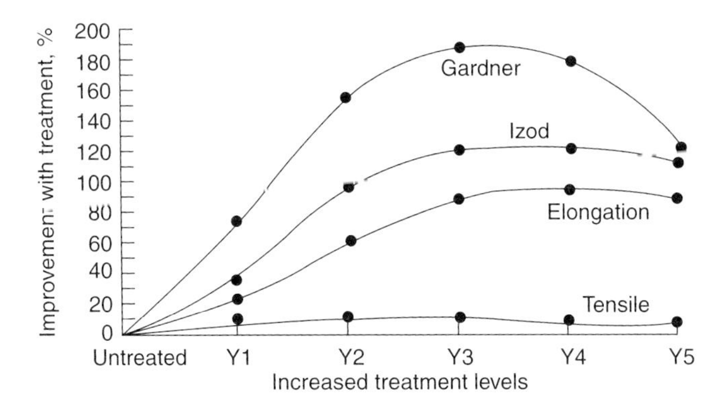

Figure 13.7 Effect of silane addition level on properties of 40 wt% kaolin-filled nylon 6,6.

especially pronounced for unnotched Izod and Gardner impact strengths, which are sensitive to the extent of dispersion in the polymer matrix. Moreover, it is demonstrated that the level of treatment should be optimized based on the cost of the silane and the level of performance sought.

Figure 13.8 shows the morphology of an aminosilane-treated 0.8 μm calcined kaolin in nylon 6,6.

As with nylons, kaolin can also improve the properties of polyolefins. A study [9] comparing silane-treated kaolin with Montana talc and ground calcium carbonate at different loadings in HDPE showed that kaolin gave higher impact and tensile strengths (Table 13.6). The talc, having the highest aspect ratio of the minerals evaluated, gave the best modulus improvement.

In another study, a highly impact-modified copolymer PP resin was extruded and then injection molded. The surface-treated kaolin doubled the flexural modulus

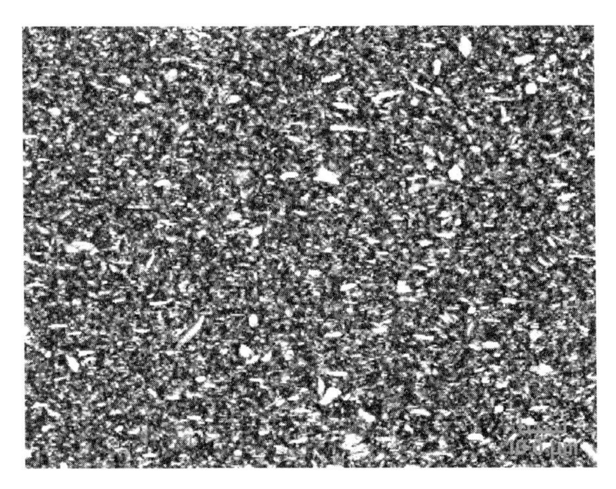

Figure 13.8 SEM micrograph of polished cross section of 40 wt% aminosilane-treated 0.8 μm calcined kaolin in nylon 6,6.

Table 13.6 Comparison of mechanical properties of mineral-filled HDPE [9].

Mineral	Tensile strength (MPa)	Tensile elongation (%)	Tensile modulus, MPa (1% secant)	Flexural modulus (MPa)	Gardner impact (J)	Notched Izod impact strength (J/m)
Unfilled	21.1	17.0	454.7	813.0	>35.2	45.6
Kaolin[a)]						
20% loading	28.3	15.0	668.3	1151	>35.2	70.5
40% loading	32.6	10.7	916.4	1778	>35.2	44.5
Kaolin[b)]						
20% loading	29.9	14.5	682.1	1240	>35.2	97.0
40% loading	33.3	10.3	1033	2101	>35.2	66.2
Calcium carbonate (3.0 μm)						
20% loading	24.1	13.6	620.1	1068	31.7	31.3
Montana talc (1.5 μm)						
20% loading	23.8	11.7	723.4	1433	8.36	23.8
40% loading	27.6	7.7	1089	2680	1.87	26.0

a) Translink® 445 (BASF).
b) Translink 555 (BASF).

over the unfilled resin at a loading of 30 wt% with a small effect on impact strength. Again, the higher tensile and flexural modulus of the talc composites are, as expected, due to the higher aspect ratio of the mineral (Table 13.7) [9].

13.7.2 Secondary Functions

13.7.2.1 Improvement of Electrical Properties

Kaolin also serves as functional filler in wire and cable insulation. As with other applications, surface-treated kaolin offers the best improvement in performance,

Table 13.7 Mechanical properties of mineral-filled polypropylene copolymer [9].

Mineral type	Mineral loading (wt%)	Tensile yield strength (MPa)	Tensile elongation (%)	Tensile modulus, MPa (1% Secant)	Flexural modulus (GPa)	Notched Izod impact strength (J/m)
Unfilled	0	20.9	37.2	337.6	730.3	779
Treated kaolin	20	20.7	23.2	509.9	1102	699
Treated kaolin	30	21.2	17.5	565.0	1378	657
Talc	20	21.1	21.8	565.0	1350	355
Talc	30	21.4	14.7	682.1	1784	207

Table 13.8 Evaluation of kaolin in cross-linked polyethylene typically used in medium voltage cable [15].

Type	Loading (phr)	Tensile strength (MPa)	Elongation at break (%)	% Retention of elongation (after 168 h at 120 °C in air)	Power factor % (after 168 h in water at 75 °C)
Unfilled	0	14.5	307	103	0.05
Satintone 5® calcined kaolin (0.8 μm)	50	16.4	131	45	1.94
Translink 77 vinylsilane treated (0.8 μm)	50	18.8	123	94	0.12

Note: Satintone 5 and Translink 77 are BASF Corp. kaolin grades.

providing a high degree of hydrophobicity and suppressing polar species at the mineral surfaces. Moreover, surface-treated grades allow better compatibility with the polymer and minimize the penetration of water through the insulation. Polymers often employed in this market are PVC, LDPE, LLDPE, X-PE, PP, and TPE, among others.

In general, volume resistivity, dielectric strength, and dielectric loss are all improved by adding kaolin. Finer particle size provides lower dielectric loss, minimizing heat buildup to the point of insulation breakdown. Vinyl functional-treated calcined kaolin outperforms untreated kaolin in terms of tensile elongation and electrical property retention in cross-linked polyethylene after high temperature aging in air and water (Table 13.8) [15]. All calcined grades give excellent initial power factor as a result of the low dielectric loss of the amorphous kaolin structure. However, after aging, the surface-treated grade performed better as a result of the hydrophobicity imparted by the vinyl functional silane (see also Chapter 4).

Metakaolin is an effective functional filler in plasticized PVC wire insulation at loadings of about 10 phr. It better protects the insulation from cracking or "treeing," a term that describes the physical breakdown of the cable polymer matrix due to moisture or other external influences. Volume resistivity greatly improves when metakaolin is incorporated compared to other minerals (see Table 13.9) [15], although the color is not as white as with higher temperature fully calcined kaolin. In addition,

Table 13.9 Effect of different fillers at 10 phr on volume resistivity of low-voltage PVC wire [15].

Filler	Volume resistivity (Ω cm)
Ground calcium carbonate	7×10^{12}
Hydrous kaolin	30×10^{12}
Hydrous kaolin silane treated	50×10^{12}
Calcined kaolin	50×10^{12}
Calcined kaolin silane treated	100×10^{12}
Metakaolin	300×10^{12}

Table 13.10 Color of injection-molded LLDPE containing TiO_2/kaolin combinations [16].

	TiO_2 % replacement	Solids %	Color "L" value	Color "a" value	Color "b" value	Opacity
Control	—	0.95	96.72	−0.51	1.64	83.3
Ultrex 96™	5	0.97	96.77	−0.56	1.67	82.1
	10	0.98	96.38	−0.60	1.58	82.0
	15	0.99	96.51	−0.58	1.70	80.7
	20	1.03	96.35	−0.64	1.45	80.1
ASP-170®	5	0.95	96.38	−0.58	1.62	79.4
	10	1.10	96.13	−0.46	1.95	79.1
	15	0.94	95.79	−0.48	1.80	79.2
	20	0.98	95.54	−0.43	2.24	78.1

Materials: LLDPE Dowlex 2553 (Dow), TiO_2 Tri-Pure® R101 (Dupont), Ultrex 96™ engineered calcined kaolin (BASF Corp.), ASP-170 specialty hydrous kaolin (BASF Corp.).
Conditions: Letdown ratio, 50 : 1; Battenfeld press, 85 tons; Barrel temperature, 230–250 °C.
Testing: Color data generated from 7.5 cm diameter injection-molded disks and opacity measured on nominal 0.575 mm thick press-outs from the letdowns.

metakaolin shows high acid solubility and therefore acts as an acid scavenger to assist in stabilizing the PVC against degradation. The kaolin types used in wire and cable generally have low soluble salts and free moisture contents.

13.7.2.2 TiO_2 Extension

TiO_2 is an excellent pigment for color and opacity in plastics. TiO_2 extension is commonly achieved with kaolin, thereby lowering cost yet still maintaining properties. A case study was conducted with a typical 50 wt% TiO_2 LLDPE concentrate. In injection-molded parts, replacement of 10% TiO_2 with calcined kaolin allowed retention of both color and opacity. At 20% replacement of TiO_2, only a slight drop in opacity occurred and color intensity was completely maintained (Table 13.10) [16].

Hydrous kaolin may also be used to replace or extend TiO_2, but not always with the same efficiency as that of calcined grades. In blown LLDPE film (Table 13.11) [17], hydrous kaolin can replace up to 10% TiO_2 with no loss in color or opacity. At 20% TiO_2 replacement, only a slight loss in opacity may occur. Another application where kaolin has delivered cost savings is in white PVC siding. Here again, TiO_2 extension can be realized up to a level of 20% [17].

An additional evaluation [18] of the extension of rutile TiO_2 by different fillers was conducted by assessing blister resistance in two types of liquid, thermosetting polyester resins (namely *ortho*- and *ortho*-neopentyl glycol (NPG). The performance of the filler/pigment portion of the cured polyester gel coat was compared after exposure to water at 65 °C for 1150 h (Table 13.12). Untreated kaolin provided better blister resistance than untreated talc or calcium carbonate. Vinylsilane surface-treated kaolin gave the best blister resistance, reducing the size and extent of blisters

Table 13.11 Color of blown LLDPE films containing TiO_2/kaolin combinations [17].

Letdown 25: 1	TiO_2 % replacement	Color "L" value	Color "a" value	Color "b" value	Opacity
Letdown 25 : 1					
Control	—	92.6	−1.0	0.1	35.7
ASP-170	5	92.4	−1.0	0.3	38.6
	10	92.4	−1.0	0.3	38.5
	20	91.9	−1.1	0.3	34.3
Letdown 100 : 1					
Control	—	91.3	−1.1	−0.1	18.5
ASP-170	5	91.3	−1.2	−0.1	19.8
	10	91.1	−1.2	−0.1	19.3
	20	91.1	−1.1	−0.1	17.9

Materials: LLDPE Dowlex 2553 (Dow), TiO_2 Tri-Pure R101 (Dupont), Kaolin ASP-170 specialty hydrous kaolin (BASF Corp.).
Conditions: Blow up ratio: 3 : 1, die diameter: 25 cm spiral, Barrel temperature: 190–245 °C. Two letdown ratios during production of blown film, 25 : 1 and 100 : 1.
Testing: Opacity was measured on nominal 0.025 mm thick films.

Table 13.12 Comparison of filler performance in polyester gel coat after aging [18].

Sample	TiO_2 %	Mineral %	Blister rating *ortho*	Blister rating *ortho*-NPG
Rutile TiO_2 control	25	0	2.5	1.1
Untreated calcined kaolin	10	15	1.0	0.7
Vinylsilane-treated Kaolin	10	15	0.8	0.0
Aminosilane-treated kaolin	10	15	0.9	0.5
Untreated talc	10	15	3.0	2.5
Wet-ground untreated calcium carbonate	10	15	2.8	2.2

Blister rating: 0 = none; 1 = slight; 2 = moderate; 3 = severe.

as a result of improved compatibility and the formation of a hydrolytically stable, strong mineral/resin interface.

References

1 Ciullo, P.A. (1996) *Industrial Minerals and Their Uses – A Handbook and Formulary*, Noyes Publication, Westwood, NJ.

2 Washabaugh, F. (1995) *Kaolins for Rubber Applications*, Engelhard Corp., Iselin, NJ.

3 Murray, H. (1991) *Appl. Clay Sci.*, **5** (3), 379–385.

4 Popson, S. (1989) Measurement and control of the optical properties of paper, Technical Paper, Technidyne Corp., New Albany, IN.

5 Finch, E. (2002) *Ind. Miner.*, **414**, 64–66.
6 Prasad, M. and Reid, K. (1990) Kaolin: processing, properties and applications, Technical Report, University of Minnesota, MN.
7 Kogel, J. *et al.* (2002) *The Georgia Kaolin – Geology and Utilization*, SME: Society for Mining, Metallurgy, and Exploration, Inc., Ann Arbor, MI, p. 41, www.smenet.org.
8 Engelhard Corporation (2002) Performance mineral reinforcements for plastics and rubber, Technical Brochure, Engelhard Corp., Iselin, NJ, www.engelhard.com.
9 Fajardo, W. (1993) Kaolin in plastic and rubber compounds, Technical Report, Engelhard Corp., Iselin, NJ.
10 Moore, P. (2003) *Ind. Miner.*, **431**, 24–35.
11 Kline & Company, Inc. (2000–2002) Extender and filler minerals North America, Kaolin Report, Kline & Company, Inc., Little Falls, NJ.
12 Bundy, W. (1993) *Kaolin Genesis and Utilization*, The Clay Minerals Society, Boulder, CO, pp. 56, 57.
13 Pickering, S., Jr. and Murray, H. (1994) Kaolin, in *Industrial Mineral and Rock*, 6th edn, SME: Society for Mining, Metallurgy, and Exploration, Inc., Ann Arbor, MI, www.smenet.org.
14 Carr, J. (1990) Kaolin reinforcements: an added dimension, Technical Report, Engelhard Corp., Iselin, NJ.
15 Khokhani, A. Kaolins in wire & cable, Technical Report, Engelhard Corp., Iselin, NJ.
16 Fajardo, W. (1997) ASP® 170 & Ultrex™ 96 Specialty Pigments, Technical Bulletin, TI2303, Engelhard Corp., Iselin, NJ.
17 Sherman, L. (1999) Stretch TiO_2, Plastics Technology Online, July 1999, http://www.plasticstechnology.com/articles/199907fa1.html.
18 Washabaugh, F. (1990) The effect of gel coat extenders on the performance of polyester laminates, Technical Bulletin, TI2180, Engelhard Corp., Iselin, N.J.

14
Wollastonite

Sara M. Robinson and Marino Xanthos

14.1
Introduction

Wollastonite is a naturally occurring acicular (needle-like) silicate mineral, named in 1822 after an English chemist and mineralogist, W.H. Wollaston. It is a calcium metasilicate ($CaSiO_3$), belonging to the family of inosilicates, with chemical similarity to the group of pyroxene minerals. Wollastonite, in its pure form is brilliant white. Contaminants or impurities from associated minerals or element substitutions may change the color to cream, pink, or gray. Wollastonite is generally described as a contact metamorphic mineral, but can be created by a magmatic process. In the primary metamorphic formation of the wollastonite crystal structure, calcite and quartz, at very high temperatures, will react to form wollastonite with emission of carbon dioxide. Wollastonite can also form by the passage of siliceous hydrothermal solutions through limestone zones. During the formation of wollastonite, other minerals including feldspars, diopsides, garnets, calcites, and quartz, may be introduced. The commercial processor uses various beneficiation techniques, such as wet cell flotation and/or high intensity magnetic separation, to remove these associated minerals and produce high-purity grades. Starting in the late 1950s, wollastonite became widely known as an important industrial mineral with applications in ceramics and paints. From the 1970s, its significantly expanded commercial use was due to the replacement of asbestos across several industries such as high performance industrial coatings, friction materials, ceramics, refractory, and metallurgy. The use of wollastonite in plastics gained momentum in the 1980s with the advent of reinforced reaction injection molding (RRIM) and bulk molding compounds (BMC). Very recently, wollastonite has also been incorporated in novel applications including biomedical composites (see Chapter 22).

Functional Fillers for Plastics: Second, updated and enlarged edition. Edited by Marino Xanthos

ISBN: 978-3-527-32361-6

14.2
Production

Production of high-purity wollastonite (at least 90%), involves first the steps of crushing and sorting. Sorting can be manual after visual inspection, or by way of an electronic optical sorter. This is followed by beneficiation, which involves either flotation or electromagnetic separation, to segregate the undesirable components. Specific milling processes are then incorporated to either preserve the inherent high aspect ratio particle forms, or to create low aspect ratio versions. The high aspect ratio grades (10 : 1–20 : 1) are typically produced via attrition mills and classifiers. By using these types of mills, aspect ratio is preserved, while "fines" are minimized. Average lengths can range from as low as 20 μm to as high as 200 μm. The low aspect ratio (3 : 1–5 : 1) grades, also known as powder grades, are produced in impact mills such as pebble or ball mills. Classifiers may also be used to control top size. Generally, the powder grades are described using high mesh values such as −325 or −400 mesh.

Surface-modified versions are commercially available with modifiers that include organosilanes, titanates, and coupling agents such as maleated polypropylene (MAPP). The modification process is based on a surface reaction via blending/coating using conventional blenders, and then drying. The plastics industry uses all forms of wollastonite and most of the end uses are determined by property requirements and polymer selection. Additional information on the production of commercial grades can be sourced from the publications [1, 2] and web sites of the two major US producers [3], and also in publications available in the US Geological Survey web site [4].

14.3
Structure and Properties

The theoretical composition of calcium metasilicate is 48.30% CaO and 51.70% SiO_2. The typical reported compositions (calculated as oxides) are 44.04–47.5% for CaO and 50.05–51.0% for SiO_2. Based on unit cell parameters, the specific gravity is calculated to be 2.96. However, the generally accepted reported value is 2.90. In commercial grades, deviations from the theoretical specific gravity are due to various impurity ions that substitute for calcium in the crystal lattice or impurity minerals such as calcite, garnet, and diopside. Typical chemical compositions of wollastonite grades for plastics from various suppliers are listed in Table 14.1.

The structure of wollastonite is characterized by chains formed from silica tetrahedra connected side-by-side through calcium in octahedral coordination [2]. These chain arrangements account for the formation of acicular crystals and preservation of their shape upon cleavage (Figure 14.1). The moderate value of wollastonite Mohs hardness (4.5–5) is attributed to the high density of the silica chains. In addition to the most important property of acicular shape, the plastics industry places high value on properties such as

Table 14.1 Typical chemical analyses of wollastonite filler grades.

Composition	wt%[a)]	wt%[b)]	wt%[c)]
CaO	47	44.0	47
SiO_2	49.5	50.0	50.0
MgO	0.2	1.5	0.3
Al_2O_3	0.6	1.8	0.3
Fe_2O_3	0.43	0.3	1.0
TiO_2	Traces	Not reported	0.05
MnO	0.29	<0.1	0.1
Na_2O	0.02	0.2	Not reported
K_2O	0.11	Not reported	0.1

a) Wolkem India Ltd (KEMOLIT® grade).
b) R.T. Vanderbilt Co. Inc. (VANSIL grade).
c) NYCO Minerals Inc. (NYAD® grade).

- white color;
- low moisture absorption;
- good thermal stability;
- low thermal expansion coefficient;
- high dielectric strength;
- low loss on ignition.

Table 14.2, based on references [5, 6] and suppliers' literature, summarizes the physical and chemical properties of wollastonite.

In engineering thermoplastics, as well as in thermoset polymers, wollastonite incorporation results in an overall improvement of the mechanical properties. This

Figure 14.1 Microphotograph of acicular wollastonite (VANSIL® WG, R.T. Vanderbilt Co., Inc.), 170×.

Table 14.2 Typical properties of wollastonite.

Property	
Color	White
Crystal system	Triclinic
Specific gravity	2.8–2.9
Coefficient of thermal expansion (K^{-1})	6.5×10^{-6}
Specific Heat (J/(kg K))	1003
Melting point (°C)	1540
Transition temperature, °C (to pseudowollastonite)	1200
Hardness (Mohs)	4.5–5
Refractive index	1.63–1.67
pH (10 wt% slurry)	9.0–11
Loss on ignition, % (950 °C)	0.1–6
Thermal conductivity (W/(m K))	2.5
Dielectric constant (10^4 Hz)	6

can be seen in increased flexural modulus, increased tensile strength, increased heat deflection temperature (HDT), and enhanced dimensional stability. More popularly, wollastonite has been used as an economic alternative/supplement to glass fibers, either chopped or milled. The surface chemistry of wollastonite provides a particular affinity for surface treatments such as silanes, titanates, polymeric esters, and other additives. Benefits of such treatments include lower compound viscosity for improved melt flow, easier dispersion for better property development, and improved physical properties. The Mohs hardness and acicular shape can be a concern for process equipment wear. This can be minimized by careful selection of metal alloys and hardened materials. On the other hand, the moderate hardness combined with appropriate grade selection can result in excellent mar and scratch resistance in certain formulations.

14.4 Suppliers/Cost

The major wollastonite producing countries are India, China, Mexico, and the United States. At one time, Finland was considered the world's fourth largest producer, but since 2004 production has been very limited. Wolkem India Ltd has claimed to be the largest global producer with an estimated production of 160 000 tons in 2002–2003 [7]. However, according to Hawley [8], China produces between 330 000 and 413 000 tons/annum from at least 60 small operations. In this reference, India is considered to be the third largest producer with a capacity of 198 000 tons/annum. In the United States, producers are NYCO Minerals Inc. and R.T. Vanderbilt Co., Inc., both mining, beneficiating, and processing wollastonite from deposits in New York state. World production likely exceeded 600 000 tons in 2007, with sales of finished grades probably near 550 000 tons [9]. In 1999, the global consumption of fillers in plastics was approximately 10 million tons. At that time, wollastonite was estimated to hold a 3% share of this consumption, or between 300 000 and 400 000 tons [10]. The estimated United States production in 2006 was approximately 110 000 metric tons [8, 9]. Approximately 30–35% of the consumption was estimated to be of plastics and rubber applications, assuming that this share percentage did not significantly change since 2003 [8]. It should be noted that one producer estimated that the worldwide consumption of wollastonite in plastics was about 45 000 metric tons [8]. A major provider of wollastonite in Europe, Quarzewerke, which did not own reserves in Europe, did enter a joint venture with S & B Industrial Minerals to mine and process wollastonite in China; this has since been discontinued. Most suppliers provide surface-treated grades and high aspect ratio forms that are the grades most commonly found in engineering plastics.

Recent prices for US produced low aspect ratio grades (200–400 mesh), exworks, ranged from US $205–290 per ton [9]. The high aspect ratio (15 : 1 +) versions were approximately US $373 per ton, exworks. Chinese wollastonite prices, FOB port, ranged from US $80–100 per ton for 200 mesh to US $90–110 per ton for 325 mesh [10]. Surface-treated grades are more costly and the differential applied can be as much as another US $1000 per ton, depending upon substrate and coating chemistries required. Quoted prices are based on cited sources at time of publication and are subject to any contractual and other conditions between selling and buying parties.

14.5 Environmental/Toxicity Considerations

At the time of publication, the current recommended exposure limits (REL) and permissible exposure limits (PEL) were as follows:

Regulatory agency	Limits (based on time-weighted average, TWA)
NIOSH	10 mg/m^3 (total particulates)
OSHA	5 mg/m^3 (respirable particulates)

Please note that at such values, similar to those of other particulates, such as kaolin, proper precautions for exposure to the mineral dust are required.

Regarding toxicity considerations, wollastonite as an acicular mineral has come under the close scrutiny of various federal agencies in the past as a possible health hazard. In a NIOSH medical survey, with respect to carcinogenic effects, "no definite association of wollastonite exposure and excess morbidity could be demonstrated." From an update of this study it was concluded that prolonged exposure to excessive wollastonite dust may affect pulmonary functions, although excessive exposure to any dust may aggravate preexisting respiratory conditions [11]. Initial exposure to the high aspect ratio grades may produce minor skin irritation. A review of *in vitro*, *in vivo*, and epidemiological studies concluded that there is no evidence to suggest that wollastonite presents a health hazard [11].

14.6
Applications Involving the Plastics Compounding Industry

Wollastonite in thermoplastic and thermoset matrices is considered a reinforcing functional filler. By this, the general benefits are overall mechanical properties' improvement with major focus on increased tensile strength, elevation of heat deflection temperature, and improved flexural modulus. With an increase in tensile strength, there can be a concomitant decrease in elongation.

The acicular, or needle-like shape of the mineral, is the principal contributing factor to the enhanced properties. Functional fillers having similar Young's moduli, but platy in shape, such as talc and mica, can provide more overall stiffness increase in different directions due to increased dissipation of stress across a wider flat plane. Impact strength may also be improved by the use of platy minerals; however, with proper selection of median diameter and control of top size of the wollastonite particles, the notched impact strength performance in certain resins can be improved over the unfilled polymer and over other fillers [12]. In order to achieve any significant improvement in flexural and tensile property values, filler incorporation needs to be above a minimum wt% level. For wollastonite, the general working belief is that this minimum is 15 wt% with preferred levels ranging from 20 to 55 wt%. The most common level is 40% for the mineral alone, or for glass fiber/mineral blends the mineral content is generally 15%. Other important factors to be considered are polymer type, filler aspect ratio, filler top size, and potential contribution of the surface coating.

In general, finished parts containing wollastonite as the major filler have low coefficient of thermal expansion, low mold shrinkage, and good dimensional stability. Other reported attributes include improved mar and scratch resistance, improved surface appearance, and cost advantage over milled or chopped glass. On the other hand, some negative effects of wollastonite that have to be addressed are warpage with thermoplastic polyolefins (TPOs) and polyesters, equipment wear, loss of some weld-line strength, and lower impact strength. Some, if not most of these issues can be mitigated by the proper selection of the surface treatment, combining with another functional platy filler (i.e., talc or mica), and wollastonite selection based on aspect ratio.

Table 14.3 summarizes the effect of wollastonite on the properties of the most common thermoplastics and thermosets.

There are other thermoplastics where wollastonite has shown reinforcing effects and these include polyvinyl chloride (PVC), linear density polyethylene (LDPE), liquid crystal polymers (LCP), and polytetrafluoroethylene (PTFE). In polyolefins, wollastonite can improve electrical properties and in PTFE, wollastonite may mitigate the abrasive nature of the polymer during processing.

In most engineering thermoplastics, from polyamide (PA) to polypropylene (PP) and thermoplastic olefins, the incorporation of coupling agents preapplied onto wollastonite has become prevalent. The development and selection of these coupling agents for both thermoplastic and thermoset composites can include reactive or nonreactive silanes. There are examples of the use of titanates as surface

Table 14.3 Effects of wollastonite on the properties of selected thermoplastics and thermosets.

Polymer	Loading (wt%)	Application examples	Expected improvements
Polyamide 6 and 6,6	30–50	Automotive parts; electric motors; gears; power tool housings; construction parts; consumer electronics	In partial replacement of glass fibers Increased tensile strength and modulus Higher impact strength Improved heat distortion temperature (HDT) Cost savings
Polypropylene	23–40	Automotive under-the-hood; furniture; construction products; recreational products	Improved HDT Improved tensile strength Improved shrinkage control Improved mar and scratch Improved impact resistance
Thermoplastic polyolefin	18–40	Automotive (interior)	Improved Izod impact strength Improved moldability Other similar to PP
Polycarbonate (PC)	15–40	Automotive exterior body applications (i.e., fascia and side cladding parts)	Improved weld-line strength Improved flex modulus Low CLTE Improved surface appearance
Polybutylene terephthalate (PBT)	15–25	Consumer products; toys; electrical and electronic components	Same as for PC
Saturated polyester, polyethylene terephthalate (PET)	15–25	Appliance panels; automotive body panels; power tool housings	Improved strength Improved heat deformation temperature Improved dimensional stability
Polyurea-reinforced reaction injection molding	25–35	Exterior automotive parts (bumpers, fascias); recreational equipment	Improved flexural modulus Improved ductility Improved tensile strength Improved notched Izod impact strength Lower CLTE vs. glass fibers Improved DOI
Unsaturated polyester	18–35	Exterior automotive/truck parts (SMC); recreational equipment; cookware; appliances; transportation repair (BMC)	Cost management substitute for glass fibers Overall strength improvement Improved surface appearance

(*Continued*)

Table 14.3 (*Continued*)

Polymer	Loading (wt%)	Application examples	Expected improvements
Phenolic resin	20–40	Friction brake pads; foundry moldings; furniture insulators; appliances	Improved high temperature performance Improved electrical properties Improved wear resistance Overall mechanical properties improvement
Epoxy	20–50	Laminates; casting compositions; construction products	Improved dimensional stability Reduced shrinkage Improved electrical properties

modifiers as well. As with other fillers, the major reasons for surface modifying pretreatments are

- improved dispersion resulting in higher filler loadings and improved mechanical properties;
- improved dispersion leading to wider processing windows and higher production rates;
- allowance for less *in situ* addition of treatments, resulting in less opportunity for error;
- reduced agglomeration;
- improved polymer–mineral bonding, yielding a stronger composite;
- reduced moisture sensitivity;
- improved electrical properties.

Table 14.4 shows the matching of the appropriate organosilane pretreatment with polymer type.

Table 14.4 Types of silane and other coupling agents for wollastonite in different polymers.

Organic reactivity	Suitable polymers
Amino	Acrylic, nylon, phenolics, PVC, urethanes, PP
Epoxy	PBT, urethanes, acrylics, polysulfides
Methacrylate	Unsaturated polyesters, acrylics, EVA, polyolefins
Vinyl	Polyolefins
Titanate	LCP, polyolefins
Methyl	PP, TPO
Phenyl	PP, TPO

14.7 Summary

In selecting wollastonite as the functional filler for plastics, most of the properties to be considered are white color, high brightness, moderate specific gravity, moderate Mohs hardness, moderate range of loose and tapped bulk densities, needle-like particle shape, high processing melt temperature, wide compatibility with silane and titanate surface treatments, and low CLTE.

The following improvements are generally expected when incorporating wollastonite in plastics: stiffness and flexural strength, impact resistance, dimensional stability, mar and scratch resistance, surface appearance, heat distortion temperature, electrical properties, and compatibility with a wide range of polymers.

Table 14.5 summarizes the expected economic benefits of using wollastonite in thermoplastics and thermosets.

14.8 Future Considerations for Use of Wollastonite in Plastics

To maximize the performance of wollastonite as plastics performance requirements increase, the following points should be considered:

Table 14.5 Anticipated economic benefits of using wollastonite as functional filler.

Features	Advantages	Benefits
Low cost reinforcing filler	Less expensive than milled or chopped glass fibers	Better economics for filled plastics while giving strength improvements
Binary reinforcement synergy	Cost effective combination with treated calcined clays	Better balance of stiffness, impact, and electrical properties in engineering resins
Moderate specific gravity	Ease of volume substitution	Smoother reformulation transition and improved cost
Lower cost than polymers	Can replace more costly resins	Maintain or improve desired properties over unfilled systems
High melting temperature	Contributes to high processing temperatures	Allows improvement of melt flow rates and yield by assisting lower melt temperature polymers to be processed at higher temperatures
Less costly than certain chemical additives	For impact modification can replace certain chemical modifiers	Better impact at lower cost

- Incorporation of very high loadings (>60%), which can be introduced via masterbatch.
- Use of newer surface treatments, especially exploration of nonreactive temperature stable compounds.
- Development of even smaller median diameter grades (1–2 μm) with top size of 5 μm.
- Combination with other performance minerals, as in binary blends, to balance properties such as stiffness and impact strength or to improve overall reinforcement capacity.

References

1 Robinson, S.M. *et al.* (2006) Wollastonite, in *Industrial Minerals and Rocks*, 7th edn (eds J.E. Kogel *et al.*), Society for Mining, Metallurgy and Exploration, Inc., Littleton, CO, pp. 1027–1037.

2 Ciullo, P.A. and Robinson, S. (2002) Wollastonite – a versatile functional filler. *Paints Coat. Ind. Mag.*, **18** (11), 50–54.

3 http://www.rtvanderbilt.com; http://www.nycominerals.com.

4 http://www.minerals.usgs.gov.

5 Hohenberger, W. (2001) Fillers and reinforcements/coupling agents, Chapter 17, in *Plastics Additives Handbook*, 5th edn (ed. H. Zweifel), Hanser Publishers, Munich, Germany.

6 Wypych, G. (2000) *Handbook of Fillers*, 2nd edn, ChemTec Publishing, Toronto, Ont., Canada, pp. 167–169.

7 Mahajan, S. (2003) Proceedings of the Functional Fillers for Plastics 2003, Intertech Corp., Atlanta, GA, October 2003.

8 Hawley, G.C. (2008) Wollastonite. *Mining Engin.*, **60** (6), 64–66.

9 Virta, R.L. (2008) Wollastonite, in *U.S. Geological Survey Minerals Yearbook 2007*, pp. 82.1–82.2, Washington, D.C., USA.

10 Anonymous (2004) Prices, Industrial Minerals 438, 31.

11 Emerson, R.J. (2004) Understanding the Health Effects of Wollastonite, accessed March 21, 2004 via www.nycominerals.com.

12 Robinson, S.M. Wollastonite-Strength in Legacy. Proceedings of the 2nd International Conference on Fillers for Polymers, Paper 10, Cologne, Germany, March 21–22, 2006.

15
Wood Flour

Craig M. Clemons

15.1
Introduction

The term "wood flour" is somewhat ambiguous. Reineke [1] states that the term wood flour "is applied somewhat loosely to wood reduced to finely divided particles approximating those of cereal flours in size, appearance, and texture." Though its definition is imprecise, the term wood flour is in common use. Practically speaking, wood flour usually refers to wood particles that are small enough to pass through a screen with 850 μm openings (20 US standard mesh).

Wood flour has been produced commercially since 1906 [2] and has been used in many and varied products including soil amendments, extenders for glues, and absorbents for explosives. One of its earliest uses in plastics was in a phenol–formaldehyde and wood flour composite called Bakelite. Its first commercial product was reportedly a gearshift knob for Rolls Royce in 1916 [3]. Though once quite prevalent as filler for thermosets, its use has diminished over the years.

In contrast to its use in thermosets, large-scale use of wood flour in thermoplastics has only occurred within the last few decades. Recent growth has been great; composites made from plastics and wood or other natural fibers have grown from less than 50 000 tons in 1995 to nearly 900 000 tons in 2007 [4]. Most of this is due to the rapid growth of wood–plastic composites in exterior building products, especially decking and railing (Figure 15.1).

Due to its low thermal stability, wood flour is usually used as filler only in plastics that are processed at temperatures lower than about 200 °C. The majority of wood–plastic composites use polyethylene as the matrix (Figure 15.2). This is, in part, due to that fact that much of the early wood–plastic composites were developed as an outlet for recycled film. Polypropylene is more commonly used in automotive applications, and polyethylene is more commonly used in exterior building applications.

Functional Fillers for Plastics: Second, updated and enlarged edition. Edited by Marino Xanthos

ISBN: 978-3-527-32361-6

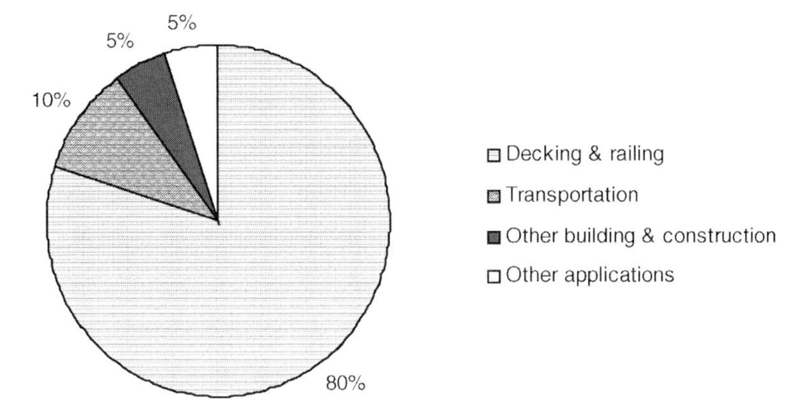

Figure 15.1 Current applications and the market size of plastics with wood flour or natural fibers [4]. "Other building and construction" include windows, doors, panels, roofing, fencing, siding, and flooring. "Other applications" include consumer products, furniture, and industrial/infrastructure.

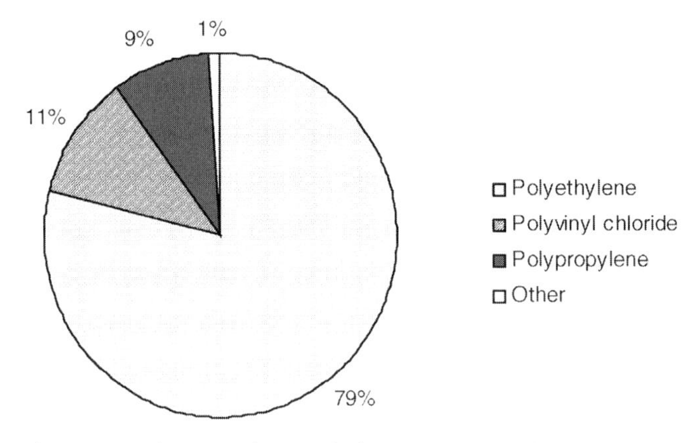

Figure 15.2 Plastics used in wood–plastic composites [4].

15.2 Production Methods

Wood flour is derived from various scrap wood from wood processors. The high quality wood flour must be of a specific species or species group and must be free from bark, dirt, and other foreign matter. Many different species of wood are offered as wood flour and are often based on the regional availability of clean raw materials from wood-processing industries. The most commonly used wood flours for plastic composites in the United States are made from pine, oak, and maple. Many reasons are given for species selection including slight color differences, regional availability, and familiarity. Some species such as red oak can contain phenolic compounds that can be oxidized and cause stains when wet [5].

Figure 15.3 Scanning electron micrograph of pine wood flour.

Though there is no standard method of producing wood flour, some generalities can be discussed. The main steps in wood flour production are size reduction and size classification. If larger raw materials are used, initial size may be reduced using equipments such as a hammer mill, hog, or chipper [2]. Once coarsely ground, the wood is pulverized by grinding between disks as in attrition mills, beating with impactors or hammers as in hammer mills, or crushing between rollers as in roller mills [2]. Other mills can also be used but are less common.

Pulverizing results in particles that contain fibers and fiber fragments. These particles typically have aspect ratios (i.e., length to diameter ratios) of only 1–5 (Figure 15.3). These low aspect ratios allow wood flour to be more easily metered and fed than individual wood fibers, which tend to bridge. However, the low aspect ratio limits the reinforcing ability [6].

Once pulverized, the wood can be classified using vibrating, rotating, or oscillating screens. Air classifying is also used especially with very finely ground wood flours [1]. Wood flour particle size is often described by mesh of the wire cloth sieves used to classify them. Table 15.1 lists the US standard mesh sizes and their equivalent particle diameters. However, different standards may be used internationally [7]. Most commercially manufactured wood flours used as fillers in thermoplastics are in the size range of 180–425 μm (80–40 US standard mesh). Very fine wood flours can cost more and increase melt viscosity more than coarser wood flours but composites made with them typically have more uniform appearance and a smoother finish. If ground too fine, fiber bundles become wood dust, fragments that no longer resemble fiber or fiber bundles.

Wood flour is commonly packaged in (i) multiwalled paper bags (approximately 23 kg or 50 lb), (ii) bulk bags (1.5 m^3 or 55 ft^3 typical), or (iii) bulk trailers [8]. Wood flour is typically supplied to the customer at moisture contents between 4 and 8% and must be dried before use in thermoplastics. Some wood flour manufacturers offer standard grades, others prefer to customize for individual buyers and applications.

Table 15.1 Conversion between US standard mesh and particle diameter.

US standard mesh [40]	Particle diameter (μm)
20	850
25	710
30	600
35	500
40	425
45	355
50	300
60	250
70	212
80	180
100	150
120	125
140	106
170	90
200	75
230	63
270	53
325	45
400	38

Specifications depend on the application but include size distribution, moisture content, species, color, and cost.

15.3 Structure and Properties

15.3.1 Wood Anatomy

As with most natural materials, the anatomy of wood is complex. Wood is porous, fibrous, and anisotropic. Wood is often broken down into two broad classes: softwoods and hardwoods that are actually classified by botanical and anatomical features rather than wood hardness. Figures 15.4 and 15.5 are schematics of a softwood and hardwood, respectively, showing the typical anatomies of each wood type. Softwoods (or *Gymnosperms*) include pines, firs, cedars, and spruces among others; hardwoods (or *Angiosperms*) include species such as the oaks, maples, and ashes.

Wood is primarily composed of hollow, elongated, spindle-shaped cells (called tracheids or fibers) that are arranged parallel to each other along the trunk of the tree [9]. The lumen (hollow center of the fibers) can be completely or partially filled with deposits, such as resins or gums, or growths from neighboring cells called tyloses [9]. These fibers are firmly cemented together and form the structural component of wood tissue. The length of wood fibers is highly variable but average about 1 mm (1/25 in.) for hardwoods and 3–8 mm (1/8 to 1/3 in.) for softwoods [9].

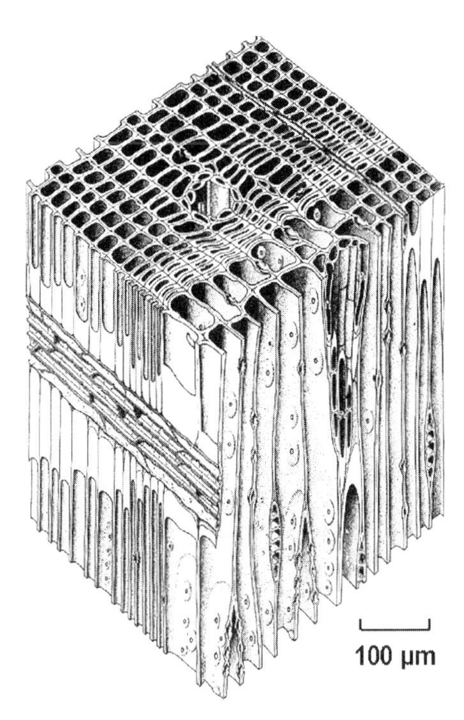

Figure 15.4 Schematic of a softwood.

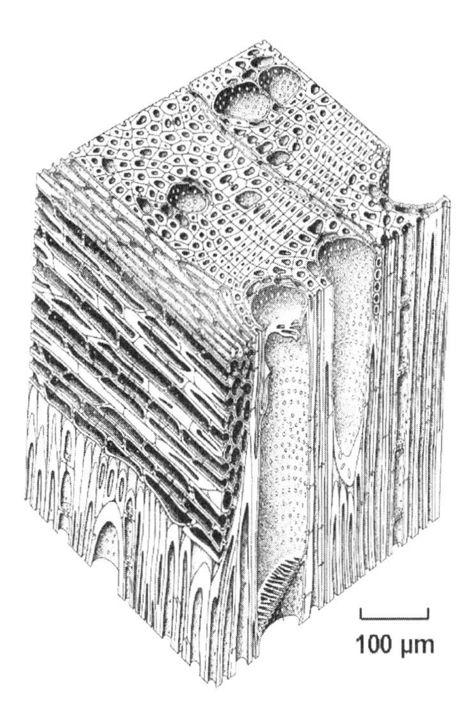

Figure 15.5 Schematic of a hardwood.

Table 15.2 Approximate chemical composition of selected woods [15].

Species	Cellulose[a)]	Hemicellulose[b)]	Lignin[c)]	Extractives[d)]	Ash
Ponderosa pine	41	27	26	5	0.5
Loblolly pine	45	23	27	4	0.2
Incense cedar	37	19	34	3	0.3
Red maple	47	30	21	2	0.4
White oak	47	20	27	3	0.4
Southern red oak	42	27	25	4	0.4

a) Alpha cellulose content as determined by ASTM D 1103 [41].
b) Approximate hemicellulose content determined by subtracting the alpha cellulose content from the holocellulose content values from Ref. [15].
c) Klason lignin content as determined by ASTM D 1106 [42].
d) Solubility in 1 : 2 volume ratio of ethanol and benzene according to ASTM D 1107 [43].

Fiber diameters are typically 15–45 μm. When wood is reduced to wood flour, the resulting particles are actually bundles of wood fibers rather than individual fibers and can contain lesser amounts of other features such as ray cells and vessel elements. Further information on wood anatomy can be found in Refs [10, 11].

15.3.2
Chemical Components

Wood itself is a complex, three-dimensional, polymer composite made up primarily of cellulose, hemicellulose, and lignin [12]. These three hydroxyl-containing polymers are distributed throughout the cell wall. The chemical compositions of selected woods are shown in Table 15.2.

Cellulose varies the least in the chemical structure of the three major components. It is a highly crystalline, linear polymer of anhydroglucose units with a degree of polymerization (n) around 10 000 (Figure 15.6). It is the main component providing the wood's strength and structural stability. Cellulose is typically 60–90% crystalline by weight and its crystal structure is a mixture of monoclinic and triclinic unit cells [13, 14]. Hemicelluloses are branched polymers composed of various 5- and 6-carbon sugars whose molecular weights are well below those of cellulose but which still contribute as a structural component of wood [15].

Lignin is an amorphous, cross-linked polymer network consisting of an irregular array of variously bonded hydroxy- and methoxy-substituted phenylpropane units [15]. The chemical structure varies depending on its source. Figure 15.7

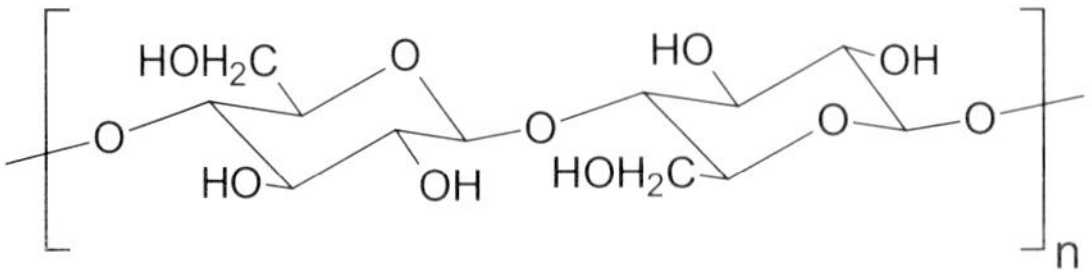

Figure 15.6 Chemical structure of cellulose [15].

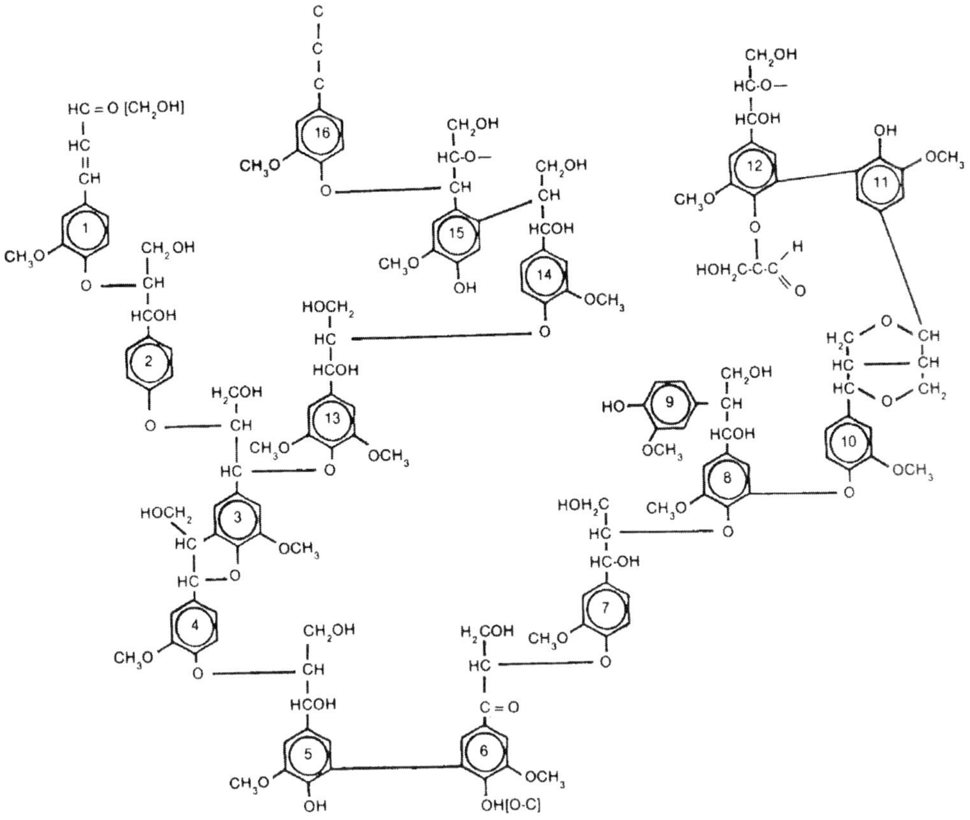

Figure 15.7 A partial softwood lignin structure [15].

represents a partial softwood lignin structure illustrating a variety of possible structural components. Lignin is more nonpolar than cellulose and acts as a chemical adhesive within and between the cellulose fibers.

Additional organic components, called extractives, make up about 3–10% of the dry wood grown in temperate climates, but significantly higher quantities are found in wood grown in tropical climates [15]. Extractives include substances such as fats, waxes, resins, proteins, gums, terpenes, and simple sugars among others. Many of these extractives function in tree metabolism and act as energy reserves or defend against microbial attack [15]. Though often small in quantity, extractives can have large influences on properties such as color, odor, and decay resistance [15]. Small quantities (typically 1%) of inorganic matter, termed ash, are also present in wood grown in temperate regions.

Cellulose forms crystalline microfibrils held together by hydrogen bonds and then cemented to lignin into the wood fiber cell wall. The microfibrils are aligned in the fiber direction in most of the cell wall, winding in a helix along the fiber axis. The angle between the microfibril and fiber axes is called the microfibril helix angle. The microfibril helix angle is typically 5–20° for most of the cell wall [16] and varies depending upon many factors including species and stresses on the wood during growth.

15.3.3 Density

The bulk density of wood flour depends on factors such as moisture content, particle size, and species, but typically is about 190–220 kg/m^3 (12–14 lb/ft^3) [8]. Because of its low bulk density, special equipment such as crammers is sometimes used to aid the feeding of wood flour.

As a filler, wood flour is unusual since it is compressible. Though the density of the wood cell wall is about 1.44–1.50 g/cm^3 [17], the porous anatomy of solid wood results in overall densities of about 0.32–0.72 g/cm^3 (20 and 45 lb/ft^3) when dry [18]. However, the high pressures found during plastics processing can collapse the hollow fibers that comprise the wood flour or fill them with low molecular weight additives and polymers. The degree of collapsing or filling will depend on variables such as particle size, processing method, and additive viscosity, but wood densities in composites approaching the wood cell wall density can be found in high-pressure processes such as injection molding. Consequently, adding wood fibers to commodity plastics such as polypropylene, polyethylene, and polystyrene increases their density.

Even these high densities are considerably lower than those of inorganic fillers and reinforcements. This density advantage is important in applications where weight is important such as in automotive components. Recently, chemical foaming agents and microcellular foaming technology have been investigated to reduce the density of wood–plastic composites [19–21].

15.3.4 Moisture

The major chemical constituents of the cell wall contain hydroxyl and other oxygen containing groups that attract moisture through hydrogen bonding [23]. This hygroscopicity can cause problems both in composite fabrication and in the performance of the end product.

Moisture sorption in wood is complex and the final equilibrium moisture content is affected by temperature and humidity. The equilibrium moisture content can also vary by up to 3–4% (although usually less), depending on whether it is approached from a higher or lower humidity (i.e., wood exhibits a moisture sorption hysteresis). Table 15.3 shows approximate equilibrium moisture contents for wood at different temperatures and humidities at a midpoint between the hysteresis curves.

Wood flour usually contains at least 4% moisture when delivered, which must be removed before or during processing with thermoplastics. Though moisture could potentially be used as a foaming agent to reduce density, this approach is difficult to control and is not common industrial practice. Commercially, moisture is removed from the wood flour: (i) before processing using a dryer, (ii) by using the first part of an

Table 15.3 Equilibrium moisture content for wood at different temperatures and humidities [18].

Temperature		Moisture content (%) at various relative humidities								
(°C)	(°F)	10%	20%	30%	40%	50%	60%	70%	80%	90%
−1.1	30	2.6	4.6	6.3	7.9	9.5	11.3	13.5	16.5	21.0
4.4	40	2.6	4.6	6.3	7.9	9.5	11.3	13.5	16.5	21.0
10	50	2.6	4.6	6.3	7.9	9.5	11.2	13.4	16.4	20.9
15.6	60	2.5	4.6	6.2	7.8	9.4	11.1	13.3	16.2	20.7
21.1	70	2.5	4.5	6.2	7.7	9.2	11.0	13.1	16.0	20.5
26.7	80	2.4	4.4	6.1	7.6	9.1	10.8	12.9	15.7	20.2
32.2	90	2.3	4.3	5.9	7.4	8.9	10.5	12.6	15.4	19.8
37.8	100	2.3	4.2	5.8	7.2	8.7	10.3	12.3	15.1	19.5

extruder as a dryer in some inline process, or (iii) during a separate compounding step (or in the first extruder in a tandem process).

Once dried, wood flour can still absorb moisture quickly. Depending on ambient conditions, wood flour can absorb several weight percents of moisture within hours (Figure 15.8). Even a compounded material often needs to be dried prior to further processing, especially if high-weight percentages of wood flour are used. Figure 15.9 shows moisture sorption curves for compounded pellets of polypropylene containing 40% wood flour at different humidities.

The hygroscopicity of wood flour can also affect the end composite. The absorbed moisture interferes and reduces hydrogen bonding between the cell wall polymers and alters its mechanical performance [22]. Moisture of up to about 30% can be adsorbed by the cell wall with a corresponding reversible increase in apparent wood

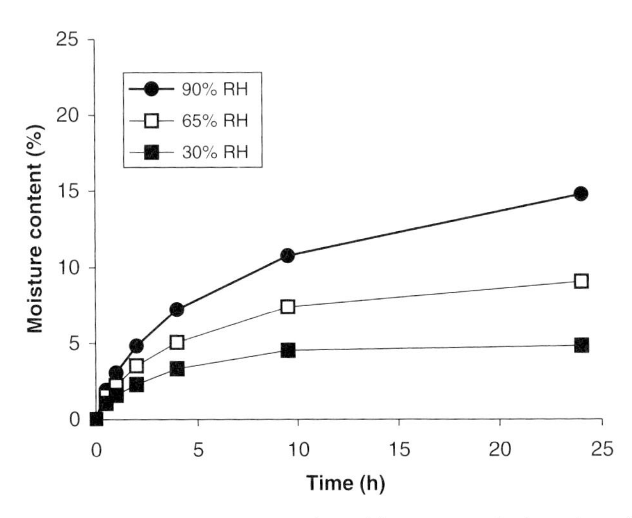

Figure 15.8 Moisture sorption of wood flour at several relative humidities and 26 °C.

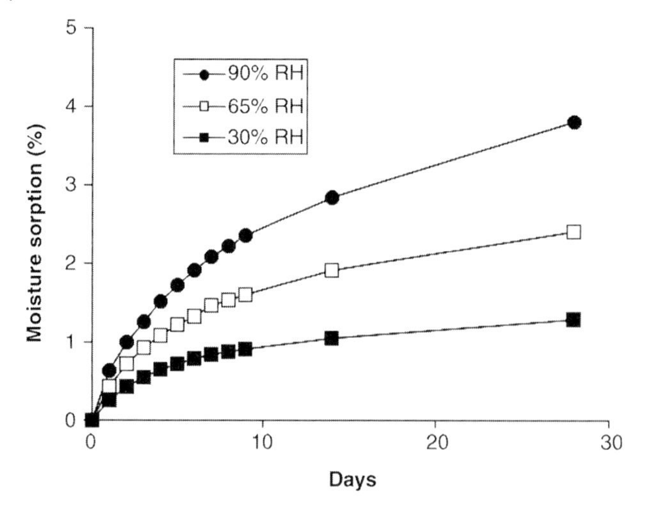

Figure 15.9 Moisture sorption of compounded pellets of polypropylene containing 40% wood flour at several relative humidities and 26 °C [49].

volume. The wood volume V_1, at a moisture content M, has been roughly approximated by [16]

$$V_1 = V_0(1 + 0.84M\varrho),$$

where V_0 is the dry volume and ϱ is the density of the wood when dry.

Volume changes due to moisture sorption, especially repeated moisture cycling, can lead to interfacial damage and matrix cracking [23]. A number of papers have discussed damage and the resulting irreversible mechanical property reductions after the composites were exposed to a humid environment or liquid water [23–25]. Water uptake depends on many variables including wood flour content, wood flour particle size, matrix type, processing method, and additives such as coupling agents. Many manufacturers of the wood–plastic composites used in exterior applications limit wood flour content to 50–60 wt% and rely on the partial encapsulation of the wood by the polymer matrix to prevent major moisture sorption and subsequent negative effects.

15.3.5
Durability

Wood will last for decades in exterior environments, especially if it is stained, painted, or otherwise protected. However, wood–plastic composites are not commonly protected. In fact, a common selling point for wood–plastic composites is that they are low maintenance material and do not require painting or staining in outdoor applications.

The surface of wood undergoes photochemical degradation when exposed to UV radiation. This degradation takes place primarily in the lignin component and results in a characteristic color change [26]. Hence, wood–plastic composites containing no pigments usually fade to a light gray when exposed to sunlight. Photostabilizers or pigments are commonly added to wood–plastic composites to help reduce this color-fade when used in exterior environments.

Mold can form on surfaces of wood–plastic composites. Mold growth has been ascribed as arising from various effects, among them moisture sorption by the wood flour, buildup of organic matter on the composite surfaces, and the lubricants used in processing of the composites. The relative contribution of these factors to mold growth is uncertain. Although mold does not reduce the structural performance of the composite, it is an aesthetic issue.

Wood is degraded biologically because organisms recognize the celluloses and hemicelluloses in the cell wall and can hydrolyze them into digestible units using specific enzyme systems [22]. If the moisture content of the wood flour in the composite exceeds the fiber saturation point (approximately 30% moisture), decay fungi can begin to attack the wood component leading to weight loss and significant reduction in mechanical performance. Figure 15.10 shows the weight loss due to exposure of a wood–plastic composite to the decay fungi *Gleophylum trabeum* in a laboratory soil block test. Decay is not found until a moisture threshold of about 15% is reached. As HDPE does not absorb moisture, the average moisture content of the wood flour in the composite would be expected to be roughly twice that shown. This suggests that when the moisture content of the wood flour reaches about 30%, approximately the fiber saturation point, significant decay begins. Additives such as zinc borate are sometimes added to wood–plastic composites to improve fungal resistance.

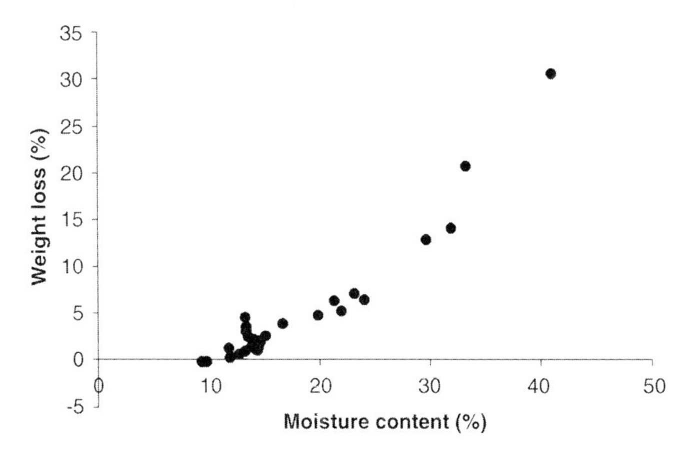

Figure 15.10 Weight loss due to fungal attack (*G. trabeum*) as a result of moisture sorption. Extruded composites of high density polyethylene containing 50% wood flour [50].

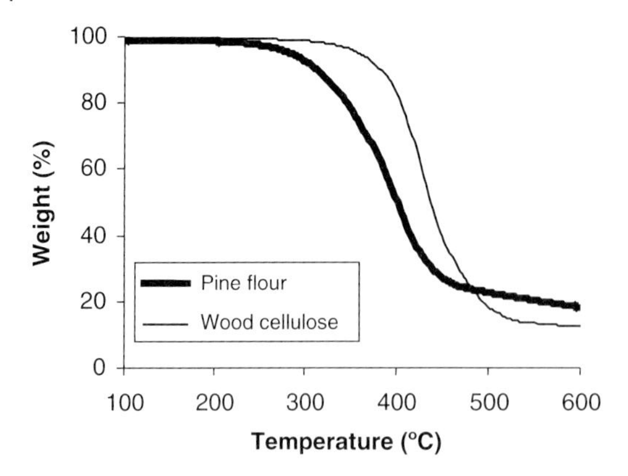

Figure 15.11 Thermogravimetric analysis of pine flour and wood cellulose.

15.3.6 Thermal Properties

Figure 15.11 shows a thermogravimetric analysis of pine flour and wood cellulose. Due to its low thermal stability, wood flour is usually used as filler only in plastics that are processed at low temperatures, lower than about 200 °C. Above these temperatures the cell wall polymers begin to decompose. High-purity cellulose pulps, where nearly all of the less thermally stable lignin and hemicelluloses have been removed, have recently been investigated for use in plastic matrices such as nylon that are processed at higher temperatures than most commodity thermoplastics [27].

Because of its practical performance, the thermal properties of wood have been extensively investigated [17]. Understandably, this work is generally performed on uncompressed wood with moisture contents typical of those found in service. Information on dry, compressed wood as might be found in a wood thermoplastic composite is lacking. Additionally, thermal properties vary depending on the chemistry and structure of wood. Factors such as extractive content, grain direction, and fibril angle are important. Though precise numbers are not known, some approximations may be made for a broad discussion on the topic.

The thermal expansion of wood is less than that of the commodity plastics commonly used as matrices. Thermal expansion coefficients for wood are directional and are roughly [17]

$$\alpha = A\varrho \times 10^{-6},$$

where α is the coefficient of thermal expansion (in K^{-1}), ϱ is the density (oven dry basis), and A is a constant with values approximately 56 in the radial direction, 81 in the tangential direction and about 5–10 times less in the fiber direction. This roughly yields an average of about $70 \times 10^{-6}\ K^{-1}$ if we assume a density of 1.5 g/cm^3. This is about half that of polypropylene ($150 \times 10^{-6}\ K^{-1}$) and 3.5 times less than that of low density polyethylene ($250 \times 10^{-6}\ K^{-1}$), both of which are commonly used as matrix materials for wood–plastic composites [28].

The specific heat of dry wood does not have a strong dependence on specific gravity and is roughly 0.324 cal/(g K) or 1360 J/(kg K) [17]. This is about half the specific heat of common polyolefins such as polypropylene and polyethylene, for which values are approximately 2–3000 J/(kg K.)

Thermal conductivity, k, of dry wood has been reported to increase linearly with density, ϱ, according to [17]

$$k = 0.200\varrho + 0.024,$$

where k is in units of W/(m K). Assuming a density of compressed wood flour in a composite of 1.5 g/cc, the thermal conductivity is calculated as 0.32 W/(m·K). This is the same order of magnitude as the values reported for polypropylene and polyethylene (0.17–0.51 W/(m·K)).

Thermal diffusivity is a measure of the rate at which a material changes temperature when the temperature of its surroundings changes. The thermal diffusivity, h, is the ratio of thermal conductivity, k, and the product of specific heat, c, and density, ϱ [17]:

$$h = k/c\varrho$$

Calculations for wood flour yield a value of $0.16 \times 10^{-6}\,\mathrm{m^2/s}$ compared to $0.11\text{–}0.17 \times 10^{-6}\,\mathrm{m^2/s}$ for polypropylene and polyethylene.

15.4 Suppliers

There is a wide range of wood flour suppliers, and they cater to a number of different industries. These are both large companies that have broad distribution networks as well as small, single source suppliers catering to single customers. Because of the varied and disperse nature of these suppliers, there are currently few good resources that list wood flour manufacturers.

Wood–plastic composite manufacturers obtain wood flour either directly from forest products companies such as lumber mills and furniture, millwork, or window and door manufacturers that produce it as a by-product, or commercially from companies that specialize in wood flour production.

With a growing number of wood flour suppliers targeting the wood–plastic composites industry, they are beginning to be listed in plastics industry resources [29]. Several major suppliers of wood flour to the North American wood–plastic composite industry include

- American Wood Fibers (Schofield, WI, USA)
- Lang Fiber (Marshfield, WI, USA)
- Marth Manufacturing (Marathon, WI, USA)
- P.J. Murphy Forest Products (Montville, NJ, USA)
- P.W.I. Industries Inc. (Saint Hyacinthe, QC, Canada).

15.5 Cost/Availability

As with most materials, wood flour costs are variable and depend on various factors such as volume, availability, particle size, and shipping distance. However, wood flour is typically about $0.11–0.22/kg ($0.05–0.10/lb) in the United States. Narrow particle size distributions and fine sizes tend to increase cost. As there are many small manufacturers and the volume is relatively lower than other wood products (solid wood, wood composites, and paper), information on wood flour availability is scarce.

15.6 Environmental/Toxicity Considerations

The environmental benefits of wood and other natural fibers have been an important influence on their use, particularly in Europe. Wood flour is derived from a renewable resource, does not have a large energy requirement to process, and is biodegradable [30].

Wood is a commonly used material and most people are very comfortable with its use. Most of the risks in using wood flour lies in the fact that (i) it has low thermal stability and can degrade and burn resulting in a fire and explosion hazard that is greater than with solid wood, and (ii) inhalation of finely ground wood flour can lead to respiratory difficulties. Some basic precautions include avoiding high processing temperatures, using well-ventilated equipment, eliminating ignition sources, and using good dust protection, prevention, and control measures. For detailed information on environmental and health risks, manufacturers should consult their suppliers and materials' safety data sheets. Regulating bodies such as the Occupational Safety and Health Association (OSHA) also have good information on health and safety information on wood dust (e.g., see www.osha.gov/SLTC/wooddust/index.html).

15.7 Applications (Primary and Secondary Functions)

15.7.1 Thermosets

In thermosetting adhesives, wood flour has been used for several functions. As an extender, it is added to reduce cost while retaining bulk for uniform spreading. Unfortunately, wood flour extenders generally also reduce durability of a given resin. As filler, it is added to thermoset adhesives to control penetration when bonding wood and to improve characteristics of the hardened film [31].

High weight percentages of wood flour have been used with thermosets such as phenolic or urea/formaldehyde resins to produce molded products. The wood flour is added to improve toughness and reduce shrinkage on curing. Wood filler may be added to the thermoset resin at elevated temperatures to form a moldable paste. This paste is then transferred to a mold and then cured under heat and pressure [32]. Alternatively, a mixture of powdered thermoset resin and wood flour is poured directly into a mold and pressed under heat and pressure. Although this type of composite was very prevalent throughout much of the twentieth century, often under the trade names such as "Bakelite," its use has diminished considerably over the years. However, a variety of products such as some salad bowls, trays, and cutting boards are still manufactured from wood thermoset composites.

15.7.2 Thermoplastics

Wood flour is often added to thermoplastics as a low cost filler to alter mechanical performance, especially the stiffness of low melt temperature, commodity thermoplastics such as polypropylene and polyethylene without increasing density excessively. Wood is much stiffer than the commodity thermoplastics usually used as matrices. Additionally, wood and pulp fibers can nucleate crystal growth in polyolefins resulting in a transcrystalline layer that can influence mechanical behavior [33, 34].

The wood flour stiffens these plastics but also embrittles them, reducing properties such as elongation and unnotched impact strength. Tensile and flexural strengths are, at best, maintained and more often decreased in the absence of a coupling agent. Many different coupling agents have been investigated for use in wood–plastic composites and are reviewed elsewhere [35, 36]. When a coupling agent is desired, maleated polyolefins are most often used commercially. However, even when a coupling agent is used, improvements in strength are limited by the low aspect ratio of the wood flour (Figure 15.12). It is often unclear how much of the strength increase is due to better wetting and dispersion and how much is due to the increased bonding. Small amounts of thermosets have also been added in wood–plastic formulations to improve mechanical performance [37].

Table 15.4 summarizes some typical mechanical property changes when various species of wood flour are added to an injection molding grade of polypropylene [44]. Both tensile and flexural moduli increase with the addition of wood flour. Composites with hardwoods (maple and oak) yield the highest flexural moduli at 60 wt% wood flour, that is, approximately four times that of the unfilled polypropylene. Heat deflection temperature is also approximately doubled by adding 60 wt% wood flour. The increase in modulus with the addition of wood flour comes at the expense of elongation, a drastic reduction in unnotched impact strength, and a general decrease in tensile strength.

The effects of wood flour particle size on the mechanical performance of filled polypropylene are summarized in Table 15.5. The largest particle size range (0.425–0.600 μm) yields the lowest performance except for notched impact strength. For the three smaller particle size ranges, properties generally decrease as the particle size is reduced except for the unnotched impact strength. Similar trends were found

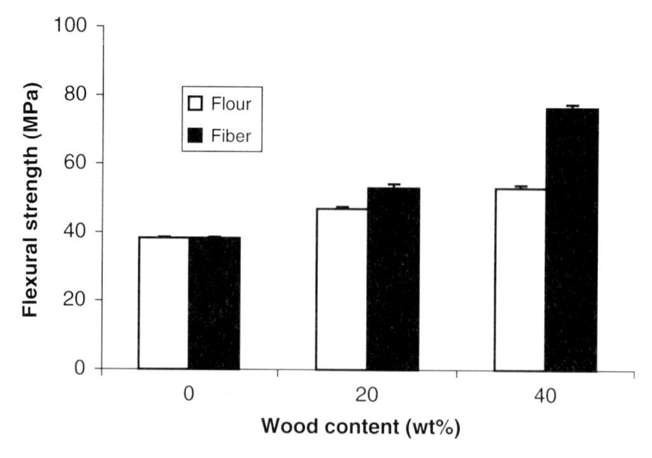

Figure 15.12 Comparison between wood flour and wood fiber as a reinforcement in polypropylene. Injection molded composites. 3% maleated polypropylene added as a coupling agent. (Derived from Ref. [52].)

when broader particle size ranges, more typical of commercially available blends, were investigated [38].

Customer and builders have a certain familiarity with wood in applications such as decking and railings (the largest wood–plastic composite market) and often desire an alternative that may have similar attributes. Mixing wood flour with plastic is seen as a way to use wood in these applications, yet improve its durability without chemical treatment or the need for painting or staining. Adding wood flour to thermoplastics results in a composite with a wood color, although often pigments and light stabilizers are added to mitigate color-fade in exterior applications. Though not nearly as stiff as solid wood, these composites are stiffer than unfilled plastics. Many of the composites do not require special fasteners or design changes such as shorter spans in applications such as deck boards. Adding wood flour also improves dimensional stability with respect to temperature changes. Although more expensive than wood, many consumers have been willing to pay for the lower maintenance required when wood–plastic composites are used.

Wood–plastic composites are heavier than wood and do not have as good mechanical performance, often having much lower stiffness [39]. The swelling of wood flour with moisture can lead to damage in its composites. Adequate dispersion of the wood flour and encapsulation by the matrix is critical and wood flour content is typically limited to approximately 50–65 wt% in exterior applications. These highly filled plastics require judicious use of additives and processing methodology. For example, the high wood content results in high viscosity and low melt strength that require significant amounts of lubricants for reasonable production rates and to prevent melt fracture. Extruders capable of removing considerable moisture from the wood flour have been developed that avoid the need for predrying. Considerable research and development is centered on economically increasing structural performance, improving durability, and decreasing weight. As improvements are being

Table 15.4 Effect of adding wood flour on the mechanical performance of injection molded polypropylene[a)] [44].

	Izod impact strength[b)]		Flexural properties[c)]		Tensile properties[d)]			
Filler content (wt %)	Notched (J/m)	Unnotched (J/m)	Flexural strength (MPa)	Modulus of elasticity (GPa)	Tensile strength (MPa)	Modulus of elasticity (GPa)	Elongation at maximum strength (%)	Heat deflection temperature[e)] (°C)
No filler								
0	15.0	600	34.7	1.03	28.5	1.31	10.4	55
Ponderosa pine[f)]								
20	15.4	128	41.6	1.89	26.5	1.99	5.7	59
30	19.0	95	43.1	2.58	24.6	3.24	3.1	76
40	20.8	76	44.2	3.22	25.5	3.71	2.3	35
50	20.5	58	41.8	3.66	23.0	4.25	1.7	39
60	21.1	41	38.8	4.04	20.1	4.56	1.4	91
Loblolly pine[f)]								
20	12.4	120	40.6	1.71	24.9	2.14	4.8	54
30	12.7	75	41.2	2.28	23.7	2.53	3.5	72
40	13.7	49	39.3	2.84	21.4	3.28	1.8	78
50	13.8	42	37.1	3.40	19.7	3.97	1.3	79
60	10.4	31	34.1	3.81	17.8	4.32	0.9	77
Maple[f)]								
20	12.8	113	46.2	2.16	27.9	2.87	4.1	59
30	15.0	87	46.5	2.47	27.1	3.33	3.3	38
40	16.5	63	45.4	3.23	25.6	4.72	2.0	104

(Continued)

Table 15.4 (*Continued*)

Filler content (wt %)	Izod impact strength[b)]		Flexural properties[c)]		Tensile properties[d)]			Heat deflection temperature[e)] (°C)
	Notched (J/m)	Unnotched (J/m)	Flexural strength (MPa)	Modulus of elasticity (GPa)	Tensile strength (MPa)	Modulus of elasticity (GPa)	Elongation at maximum strength (%)	
50	17.7	49	42.1	4.16	24.0	5.20	1.4	111
60	17.9	44	38.0	4.35	19.9	4.77	1.1	110
Oak[f)]								
20	14.6	87	44.1	1.83	27.2	2.48	4.7	74
30	17.5	63	45.9	2.87	25.7	3.81	2.4	98
40	18.6	68	44.8	3.39	25.2	4.19	2.1	100
50	20.9	46	42.8	3.99	23.4	4.79	1.5	112
60	18.8	33	38.1	4.60	19.8	5.05	1.2	114
Average COV[g)]	12	6	2	4	2	8	10	2

a) Fortilene 3907, polypropylene homopolymer, Solvay Polymers, Deer Park, TX, USA.
b) ASTM D 256 [45].
c) ASTM D 790 [46].
d) ASTM D 638 [47].
e) ASTM D 648 [48].
f) Wood flours were commercial grades from American Wood Fibers, Schofield, WI, USA.
g) COV, the coefficient of variation is 100 × (standard deviation)/(average).

Table 15.5 Mechanical properties of composites made from polypropylene[a] filled with 40 wt% wood flour [38].

	Izod impact strength[b]		Flexural properties[c]		Tensile properties[d]		
Particle size range (μm)	Notched (J/m)	Unnotched (J/m)	Flexural strength (MPa)	Modulus of elasticity (GPa)	Tensile strength (MPa)	Modulus of elasticity (GPa)	Elongation at maximum strength (%)
No filler							
—	15.0	600	34.7	1.03	28.5	1.31	10.4
Composites with 40 wt% wood flour[e]							
425–600	22	54	38.7	2.69	21.8	3.20	2.3
180–250	20	79	42.6	3.15	25.5	3.61	2.3
106–150	19	84	42.9	3.00	24.9	3.47	2.2
53–75	16	91	41.4	2.89	24.3	3.46	2.1
Average COV[f]	5	10	1	2	1	7	6

a) Fortilene 3907, polypropylene homopolymer, Solvay Polymers, Deer Park, TX, USA.
b) ASTM D 256 [45].
c) ASTM D 790 [46].
d) ASTM D 638 [47].
e) Specially screened wood flour from American Wood Fibers, Schofield, WI, USA.
f) COV, the coefficient of variation is 100 × (standard deviation)/(average).

made, wood plastic composites' use is extending to other exterior building applications such as roofing, fencing, and siding. It is also increasingly being used in furniture and injection molded consumer products.

Environmental considerations can influence the use of wood flour. Wood flour is manufactured from industrial byproducts, mitigating a disposal issue. It is also lighter than inorganic fillers, which can lead to benefits such as fuel savings when its composites are used in transportation and packaging applications. With changing consumer perceptions, some manufacturers have used the natural look of wood flour composites as a marketing tool. Others add wood flour to increase bio-based material content. Though not specifically added to plastics to impart biodegradability, wood flour can be used as a filler in biodegradable polymers where its biodegradability is an attribute rather than the detriment; it is sometimes considered to be in more durable composites.

References

1 Reineke, L.H. (1966) Wood flour, U.S. Forest Service Research Note FPL-0113, USDA Forest Service, Forest Products Laboratory, Madison, WI, USA, pp. 13.

2 Panshin, A.J. *et al.* (1962) Wood Flour, Chapter 13, in *Forest Products: Their Sources, Production, and Utilization*, McGraw-Hill, New York, pp. 265–271.

3 Gordon, J.E. (1988) *The New Science of Strong Materials (or why you don't fall through the floor)*, 2nd edn, Princeton University Press, Princeton, NJ, p. 179.

4 Morton, J., Quarmley, J., and Rossi, L. (2003) Current and emerging applications for natural and wood fiber composites. Proceedings of the 7th International Conference on Woodfiber-Plastic Composites, Madison, WI, Forest Products Society, pp. 3–6.

5 Rowe, J.W. and Conner, A.H. (1979) Extractives in Eastern Hardwoods – a review, General Technical Report FPL 18, USDA Forest Service, Forest Products Laboratory, Madison, WI, USA, p. 17.

6 Bigg, D.M. *et al.* (1988) High performance thermoplastic composites. *J. Thermoplast. Compos.*, **1**, 146–161.

7 International Sieve Chart, www.reade.com/Sieve/international_sieve.html ©1997 Reade Advanced Materials.

8 Wood Flour: Discover the Possibilities, product literature, PJ Murphy Forest Products Corp., Montville, NJ.

9 Miller, R.B. (1999) Structure of wood, Chapter 2. *Wood Handbook: Wood as an Engineering Material*, General Technical Report, FPL-GTR-113, USDA Forest Service, Forest Products Laboratory, Madison, WI, USA, pp. 463.

10 Panshin, A.J. and de Zeeuw, C. (1980) *Textbook of Wood Technology*, 4th edn, McGraw-Hill, New York.

11 Hoadley, R.B. (1990) Identifying wood, *Accurate Results with Simple Tools*, Taunton Press, Newtown, CT.

12 Rowell, R.M. (1983) Chemical modification of wood. *For. Prod. Abstr.*, **6** (12), 363–382.

13 Imai, T. and Sugiyama, J. (1998) Nanodomains of Iα and Iβ Cellulose in Algal Microfibrils. *Macromolecules*, **31**, 6275–6279.

14 Wada, M. *et al.* (1994) The monoclinic phase is dominant in wood cellulose. *Mokuzai Gakkaishi*, **40** (1), 50–56.

15 Pettersen, R.C. (1984) The chemical composition of wood, Chapter 2, in *The Chemistry of Solid Wood* (ed. R.M. Rowell), American Chemical Society, Washington, DC, pp. 76–81.

16 Parham, R.A. and Gray, R.L. (1984) Formation and structure of wood, in *The Chemistry of Solid Wood* (ed. R.M. Rowell), American Chemical Society, Washington, DC, pp. 1–56.

17 Kellogg, R.M. (1981) Physical properties of wood, in *Wood: Its Structure and Properties*

(ed. F.F. Wangaard), Pennsylvania State University, College Park, PA, pp. 191–223.
18 Simpson, W. and TenWolde, A. (1999) Physical properties and moisture relations of wood, Chapter 3. *The Wood Handbook: Wood as an Engineering Material*, General Technical Report, FPL-GTR-113, USDA Forest Service, Forest Products Laboratory, Madison, WI, USA, pp. 2-1–2-4.
19 Karayan, V. (2001) The effect of chemical foaming agents on the processing and properties of polypropylene-woodflour composites. Proceedings of the Sixth International Conference on Woodfiber-Plastic Composites, Forest Products Society, Madison, WI, pp. 249–252.
20 Turng, L.-S. *et al.* (2003) Applications of nano-composites and woodfiber plastics for microcellular injection molding. Proceedings of the Seventh International Conference on Woodfiber-Plastic Composites, Forest Products Society, Madison, WI, pp. 217–226.
21 Matuana-Malanda, L. *et al.* (1996) Characterization of microcellular foamed PVC/cellulosic fiber composites. *J. Cell. Plast.*, **32** (5), 449–469.
22 Winandy, J.E. and Rowell, R.M. (1984) The chemistry of wood strength, Chapter 5, in *The Chemistry of Solid Wood* (ed. R.M. Rowell), American Chemical Society, Washington, DC, p. 218.
23 Peyer, S. and Wolcott, M. (2000) Engineered Wood Composites for Naval Waterfront Facilities. *2000 Yearly Report to Office of Naval Research*, Wood Materials and Engineering Laboratory, Washington State University, Pullman, WA, pp. 14.
24 Stark, N.S. (2001) Influence of moisture absorption on mechanical properties of wood flour-polypropylene composites. *J. Thermoplast. Compos.*, **14** (5), 421–432.
25 Balatinecz, J.J. and Park, B.-D. (1997) The effects of temperature and moisture exposure on the properties of woodfiber thermoplastic composites. *J. Thermoplast. Compos.*, **10**, 476–487.
26 Rowell, R.M. (1984) Penetration and reactivity of cell wall components, Chapter 4, in *The Chemistry of Solid Wood* (ed. R.M. Rowell American Chemical Society, Washington, DC, p. 176.
27 Sears, K.D. *et al.* (2001) Reinforcement of engineering thermoplastics with high purity cellulose fibers. Proceedings of the Sixth International Conference on Woodfiber-Plastic Composites, Forest Products Society, Madison, WI, pp. 27–34.
28 Osswald, T.A. and Menges, G. (1996) *Materials Science of Polymers for Engineers*, Carl Hanser Verlag, New York City, NY, pp. 458–459.
29 Anonymous (2003) Special report: who's who in WPC, *Natural and Wood Fiber Compos.*, **2** (8), p. 4.
30 Suddell, B.C. and Evans, G. (2003) The increasing use and application of natural fibre composite materials within the automotive industry. Proceedings of the 7th International Conference on Woodfiber-Plastic Composites, Madison, WI, pp. 7–14.
31 Marra, A.A. (1992) *Technology of Wood Bonding: Principles in Practice*, Van Nostrand Reinhold, New York, NY, p. 438.
32 Dahl, W.S. (1948) *Woodflour*, The Mercury Press, The Parade, Northampton, England, p. 119.
33 Quillin, D.T. *et al.* (1993) Crystallinity in the polypropylene/cellulose system. I. Nucleation and crystalline morphology. *J. Appl. Polym. Sci.*, **50**, 1187–1194.
34 Lee, S.Y. *et al.* (2001) Influence of surface characteristics of TMP fibers to the growth of TCL on the linear fibers surface (I): aspect in the surface roughness. Proceedings of the Sixth International Conference on Woodfiber-Plastic Composites, Forest Products Society, Madison, WI, pp. 107–118.
35 Lu, J.Z. *et al.* (2000) Chemical coupling in wood fiber and polymer composites: a review of coupling agents and treatments. *Wood Fiber Sci.*, **32** (1), 88–104.
36 Bledski, A.K. *et al.* (1998) Thermoplastics reinforced with wood fillers: A literature review. *Polym. Plast. Technol. Eng.*, **37** (4), 451–468.
37 Wolcott, M.P. and Adcock, T. (2000) New advances in wood fiber-polymer formulations. Proceedings of the Wood-Plastic Conference, Baltimore, MD, Plastics Technology Magazine and

Polymer Process Communications, pp. 107–114.

38 Stark, N.M. and Berger, M.J. (1997) Effect of particle size on properties of wood flour reinforced polypropylene composites. Proceedings of the Fourth International Conference on Woodfiber-Plastic Composites, Forest Products Society, Madison, WI, pp. 134–143.

39 English, B.W. and Falk, R.H. (1995) Factors that affect the application of woodfiber-plastic composites. Proceedings of the Woodfiber-Plastic Composites: Virgin and Recycled Wood Fiber and Polymers for Composites Conference, The Forest Products Society, Madison, WI, pp. 189–194.

40 ASTM International (2001) Test Method ASTM E11-01. *Standard Specification for Wire Cloth and Sieves for Testing Purposes, Annual Book of ASTM Standards*, vol. **14.02**, ASTM International, West Conshohocken, PA.

41 ASTM International (2001) Test Method ASTM D1103-60 (1977). *Method of Test for Alpha-Cellulose in Wood, Annual Book of ASTM Standards*, vol. **04.10**, ASTM International, West Conshohocken, PA

42 ASTM (2001) Test Method ASTM D1106-96. *Standard Test Method for Acid-Insoluble Lignin in Wood.*

43 ASTM (2001) Test Method ASTM D1107-96. *Standard Test Method for Ethanol-Toluene Solubility of Wood.*

44 Berger, M.J. and Stark, N.M. (1997) Investigations of species effects in an injection-molding-grade, wood-filled polypropylene. Proceedings of the Fourth International Conference on Woodfiber-Plastic Composites, Forest Products Society, Madison, WI, pp. 19–25.

45 ASTM International (2002) Test Method D256-02e1. *Standard Test Methods for Determining the Izod Pendulum Impact Resistance of Plastics, Annual Book of ASTM Standards*, ASTM International, West Conshohocken, PA.

46 ASTM International (2002) Test Method D790-03. *Standard Test Methods for Flexural Properties of Unreinforced and Reinforced Plastics and Electrical Insulating Materials, Annual Book of ASTM Standards*, ASTM International, West Conshohocken, PA.

47 ASTM International (2002) Test Method D638-02a. *Standard Test Method for Tensile Properties of Plastics, Annual Book of ASTM Standards*, ASTM International, West Conshohocken, PA.

48 ASTM International (2002) Test Method D648-01. *Standard Test Method for Deflection Temperature of Plastics Under Flexural Load in the Edgewise Position, Annual Book of ASTM Standards*, ASTM International, West Conshohocken, PA.

49 English, B. *et al.* (1996) Waste-wood derived fillers for plastics. General Technical Report FPL-GTR-91, U.S. Department of Agriculture, Forest Service, Forest Products, Laboratory, Madison, WI, pp. 282–291.

50 Clemons, C.M. and Ibach, R.E. (2004) The effects of processing method and moisture history on the laboratory fungal resistance of wood-HDPE composites. *Forest Prod. J.*, **54**, (4), 50–57.

51 Nakagawa, S. and Shafizadeh, F. (1984) Thermal properties, in *Handbook of Physical and Mechanical Testing of Paper and Paperboard* (ed. R.E. Mark), Marcel Decker, New York.

52 Stark, N.M. (1999) Wood fiber derived from scrap pallets used in polypropylene composites. *Forest Prod. J.*, **49** (6), 39–45.

16
Calcium Carbonate

Yash P. Khanna and Marino Xanthos

16.1
Background

Calcium carbonate is the most common deposit formed in sedimentary rocks. Natural $CaCO_3$ used as filler in plastics is produced from chalk, limestone, or marble found in the upper layers of the earth's crust to a depth of about 15 km. Chalk is a soft textured, microcrystalline sedimentary rock formed from marine microfossils. Limestone is also of biological origin but it is harder and denser than chalk, having been compacted by various geological procedures. Marble is even harder, having been subjected to metamorphosis under high pressures and temperatures, which results in recrystallization with the separation of impurities in the form of veins. The origins of calcium carbonate in deposits and in living organisms are discussed in Ref. [1].

The sedimentary rocks consisting mainly of calcite crystals are processed by standard mining procedures and then subjected to grinding and classification. In addition to natural ground calcium carbonate (GCC), there also exists a chemically produced form known as precipitated calcium carbonate (PCC), which may be finer and of higher purity, but also more expensive than the natural one.

Calcium carbonate is an abundant, largely inert, low cost, white filler with cubic, block-shaped, or irregular particles of very low aspect ratio (see Tables 1.1–1.3). Generally speaking, its usage reduces costs in a variety of thermoplastics and thermosets, with moderate effects on the mechanical properties. As a filler in plastics, calcium carbonate leads to increased modulus with minimal effect on the impact strength. These benefits are accompanied by shrinkage reduction and improved surface finishing. As a value-added functional filler, calcium carbonate may act as a surface property modifier, as a processing aid, as a toughener, and as a stress concentrator by introducing porosity in stretched films. In addition, the value-added functionality of calcium carbonate lies in improved productivity from a combination of high thermal conductivity and lower specific heat in comparison to the plastics matrices. These benefits may be enhanced or modified by selecting a grade with an appropriate particle size distribution, in particular, the median particle

Functional Fillers for Plastics: Second, updated and enlarged edition. Edited by Marino Xanthos

ISBN: 978-3-527-32361-6

size and the topcut, or via suitable surface treatments with hydrophobic chemical agents such as stearic acid (SA).

16.2 Production Methods

More than 90% of $CaCO_3$ used in plastics is processed by the conventional grinding methods. The following processes are used to produce GCC [1, 2]:

1) Dry grinding using a single stage milling process leads to a relatively coarse powder with the finest ground material having a median diameter of $\sim$12 μm and a broad size distribution. It is suitable for inexpensive dark floor tiles and vinyl foam carpet backing. Multiple-stage milling using air classifier(s) leads to products with a median particle size in the 3 μm range for high-end applications.
2) Wet grinding processes, generally speaking, provide products of higher purity and a much finer median particle size in the 1 μm range for demanding applications. In the wet grinding – high solids process, a dispersant is added to facilitate deagglomeration, which can affect the moisture pick-up; however, the product can have a median particle size of as low as about 0.7 μm. In the wet grinding – low solids process, no dispersant is added and the product is somewhat superior in moisture resistance, although the median particle size is slightly higher than about 1 μm. After the wet grinding and classification (screens/centrifuge) steps, the material is either flash or spray dried.
3) Surface modified GCC. The GCCs made by the dry or wet processes for plastics applications undergo a surface treatment to impart hydrophobicity, stearic acid being the coating of choice.

The grinding costs increase as the degree of fineness increases, until wet milling becomes more economical than dry grinding in spite of the additional drying costs [3]. If the wet grinding process is used, the material may often be delivered to the customer in a slurry form, at least for applications that can handle slurries, for example, paints and coatings. The typical particle morphology of the ground materials is rhombohedral.

Precipitated calcium carbonate is also known as synthetic $CaCO_3$ since several chemical operations may be involved in its manufacturing. In the earlier synthetic processes, it was obtained as a by-product during the manufacture of Na_2CO_3 by the ammonia process or during the manufacture of NaOH by the soda-lime process [2]. These methods have been replaced by a direct synthetic process that involves calcination of $CaCO_3$ at 900 °C to produce limestone, CaO, conversion into lime, $Ca(OH)_2$, by mixing with water, and reaction with CO_2, the latter being recovered from the calcination process. Following carbonation, the suspension is filtered, and the collected solids are dried and deagglomerated in grinders. Depending on the process conditions, various morphologies (spherical, discrete, or clustered acicular, prismatic, rhombohedral, scalenohedral, orthorhombic) and crystal forms (calcite, aragonite) are possible. A schematic of the process may be found in Ref. [1] and photographs of the obtained crystals accessed via the supplier's web site, for example [4].

PCC products are of high purity, with a very fine, regular particle size, a narrow particle size distribution, and high surface area. An advantage of the process is that there is no coproduct to separate, thus, eliminating the need for any additional steps. However, it is a higher cost process and more energy intensive than that involved in GCC production. Typically, the median particle size is 0.7–2.0 μm and as low as 0.02 μm (20 nm) with newer grades. Obviously, these fine particles must always be coated with a hydrophobic coating to reduce particle agglomeration. The high specific surface area may adversely affect rheology at high loadings through excessive adsorption of stabilizers/plasticizers in PVC, and can cause excessive viscosity increase in unsaturated polyesters.

Due to the hydrophilic nature of its surface, calcium carbonate is incompatible with the most common hydrophobic polymers such as polyethylene (PE) and polypropylene (PP) of low surface energy of about 35 mJ/m^2. While larger particles having, for example, a median particle size (d_{50}) $\geq 3\,\mu m$ can be incorporated into polymers, the smaller particles by virtue of enhanced particle–particle interactions tend to agglomerate, thus leading to dispersion and performance problems. Moisture pick-up by $CaCO_3$ (equilibrium uptake ≅0.5–3.0% depending on GCC versus PCC, surface area, process aids, etc.) poses additional problems during handling and processing. As a result, it is necessary to render the $CaCO_3$ surface hydrophobic via treatment with surface modifiers, by far the most common being stearic acid. Surface coating is typically carried out by dry methods in high intensity mixers above the melting temperature of SA using a monolayer amount, for example, 0.5–1.5 wt% depending upon the surface area of the grade of calcium carbonate.

A vast literature exists on the surface modification of $CaCO_3$ mineral with SA [3] including information in Chapter 6. Recently, Khanna *et al.* [5] have investigated the nature of surface interactions between $CaCO_3$ and SA. Using multiple analytical techniques, they have been able to identify and quantify as many as four different phases of stearic acid coating applied at the monolayer concentration:

1) The majority of the monolayer SA (≥80%) is adsorbed on the surface of $CaCO_3$ particles via a chemical reaction leading to the formation of calcium monostearate (CaSt).
2) A minor reaction product, calcium stearate ($CaSt_2$), can also form (0–10%) depending upon the process conditions and exists as a precipitate in the bulk.
3) A small but significant fraction of the SA exists as the unreacted material (≅10%) of which a minor component is bound, most likely, trapped within the CaSt chains on the surface of $CaCO_3$ particles.
4) Most of the unreacted SA (≅10%) exists as a residual impurity. The challenge is to convert as much of the coating to the surface reacted form, reduce $CaSt_2$, and minimize unreacted/free SA, the latter being detrimental for the downstream operations.

As far as other chemical modifiers are concerned, titanates/zirconates (Chapter 5) along with carboxylated polyolefins (Chapter 6) have been used, although SA remains by far the coating of choice [6, 7].

Table 16.1 Properties of ground calcite [1, 2, 7].

Property	Value
Loss on ignition (950 °C)	43.5
Density (g/cm^3)	2.7
Hardness (Mohs)	3
Water solubility (g/100 ml)	0.0013
Acid solubility	High
Young's modulus (MPa)	35 000
Thermal conductivity (W/(m K))	2.5
Thermal expansion coefficient (K^{-1})	10^{-5}
Volume resistivity (Ω cm)	10^{10}
Dielectric constant at 10^4 Hz	8–8.5
Refractive indices	1.48, 1.65
pH in 5% water slurry	9.0–9.5

16.3 Structure and Properties

Calcium carbonate occurs in different crystalline forms. The most widespread is calcite, which has either a trigonal-rhombohedral or a trigonal-scalenohedral crystal lattice. The fundamental properties are shown in Table 16.1. Another form is the orthorhombic aragonite, which is less stable and can be converted by heat to calcite. Vaterite, a third form is unstable and overtime will transform into the other two forms. Aragonite has a higher density (2.8–2.9 g/cm^3), a higher single refractive index (1.7), and a somewhat higher Mohs hardness (3.5–4) than calcite. Its other properties are very similar. Both minerals are white and their refractive indices are not high enough to interfere with effective coloration.

Commercial GCC grades, wet or dry ground, contain 94–99% $CaCO_3$, $MgCO_3$ being the major impurity, and alumina, iron oxide silica, manganese oxide the minor ones. PCC may contain 98–99% $CaCO_3$ or even higher concentrations for pharmaceutical grades.

16.4 Suppliers

Omya and *Imerys* are among the largest suppliers of calcium carbonate worldwide [8]. Other European suppliers include *Provençale S.A.* (France), *Reverté S.A.* (Spain), *Alpha Calcit* (Germany), *Mineraria Sacilese* (Italy), *Dankalk* (Denmark), *Solvay* (Belgium), and *Schaefer Kalk* (Germany). Major US suppliers in addition to *Omya* and *Imerys* include *J.M. Huber* (GA), *Specialty Minerals, Inc.* (PA), *Columbia River Carbonates* (WA), *Global Stone PenRoc, Inc.* (PA), *Franklin Industrial Minerals* (TN), *Old Castle Industrial Minerals* (GE), and *Unimin* (CT). All these suppliers list their products on their respective web sites and some of them also provide PCC grades.

Numerous other suppliers exist in the Asia-Pacific region and other European countries.

GCC grades available for plastics are classified based on their particle size distribution (psd) and type of surface treatments. Ultrafine grades have a median particle size of about 0.7–2 μm. Fine grades are between 3 and 7 μm and surface treated have median particle size (based on untreated material) ranging from 1 to 3 μm [9]. Grades of different particle sizes are recommended depending on the polymer, the intended application, and the fabrication method. PCC grades can be submicron in median particle size.

16.5 Cost/Availability

Prices vary depending on degree of fineness and surface treatment with coated grades being more expensive [10]. As of 2008, given the volatility in commodity prices, GCC prices are estimated to be about US $200/ton for untreated 3–7 μm grades and about US $300/ton for 1–3 μm surface-treated grades. PCCs, on a relative basis, are significantly more expensive.

Global production of ground calcium carbonate is estimated to be 70 million tons in 2007 [8]. This is expected to grow by no more than 2%/year through 2012, limited by the 2008 economic recession. The paper industry is the most important consumer of GCC, representing about 30–50% share of consumption, dependent on the specific region. Plastics and paints each consume about 15% share, and the remainder is consumed by various industries including construction (joint compounds), adhesives, food, rubber, and personal care, to name a few. The GCC market represents more than 90% of the total calcium carbonate industry, while the small remainder is primarily PCC (considered a higher per unit value than GCC).

World production is almost equally split between Europe/North America and Asia-Pacific where production is dominated by China and Japan [8]. As mentioned earlier two large companies, *Omya* and *Imerys*, with worldwide operations are the major producers accounting for a significant proportion of the global supply. A number of other medium sized suppliers have operations in more than one country. Over the years there has been a considerable consolidation of both producers and consumers of GCC and this trend is expected to continue.

16.6 Environmental, Toxicity, and Sustainability Considerations

Calcium carbonate is generally regarded as safe (GRAS) without any known occupational diseases associated with its handling. It is nontoxic, nonhazardous and most grades meet United States FDA codes. The United States OSHA regulates the substance under the generic total particulate limit of 15 mg/m^3. Specific exposure limits for the various forms of natural $CaCO_3$ dust containing less than 1% quartz by different US regulatory agencies are

- OSHA PEL, 8 h TWA 15 mg/m^3 (total) TWA 5 mg/m^3 (respiratory);
- NIOSH REL, TWA 10 mg/m^3 (total) TWA 5 mg/m^3 (respiratory);
- ACGIH 8 h TLV, TWA of 10 mg/m^3 dust.

Further details on environmental/toxicity considerations can be found in suppliers Material Safety Data Sheets (MSDS).

With growing concern of environment and sustainability, Imerys has taken an initiative to evaluate the sustainability aspects of the calcium carbonate mineral. Life-cycle inventory (LCI) is a key part of the International Environmental Management System standards 14044:2006. This tool is used by companies in the plastics supply chain to understand the environmental impact of raw materials critical for plastics production. LCI analysis of treated ground calcium carbonate compared to base resins was carried out since treated GCC is used to replace some of the resin volume in plastics production. The production of treated GCC was found to have a lower impact on the environment than the production of various resin systems, including PVC, PE, and PP. Raw material and energy use inputs were included in the study, as were environmental emissions to land, air, and water [11].

16.7 Applications

16.7.1 General

The primary use of calcium carbonate as a filler is to lower costs, while having moderate effects on mechanical properties. However, depending on the particular polymer system, it may also be considered as a multifunctional filler with a variety of specific effects on rheology, processing, and morphology. In any case, the introduction of surface-treated calcium carbonate and ultrafine grades has undoubtedly led to the development of new applications. Examples of the importance of surface treatments on the rheological, mechanical, and other properties of $CaCO_3$ filled plastics are presented in Chapters 5 and 6. The sections below highlight some of the traditional and novel applications in specific polymers.

16.7.2 Polyvinyl Chloride

PVC is responsible for about 50% of plastics consumption of calcium carbonate [8]. In filled PVC, 80% of the mineral market involves $CaCO_3$ (ground 70%, precipitated 10%). Important characteristics of GCC are its ready availability in different particle sizes and its low cost. It imparts increased stiffness, often without adversely affecting the impact strength in rigid formulations, and results in low shrinkage and improved dimensional stability.

Filler levels in flexible PVC are usually in the range of 10–60 phr (parts per hundred resin) of largely 3 μm particle size in such applications as upholstery coverings, hose extrusion compounds, and so on [2]. In general, finer grades have a less detrimental effect on physical properties. There are special requirements for the use of $CaCO_3$ (up to 60 phr) in vinyl electrical wire insulation compounds, where ionic impurities need to be carefully controlled in order to meet applicable volume resistivity standards. In PVC plastisols and organosols, loading is usually 20–100 phr and a wide range of particle sizes is used, from coarse grades in carpet backing to ultrafine precipitated grades and coated grades that control rheological properties. Loadings up to 400 phr of fine GCC grades may be used in PVC floor tiles, where the installation requirements of easy break of a notched tile is achieved by the high filler loadings.

PVC pipe is a major application, and grades with an average particle size of 3 μm or less are generally used. Pipe loadings vary from about 5 to 20 phr, depending on the end use. For example, loadings for conduit are approximately 20 phr, loadings for sewer and drainpipe are about 8–10 phr, and loadings for potable water are 5 phr. The National Sanitation Foundation (NSF) approves grades of GCC for use with potable water and prescribes limits on both loadings and particle size. Ultrafine GCC products are used to improve impact resistance and resin dispersion in thin-walled rigid PVC applications such as downspout and drainpipe, high-pressure pipe, and siding. Flexible PVC applications for wire and cable may also require in general, ultrafine GCC.

In rigid PVC extrusion, injection molding, and siding compounds, concentrations of a 3 μm $CaCO_3$ ranging from 10 to 40 phr usually result in decreased impact and tensile strengths. However, use of ultrafine (about 0.7–1.1 μm) surface-treated filler does not adversely affect the impact strength (Izod or Gardner) or the tensile strength, when properly formulated and processed [12]. Significant technical information on $CaCO_3$ filled PVC may be found in trade literature of PVC resin manufacturers, $CaCO_3$ suppliers, and stabilizer manufacturers.

PCC is characterized by fine particle size and a narrow size distribution. In high value rigid PVC applications such as sidings, vertical blinds, or pipe, it imparts specific properties such as improved mar resistance, increased surface gloss, reduced flex whitening, and reduced plate-out. These benefits are in addition to increased impact strength and higher modulus. The improved impact strength of rigid PVC containing coated ultrafine PCC at concentrations 5–10 phr is maintained even at temperatures below 0 °C; this may allow a reduction in the concentration of expensive elastomeric additives required to maintain ductility [13].

PCC can be considered as a processing aid for the extrusion of rigid PVC. Its finer size is compatible with the PVC primary particles, improving dispersion of the formulation components. This may result in shorter fusion/gelation time. More complete gelation may provide a matrix that has fewer defect sites than in the case of coarser $CaCO_3$. Further advantages claimed by material suppliers [13] are elimination of plate-out; elimination of surface defects, thereby improving surface finish; and increased output. PCC may also act as acid acceptor for the secondary stabilization of PVC, neutralizing chloride ions and as rheology modifier promoting uniform cell structure in PVC foams; similar mechanisms are valid for GCC, but PCC with its higher surface area might be more effective.

16.7.3
Glass Fiber-Reinforced Thermosets

Significant uses of GCC in unsaturated polyesters are in bulk molding compounds (BMC), sheet molding compounds (SMC), lay-up/spray-up, glass mat transfer, pultrusion, and cultured/densified marble applications. Typical particle sizes range from 3 to 80 μm at concentrations from 150 to 220 phr depending on the filler size. In addition to cost reduction, the fillers improve surface finish by reducing shrinkage and modify rheology to prevent segregation of the glass fibers. The key requirements for these grades are low cost, low moisture, low resin demand (through excellent particle packing), and absence of contaminants that interfere with the thickening reaction in SMC [2, 9, 14]. Coarser grades are used in "hand-lay up" and "spray-up" applications and even coarser grades (up to 60 μm) with low resin demand are used for cultured marble made by casting at loadings up to 300 phr. Table 6.1 contains an example of improved adhesion in filled UP through the use of unsaturated acids.

16.7.4
Polyolefin Moldings

When compared to other fillers of different shape and aspect ratios (glass fibers, mica flakes, talc), calcium carbonate in polyolefins usually provides higher ductility and unnotched impact, though at the expense of modulus, strength, and HDT. These effects are shown in Table 16.2 for injection molded composites containing a predominantly HDPE matrix that was designed to simulate a recyclable composition [15]. Note that the lower aspect ratio $CaCO_3$ and talc are used at 40% whereas the higher aspect ratio glass and mica are used at lower concentrations (20–25%).

Effects of surface treatment in $CaCO_3$ filled polyolefins are presented in Chapters 5 and 6. With respect to titanate treatment, data on the improvement of dispersion, rheology, processabilty, and certain mechanical properties are provided in Tables 5.6–5.8. The effect of deagglomeration afforded by titanates is shown in Figure 5.2. Figure 5.3 shows the effects of titanate treatment on allowing higher loadings in thermoplastics through lower viscosities. Figure 5.4 demonstrates the flexibility imparted to a 70 wt% $CaCO_3$-treated filled PP homopolymer through the presence of 0.5% titanate coupling agent.

Fatty acid (stearic, isostearic) and other polar coupling agent effects on rheology and properties of $CaCO_3$ filled polyolefins are discussed in detail in Chapter 6. The beneficial effects of maleated PP in filled PP compounds are shown in Table 6.9. Modification of the polyolefin matrix of Table 16.2 by maleic anhydride grafting through reactive extrusion increases the tensile and flexural and unnotched impact

Table 16.2 Properties of injection molded polyethylene containing different fillers [15].

Property	Unfilled control	20% glass	25% mica	40% $CaCO_3$ uncoated	40% $CaCO_3$ coated	40% talc
Tensile yield strength (MPa)	24.8	—	—	23.4	20.8	26.9
Tensile break strength (MPa)	10.8	36.5	25.0	21.8	18.3	26.7
Tensile yield elongation (%)	9.2	—	—	5.5	5.3	3.0
Tensile break elongation (%)	23	3.5	2.8	8.7	13	3.3
Flexural modulus (MPa)	1030	2620	2480	1800	1590	2380
Flexural strength (MPa)	30.3	49.4	36.5	33.9	30.6	35.0
Izod impact notched (J/m)	48.0	53.4	58.7	48.0	53.4	37.4
Izod impact unnotched (J/m)	1014	208	139	534	406	245
HDT at 1.82 MPa (°C)	47	79	65	49	50	53

Note: Fiber glass OCF457AA chopped strands 46 mm, Owens Corning Fiberglas Corp.; Suzorite mica 200H-K, Suzorite Mica Products; $CaCO_3$ Atomite/Microwhite 25, Uncoated ECC Intern.; $CaCO_3$ Kotamite, Coated ECC Intern.; Talc, Microtalc MP1250, Pfizer; Matrix, polyolefin matrix based on at least 80% HDPE.

strength of the untreated $CaCO_3$ compositions without any significant effect on modulus, HDT, or elongation at break. Increased tensile or flexural strength is accompanied by the disappearance of a yield point and a decrease in unnotched impact strength (Table 16.3) [15]. Similar effects are observed when comparing the unmodified with the modified matrix in composites containing surface-treated $CaCO_3$. Comparison of the fracture surfaces of Figures 16.1 and 16.2 is indicative of improved coating of the particles with the maleated matrix, which presumably results in improved adhesion. An interesting example of the multifunctional characteristics of $CaCO_3$ is included in Table 5.5. In LLDPE, 44% loading of filler treated with pyrophosphato titanates can lead to self-extinguishing characteristics.

Table 16.3 Properties of injection molded maleated polyethylene containing different fillers [15].

Property	20% glass	25% mica	40% uncoated $CaCO_3$	40% coated $CaCO_3$	40% talc
Tensile yield strength (MPa)	—	—	—	—	29.4
Tensile break strength (MPa)	42.1	26.9	33.1	34.0	29.2
Tensile yield elongation (%)	—	—	—	—	3.0
Tensile break elongation (%)	4.8	3.7	8.2	7.8	3.4
Flexural modulus (MPa)	2340	2620	1850	2020	2290
Flexural strength (MPa)	51.3	40.7	38.2	39.5	37.4
Izod impact notched (J/m)	80.1	58.7	80.1	58.7	58.7
Izod impact unnotched (J/m)	272	208	422	315	251
HDT at 1.82 MPa (°C)	84	65	49	52	60

Note: Fiber glass OCF457AA, Owens Corning; Suzorite mica 200H-K; $CaCO_3$ Atomite/Microwhite 25, ECC Intern.; $CaCO_3$ Kotamite, Coated ECC Intern.; Talc, Microtalc MP1250, Pfizer; Matrix, polyolefin matrix based on at least 80% HDPE, modified with peroxide/maleic anhydride in twin-screw extruder.

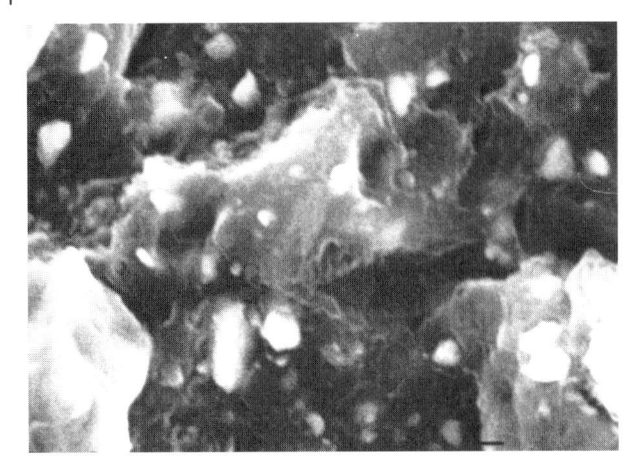

Figure 16.1 SEM fracture surface of injection molded specimen showing poor adhesion between the unmodified polyethylene matrix and the untreated calcium carbonate [15].

Significant applications, particularly in automotive items (automotive interior, exterior, under the hood), household items, and furnitures exist for molded filled PP. Particle size usually ranges from 0.7 to 3 μm at 20–40% filler content. Primary effects with respect to the unfilled resin are increased stiffness and heat distortion temperature, retention or improvement of impact strength, and improved dimensional stability. Shrinkage decreases with increasing concentration. $CaCO_3$ also provides more isotropic shrinkage and lower warpage than other filler shapes and may also act as crystal nucleator in PP. A stearate-coated PCC ultrafine (mean particle size 70–40 nm) grade has been shown to promote the formation of higher amounts of the β-crystalline phase that has been reported to be tougher than the usually

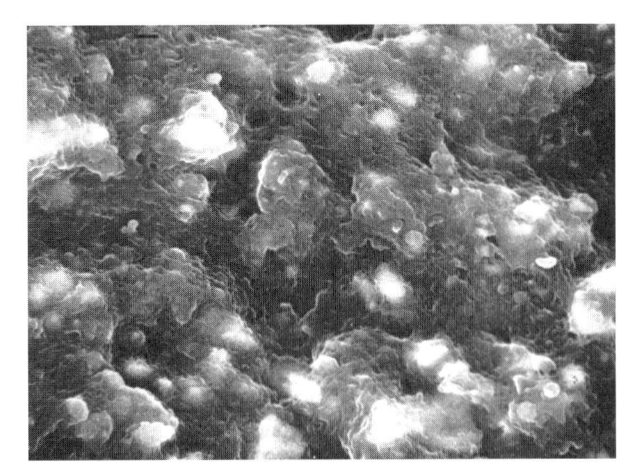

Figure 16.2 SEM fracture surface of injection molded specimen showing improved adhesion between the maleated polyethylene matrix and the untreated calcium carbonate [15].

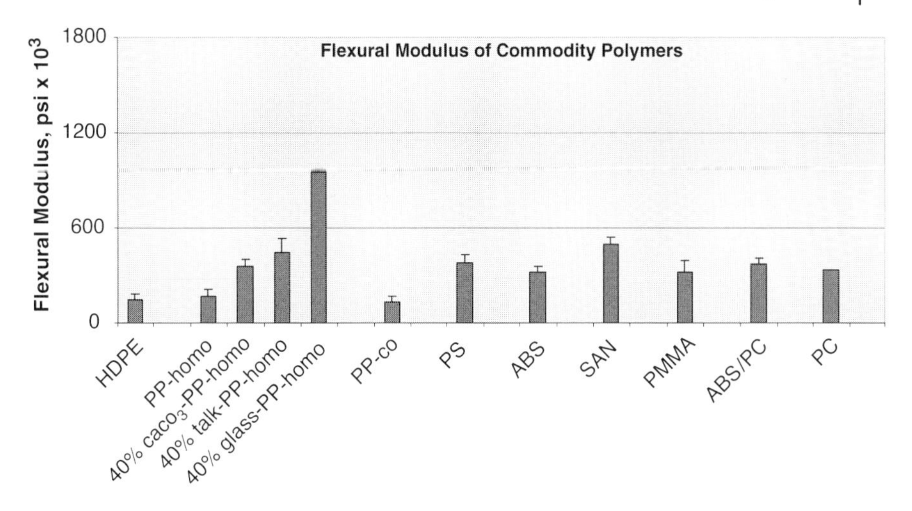

Figure 16.3 Comparison of flexural modulus of $CaCO_3$ filled PP with unfilled polymers; to convert psi into Pa multiply by 6895. (Replotted data from CAMPUS database courtesy of Dr. S.K. Dey, Sonoco Corp.)

prevailing PP α-phase [16]. In PP sheet thermoforming, 30% filler incorporation increases stiffness and can yield approximately the same performance as PS or PVC with the same shrinkage characteristics. In Figure 16.3, $CaCO_3$ filled PP is compared in terms of stiffness with several unfilled polymers and PP containing other fillers. In polymer processing operations such as injection molding, film extrusion, sheet formation, and thermoforming, the presence of the filler can also increase productivity due to more rapid cooling by virtue of the increased thermal conductivity. PP molding compounds containing 10–40% $CaCO_3$ are available by a plethora of compounders.

16.7.5 Polyolefin Films

Specific grades of calcium carbonate are available for polyolefin film applications where high clarity is not required. It improves stiffness, dart impact resistance, and tear strength in LLDPE and HMW HDPE blown films and biaxially oriented PP (BOPP) films. This may allow down gauging and, hence, lower overall materials' cost. As mentioned above, it can further reduce costs as it improves processing characteristics by increasing thermal conductivity, thus, enabling the part to be cooled faster; it also reduces specific heat, allowing more rapid temperature increases and overall increased line speeds and productivity [17, 18]. $CaCO_3$ levels of 20–35 wt% may increase output by as much as 20–40%. Grades for LLDPE films are very specifically produced for this application and are wet ground, surface-treated materials with median diameter of 1.4 μm added as a masterbatch.

An additional use of $CaCO_3$ in polyolefin and other films (polyester and cellulose acetate) is based on its ability to increase surface roughness at low filler content and minimize "blocking," that is, the tendency of adjacent film surfaces to stick together under pressure (see also Chapter 19). Examples of the use of $CaCO_3$ as an antiblock in LLDPE can be found in Refs [1, 6]. Related to its surface roughening capacity is the ability of $CaCO_3$ to improve printability in extrusion coating, allowing better flow of the printing ink on the film surface.

16.7.6 Polyolefin Microporous Films

Microporous films have rapidly developed as a large majority of the product lines in infant and adult health care, construction, protective apparel, and medical sectors consuming large quantities of GCC [19]. Such films are manufactured from highly filled, nonporous precursor polyolefin films containing up to 60% of specifically made GCC by stretching on- or off-line [20]. Porosity down to 1 μm is induced by interfacial debonding of the $CaCO_3$ particles, which act as stress concentrators in the stretching step. Figure 16.4 shows the surface characteristics of an experimental LLDPE/60% $CaCO_3$ stretched film with an average pore size of <2 μm [21]. The films are termed "breathable" since, after stretching, an interconnected pathway is created for the transport of water vapor but not of liquid water.

Manufacturing involves the following four steps:

- Extrusion compounding, usually in a TSE.
- Film extrusion by casting or blowing.
- Rapid stretching to thin films $\leq$25 μm, either uniaxially or biaxially, to produce uniform microvoids.

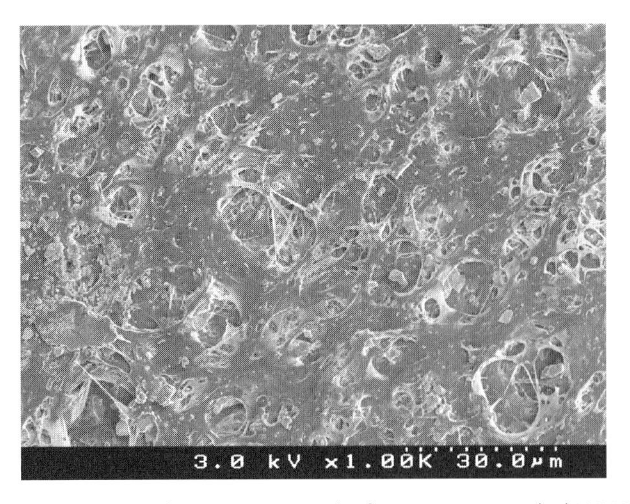

Figure 16.4 Surface SEM micrograph of microporous stretched LLDPE/60% $CaCO_3$ film [21].

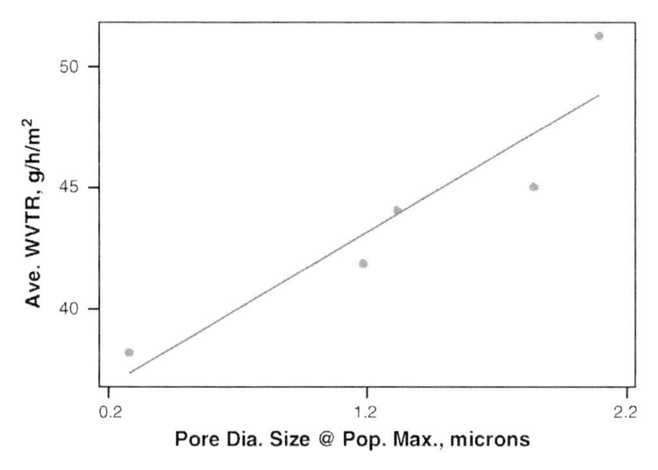

Figure 16.5 WVTR versus pore diameter for microporous stretched LLDPE/60% $CaCO_3$ films containing different filler grades [21].

- Combination with a nonwoven to provide a cloth-like product with good strength characteristics.

In order to provide maximum dispersion and deagglomeration at these high loadings, extrusion compounding usually involves addition of the filler downstream after the polymer has fully melted. Appropriate selection of the screw configuration and the removal of trapped air are important, as also discussed in Chapter 3.

Important characteristics of the stretched, porous films are microstructure, mechanical properties, porosity, and water vapor transmission rate (WVTR). Porosity and average pore size have been shown to correlate fairly well with WVTR and can be used to optimize GCC grade characteristics (Figure 16.5). Important characteristics of the selected GCC grade are median particle size (1–2 μm), particle size distribution, surface area, stearic acid coating levels, and very low moisture content. Figure 16.6 shows comparative data of modulus in the machine (MD) and transverse (TD) directions, along with WVTR data of six microporous LLDPE films containing different grades of GCC, all at 60 wt% [21, 22]. As expected, an increase in porosity, as manifested by increased WVTR, results in decreased modulus. These effects depend significantly on the surface area of calcium carbonate, its particle size, and distribution, as well as on the quality of the stearic acid coating.

16.7.7 Nonwoven Fabrics

As mentioned earlier, the majority of commodity thermoplastic products marketed today include some inexpensive mineral fillers in their formulation for cost as well as performance improvement purposes. Even though nonwovens are in a similar cost sensitive market, products such as spunbond polypropylene have not followed this trend. Recent work by McAmish [23–25] has shown that certain grades of $CaCO_3$

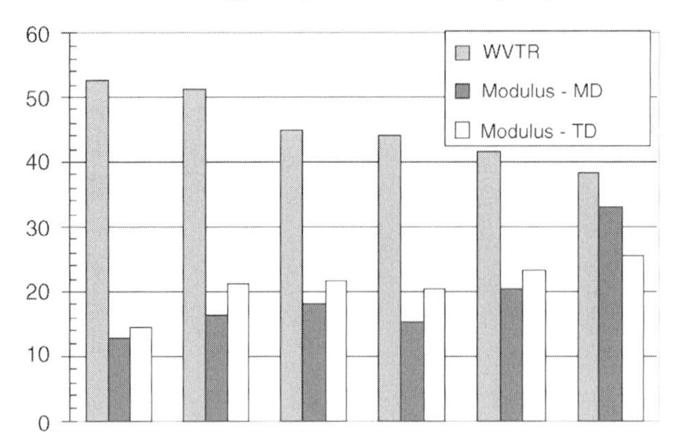

Figure 16.6 Average WVTR versus tensile modulus of microporous stretched LLDPE/60% $CaCO_3$ films containing different filler grades [21].

could be used successfully in fibrous products. Experiments were conducted on pilot equipments at the University of Tennessee (Tandec), North Carolina State University (NCSU), and at the School of Materials Science at Clemson University. The pilot lines were producing spunbond, melt blown, and monofilament products. The results from these experiments demonstrate that the process parameters are unaffected, along with little impact on the physical characteristics of the end products up to 40% filler loadings. Figure 16.7 demonstrates a uniform dispersion of calcium carbonate in a spunbond PP fiber.

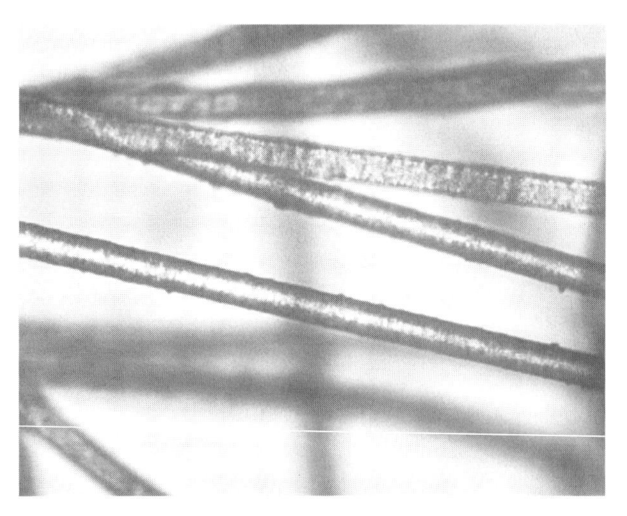

Figure 16.7 SEM photomicrograph showing the dispersion of $CaCO_3$ in a fiber of spunbond polypropylene.

16.7.8
Bioactive Composites

Among the various fillers evaluated for bone regeneration, a pure form of vaterite has been shown to promote bioactivity in degradable polymers (polylactic acid) through apatite formation. Figure 22.3 shows the formation of an apatite structure through contact with simulated body fluids. Bioactive fillers may be a very promising area for expanded use of highly purified forms of calcium carbonate.

References

1 Wypych, G. (2000) *Handbook of Filllers*, ChemTec Publishing, Toronto, Ont., Canada, pp. 48–58, 800.
2 Katz, H.S. and Milewski, J.V. (eds) (1978) Chapter 5, in *Handbook of Fillers and Reinforcements for Plastics*, Van Nostrand Reinhold Co., New York.
3 Rothon, R.N. (1999) Mineral fillers in thermoplastics: filler manufacture and characterization, in *Advances in Polymer Science*, vol. 139, Springer-Verlag, Berlin, Heidelberg, pp. 67–107.
4 Specialty Minerals Inc. (SMI) at www.mineralstech.com.
5 Khanna, Y.P., Taylor, D.A., Paynter, C.D., and Skuse, D.S. (2008) Surface modification of calcium carbonate. I. Characterization of physico-chemical phases of stearic acid coating. *J. Mater. Sci.*, submitted to publication.
6 Solvay Chemicals US, Coated precipitated calcium carbonate, MSDS No Winnofil-1003, Revised 10-2003.
7 Zweifel, H. (ed.) (2001) Chapters 7 and 17, in *Plastics Additives Handbook*, Hanser Publishers, Munich.
8 Blum, H. (2008) Kline and Company, Little Falls, NJ, www.klinegroup.com.
9 Lamond, T.G. (1995) Proceedings of the Functional Fillers 95, Intertech Corp., Houston, TX, December 1995.
10 Industrial Minerals Prices, Ind. Minerals, February 2004, pp. 72–73.
11 Anderson, R., Cribb, A., and Ngo, V. (2008) Proceedings of the Functional Fillers Conference, Intertech Corp., Atlanta, GA., September 2008.
12 Crowe, G. and Kummer, P.E. (1978) Plastics compounding, September/October 1978, 14–23.
13 Solvay precipitated calcium carbonate, Technical information at www.solvaypcccom/market/application.
14 Champine, N. (1995) Proceedings of the Functional Fillers 95, Intertech Corp., Houston, TX, December 1995.
15 Xanthos, M. *et al.* (1995) Thermoplastic composites from maleic anhydride modified post-consumer plastics. *Polym. Compos.*, **16** (3), 204.
16 Kotek, J. *et al.* (2004) Tensile behavior of isotactic polypropylene modified by specific nucleation and active fibers. *Eur. Polym. J.*, **40**, 679.
17 Hancock, M. (1998) Proceedings of the Functional Fillers & Fibers for Plastics 98, Intertech Corp., Beijing, PR China, June 1998.
18 Guy, A.R. (2003) Proceedings of the Functional Fillers for Plastics 2003, Intertech Corp., Atlanta, GA, October 2003.
19 Moreiras, G. (2001) GCC in microporous films. Proceedings of the 3rd Minerals in Compounding Conference, Cologne, Germany, April 2001, *Ind. Minerals*, July 2001, also available at www.omya.com.
20 Clemensen, P.D. (2001) Proceedings of the Functional Fillers for Plastics 2001, Intertech Corp., San Antonio, TX, September 2001.
21 Zhang, Q., Xanthos, M., Freeman, M.G., and Ashton, H. (2002) Highly filled calcium carbonate/polyethylene porous films for water vapor breathable

applications. Proceedings of the 60th SPE ANTEC, vol. 48, p. 2840.

22 Xanthos, M. and Wu, J. (2004) Microporous films from polymer blends and composites – processing/structure/property relationships. Proceedings of the 62nd SPE ANTEC, vol. 50, p. 2641.

23 McAmish, L. and Skelhorn, D.Spunlaid fibers comprising coated calcium carbonate, processes for their production, and nonwoven products, WO 2008/077156 A2.

24 McAmish, L. (2007) Proceedings of the TAPPI – International Nonwovens Tech. Conference, Atlanta, GA, USA, September 24–27, 2007.

25 McAmish, L. (2008) Proceedings of the TAPPI– International Nonwovens Tech. Conference, Houston, TX, USA, September 8–11, 2008.

C Specialty Fillers

Functional Fillers for Plastics: Second, updated and enlarged edition. Edited by Marino Xanthos

ISBN: 978-3-527-32361-6

17
Fire Retardants

Henry C. Ashton

17.1
Introduction

The chemistry and dynamics of combustion have been well described in many publications, but for the purpose of this chapter a short overview is helpful. Combustion may be defined as the rapid uncontrolled reaction of oxygen with a substrate. A feature of this process is that it is generally exothermic and as a consequence of the heat of combustion, the reaction of the fuel with oxygen is accelerated resulting in a self-propagating reaction. Although the reaction can be viewed as an extreme case of oxidation, as with thermal oxidative degradation of polymers, once the process starts it is extremely difficult to control or stop. Thus, any fire-retardant strategy should focus on the prevention of initiation of a combustion reaction or on rapid intervention if combustion has been initiated.

This chapter will focus on the combustion of polymeric materials and the strategies used to counteract fire. In many applications such as electrical, electronic, transport, building, and so on, the use of polymers is restricted by their flammability despite the other advantages that their use brings. At this point, it is worth noting that polymers display a wide range of flammabilities and smoke generation. Some polymers such as polyvinyl chloride (PVC) are, by their nature, inherently fire retardant (FR). The generation of halogen free radicals during combustion provides an active species that can intercept oxygen before it can exothermically combine with the burning polymer in the solid state. A dominant thermooxidative degradation mechanism of polyethylene (PE) is intermolecular cross-linking. Such processes can assist in the formation of a fire-retarding char. Polypropylene (PP) with an empirical formula very similar to PE does not behave in this way. The dominant degradation mechanism is intramolecular chain scission resulting in shorter molecular chains. As a consequence, PP tends to drip during combustion and in doing so it spreads flames from the burning to the nonburning surfaces.

Functional Fillers for Plastics: Second, updated and enlarged edition. Edited by Marino Xanthos

ISBN: 978-3-527-32361-6

17.2
Combustion of Polymers and the Combustion Cycle

The increasing diffusion of synthetic polymers into the marketplace has greatly increased the "fire risk" and the "fire hazard," that is, respectively, the probability of fire occurrence and its consequences either for humans or for structures.

To enable an effective use of polymers, flame-retardant materials have to be added to polymer-based formulations. The role of these additives is to

- slow down polymer combustion and degradation (fire extinction);
- reduce smoke emission;
- avoid flame spread by dripping of hot or burning material or by other mechanisms.

A key safety element of a fire situation is to allow people the maximum amount of time possible to escape the fire. Fire safety regulations are developed with this factor in mind. The stringency of the regulations often depends on the time needed to safely escape the environment of a fire.

As shown in Figure 17.1, the three required elements to sustain combustion include a source of fuel, heat, and oxygen. Removal of any one of these elements from the cycle can prevent combustion. It takes heat to initiate combustion (ΔH 1) and once combustion begins heat is liberated (ΔH 2). This would correspond to the total heat released as measured by cone calorimetry (see Section 17.8)

Several techniques are available to break down this combustion cycle. Among materials that can be used as fire-retarding agents, some function as absorbers of heat, thus lowering the ambient temperature below the critical temperature for ignition. Another approach offering a barrier to combustion is to use materials that will cross-link or intumesce providing a networked structure that prevents the ingress of oxygen to the site of combustion. These are examples of the so-called condensed phase fire-retarding mechanisms. Others function by providing a vapor

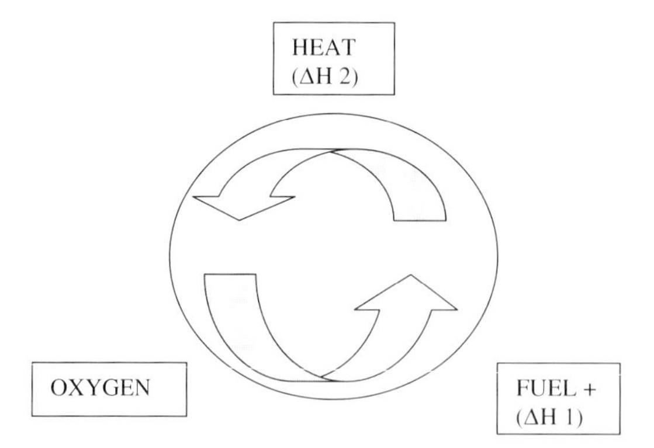

Figure 17.1 The combustion cycle.

barrier that dilutes the concentration of the incoming oxygen required to feed the fire and, thus, slow down the kinetics of combustion. The generation of active nonpolymeric free radical species in the flame (e.g., halogens) that can intercept the reactive oxygen molecule prior to reaction with the fuel is a well-practiced approach to fire retardance. This mechanism, in effect, decreases the free-radical character of oxygen, thus rendering it less reactive with the fuel. The last two approaches are vapor phase or noncondensed phase mechanisms. An effective fire retardance strategy can often involve the use of more than one of the above approaches depending upon the substrate polymers and the FR standards that have to be passed.

Dilution of the available fuel (polymer system) can be accomplished by using materials such as inorganic fillers that allow the use of diminished amounts of the more combustible organic polymers. Both the physical characteristics of materials and their chemical composition influence combustion behavior. Factors such as surface area, availability of air (oxygen) to impinge on the site of ignition, and heat transfer to and from the flame, *inter alia*, can influence the rate and the extent of combustion.

17.3 Fuel

There is a wide and structurally complex variety of polymers. However, all of them will burn under appropriate conditions, reacting with oxygen from the air, releasing heat, and generating combustion products. The oxidation reaction in a flame takes place in the gas phase. Hence, for liquids or solids to burn, there must be conversion to gaseous species. In the case of liquids, this can be accomplished by evaporation or by boiling from the liquid surface. For most solids, decomposition by heating (pyrolysis) is necessary to afford products that have a molecular weight that is sufficiently low to volatilize. This requires that the surface temperature of the solid be sufficiently high to generate species with energies higher than that required for evaporation alone. Typically, the surface temperature of burning polymeric materials has to be close to the temperatures at which carbon–carbon and carbon–hydrogen bond scission will take place, that is, at approximately 400 °C [1, 2].

17.4 Smoke

Perhaps the best working definition of smoke is given by the National Fire Protection Association, in NFPA 92B: "The airborne solid and liquid particulates and gases evolved when a material undergoes pyrolysis or combustion together with the quantity of air that is entrained or otherwise mixed into the mass."

Ideally, the combustion of a linear hydrocarbon can be represented by the equation

$$C_nH_{2n+2} + [(3n+1)/2]O_2 \rightarrow nCO_2 + (n+1)H_2O$$

Under these conditions the only products expected are carbon dioxide and water. However, in most cases combustion involves the generation of more than these species as products. In cases of incomplete combustion, the formation of carbon monoxide often occurs with the concomitant formation of lower molecular weight or oligomeric species.

In general, the volatile components make a complex mixture whose composition ranges from simple molecules such as hydrogen and low molecular weight hydrocarbons (e.g., ethylene) to higher molecular weight species that can volatilize only at higher temperatures.

Smoke particles are mainly of two kinds:

1) Carbonaceous solid particles that produce black smoke, often called soot.
2) Liquid droplets that form as some gas molecules cool and condense producing light-colored smoke.

The droplets described above may also contain some finely divided solid materials such as minerals originally present in the combusting material; gases can also be entrained. The relative amounts of solids, liquids, and gases depend on ambient conditions, particularly temperature.

There are two characteristics of smoke that deserve mention: toxicity and visibility.

17.4.1 Toxicity

Toxic potency of smoke or its components gases is defined by American Society for Testing and Materials (ASTM) as "a quantitative expression relating concentration of smoke or combustion gases and exposure time to a particular degree of adverse physiological response, for example, death on exposure of humans or animals" [3]. The toxic potency of smoke from any material product or assembly is related to the composition of that smoke that in turn depends upon the conditions under which the smoke is generated. LC_{50} is a common end point used in laboratories to assess toxic potency. The lower the value, the more the toxic the species are.

In the 1980s and 1990s, considerable work was done at the National Institute of Standards and Technology (NIST) and the Southwest Research Institute (SwRI) to establish the extent to which the toxicity of a material's combustion products could be explained and predicted by the interaction of the major toxic gases. This work led to the development of the N-Gas models [4–6]. The underlying hypothesis is that the toxicity of smoke in a fire is due to a small number "*N*" of constituent gases.

The Six- and Seven-gas models (N-gas) were empirical relationships developed to calculate the "N-Gas Value," which is an index of toxicity. The Six-gas model includes terms for CO, CO_2, oxygen (depletion), HCN, HCl, and HBr concentrations as well as the LC_{50} for selected animals exposed to the constituent components. These models are based on exposure data both for individual gases and for complex gas mixtures. In this way, the effect of synergistic or antagonistic interactions between the constituent gases in a smoke can be factored into the determination of toxicity [7, 8].

The Seven-gas Value includes terms for NO_2 whose toxic effects are considered representative of the toxic effects of the five known gases represented by NO_x. CO_2 and NO_2 show synergistic toxicological effects. Five percent of CO_2 added to NO_2 has been shown to double its toxicity [9]. HCN and NO_2 exhibit antagonistic toxicological effects. NO_2 is known to react with H_2O to form nitrous (HNO_2) and nitric (HNO_3) acids [10]. In the blood, HNO_2 can dissociate to form the NO_2^- anion. This will react with oxyhemoglobin to form methemoglobin, which is a CN^- antidote; through its reaction with CN^-, it forms cyanmethemoglobin, thus preventing the entry of CN^- to the cells [11].

Every year approximately 48 000 people die from the toxic effects of smoke in the United States [12, 13]. Carbon monoxide is recognized as the major intoxicant in many cases. Carbon monoxide converts blood hemoglobin to carboxyhemoglobin and in so doing prevents the formation of the oxygen complex oxyhemoglobin that is required for cellular metabolism. Exposure to carbon monoxide at a level of 2% in air can incapacitate a person in 2 minutes and unless there is immediate medical attention it is likely that death will follow in a matter of minutes [14]. Some attempts have been made to quantitatively predict the toxic effects of fire, but overall accurate predictions are difficult. The conditions of burning can influence the relative amounts of gases and particulates and the after-flame conditions can also affect the composition of gases produced [15–17].

Another intoxicant is hydrogen cyanide, which is also found in the combustion products of nitrogen-containing polymers, including wool, acrylonitrile, polyurethanes, polyamides, and urea–formaldehyde resins. This is estimated to be 10–40 times more toxic than carbon monoxide. Hydrogen cyanide functions by rendering cells incapable of accepting oxygen from oxyhemoglobin. Levels below 1 ppm in human blood can lead to toxicity.

Carbon dioxide is not highly toxic but exposure to it can render the effects of other toxic agents more severe because it causes increased breathing rates as a result of depletion of the available oxygen in the air. In extreme cases of exposure, it can cause suffocation. Other toxic agents that deserve mention are hydrogen chloride and acrolein. Hydrogen chloride is a severe irritant and causes lung damage [18]. Acrolein is a by-product of the combustion of materials as diverse as wood, paper, polyethylene, and cotton and has toxicity comparable to that of SO_2. Prolonged exposure (20 min or more) can be lethal.

Hydrogen bromide (HBr) and hydrogen fluoride (HF) are also significant toxins. Hydrogen fluoride, in particular, can result in bone damage upon exposure. HF is unique among acids in that it easily penetrates tissue. The high electronegativity of the fluoride anion results in a weak acid that exists predominantly in the nondissociated state compared to other acids (10^3 times less than HCl). The insidious nature of HF exposure means that injuries can often seem minor initially, but over time (hours) serious metabolic derangement can occur including corneal opacification or sloughing, skin damage, edema, or cardiac arrest [19]. While exposure to HF is relatively rare, great care must be taken in such cases to ensure prompt medical treatment [20–23].

Oxides of nitrogen, in particular, NO and NO_2 are produced during the combustion of nitrogen-containing polymers such as polyurethanes. The symptoms following the inhalation of NO_x are mostly caused by nitrogen dioxide (see above). Inhalation of nitric oxide (NO) causes the formation of methemoglobin that can carry oxygen but is unable to release it for cellular metabolism [24]. NO_2 inhalation leads to a fall in blood pressure, production of methemoglobin, and cellular hypoxia. High concentrations will likely cause rapid death. Even milder yet still severe exposures may result in death with production of yellow frothy fluid in the airways.

A full treatment of toxic effects of products of combustion is given by Purser [25].

17.4.2
Visibility

The most common optical method for measuring the density of smoke particles is to make use of the Beer–Lambert law that is described by the following equations:

$$I/I_0 = \exp(-\alpha lc) \text{ and } A = \log(I/I_0), \tag{17.1}$$

where α is absorption coefficient, A is absorbance, ε is molar absorptivity, I_0 is intensity of the incident light, I is intensity of light after passing through the material, l is path length of light, and c is concentration of absorbing species. The extinction coefficient, defined as the fraction of light lost to scattering and absorption per unit distance in a participating medium, is a measure of the opacity of the smoke and can be calculated using the above relationships.

Optical methods are favored for measuring smoke density because they provide information relevant to visibility through smoke. Other methods involve weighing particulates and calculating the weight loss on combustion relative to the weight loss of the combustible. This method correlates with smoke measurements by optical methods [26–28].

17.5
Flammability of Polymers

Flammability of polymers can be determined by their reactivity with oxygen. The limiting oxygen index (LOI), defined as the minimum amount of oxygen required to sustain combustion under specified conditions, is a quantitative measure of the tendency of materials to burn (Table 17.1). In a flame-retarded system, the study of the LOI can provide information on the effectiveness of flame-retardant materials [29].

Depending on the polymer and the applicable fire retardance standards, flame retardants are chosen to interfere with one or more stages of the combustion process: heating, decomposition, ignition, flame spread, and smoke density. Fire retardants have to inhibit or even suppress the combustion process.

Fire retardants can be divided into halogen containing and nonhalogen containing. This is a useful categorization as it leads into an explanation of their different modes of actions. Chief among the nonhalogen fire retardants are aluminum trihydrate

Table 17.1 Limiting oxygen index values for selected polymers [29].

Material	LOI	Material	LOI
Polyoxymethylene	15.7	Neoprene	26.3–40
Polyurethane foam (flexible)	16.5	Nomex	26.7–28.5
Natural rubber foam	17.2	Modacrylic fibers	26.8
Polymethylmethacrylate	17.3	Leather	34.8
Polyethylene	17.4	Phenol formaldehyde resin	35
Polypropylene	17.4	PVC	37–42
Polyacrylonitrile	18	Polyvinylidene chloride	60
ABS	18.3–18.8	Polytetrafluoroethylene	95
Cellulose	19	Polystyrene	18–19
Nylon	20.1–26	Styrene–acrylonitrile	19
Wood (birch)	20.5	Polyethylene terephthalate	20–23
Polycarbonate	22.5–28	Epoxy	21–25
Wood (red oak)	23	Polyvinylidene fluoride	44
Wool	23.8	Fluorinated ethylene propylene copolymer	>95

(ATH) and magnesium hydroxide (MOH) as well as materials such as melamine and derivatives, ammonium polyphosphates (APPs), antimony oxide, and boron-containing materials such as zinc borate and others.

17.6 Mechanisms of Fire-Retardant Action

Flame retardants can act *chemically* and/or *physically* in the condensed phase and/or in the gas phase. In reality, combustion is a complex process occurring through simultaneous multiple paths that involve competing chemical reactions. Heat produces flammable gases from the pyrolysis of polymers and if the required ratio between these gases and oxygen is attained, ignition and combustion of the polymer will take place.

17.6.1 Condensed Phase

In the condensed phase, three types of processes can take place:

1) Breakdown of the polymer, which can be accelerated by flame retardants, leads to its pronounced flow that decreases the impact of the flame.
2) Flame retardants can cause a layer of carbon (charring) on the polymer's surface. This occurs, for example, through the dehydrating action of the flame retardant generating double bonds in the polymer. These processes form a carbonaceous layer as a result of cyclization and cross-linking (see Figure 17.2).

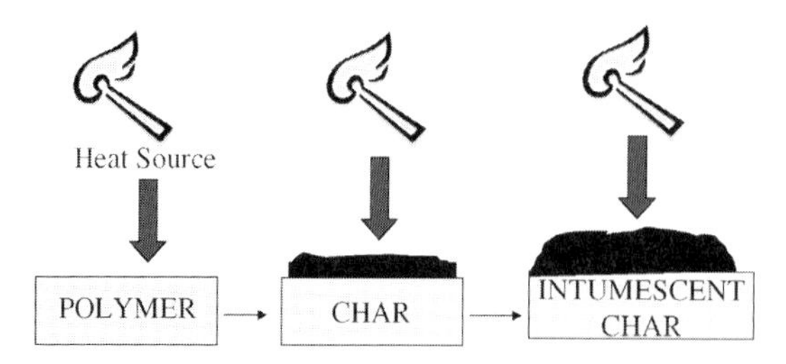

Figure 17.2 Char and intumescence formation.

3) Heat absorption through materials such as ATH that have a very high heat capacity.

Flame retardancy by intumescence is essentially a special case of a condensed phase activity without apparent involvement of radical trap mechanisms in the gaseous phase. Intumescence involves an increase in the volume of the burning substrate as a result of network or char formation. This char serves as a barrier to the ingress of oxygen to the fuel and also as a medium through which heat can be dissipated (Figure 17.2).

In intumescence, the amount of fuel produced is also greatly diminished and char, rather than combustible gases, is formed. The char constitutes a two-way barrier, both for the hindering of the passage of the combustible gases and molten polymer to the flame and the shielding of the polymer from the heat of the flame. Many intumescent systems have been developed in the past 25 years. These can be generally formulated to consist of following three basic ingredients:

- a "catalyst" (acid source) promoting charring,
- a charring agent,
- a blowing agent (spumific compound).

The catalyst or acid source can consist of ammonium phosphate or polyphosphate salts, phosphoric acid-derived amides or alkyl or halo-alkyl phosphates. Charring agents are based on molecular structures that can form cross-linked networks such as pentaerythritol, sorbitol, melamine, and phenol–formaldehyde resins. Other polymeric systems capable of intumescence are some polyamides and polyurethanes. Blowing agents help form a porous structure in the char and can facilitate its formation. Common blowing agents are based on urea and urea–formaldehyde resins, melamines, and polyamides that can liberate moisture.

17.6.2
Chemical Effects in the Gas Phase

In the gas phase, the process of combustion is slowed down by reactive species that interfere chemically with the propagation process of the fire. The flame retardants

Table 17.2 Components of polymer combustion.

Gas-Phase components of polymer combustion	
Ever-increasing concentration of the excited state reactive OH^*, O^*, and H^* free radical species	The excited state free radicals have been removed and rendered unavailable to the combustion cycle
CO_2 OH^* O_2 O^* CO OH H^*	CO_2 CO HOH HX HX HH
Polymer without RX flame retardant	Polymer with flame retardant RX

themselves or species derived from them interfere with the free radical mechanism of the combustion process. This slows or stops the exothermic processes that occur in the gas phase and results in cooling of the system and reduction in the supply of flammable gases. Hydrogen halides, HX (X = Br or Cl), produced by the reaction of halogenated organic compounds, R–X, with a polymer P–H, can react with the excited state $HO^{\bullet}$ and $H^{\bullet}$ radicals to produce the less reactive halogen free radicals $X^{\bullet}$ leading to an overall decrease in the kinetics of the combustion as shown below:

$$R{-}X + P{-}H \rightarrow HX + RP$$

$$HX + H^{\bullet} \rightarrow H_2 + X^{\bullet}$$

$$HX + OH^{\bullet} \rightarrow H_2O + X^{\bullet}$$

Table 17.2 shows the components of polymer combustion. In the presence of halogenated fire retardants, the active radical species such as $OH^{\bullet}$, $O^{\bullet}$, and $H^{\bullet}$ can be quenched in the gas phase to form species such as H_2O, H_2, and HX that are relatively less reactive in the combustion cycle.

A schematic summary of the mechanism of action of flame retardants is given in Figure 17.3. The relationship between condensed phase and gas phase processes is indicated by referring to specific flame retardants. The influence of both chemical and physical processes should also be noted.

17.7 Classification of Flame Retardants

17.7.1 Metal Hydroxides

Metal hydroxides, particularly aluminum trihydrate and magnesium hydroxide, contribute to several fire-retardant actions. They first decompose endothermically and release water. The endothermic decomposition serves to remove heat from the surroundings of the flame and, thus, cool the flame. This is often referred to as the "heat sink" phenomenon. Pyrolysis decreases in the condensed phase as a result of

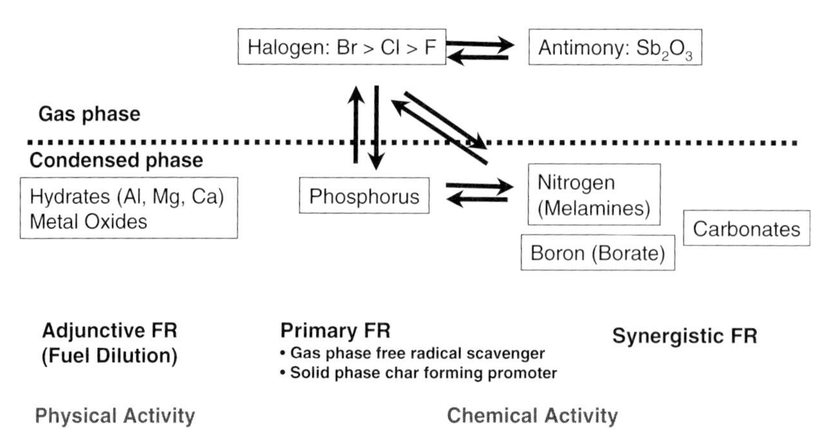

Figure 17.3 Chemical mechanism of flame retardance.

this action. The release of water dilutes the amount of oxygen capable of ingress to the flame and avoids the critical fuel/oxygen ratio (physical action in the gas phase). Both mechanisms combat ignition. In some fire tests used for electric/electronic, cable applications, the ignition has to be significantly delayed. Thus, metal hydroxides are suitable for these applications. Moreover, after the degradation, a ceramic-based protective layer is created that improves insulation (physical action in the condensed phase) and gives rise to a smoke suppressant effect (chemical action in the condensed phase). The ceramic-based protective layer ensures an efficient protection of the polymer during combustion leading to a severe decrease in the heat released.

17.7.1.1 **Aluminum Trihydrate**

Aluminum trihydrate is well known as the largest volume flame retardant used in the world with consumption of around 200 000 tons/year, representing 43% of all flame-retardant chemicals in volume (about 29% in value). Grades available include general purpose (particle sizes of 5–80 μm), high loading (5–55 μm), superfine (controlled surface areas from 4 to 11 m^2/g), ultrafine (15–35 m^2/g), and low electrolyte level. Also known as ATH or "hydrated alumina," it is technically aluminum hydroxide, with the chemical formula $Al(OH)_3$. The term "hydrate" became part of the common name because chemically combined water is released during its decomposition in fire.

ATH came into wide use as a flame retardant in the 1960s, primarily as a result of demands of consumer-driven safety legislation for carpet backing and fiberglass-reinforced polyester products. For these applications, ATH imparts flame retardance and smoke suppression. End-use markets for ATH include transportation, construction, cast polymers, electrical/electronic, wire and cable, leisure and appliances. Specific polymer applications range from thermosets such as "solid surface," sheet molding compounds (SMCs) and bulk molding compounds (BMCs) to wire and

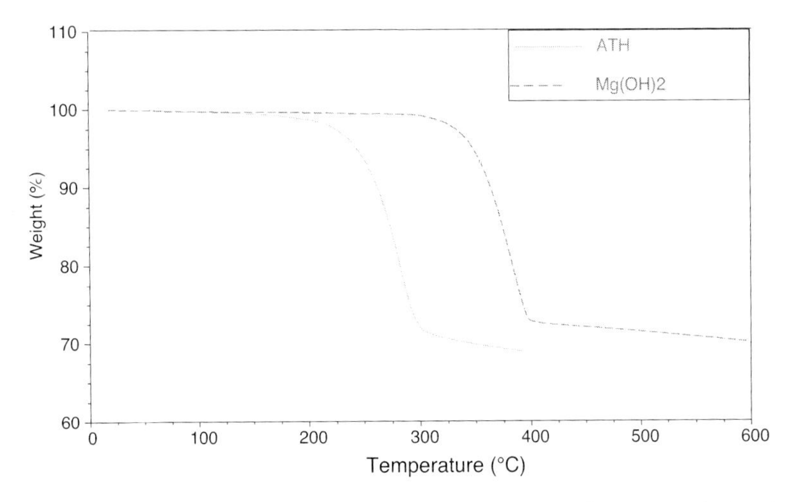

Figure 17.4 Comparison of the thermal degradation properties of magnesium hydroxide and aluminum trihydrate.

cable and other thermoplastic applications where it is compounded with PVC, polyolefins, or ethylene propylene diene (EPDM) rubber.

Thermodynamic properties and flame retardance At room temperature, ATH is very stable but when the temperature exceeds 205 °C, as could happen in a fire situation, it begins to undergo an endothermic decomposition with a reaction heat of −298 kJ/mol. This decomposition, as shown below, is kinetically slow between 205 and 220 °C, but above 220 °C (see Figure 17.4 comparing ATH and MOH) it becomes very rapid:

$$2Al(OH)_3 \rightarrow Al_2O_3 + 3H_2O$$

Water that is released during this decomposition dilutes the gases that feed combustion. This dilution slows down the rate of polymer combustion by forming a vapor barrier that prevents oxygen from reaching the flame. The combustion of polymers is also retarded since ATH acts as a heat sink (heat capacity of $Al_2O_3{\cdot}3H_20 = 186.1$ J/K mol) and absorbs a portion of the heat of combustion. It is also believed that aluminum oxide (Al_2O_3) formed during this decomposition aids in the formation of an insulating barrier on the surface of the burning polymer, which acts to insulate the polymer from fire.

Ground versus precipitated ATH ATH can be produced in a variety of particle sizes and particle size distributions that are controlled by grinding or precipitation processes. Both precipitated and ground ATH products can be used as long as processing temperatures stay below the onset of decomposition temperature at 200 °C. Mechanically ground ATH has a wide particle size distribution and a better packing fraction than precipitated ATH, and this contributes to less dusting, faster

Table 17.3 Chemical analysis of wet ground and precipitated ATH.

Parameter	Precipitated ATH	Wet-ground ATH
$Al(OH)_3$ (%)	99.2	99.2
SiO_2 (%)	0.05	0.005
Fe_2O_3 (%)	0.035	0.007
Na_2O (%) soluble	0.01	0.005

incorporation into the polymer matrix, and lower compound viscosity (by as much as 50%) compared to precipitated grades. Table 17.3 shows a chemical analysis of wet ground and precipitated ATH.

The most significant difference between the two grades is in their morphology as illustrated in Figure 17.5. Precipitated ATH, which is allowed more time to crystallize, displays a hexagonal platelet morphology and a more homogeneous particle size distribution. This can be a benefit in some electrical insulation applications where mineral homogeneity can influence dielectric properties. However, this homogeneity may be offset by difficulty in processing compared to ground ATH; in the latter, packing fractions are controlled allowing ease of processing and, thus, an improved distribution of the mineral in the polymer matrix.

The relative ease of processing of ground ATH can allow higher levels to be incorporated into a polymer composite and this, in turn, enables FR standards requiring high loading levels to be more easily attained. The relative ease of incorporation and processing of the ground ATH compared to the precipitated product is manifested by lower viscosities; for example, in a 55% ATH loaded EPDM compound, Mooney viscosities were respectively 28 and 52 for the wet ground and precipitated grades

A comparison of the combustion properties of EPDM composites containing wet ground and precipitated ATH (Table 17.4) shows significant differences. It is likely that the lower average heat release rate (HRR) and average mass loss rate for the wet

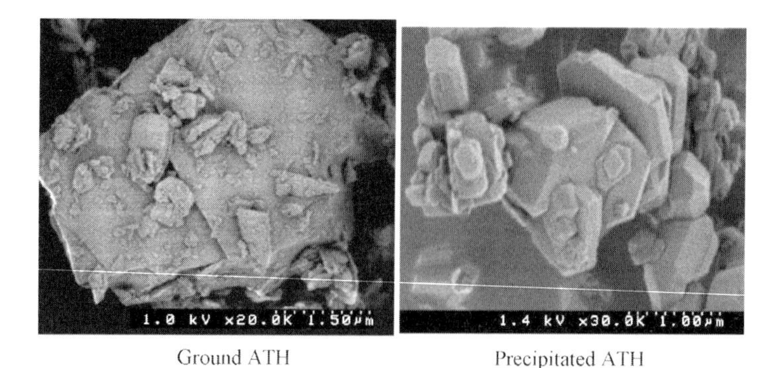

Ground ATH Precipitated ATH

Figure 17.5 Scanning electron micrographs of ground (20 000×) and precipitated ATH (30 000×).

Table 17.4 Cone calorimetric data comparison of wet-ground and precipitated ATH in EPDM.

Cone calorimetry data at 35 kW/m^2	Precipitated ATH	Wet-ground ATH
Peak HRR (kW/m^2)	164	111
Peak HHR (s)	489	272
Average HHR (kW/m^2)	123	84
Total heat (mJ/m^2)	80	85
Average effective heat of combustion (MJ/kg)	29	34
Average mass loss rate ($g/s/m^2$)	4.4	2.9

ground ATH composite may derive from a more homogeneous distribution of the ATH in the polymer. This is a good example of the influence of physical properties of a fire-retarding material on the performance in combustion testing.

It deserves to be stated that there have been recent significant advances in the control of morphology and purity of precipitated ATH [34]. New methods of precipitation yield products with better bulk flow feeding properties and thermal stability that compound faster with greater ease and produce lower compounding viscosity than typical precipitated ATHs. The nature of the precipitation processes is such that they allow a greater manipulation of conditions and compositions such as concentration, time, agitation, doping, electrolyte removal than typical grinding processes.

Applications In *reinforced thermoset* applications, the principal properties that ATH imparts are flame retardance, smoke suppression, thermal conductivity, optical translucency, and chemical stability. Aside from the benefits imparted in the above properties, the use of ATH as a resin extender also affords lower cost formulations with an acceptable balance of performance properties.

Finely divided ATH or MOH produced in precipitation processes or finely ground and surface-treated ATH grades can be used in wire and cable applications, most notably in PVC and EPDM. The FR tests that are relevant to this application are UL 94/horizontal burn, VW-1, UL-1685, UL-1666, and UL-910. The ATH imparts the following fire retarding processes:

- Cooling of the polymer and reduction of pyrolysis products.
- Insulation of the substrate through a combination of aluminum oxide (derived from ATH when it loses water) with the char formed.
- Dilution of combustion gases by release of water vapor.

Furthermore, it is

- unleachable owing to its insolubility in water;
- nontoxic;
- noncorrosive.

The so-called "solid surface" is a solid, nonporous surfacing composite material derived from ATH and a polyester or acrylic resin that is cured either thermally or at room temperature. The color and pattern run throughout its thickness with a soft, deep translucence and a natural feeling of warmth. "Solid surface" resists heat, stains, mildew, and impact. Incidental damage to the surface can be easily repaired to maintain the appearance. It can be easily cut, routed, and shaped so fabricators can create distinctive designs and customized patterns. As a result of the high loading levels of ATH that often exceed 60 wt%, the material easily resists fire even if contact is made by red-hot metal. For this reason, the material finds use in kitchens and homes as counter tops and bathroom fixtures.

Surface-treated ATH To enhance the overall performance of ATH-filled compounds, a wide selection of chemical modifications of the ATH surface can be carried out with surfactants, stearates, and organofunctional silanes. As also discussed in Chapters 4–6, these chemical treatments can aid processing and improve mechanical properties, chemical resistance, electrical performance, and flame retardance. Although many of the surface treatments used in ATH or MOH are organic materials that contain combustible functional groups, the effect of these treatments on FR properties is usually negligible. In fact, as a consequence of the increased loading levels and improved dispersion that these treatments afford compared to the nontreated counterparts, the FR performance of composites is often enhanced by using treated ATH or MOH.

Table 17.5 shows the various surface treatments that are suitable for a variety of polymers and the associated impact on end-use properties. Surface treatments can be classified under three broad headings:

1) Encapsulating and nonreactive
2) Reactive with the ATH but noncoupling to the polymer matrix
3) Reactive and coupling [30]

It is clearly evident that the end-use properties are significantly influenced by the choice of ATH and the chemical surface treatment. The ATH products of the future will include new combinations of morphology, purity (low electrolyte levels), and surface treatments as a way to optimize the desired combination of end-use properties. Table 17.6 shows a survey of surface treatment options for ATH in wire and cable applications using a variety of polymers where flame retardance and smoke suppression are required.

Summary ATH is used in a variety of polymers in different areas of applications. The products of the future will be tailored to conform to specific end-use requirements of the customer. This will be done by simultaneous control of morphology and optimization of surface treatments as well as optimization of the end-use formulations. Users of ATH will continue to seek lower processing costs through higher throughput, ease of processing, and polymer compatibility that can be achieved and optimized by the methods outlined above, or quite probably by newer products from synthetic processes, such as precipitation, that have morphologies more suited to high throughput and ease of processing [31–33]. Table 17.7 summarizes the various flammability and electrical property standards where ATH plays a major role.

Table 17.5 Chemical surface treatments applied to ATH for specific polymers.

Surface modification	Polymer type	Properties affected
AX 5102, a blend of anionic and nonionic surfactants	Epoxy, phenolic, polyolefins, Urethane	Lower viscosity, improved dispersion, higher loading potential, improved surface profile
ST, proprietary surfactant	Latex, polyester, polyolefins, urethane	Lower viscosity, improved suspension, better air release properties, faster dispersion, improved wet-out
SL, isostearic acid	EVA, polyolefins	Improved processing and dispersion, higher loading potential
Hyflex silane, Y-5889 OSi product, ethoxylated propylsilane	Epoxy polyester, urethane	Lower viscosity, higher loading potential, faster dispèrsion, improved physical properties, high polymer compatibility
SP, RC-1 (vinylsilane + methylsilane oligomer mix)	Acrylic, EPDM, EVA, neoprene, polyester, polyolefins, PVC, SBR	Improved mechanical properties, better processing, increased water resistance
Isobutyltrimethoxysilane	Epoxy, polyester, urethane	Lower viscosity, higher loading potential, faster dispersion, enhanced physical properties, high polymer compatibility
SA, AMEO 3-aminopropyltriethoxysilane	EVA, nitrile, phenolic, PVC, urethane	Improved mechanical properties and processing
SM, A-189	EPDM, neoprene, nitrile	Greater abrasion resistance, higher mechanical properties, improved processing
3-Mercaptopropyltrimethoxysilane		
SE, A-187, 3-glycidoxypropyltrimethoxysilane	Epoxy, phenolic	Improved processing, improved mechanical properties, improved water resistance
SH, A-174, methacryloxypropyltrimethoxysilane	Acrylic, EPDM, EVA, polyester, polyolefins, SBR	Improved mechanical properties and processing, improved water resistance

Table 17.6 ATH surface treatments used in a variety of polymers.

Polymer	Stearic acid (SL)	Aminosilane (SA)	Vinylsilane (SP)	Proprietary silane
XL-EVA	+	+	+	+
TP-EVA	+	+		+
XL-polyolefin	+	+	+	+
TP-polyolefin	+	+	+	+
PVC	+	+		

Table 17.7 Standards met with ATH use.

Standard	Description	Target
UL 94	Vertical	V0
ASTM E 84 (UL 723)	Tunnel test	<450 Smoke, <25 flame spread
ASTM D 495	Arc resistance	>200 s
ASTM E 662	NBS smoke chamber	<200 at 4 min
ASTM E 1354	Cone calorimeter	
ASTM E 648	Radiant panel	Critical heat flux of floor coverings using a radiant panel
ASTM D 635	Horizontal burn	
ASTM D 2863	Limiting oxygen index	
ASTM D 149	Dielectric strength	
ASTM D 150	Dielectric constant	
ASTM D 257	Volume resistivity	
ASTM E 906	OSU test	
ANSI Z 124	Torch test	
MVSS-302	Auto interior flammability	
UBC 17-5	Corner room burn test	
CAL 133	Contact furniture flammability	

17.7.1.2 **Magnesium Hydroxide**

Magnesium hydroxide functions in a manner similar to ATH. It liberates water from crystallization in a fire situation and in so doing it forms a barrier to the ingress of oxygen required for combustion. The material is also capable of absorbing heat, although the heat capacity value of 77.0 J/K mol (crystalline) is lower than that for ATH.

Magnesium hydroxide is produced by extraction from ores, such as magnesite, dolomite, or serpentinite, and brine and seawater. The material used as flame retardant is generally of high purity (>98.5%). It is most often obtained from seawater or brine, although other ore-derived products can also be very pure. Three major processes for magnesium hydroxide are isolation from seawater and brine, the Aman process that employs thermal decomposition of brine to yield a high-purity magnesia that is converted to magnesium hydroxide, and the Magnifin® process in which magnesium oxide is isolated from serpentine ore and converted to magnesium hydroxide.

Magnesium hydroxide is a white powder. Median particle sizes range from 0.5 to 5 μm and specific surface areas typically range from 7 to 15 m^2/g, depending on size and morphology. It is used at high loading levels, usually between 40% to 65 wt%. A key difference from ATH is the higher decomposition temperature of about 320 °C (Figure 17.4). This affords an advantage over ATH in that it may be used in thermoplastics, engineering thermoplastics, and thermoset resin systems processed at temperatures above 200 °C. From a chemical standpoint, magnesium hydroxide is a much more alkaline material than ATH. Since slurries in water (~5% by weight) can yield pH values around 10.5 or more, this property may need to be considered in some formulations.

Magnesium hydroxide finds use in applications such as wire and cable, appliance housings, construction laminates, roofing, piping, and electrical components. Different grades are developed for different applications. The use of surface-treated grades is increasing as market demands move toward increased product and performance differentiation [35, 36].

17.7.1.3 Zinc Stannates

One of the disadvantages of using ATH or MDH is the high levels of the filler required for adequate flame retardancy. Difficulties in incorporation increase as the system becomes more rigid as in the case of PVC. This can result in processing difficulties and significant deterioration of critical polymer properties including physical, mechanical, and electrical. With the increase in energy costs, since 2006, there is increasing pressure on formulators to develop lower weight compounds at competitive cost without the loss of key properties. The low-toxicity inorganic tin compounds, zinc stannate [$ZnSnO_3$] and zinc hydroxystannate [$ZnSn(OH)_6$], have been developed as flame retardants for flexible PVC [37–39] resulting in enhancement of the limiting oxygen index, when only marginal impact is seen with ATH and MDH. Furthermore, the char yields obtained with the tin compounds are significantly higher than in the case of ATH and MDH, although MDH did contribute to an increase in the char yield. As mentioned above, ATH and MDH are not recommended for smoke suppression in rigid PVC formulations mainly due to difficulty in incorporation. However, the use of zinc hydroxystannate as a smoke suppressant, synergistic with antimony trioxide in PVC, has recently been described [40]. Testing also showed that the incorporation of the zinc hydroxystannate as an effective and cost-efficient fire retardant and smoke suppressant of PVC did not affect its heat stability, color, or impact strength at 3 phr levels [41].

17.7.2 Halogenated Fire Retardants

17.7.2.1 Mechanism of Action

The mode of action of halogenated fire retardants depends on the reaction of halogen-based free radicals with excited state fire-propagating components in the gas phase resulting in lower system temperature and a decreased reaction rate for the fire-sustaining chemical processes.

Brominated and chlorinated organic compounds are generally used as fire retardants. Iodine-containing materials are avoided as they tend to be more unstable, in general, than their chlorinated or brominated counterparts. The choice of retardant depends on several factors such as the polymer type, the behavior of the halogenated fire retardant under processing conditions (stability, melting, distribution, etc.), and/or its effect on properties and long-term stability of the resulting material. It is advantageous to use an additive that introduces a halogen species to the flame in the same temperature range as that in which the polymer degrades into volatile combustible products. Then, the fuel and inhibitor would both reach the gas phase according to the "right place at the right time" principle.

The most effective fire-retardant polymeric materials are halogen-based polymers (e.g., PVC, chlorinated PVC, polyvinylidene fluoride (PVDF)) and additives (e.g., chlorinated paraffins (CPs), tetrabromobisphenol A (TBBA)). However, the improvement in fire performance depends on the type of fire tests, that is, the application.

Halogenated organic fire retardants may rather be classified as additives than as functional fillers. Many are liquids, others are high molecular weight polymers, and others exhibit a certain degree of miscibility with the polymer matrix; thus, they do not fall within this book's scope of solid (mostly inorganic) fillers forming a distinct dispersed phase. They are, however, briefly discussed since they are commonly used synergistically in combination with more traditional inorganic fillers such as antimony oxide, molybdenum oxide, or occasionally with aluminum trihydrate.

17.7.2.2 **Synergy with Antimony Trioxide**

Antimony oxide (Sb_2O_3) is often used as a synergist in combination with halogenated fire retardants. For efficient action, trapping of excited state free radicals occurs in the gas phase in the flame and hence the radical trapping agent needs to reach the flame in the gas phase. Addition of Sb_2O_3 allows formation of volatile antimony species (halides or oxyhalide) that interrupt the combustion process by inhibiting H^* radicals via a series of reactions shown in Figure 17.6. This mechanism is an illustration of the synergy between halogenated flame retardants and the antimony oxide.

Loading levels of antimony trioxide can often be quite high and in some cases this has led to toxicity concerns. In recent years, this issue has received considerable attention. The aircraft manufacturer Airbus has recently adopted a policy that prohibits the use of any new product or material in its aircraft that contains antimony trioxide. This may require significant changes in the design of products for use in aircraft interiors, for example, flooring, window shades, or seat components. From a designer's perspective, this may tend to limit the use of PVC unless suitable nonantimony-containing FR packages can be developed that will provide the same or improved level of performance at an acceptable cost.

17.7.2.3 **Bromine-Containing Fire Retardants**

Bromine-containing fire retardants may be used in many polymeric applications as effective gas phase flame retardants. They may be broadly classified into aliphatic and aromatic materials. In general, the aliphatic materials are more effective flame

$Sb_2O_3 + 6\,HX$	$\rightarrow$	$2\,SbX_3$
$SbX_3 + H^*$	$\rightarrow$	$SbX_2 + HX$
$SbX_2 + H^*$	$\rightarrow$	$SbX + HX$
$SbX + H^*$	$\rightarrow$	$Sb + HX$
$Sb + O^*$	$\rightarrow$	SbO^*
$SbO^* + H^*$	$\rightarrow$	$SbOH$
$SbOH + H^*$	$\rightarrow$	$SbO^* + H_2$

Figure 17.6 Mechanism of action of antimony oxide as a flame retardant [11].

retardants by virtue of their lower decomposition temperature than the aromatic bromine FRs that have a high degree of thermal stability. In some systems, the decreased chemical lability of the aromatic systems can allow action over a longer time cycle. This can be advantageous in the case of longer burning fires or cases where a fire may have taken hold and the more kinetically labile retardants have become exhausted [42].

The mechanism of action of halogenated flame retardants has been outlined in Table 17.2 and Figure 17.3. These materials function mainly in the gas phase by liberating volatile compounds of bromine such as hydrogen bromide, HBr, which inhibit the reactions that propagate the flame. Hydrogen bromide reacts readily with excited state radicals of oxygen, hydroxyl, and hydrogen to generate molecular species that have little or no tendency to propagate the flame. One important reaction in this process is the reaction of HBr with HO^* to form H_2O Br^*. This prevents the reaction of HO^* with CO to form CO_2, which is a highly exothermic process and assists the propagation of combustion.

A wide range of brominated flame retardants is available, and the choice of material is often made with consideration of the polymer processing operations and conditions. Ideally, the flame-retardant material should remain stable during normal processing of the polymer. Thermogravimetric weight loss studies are often used to determine the stability of many polymer additives, including flame retardants, over the projected range of processing conditions. In addition to low MW compounds (see examples in Table 17.8), higher MW materials such as brominated polystyrene (MW 200 000) are used when nonblooming characteristics are desirable. Brominated epoxy fire retardants are also available as oligomers with MWs in the range of 1600–3600 g/mol or as more defined polymers with MWs in the range of 10 000–600 000 g/mol. These materials can be either terminated with epoxy groups or with brominated aromatic rings. The epoxy-terminated materials are suitable for high-end engineering polymer applications providing high thermal and UV stability, do not bloom, and are easily processed. The brominated phenyl group-terminated polymers are used in HIPS and ABS where they provide good processability with low migration, high thermal and UV stability, good HDT, impact strength, and low metal adhesion. The choice of the brominated epoxies depends on the molecular weight that correlates well with the onset of softening. In general, the lower the molecular weight the greater the ease of incorporation into polymers. However, this advantage can be offset by greater mobility of the additive.

17.7.2.4 **Chlorine-Containing Flame Retardants**

The use of chlorinated additives as flame retardants is often in conjunction with other materials such as phosphorus or antimony-containing additives. The choice of a chlorinated additive depends on the stability of the material, the volatility, chlorine content, and kinetics of reaction in the gas phase during combustion. The use of such materials, however, is decreasing owing to the toxicity of polychlorinated biphenyls (PCBs) and products with related molecular structures.

Examples of chlorinated additives include chlorinated paraffins derived from chlorination of petroleum distillates and represented by the general formula

Table 17.8 Examples of aliphatic and aromatic brominated flame retardants.

Material	CAS no.	MW	Br content (%)	Melting point (°C)	Advantages	Application
Aliphatic						
Trisbromoneopentyl phosphate	19186-97-1	1018	67	230	UV, heat stability	PP, ABS
Trisbromoneopentyl alcohol	36483-57-5	325	73	65	Low leaching, heat stability	Polyurethanes
Dibromoneopentyl glycol	3296-90-0	262	60	110	High FR performance transparency	UPEs
Hexabromocyclo-dodecane	25637-99-4	642	73	180	High purity and efficiency	EPS, XPS
Aromatic						
Tris(bromophenyl)triazine	25713-60-4	1067	67	230	Nonblooming	PE, ABS epoxy
Tetrabromobisphenol A bis(2,3-dibromo propyl ether)	21850-44-2	943	68	115	High efficiency	PP, ABS, PS
Octabromo-diphenyloxide	32536-52-0	801	78	120–185	High surface quality	ABS, styrenic copolymers
Decabromo-diphenyloxide	1163-19-5	959	83	305	Excellent thermal stability	General purpose
Tetrabromo bisphenol A	79-94-7	544	58	181	High reactivity and efficiency	PC, epoxy, phenolic, ABS
Brominated triethylphenyl indane	155613-93-7	857	73	240–55	Good flow and impact	PA, styrenics
Pentabromotoluene	87-83-2	487	82	299		PE, PP, rubber

$C_nH_{2n+2-m}Cl_m$. They typically contain chlorine in the range of 30–70% by weight and range in length from 10 to 30 carbon units and they may be liquid or solid. These materials function by releasing hydrogen chloride that reacts in the gas phase to decrease the concentration of the reactive hydroxyl, oxygen, or hydrogen radical species that propagate the flame. Chlorinated resins arc mainly used as plasticizers for flexible PVC in combination with dioctyl phthalate or diisononyl phthalate improving flame-retardant properties in applications such as flooring and cables. Solid grades with high chlorine content are used in thermoplastics such as low-density polyethylene (LDPE) in cable jacketing in combination with Sb_2O_3.

Other examples of chlorinated additives include chlorinated alkyl phosphates with main applications in rigid and flexible polyurethane foams and chlorinated cycloaliphatics such as dodecachlorodimethanodibenzocyclooctane. The latter is used with various synergists such as antimony trioxide and zinc borate in numerous polymers including polyamide, polyolefins, and polypropylene.

17.7.3 Zinc/Boron Systems

Zinc borate is a boron-based fire retardant available as a fine powder with a chemical composition of $(ZnO)_x(B_2O_3)_y(H_2O)_z$. The most commonly used grades have the structure $2ZnO{\cdot}3B_2O_3{\cdot}zH_2O$.

Zinc borate can be used as a fire retardant in PVC, polyolefins, elastomers, polyamides, and epoxy resins. In halogen-containing systems, it is used in conjunction with antimony oxide, while in halogen-free systems it is normally used in conjunction with other FRs such as aluminum trihydrate, magnesium hydroxide, or red phosphorus. In a small number of specific applications, zinc borate can be used alone.

17.7.4 Melamines

Melamine (2,4,6-triamino-1,3,5-triazine) or melamine-derived flame retardants represent a small but fast growing segment in the flame-retardant market. In this family, three chemical groups can be defined; pure melamine, melamine derivatives, that is, salts with organic or inorganic acids such as boric acid, cyanuric acid, phosphoric acid or pyro/poly-phosphoric acid, and melamine homologues containing multiring structures. Melamine-based flame retardants show excellent flame-retardant properties and versatility in use because of their ability to employ various modes of flame-retardant action. Melamines are only briefly mentioned in this chapter since they may be considered as additives rather than true functional fillers.

17.7.5 Phosphorus-Containing Flame Retardants

The element phosphorus (red phosphorus) is known to be an effective inhibitor of combustion. A wide variety of phosphorus-containing materials are used as fire retardants in a broad range of polymers from elemental red phosphorus to both inorganic and organic compounds such as ammonium polyphosphates, organophosphates, and phosphinates, phosphorus-nitrogen compounds such as melamine phosphates, phosphorus-halogen compounds, and phosphorus-containing polyols that are used to produce polyurethane foams. Phosphate esters, with or without halogen, are widely used in PVC or in polymers that have high hydroxyl group content. The level of phosphorus in different chemicals can vary from below 10% to close to 50% in the case of red phosphorus. The oxidation state of the phosphorus can range from 0 to +5.

17.7.5.1 **Mechanism of Action**

While the details of the chemical mechanisms involving phosphorus-containing fire retardants are complex, the mechanisms can be broadly described as condensed phase and vapor phase depending to a large extent on the chemical environment and the host polymer. In a fire situation, phosphorus-containing fire retardants can act in the condensed phase or the vapor phase, and sometimes both. In the condensed phase, the oxidation of phosphorus to phosphoric acids leads to the formation of a carbonaceous char, sometimes accompanied by foaming or intumescence and formation of a protective layer on the polymer surface that inhibits the release of volatile species from the burning polymer. This mechanism tends to lower the level of smoke in a fire situation. Under conditions when there are no heteroelements present, other than oxygen, phosphoric or phosphonic acids are formed. These accelerate the loss of volatile groups from the burning polymer chain and provide a less combustible vapor barrier. The loss of volatile groups also has a cooling effect on the host polymer.

There is a correlation between char yield and fire resistance [43]. Char is formed at the expense of combustible gases. Char inhibits flame spread by acting as a thermal barrier around the unburned polymer. Acids of phosphorus or their derived anions can dehydrate to form oligomeric molecules and lose water in the process. This water may act as a barrier to the ingress of oxygen, in effect as a diluent for combustion gases [44].

The mode of action in the gas phase involves the formation of reactive phosphorus-based species such as PO^*, P^*, and HPO that are capable of removing the free radicals that can drive combustion. The reaction scheme below for a triaryl phosphate is indicative of the underlying chemical mechanism of phosphorus-based flame retardance:

$$R_3P = O \rightarrow PO^*, P^*$$

$$H^* + PO^* + M \rightarrow HPO + M$$

$$HO^* + PO^* \rightarrow HPO + \frac{1}{2}O_2$$

$$HPO + H^* \rightarrow H_2 + PO^*$$

17.7.5.2 Red Phosphorus

Elemental red phosphorus is a very efficient fire retardant for a variety of polymers including polyesters, polycarbonates, epoxies, polyurethanes and, in particular, polyamides and phenolics. Red phosphorus and atmospheric moisture react to form the highly toxic gas phosphine and the material will readily ignite in air to form phosphine dimer.

$$2P + 3H_2O \rightarrow 2PH_3 + \frac{3}{2}O_2$$

Consequently, the commercial product is encapsulated with a high loading of polymer to give concentrates of up to 50% phosphorus.

The mechanism of action is believed to be due to the formation by oxidation during combustion of phosphoric acid or phosphorus pentoxide. These species can form a carbonaceous char in the condensed phase. Red phosphorus can act as a flame retardant without coagents or synergists in oxygen and/or nitrogen-containing polymers. In polymers that have no oxygen functionality, coagents are required [45–47].

17.7.5.3 Phosphorus-Containing Organic Compounds

Similar to halogenated organic fire retardants, many phosphorus-containing organics are liquids and, thus, may be classified as additives rather than as fire-retardant solid functional fillers. Although they do not fall within the scope of this book, they are briefly mentioned since they may synergistically act in combination with more traditional fire retardants. They include the following:

- Aryl phosphates [48, 49] such as triphenylphosphate, resorcinol bis-diphenylphosphate, and bisphenol-A bis-diphenylphosphate with applications in engineering thermoplastics, and tricresylphosphate as a PVC flame-retardant plasticizer.
- Alkyl phosphonates with a general structure of $RP(=O)(OR^1)(OR^2)$. They are generally viscous liquids with applications in rigid PUR, highly filled polyesters, and PET fibers.
- Phosphinates denoted by the general formula $[R^1{\cdot}R^2{\cdot}PO(=O)]^- M^+$ (M = metal or NH_4^+), are efficient fire retardants for engineering thermoplastics such as glass fiber-reinforced polyamides and polyesters. Synergistic effects with nitrogen-containing materials such as melamine have been reported [50, 51].
- Chlorinated alkyl phosphates such as tris(2-chloroethyl)phosphate are sometimes used in rigid and flexible polyurethane foams.

The polyurethane industry is searching for a more environmentally acceptable alternative to pentabromodiphenyl ether, which has been banned in Europe and voluntarily withdrawn in the United States. The preference appears to be in favor of phosphorus-containing materials.

17.7.5.4 Ammonium Polyphosphate

This material is an inorganic salt of polyphosphoric acid and ammonia with the general structure $[NH_4PO_3]_n$. There are two main families of ammonium polyphosphate as shown below:

Table 17.9 Solubility and properties of some commercial grades of APP I.

Monomer units in chain (*n*)	pH	Solubility in water (g/100 ml)	Median particle size (μm)
20	5–6	5	12
40	5–6	4	14
60	5–6	3	16
80	5–6	2	20

$$\text{APP I} \quad NH_4{}^{+\,-}OP(=O)O-[NH_4PO_3]_n-OP(=O)O^{-\,+}NH_4$$

$$\text{APP II} \quad NH_4{}^{+\,-}OP(=O)O-[NH_4PO_3]_m-P(=O)O-[PO_3NH_4]_nOP(=O)O^{-\,+}NH_4$$

Crystal-phase I APP (APP I) is characterized by a variable linear chain length, showing a lower decomposition temperature (approximately 150 °C) and a higher water solubility than crystal phase II (APP II). APP II has a cross-linked and branched structure. The molecular weight is much higher than that of APP I with $n > 1000$. APP II has a higher thermal stability and lower water solubility than APP I (see Tables 17.9 and 17.10).

APP is a stable, nonvolatile material. It slowly hydrolyzes in contact with water to produce monoammonium phosphate (orthophosphate). This process is autocatalytic and is accelerated by higher temperatures and prolonged exposure to water. The decomposition temperature of APP is related to the length of the polymer chain. Long-chain APP starts to decompose at temperatures above 300 °C to polyphosphoric acid and ammonia. Short-chain APP, defined as that having a chain length below 100 monomer units, begins to decompose at temperatures above 150 °C.

APP and APP-based systems are very efficient halogen-free flame retardants mainly used in polyolefins (PE, PP), epoxies, polyurethanes, unsaturated polyesters, phenolic resins, and others. APP is a nontoxic, environment friendly material and it does not generate additional quantities of smoke due to intumescence. Compared to other halogen-free systems, APP requires lower loadings. In thermoplastic formulations, APP exhibits good processability, retention of good mechanical properties,

Table 17.10 Solubility and properties of some commercial grades of APP II.

Material type 2 APP	pH	P_2O_5 (%)	Solubility in water (g/100 ml)	Median particle size (μm)
Material 1	5–6	72	0.5	18
Material 2	5–6	72	0.5	15
Material 3	5–6	72	0.5	7
Material 4	5–6	72	0.5	4

Table 17.11 Benefits of APP in several polymer systems.

Applications	Benefits
Thermoplastics	Inorganic polymer
Polyethylene	High FR efficiency
Polypropylene	Halogen free
Thermosets	Environment friendly
Polyurethane	Excellent processability
Epoxy resin	Good mechanical properties
Unsaturated polyester	Good electrical properties
Phenolic resin	Less smoke generated

and good electrical properties. In thermosets, APP can be used in combination with ATH to obtain a significant reduction of the total FR filler. Such combinations are used in construction and electrical applications. The processability can be improved when the total filler level is reduced. Table 17.11 summarizes some benefits of APP in several polymer systems.

Ammonium polyphosphate acts as a flame retardant by intumescence as shown in Figure 17.7.

When plastic or other materials that contain APP are exposed to an accidental fire or heat, the flame retardant starts to decompose commonly into polymeric phosphoric acid and ammonia. The polyphosphoric acid reacts with hydroxyl or other groups of a synergist to form an unstable phosphate ester. Dehydration of the phosphate ester then follows. Carbon foam is built up on the surface against the heat source (charring). The carbon barrier acts as an insulation layer, preventing further decomposition of the material.

$$(NH_4PO_3)_n + \text{Heat } (>250\,^\circ C) \rightarrow (HPO_3)_n + nNH_3$$

$$(HPO_3)_n + \text{Polymer} \rightarrow \text{Carbon Char} + H_3PO_4 - H_2O$$

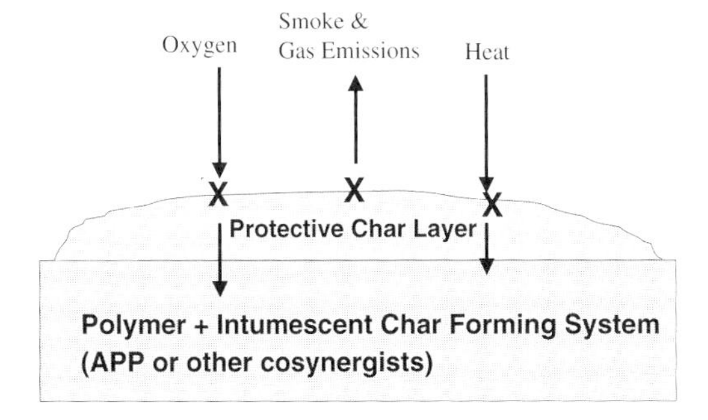

Figure 17.7 Char formation in polymer-containing ammonium polyphosphate as flame retardant.

Cosynergists may also be used. Addition of synergetic products such as pentaerythritol derivatives, carbohydrates, and spumific agents will significantly improve the flame-retardant performance of APP. Red phosphorus and APP are used in intumescent coatings and paints suitable for materials such as wood and steel as well as polymer systems.

17.7.6
Low Melting Temperature Glasses

Low melting temperature phases (glasses) have long been considered as flame-retardant polymer additives. Low melting temperature glasses can improve the thermal stability and flame retardance of polymers by

- providing a thermal barrier for both polymer and char that may form as a combustion product;
- providing a barrier to retard oxidation of the thermally degraded polymer and combustion of char residue;
- providing a "glue" to maintain structural integrity of the char;
- providing a coating to cover over or fill in voids in the char, thus providing a more continuous external surface with a lower surface area;
- creating potential useful components of intumescent polymer additive systems.

A low melting temperature glass flame-retardant system for PVC consisting of $ZnSO_4$–K_2SO_4–Na_2SO_4 may be used. This system is an excellent char former and a smoke suppressant. Low melting glasses have also been tested with transition metals such as Ni, Mn, Co, and V and with main group metals such as Al, Ca, and Ce. It has been shown that silica gel in combination with potassium carbonate is an effective fire retardant (at a mass fraction of only 10% total additive) for a wide variety of common polymers such as PP, polyamide, PMMA, polyvinylalcohol, and cellulose and to a less extent for PS and SAN copolymers. The mechanism of action for these additives seems to involve the formation of a potassium silicate glass during the combustion.

The action of ammonium pentaborate (APB) in polyamide-6 as glass former has been studied. The degradation of APB into boric acid and boron oxide provides a low melting glass at temperatures of interest and then the char formed by the degradation of polyamide-6 is protected by the glass that induces the fire-retardant effect.

Zinc phosphate glasses, with glass transition temperatures in the range of 280–370 °C, have recently been developed. These may be compounded with engineering thermoplastics such as polyarylether ketones, aromatic liquid polyesters, polyarylsulfones, perfluoro-alkoxy resins, or polyetherimides to afford a significant increase in LOI compared to the unfilled polymer. The incorporation of low-temperature glasses as conventional fillers in commodity polymers can, however, produce measurable flame retardance effects as well. Nevertheless, the loading must be high enough (>60 wt% in polycarbonate) to significantly increase the LOI.

17.7.7
Molybdenum-Containing Systems

Molybdenum oxide (MoO_3), ammonium dimolybdate ($(NH_4)_2Mo_2O_7$), and ammonium octamolybdate ($(NH_4)_4Mo_8O_{26}$) are commonly used in thermoset polyesters and in flexible and rigid PVC as smoke suppressants [52]. Molybdenum compounds can be used as coadditives with Sb_2O_3, ATH, organophosphates, and zinc borate [53–57]. Although they are expensive materials, they are very efficient at low loading levels and this may prove advantageous in applications such as transportation and aviation requiring lower weight products.

MoO_3 is known as a good char-former. Molybdenum appears to work in the condensed phase preventing the formation of volatile organic compounds that could contribute to smoke formation [58]. Two mechanisms have been proposed: a redox mechanism in which molybdenum in a low oxidation state reacts to form a higher oxidation-state molybdenum chloride species and facilitates coupling of alkyl groups leading to less volatile species [59]. The second mechanism involves the conversion of linear *cis*-polyenes to *trans*-polyenes catalyzed by molybdenum. The *cis*-isomers can easily convert to aromatic and cyclic structures such as benzene or substituted benzene compounds that will contribute to smoke. The *trans*-isomers cannot convert so readily to volatile cyclic species [60].

Some interesting work has been reported using Zn and Mo systems in plenum cable formulations. The use of a zinc molybdenum complex in place of ammonium octamolybdate is described in Ref. 61 and the use of ammonium octamolybdate in combination with zinc stannate as a smoke suppressant in plenum cable compounds is discussed in Ref. [62].

17.7.8
Nanosized Fire Retardants

With the need to improve both performance and toxicity of fire retardants, there is an emerging focus on nanomaterials. The chemical mechanisms of fire retardance using nanomaterials are a subject with a great potential for exploration in the coming years.

Fire-retardant coatings and foams that contain nanomaterials is one area with emerging developments. Most fire-retardant coatings work by suppressing flame through intumescence, solidifying into foam on exposure to flame or extreme heat. This foam insulates the substrate and reduces flame spread on the material surface. The relevant test is the fire chamber test using ASTM method E 84. Recent work describes the development of a variety of systems including two-part epoxy coatings with nanomaterials that have zero VOCs and are nonhazardous [63]. Other reported fire-retardant systems of interest are nanoclay-containing PUR foams [64], nanoclay/polyolefin composites [65, 66], laponite and montmorillonite/polyamide-6 nanocomposites [67], and carbon nanotube/PMMA [68], to cite a few examples.

One of the difficulties in using nanomaterials is that they generally do not incorporate easily into polymers with a high degree of dispersion and/or exfoliation (see also Chapter 9). However, there has been considerable progress in this area using functionalized montmorillonites that can be processed in compounding extruders [69, 70]. The aim is to create an interconnected but well-dispersed and distributed network of the nanomaterials within the host polymer. It is theorized that this nanonetwork can prevent the polymers from breaking down into small molecular fragments that vaporize and mix with air [71, 72]. The formation of an interconnected network also helps the process of char formation. There are several literature examples of nanomaterials being a key component of char formation on combustion. A description of the burning behavior and, in particular, the dripping behavior of nanoclay/nylon composites is provided in Ref. [73].

There have been recent commercial developments at some major FR producers where nanoclays have been combined with metal hydroxide fire retardants acting synergistically. As nanoclays are added at relatively small amounts, typically 1–5 wt%, and metal hydroxides traditionally at much higher loadings, these synergies may be exploited to allow lower overall loading levels of nanoclay and metal hydroxide.

Recent research on alternative nanomaterials includes double-layered hydroxides, and polyhedral oligomeric silsesquioxanes (see Chapters 20 and 23). The potential of nanoclays as intumescent fire retardants to protect materials such as wood that are easily combustible in contact with air is also exploited. It is expected that the pace of this type of research will increase in the coming years [74–77].

17.8 Tools and Testing

While a complete survey of the testing techniques for flame retardants is beyond the scope of this chapter, testing methods such as cone calorimetry, the requirements of the UL 94 testing protocols, and radiant heat panels deserve mention here.

The cone calorimeter (Figure 17.8) is perhaps the most valuable research instrument in the field of fire testing. It functions by burning sample materials cut to specific geometry and measuring the evolved heat using the technique of oxygen depletion calorimetry. The basis of this technique is that the heat released during combustion is directly proportional to the quantity of oxygen used in the process. This name was derived from the shape of the truncated conical heater that is used to irradiate the test specimen with fluxes up to 100 kW/m^2. Heat release is a key measurement in the progress of a fire [78–80]. The FTT Cone Calorimeter has been produced to meet all existing standards (including ISO 5660, ASTM E 1354, ASTM E 1474, ASTM E 1740, ASTM F 1550, ASTM D 6113, NFPA 264, CAN ULC 135, and BS 476 Part 15).

The UL 94 standard provides a preliminary indication of a polymer's suitability for use as part of a device or appliance with respect to its flammability. It is not intended to reflect the hazards of a material under actual fire conditions. UL 94 contains the following tests: 94HB, 94V, 94VTM, 94-5V, 94HBF, 94HF, and Radiant Panel (ASTM D 162). The 94HB test describes the horizontal burn method. Methods 94V and 94VTM are used for vertical burn, a more stringent test than 94HB. The 94-5V

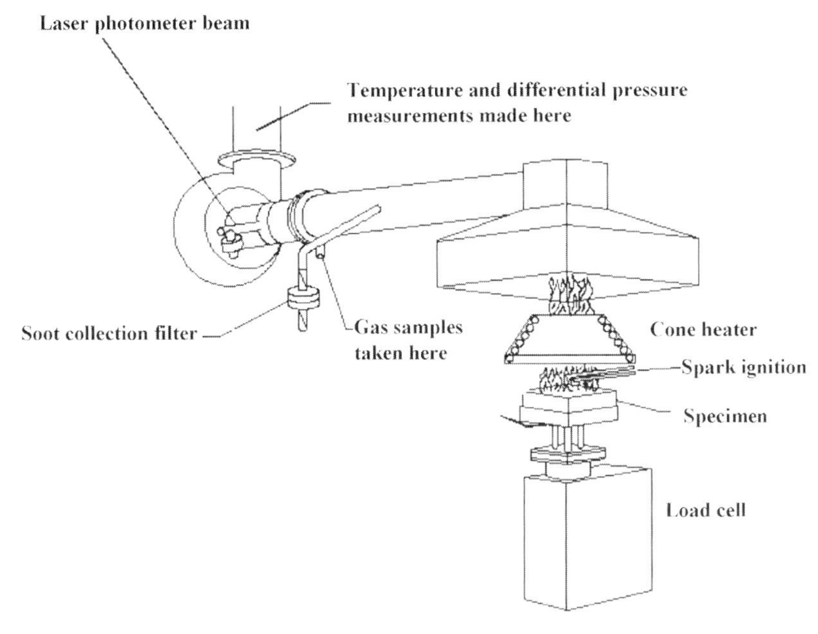

Figure 17.8 Cone calorimeter schematic.

test is for enclosures, that is, for products that are not easily moved or are attached to a conduit system. The 94HBF and HF are used for nonstructural foam materials, that is, acoustical foam. Radiant panel test is an ASTM (E 162) test to determine the flame spread of a material that may be exposed to fire.

Flame spread, critical heat flux, and smoke production rate of floor coverings and floor coatings are measured using a radiant heat source according to DIN 4102-14 and EN ISO 9239 for classification into DIN 4102 and Euroclasses.

17.9 Toxicity

In considering the toxicity of flame retardants, one has to consider both the toxicity of materials liberated during combustion from originally nontoxic materials and the toxicity of the FR materials. The most toxic materials typically liberated from combustion polymers are carbon monoxide, hydrogen cyanide (from nitrogen-containing polymers) among others. When considering the production of toxic chemicals that can arise from the use of flame retardants such as halogenated flame retardants, the negative effects have to be considered in many cases as offsetting, by orders of magnitude, the potential toxicity from gas evolution that would happen in a nonflame retarded polymer. This is not to say that the toxicity effects, for example, of halogenated fire retardants should be ignored and efforts to minimize their toxicity should not be undertaken, but the danger has to be seen in the context of their use.

17.9.1 Metal Hydroxides

Aluminum trihydrate is not considered a toxic material and is classified as GRAS (generally recognized as safe) in paper and is specifically listed for use in polymers. It is not approved for use in direct food contact but in indirect food additives such as polymers, cellophane, and rubber articles intended for repeated use, paper and paperboard components, and also in defoaming agents used in the manufacture of paper and paperboard. The human therapeutic category is as an antacid and an antihyperphosphatemic [81]. From a hygiene perspective, ATH is a nuisance dust and is mildly irritating.

Magnesium hydroxide meets the specifications of the Food Chemicals Codex [82]. It is used in food with no limitation other than current good manufacturing practice. The affirmation of this ingredient is as GRAS as a direct human food ingredient. It is permitted to be used for indirect use in polymers. It is a nontoxic material and the human therapeutic category is as an antacid and a cathartic [83]. The material can be irritating owing to its potential to develop alkalinity on contact with water.

17.9.2 Antimony Trioxide

Antimony trioxide is a material that is suspected of being a human carcinogen but has not been proven to be so. Epidemiological studies suggest that occupational exposure to antimony trioxide could be associated with an excess of lung cancer [84]. Concerns about human carcinogenicity have arisen because the material has been shown to be carcinogenic in animal studies [85, 86]. Antimony trioxide is carcinogenic in female, but not male, rats after inhalation exposure, producing lung tumors. The studies on rats tend to indicate that there is an increased risk of cancer developing from long-term chronic exposure than from acute or occasional exposure [87–91]. Antimony trioxide has been reported to induce DNA damage in bacteria. In summary, antimony trioxide is often treated as though it were carcinogenic in humans, a position that is supported in practice by many toxicologists. Regardless of the defined regulatory status, appropriate environmental and engineering controls are recommended in handling this material. Antimony trioxide most likely exerts its effect as a contact irritant, since patch tests for allergy usually produce no reaction. Further information on its effects on humans can be found in Refs [92, 93].

17.9.3 Brominated Fire Retardants

At present, there are studies underway to determine the toxicological characteristics and effects on hygiene of a number of halogen-containing FRs. Work has been published in Sweden suggesting the possibility of environmental accumulation resulting in the presence of these materials in the food chain from fish to humans (breast milk) [94]. The use of some brominated FRs such as polybrominated biphenyls and brominated diphenylether either has been phased out or has been limited in Europe. There is also increasing focus on higher polybrominated diphenylethers that, although not believed to bioaccumulate or be toxic, are suspected to degrade to the more problematic lower brominated diphenyl ethers. Without prejudice to the outcome of these or other studies, it is fair to say that concern over the toxicity of halogenated FRs has led many consumers to consider alternative FRs that do not contain halogens. Public concern is driven in part by anxiety over the health effects of dioxins and dioxin-like materials.

Regarding brominated flame retardants, Cullis [95] stated that unless suitable metal oxides or metal carbonates are also present, virtually all of the available bromine is eventually converted to gaseous hydrogen bromide. This is a corrosive and powerful sensory irritant. In a fire situation, however, it is always carbon monoxide (CO) or hydrogen cyanide (HCN), rather than an irritant, that causes rapid incapacitation. Owing to its high reactivity, hydrogen bromide is unlikely to reach dangerously high concentration levels [95].

The toxicity profile of brominated flame retardants, in contrast to other flame retardants, is relatively well understood. Studies have been carried out by the World Health Organization (WHO), the Organization for Economic Development and Cooperation (OECD), and the European Union (EU). The World Health Organization concluded in the early 1990s [96] that any risk presented by the major brominated flame retardants such as decabromodiphenyl oxide and tetrabromobisphenol A was minimal and manageable. Toxicology testing is ongoing on several brominated flame retardants from the diphenyl ether family and hexabromocyclododecane in accordance with industry's commitments under the OECD Voluntary Industry Commitment and EU risk assessment programs [97, 98].

17.9.4 Chlorinated Fire Retardants

A major concern with respect to the use of chlorinated FRs is their potential to form known hazardous materials, particularly dioxins or dioxin-like materials. As in the case of the brominated analogues, the most likely product of combustion of chlorinated FRs is HCl. However, the potential for formation of other potentially hazardous materials is real and has to be considered carefully [99].

The number of mixed halogen congeners based on dibenzodioxins or dibenzofurans is in excess of 4000. Interestingly, at present only a relatively small number of these substances are regulated under US or European laws. Among all these substances, only the 2,3,7,8-substituted polychlorinated, polybrominated, and mixed halogenated dibenzodioxins and dibenzofurans are of concern in terms of toxicity because dioxin toxicity is a receptor-mediated event and only these congeners have the proper conformation to interact with the biological receptors. They also accumulate in the fat of animals and humans [100, 101].

Some countries (e.g., Germany) have developed rules for the maximum content of some 2,3,7,8-substituted polychlorinated dibenzo-*para*-dioxins and dibenzofurans in products. The availability of relevant data on flame retardants in the open literature is limited, especially for some existing chemicals produced before regulations for commercialization were strengthened in several countries. The International Program for Chemical Safety (IPCS) has issued evaluations for some flame retardants and is preparing evaluations for others.

17.9.5 Boron-Containing Fire Retardants

There is little information on the health effects of long-term exposure to boron. Most of the studies are on short-term exposures. The Department of Health and Human Services, the International Agency for Research on Cancer, and the US Environmental Protection Agency (EPA) have not classified boron with regard to its human carcinogenicity. One animal study found no evidence of cancer after lifetime exposure to boric acid in food. No human studies are available [102].

Zinc borate is also used commonly as a fire retardant. The toxicity of zinc is low [103] and data indicate that relatively large amounts of zinc may pass for years through the kidneys and gastrointestinal tract in humans without causing any detectable clinical damage [104]. Some zinc salts are irritants and corrosive on skin contact and when ingested they can act as emetics [105].

17.9.6 Phosphorus-Containing Fire Retardants

Organic phosphate esters are commonly used as fire retardants in many applications including housings for personal computers and video display monitors. While there have been no definitive findings to date that would directly negate the use of these materials, the European Committee for Standardization has established a working group to prepare a standard addressing the risk associated with the presence of organic chemical compounds in toys [106]. Their priority list of chemicals include more than 50 flame retardants – most of them chlorine, bromine, or phosphorus based. The Electronics Industry Alliance (EIA) is currently working on a "Material Declaration Guide" [107]. All listed "materials of interest" must be disclosed if they are incorporated into parts in amounts greater than 0.1%. FRs are listed in this material category, including organophosphorus FRs [94].

There is an evolving concern about the use of some particular organophosphates such as tricresyl phosphate, tri-*n*-butyl phosphate, and triphenyl phosphate [94, 108–110]. Animal studies have shown that these phosphate esters can cause allergies, learning problems, and deterioration in sperm production and can influence the white and red blood cell counts of humans [109]. In a recent publication [110], levels of flame retardants in indoor air at electronics scrap recycling plant and other work environments were considered; 15 brominated FRs and 9 organophosphorus compounds were analyzed. Although the measured concentrations found in all these studies were very low and orders of magnitudes below any occupational limit values (and so the levels of exposure provided very significant margins of safety), health-related concerns were still raised by researchers.

There are few demonstrated toxicological concerns associated with the handling and use of APP. Thus, APP does not require hazard warning labels and has been shown to cause acute oral toxicity in rats only at a level of 5000 mg/kg. APP is capable of causing mild skin irritation. As in the case of most phosphorus-containing materials, APP should be prevented form entering surface waters as it may contribute to eutrophication of stagnant water.

17.9.7 Molybdenum-Containing Fire Retardants

Molybdenum is a transition metal in Group VI of the Periodic Table between chromium and tungsten. Although molybdenum is sometimes described as a "heavy metal," its properties are very different from those of the typical heavy metals, mercury, thallium, and lead. Its position in the periodic table belies its toxicity. It is much less toxic than these and other heavy metals. Molybdenum's low toxicity makes it an attractive substitute for toxic metals in a number of applications, for example, in place of chromium in corrosion inhibitors and antimony in smoke suppressants [111].

17.9.8 Nanosized FRs

With the development of nanomaterials concerns exist about their toxicity and health impacts. Because of the size of nanoparticulates, concerns arise about their ability to penetrate tissue and cause harmful effects ranging from mild irritation to more serious tumor developments. In the case of nanoclays, one has to consider that nanoclays only become nanosized after incorporation and exfoliation in a polymer matrix. In a practical sense, the presence of crystalline silica, a material that may be found in clays, is probably more of a real hazard. Crystalline silica is regulated to extremely low levels in commercial clay materials [112] (see also Chapter 19).

The situation with carbon nanotubes is somewhat less clear. Many studies have explored the toxicity of carbon nanotubes, some concluding that the nanotubes are acutely toxic and some not (see also Chapter 10). Reports suggest that soot-containing nanotubes did not present any significant health hazards [113, 114]. Cellular uptake of carbon nanotubes and the potential for interaction with cellular proteins, organelles, and DNA has been discussed in Refs [115–117]. In 2007, a 3-year study on the toxicity of carbon nanotubes was announced by CNRS in France. The project is to address three main subjects: the polluting effects of nanotubes in the environment; toxicity in humans, and how to produce nanotubes using cleaner methods [118]. While no studies to date have shown a proven linkage between carbon nanomaterials

Table 17.12 Major global producers of fire retardants.

Aluminum trihydrate	ALCAN
	Alcoa
	Martinswerk
	J.M. Huber Corp.
	Custom Grinders
	R.J. Marshall
	Shandong Corp.
	Franklin Minerals
	Aluchem
	Nabaltech GmbH
	Albemarle Inc.
Magnesium hydroxide	Martin Marietta
	Kisuma
	Incemin
	Dead Sea Bromine Company
	Peñoles
	Magnesium Products Inc.
	Cimbar
	Albemarle
	J.M. Huber Corp.
	Martinswerk
Antimony compounds	Great Lakes Chemical Corp.
	Albemarle Corp.
	Campine
	Sica
	Jiefu Corporation
	GFS Chemicals
	ACI Alloys
	Production Minerals Inc.
	Chemtura
	Nyacol
	Chemico Chemicals
Halogenated FRs	Great Lakes Chemical Corp.
	Albemarle Corp.
	Clariant Inc.
	Akzo Nobel
	Eurobrom
	Dover Corp.
	Ineos Chlor
	Caffaro
	Dead Sea Bromine Company
	Occidental Chemical Corp.
	Chemtura
	Campine
	Ferro
Melamines	DSM Melamine
	Buddenheim Iberica
	Akzo Nobel
	Argolinz
	Chemie Linz

Table 17.12 (*Continued*)

Phosphorus-containing FRs	Clariant
	Buddenheim Iberica
	Great Lakes Chemical Corp.
	Songwon
	Albright and Wilson Americas
	Hoechst Celanese
	Akzo Nobel
Boron compounds	US Borax Inc.
	Etibank
	Anzon
	NACC
	Joseph Storey
	William Blythe
	Rio Tinto
	Borax Europe Ltd
	Buddenheim Iberica
Molybdenum compounds	Sherwin Williams
	Climax Molybdenum
	Shanghai Nanjing Chemical Products Co. Ltd
	Guangzhou Wade Chemical Industry

and disease in humans, relatively little work has been done with regard to oral and dermal toxicity of carbon nanotubes; thus, it makes sense to be prudent and exercise caution when handling these materials.

17.10 Manufacturers of Fire Retardants

Table 17.12 shows a list of the major global producers of fire retardants. Statistics on the production of fire retardants may be obtained from the Flame Retardant Chemicals Association (FRCA) in the United States (www.fireretardants.org) and from the European Flame Retardants Association (EFRA) that may be contacted through www.cefic-efra.com.

17.11 Concluding Remarks

Fire retardants have proven benefits. Their use has saved an innumerable number of people from death and injury. Their use has enabled the proliferation of plastics to the present extent in society. Nevertheless, there are valid concerns about the safety of individual fire retardants that need to be addressed on an individual basis. There is also a need to collect data to fill the gaps that exist for some existing fire-retardant materials. Only through the acquisition of valid data in a thorough scientific manner can many of the current and emerging concerns about fire-retardant materials be conclusively addressed and the benefits of fire retardance are realized to the greatest possible extent [119–122].

Acknowledgments

The author thanks Schneller LLC for permission to publish this chapter. The following colleagues are acknowledged: Dr Edward Weil, Polytechnic, NYU, NYC, for information on smoke suppression and flame retardants in PVC and Dr John Mara of Songwon International for information on phosphorus-containing fire retardants.

References

1 Kasahiwagi, T. (1994) Proceedings of the 25th International Symposium on Combustion The Combustion Institute, Pittsburgh PA, pp. 1423–1437.

2 Beyler, C.M. and Hirschler, M.M. (1988) Thermal decomposition of polymers, in *SFPE Handbook of Fire Protection Engineering*, 1st edn (eds P.J. Di Nenno *et al.*), National Fire Protection Association, Quincy, MA, pp. 1–119.

3 American Society for Testing and Materials (ASTM) (2004) ASTM E176-04, *Standard Terminology of Fire Standards, Annual Book of ASTM Standards*, ASTM, West Conshohocken, PA.

4 Levin, B.C., Braun, E., and Paabo, M. (1995) Further development of the N-Gas model: fire and polymers II, in *Materials and Tests for Hazard Prevention*, ACS Symposium Series 599 (ed. G.L. Nelson), ACS, Washington, DC, pp. 293–311.

5 Braun, E. and Levin, B.C. (1986) Polyesters: a review of literature on products of combustion and toxicity. *Fire Mater.*, **10**, 107–123.

6 Braun, E. and Levin, B.C. (1987) Nylons: a review of literature on products of combustion and toxicity. *Fire Mater.*, **11**, 71–78.

7 American Society for Testing and Materials (ASTM) (2002) ASTM E1678-02, *Standard Test Method for Measuring Smoke Toxicity for use in Fire Hazard Analysis, Annual Book of ASTM Standards*, ASTM, West Conshohocken, PA

8 American Society for Testing and Materials (ASTM) (2001) ASTM E800-01, *Standard Guide for Measurement of Gases Present or Generated During Fires, Annual Book of ASTM Standards*, ASTM, West Conshohocken, PA.

9 Levin, B.C. *et al.* (1989) Synergistic effects of nitrogen dioxide and carbon dioxide following acute inhalation exposure in rats. NISTIR 89-4105, NIST, Gaithersburg, MD.

10 Cotton, F.A. and Wilkinson, G.L. (1972) *Advanced Inorganic Chemistry*, 3rd edn, Wiley Interscience, p. 358.

11 Levin, B.C. and Kuligowski, E.D. (2006) Toxicology of fire and smoke, Chapter 10, in *Inhalation Toxicology* (eds H. Salem and S.A. Katz), CRC Press, Taylor and Francis, Boca Raton, FL, pp. 205–228.

12 Committee on Fire Toxicology, Dubois, A.B. Chair, National Research Council – National Academy of Sciences (1986) *Fire and Smoke, Understanding the Hazards*, National Research Council – National Academy of Sciences, Washington, DC.

13 Karter, M.J., Jr. (2003) 2002 U.S. fire loss. *NFPAJ*, **97**, 5, 59–63.

14 Hartzell, G.E. (1988) Fire and life threat. *Fire Mater.*, **13**, 53–60.

15 Gottuk, D.T. *et al.* (1995) The role of temperature on carbon monoxide production in compartment fires. *Fire Safety J.*, **24**, 315–331.

16 Pitts, W.M. (1995) Gobal Equivalence Ratio Concept and the Formation Mechanism of Carbon Monoxide in Exclosure Fires. *Prog. Energy Combust. Sci.*, **21**, 197–237.
17 Pitts, W.M. (1997) Proceedings of the 5th International Symposium, Int.' Assoc. for Fire Safety Science Interscience Communications Ltd., Melbourne, Australia, pp. 535–546.
18 Hartzell, G.E. *et al.* (1990) Toxicity of smoke containing hydrogen chloride, in *Fire and Polymers, Hazard Identification and Prevention*, ACS Symposium Series 425 (ed. G.L. Nelson), American Chemical Society, Washington, DC, pp. 12–20.
19 Bosse, G.M. and Matyunas, N.J. (1999) Delayed toxidromes. *J. Emerg. Med.*, **17** (4), 679–690.
20 Bertolini, J.C. (1992) Hydrofluoric acid: a review of toxicity. *J. Emerg. Med.*, **10**, 163–168.
21 Kirkpatrick, J.J.R. *et al.* (1995) Hydrofluoric acid burns: a review. *Burns*, **21** (7), 483–493.
22 Sheridan, R.L. *et al.* (1995) Emergency management of major hydrofluoric acid exposures. *Burns*, **21** (1), 62–64.
23 Minnesota Poison Control System, http://www.mnpoison.org.
24 Bloom, J.(1997) Emergency Medicine: Concepts and Clinical Practice. *Comp. Toxicol.*, **4**, 62–66.
25 Purser, D.A. (1995) Toxicity effects of combustion products, in *SFPE Handbook of Fire Protection Engineering*, 2nd edn (eds P.J. Di Nenno*et al.*), National Fire Protection Association, Quincy, MA, pp. 85–146.
26 Tewarson, A.Improved Fire and Smoke Resistant Materials in Ref. [2].
27 Robertson, A.F. (1975) Estimating Smoke Production During Building Fires. *Fire Technol.*, **11**, 80.
28 Fenimore, C.P. and Jones, G.W. (1966) Flammability of Polymers. *Combust. Flame*,**10**, 295.
29 Cote, A.E. (1997) *Fire Protection Handbook, Table A-5*, 18th edn, National Fire Protection Association, Quincy MA.
30 Ashton, H.C. *et al.* (2003) Proceedings of the SPE RETEC Polyolefins 2003, Houston, TX, February 24–26, 2003 pp. 395–407.
31 Ashton, H.C. (2001) Proceedings of the Functional Fillers for Plastics 2001, Intertech Corp, San Antonio, TX, September 11–13, 2001
32 Green, D.W. and Dallavia, A.J., Jr. (1991) Alumina trihydrate in polyester resin-viscosity reduction via particle size blends and surface modification. Session 7-E, p1, Proc. 46th Ann. Conf. SPI-Composites Institute, The Society of the Plastics Industry Inc., February 18–21, 1991.
33 Green, D. and Dallavia, A., Jr. (1989) Proc. of 46th SPE ANTEC, vol. 34, p. 619.
34 Naitove, H.M. (2008) Additives and Colorants score dramatic Advances, Plastics Technology (online); http://www.ptonline.com/articles/200803fa2.html.
35 Robertson, A.F. (1975) *Surface Flammabiliity Measurements by the Radiant Panel Method*, ASTM Spec. Tech. Publ. No. 344, ASTM International, West Conshohocken, PA, p. 33.
36 Gerards, T. (2002) Proc. of the FILPLAS 1992, Manchester, UK, May 19–20, 2002.
37 Tain, C.M., Qu, H.Q., Wu, W.H., Guo, H.Z., and Xu, J.Z. (2005) Metal Chelates as synergistic Flame Retardants for Flexible PVC. *J. Vinyl. Addit. Technol.*, **11**, 70.
38 Cross, M.S., Cusack, P.A., and Hornsby, P.R. (2003) Fire Retardancy of Polymers: the use of Zinc Hydroxystannate in Ethylene-Vinyl acetate Copolymer. *Polym. Degrad. Stab.*, **79**, 309.
39 Cusack, P.A., Heer, M.S., and Monk, A.W. (1997) Zinc Hydroxystannate as an Alternative Synergist to Antimony Trioxide in Polyester Resins. *Polym. Degrad. Stab.*, **58**, 229.
40 Daniels, C., Herbert, M.J., and Rai, M. (2001) Halogenated polymeric formulation containing divalent (hydroxy) stannate and antimony compounds as synergistic flame retardants, US Patent 6,245,846.
41 Thomas, N.L. (2003) Zinc compounds as flame retardants and smoke

suppressants for rigid PVC. *Plastics, Rubber Compos.*, **32** (8–9), 413–419.

42 Bar Yaakov, Y. *et al.* (2000), Proc. Flame Retardants Conference, London, UK, February 8–9, 2000, pp. 87–97.

43 Lomakin, S.M. and Zaikov, G.E. (2003) *Modern Polymer Flame Retardancy*, VSP Publ., Utrecht, The Netherlands, Boston, MA.

44 Cotton, F.A. and Wilkinson, G. (1972) *Advanced Inorganic Chemistry*, 3rd edn, John Wiley & Sons, Ltd, London, UK, pp. 396–397.

45 Braun, U. and Scharter, B. (2003) Fire retardance mechanisms of red phosphorus in thermoplastics. Proc. of the Additives 2003 Conference, San Francisco, CA, USA, April 6–9, 2003.

46 Levchik, G.F., Vorobyova, S.A., Gobarenko, V.V., Levchik, S.V., and Weil, E.D. (2000) Some mechanistic aspects of the fire retardant action of red phosphorus in aliphatic nylons. *J. Fire Sci.*, **18** (3), 172–182.

47 Levchik, S.V. and Weil, E.D. (2006) A review of recent progress in phosphorus-based flame retardants. *J. Fire Sci.*, **24** (5), 345–436.

48 Murashko, E.A., Levchik, G.F., Levchik, S.V., Bright, D.A., and Dashevsky, S. (1999) Fire retardant action of resorcinol bis(diphenyl phosphate) in PC ABS Blend. II. Reactions in the condensed phase. *J Appl. Polym. Sci.*, **71** (11), 1863–1872.

49 Bright, D.A., Dashevsky, S., Moy, P.Y., and Williams, B. (1997) Resorcinol bis(diphenylphosphate). A non-halogen flame retardant additive. Proc. 55th Ann. Technical Conference Society of Plastics Engineers, SPE, vol. 43, p. 2936.

50 http://www.plastemart.com/upload/Literature/Phosphorus-based-flame-retardant-characteristics.

51 Horold, S. Phosphinates, the flame retardants for polymers in electronics, http://www.flameretardants-online.com/news/downloads/over_english/phosphinate.pdf.

52 Cook, P.M. and Musselman, L.L. (2000) Proc. Flame retardants 2000, Conf. Interscience Communications, London, UK, pp. 69–76.

53 Stames, W.H., Wescott, L.D., Jr., Reents, W.D., Jr., Cais, R.E., Villacorta, G.M., Plitz, I.M., and Anthony, L.J. (1984) *Polymer Additives* (ed. J.L. Kresta), Plenum Press, New York, pp. 237–248.

54 Di Pietro, J. and Stepniczka, H.J. (1971) A Study of Smoke Sensity and Oxygen Index. *J. Fire Flammability*, **2**, 36.

55 Pitts, J.J. (1972) Antimony Halogen Synergistic Reactions in Fire Retardance. *J. Fire Flammability*, **3**, 235.

56 Moore, F.W. and Tsingidos, G.A. (1981) Advances in the Use of Molybdenum Compounds as Smoke Suppressants for PVC. *J. Vinyl Technol.*, **3** (2), 139.

57 Lattimer, R.P. and Kroenke, W.J. (1981) The Functional Role of Molybdenum Trioxide as a Smoke Retarder Additive in Rigid Poly(vinyl chloride). *J. Appl. Polym. Sci.*, **26**, 1191.

58 Levchik, S.V. and Weil, E.D. (2005) Overview of the recent Literature on flame retardancy and smoke suppression in PVC. *Polym. Adv. Technol.*, **16**, 707–716.

59 Lattimer, R.P. and Kroenke, W.J. (1981) Metal smoke retarders for poly(vinyl chloride). *J. Appl. Polym. Sci.*, **26**, 1167.

60 Stames, W.H., Wescott, L.D., Jr., Reents, W.D., Jr., Cais, R.E., Villacorts, G.M., Plitz, I.M., and Anthony, L.J. (1984) *Polymer Additives* (ed. J.L. Kresta), Plenum Press, New York, pp. 237–248.

61 Linsky, L.A., Andries, J.C., Ouellette, D., Buono, J.A., and Tao, T. (1995) U.S. Patent 5,886,072.

62 Brown, S. (1999) Trends and developments in flame retardants for cables. IEEE Colloquium on Developments in Fire Performance Cables for energy, Ref. No. 1999/074, London, UK, May 27, 1999, pp. 3/1–3/17.

63 http://www.advancedepoxycoatings.com/library.html.

64 White, L. (2006) Nano-clay for FR foams: nano materials may be alternative to harmful flame retardants (newslines), Urethanes technology, February 2006.

65 Razdan, S., Petra, P.K., and Warner, S. (2003) Morphology and thermal stability

of PP nanocomposites. *Polym. Mater. Sci. Eng.*, **69**, 722.

66 Razdan, S., Petra, P.K., Warner, S., and Kim, Y.K. (2003) Effect of nanofillers on thermal stability of polypropylene. Proc. Tenth Ann. International Conference on Composite/NanoEngineering (ICCE-10), July 20–26, 2003.

67 Inan, G., Patra, P.K., Kim, Y.K., and Warner, S.B. (2003) Flame retardancy of laponite and montmorillonite-based nylon-6 nanocomposites. *Mater. Res. Soc. Symp. Proc.*, **L8.**, 46–49.

68 Kashiwagi, T. (2007) Flame retardant mechanism of the nanotubes-based nanocomposites, Final Report, NIST GCR 07-91, Gaithersburg, MD.

69 Cho, J.W., Logsdon, J., Omachinski, S., Qian, G., Lan, T., Womer, T.W., and Smith, W.S. (2002) Nanocomposites: a single screw mixing study of nanoclay-filled polypropylene. Proc. 60th Ann. Technical Conference Society of Plastics Engineers, SPE, vol. 48, Session T28.

70 Adanur, S. and Ascioglu, B. (2005) Challenges and opportunities in nano-fiber manufacturing and applications. 2nd International Technical Textiles Congress, Istanbul, Turkey, July 12–15, 2005.

71 Betts, K.S. (2008) New thinking on flame retardants. *Environ. Health Perspect.*, **116** (5), A210–A213, http://foodcinsumer.org/7777/8888/M_edicare_54/050609212008.

72 See Ref. [68].

73 Kashiwagi, T., Fagan, J., Douglas, J.F., Yamamoto, K., Heckert, A.N., Leigh, S.D., Obrzut, J., Du, F., Lin-Gibson, S., Mu, M., Winey, K.I., and Haggenmueller, R. (2007) Relationship between dispersion metric and properties of PMMA/SWNTnanocomposites. *Polymer*, **48**, 4855–4866.

74 Patra, P.K., Warner, S.B., Kim, Y.K., Fan, Q., Calvert, D., and Adanur, S. (2005) NTC Project No: M02-MD08 (formerly M02-D08), Nano Engineered Fire Resistant Composite Fibers, National Textile Center (NTC) Annual Report, November 2005, Spring House, PA.

75 Morgan, A.B. and Wilkie, C.A. (eds) (2007) *Flame Retardant Polymer Nanocomposites*, John Wiley & Sons, Inc., Hoboken, NJ.

76 Nelson, G.L. and Wilkie, C.A. (eds) (2001) *Fire and Polymers: Material Solutions for Hazard Prevention, ACS Symposium Series #797*, American Chemical Society, Washington, DC.

77 Gou, G., Park, C.B., Lee, Y.H., Kim, Y.S., and Sain, M. (2007) Flame retarding effects of nanoclay on wood-fiber composites, in *Flame Retardant Polymer Nanocomposites* (eds A.B. Morgan and C.A. Wilkie), John Wiley & Sons, Inc., Hoboken, NJ.

78 Babrauskas, V. *The Cone* Calorimeter, *Section 3/Chapter 3*, pp. 3-37–3-52 in Ref. [25].

79 ASTM (2001) Test Method E 1354, *Standard Test Method for Heat and Visible Smoke Release Rates for Materials and Products Using an Oxygen Consumption Calorimeter, Annual Book of ASTM Standards*, vol. 04.07, ASTM International, West Conshohocken, PA.

80 Kaplan, H.L. *et al.* (1983) *Combustion Toxicology, Principles and Test Methods*, Technomic Publishing Company Inc., Lancaster, PA, pp. 7–49.

81 Mak, R.H.K. *et al.* (1985) *Brit. Med. J.*, **291**, 623 referenced in (1989) *The Merck Index, An Encyclopedia of Chemicals, Drugs and Biologicals*, 11th edn, Merck and Co., Rahway, NJ, p. 57.

82 (1996) *Food Chemicals Codex*, 4th edn, National Academy Press.

83 See Ref. [23], p. 892.

84 Web site: http://toxnet.nlm.nih.gov/cgi-bin/sis/search/f?./temp/~AAAXEa4wa.

85 Groth, D.H. *et al.* (1986) Carcinogenic effects of antimony trioxide and antimony ore concentrate in rats. *J. Toxicol. Environ. Health*, **18**, 607–626.

86 International Agency for Research on Cancer, IARC (1989) *Monographs on the Evaluation of the Carcinogenic Risk of Chemicals to Man, 1972 – Present* vol. 47, World Health Organization, Geneva, p. 296.

87 Patty, F. (ed.) (1963) *Industrial Hygiene and Toxicology: Volume II: Toxicology*, 2nd edn, Interscience Publishers, New York, p. 995.

88 Clayton, G.D. and Clayton, F.E. (eds) (1981–1982) *Patty's Industrial Hygiene and Toxicology: Volume 2A, 2B, 2C: Toxicology*, 3rd edn, John Wiley & Sons, Inc., New York, p. 1510.

89 ACGIH (1980) *Documentation of the Threshold Limit Values and Biological Exposure Indices*, 4th edn, American Conference of Governmental Industrial Hygienists, Inc., Cincinnati, OH, p. 20.

90 See Ref. [28], pp. 47, 302.

91 ACGIH (2002) Threshold limit values for chemical substances and physical agents and biological exposure indices for 2002, A2; suspected human carcinogen/antimony trioxide production/TLVs & BEIs: 15. American Conference of Governmental Industrial Hygienists, Cincinnati, OH.

92 Rom, W.N. (ed.) (1992) *Environmental and Occupational Medicine*, 2nd edn, Little, Brown and Company, Boston, MA, p. 815.

93 Prager, J.C. (1996) *Environmental Contaminant Reference Databook*, vol. 2, Van Nostrand Reinhold, New York, p. 90.

94 Carlson, H. *et al.* (2000) Vido display units: on emmition [sic] source of the contact allergenic flame retardant triphenyl phosphate in the indoor environment. *Environ. Sci. Technol.*, **34**, 3885–3889.

95 Cullis, C.F. (1987) Bromine compounds as flame retardants. Proc. International Conf. on Fire Safety, Product Safety Corporation, Sunnyvale, CA, 12, pp. 307–323.

96 World Health Organization (1994) Brominated Diphenyl Ethers, IPCS Environmental Health Criteria 162, Geneva, Switzerland.

97 EEC (1993) Council regulation EEC 793/93 of 23 March 1993 on the evaluation and control of the risks of existing substances. *EEC Official J.*, **L84**, 1–75.

98 Hardy, M.L. (1997) Regulatory status and environmental properties of brominated flame retardants undergoing risk assessment in the EU: DBDPO, OBDPO, PEBDPO and HBCD. 6th European Meeting on Fire Retardancy of Polymeric Materials, Lille, France, September 24–26, 1997.

99 Pomerantz, A. *et al.* (1978) Chemistry of PCBs and PBBs–Final Report of the Subcommittee on Health Effects of PCBs and PBBs. *Environ. Health Perspect.*, **24**, 133–146.

100 Kurz, R. (1998) An introduction to brominated flame retardants. Ricoh presentation, Paper presented at the Bromine Science and Environmental Forum Japan Seminar, Tokyo, November 6, 1998.

101 Brominated Flame Retardants/CEM Working Group (1989) Polybrominated dibenzodioxins and dibenzofurans (PBDDs/PBDFs) from flame retardants containing bromine: assessment of risk and proposed measures. Report No. III/4299/89, German to the Conference of Environmental Ministers, Bonn, Germany, September 1989.

102 Agency for Toxic Substances, Disease Registry (ATSDR) (1992) *Toxicological Profile for Boron*, U.S. Department of Health and Human Services, Public Health Service, Atlanta, GA.

103 ACGIH (1986) *Documentation of the Threshold Limit Values and Biological Exposure Indices*, 5th edn, American Conference of Governmental Industrial Hygienists, Cincinnati, OH, p. 645.

104 Browning, E. (1969) *Toxicity of Industrial Metals*, 2nd edn, Appleton-Century-Crofts, New York, p. 353.

105 See Ref. [30], p. 2039.

106 European Commitee for Standardization (CEN) Report (2003) On the risk assessment of organic chemicals in toys. Final report of the work of CEN/TC 52/WG 9. Published by the European Commission, Health and Consumer Protection Directorate General, Brussels, C7/VR/csteeop/CEN/12-13112003 D(03).

107 Electronic Industry Alliance (EIA) (2000) Material Declaration Guide, p. 25.

108 Carlson, H. *et al.* (1997) Organophosphate ester Flame Retardants and Plasticizers in the Indoor Environment: Analytical Methodology and Occurance. *Environ. Sci. Technol.*, **31**, 2931–2936.

109 Svenska Dagbladet, "Nytt plastämne kann ge allergi (New plastic material causes allergies)," November 11, 1999.

110 European Flame Retardants Association (EFRA), Position Paper on Triphenyl Phosphate (TPP), 2000; http://efra.cefic.org.

111 http://.wwwmoly.imoa.info/Default.asp?Page=32.

112 Southern Clay Products, Material Safety Data Sheet, Cloisite® 20A, 2003.

113 Huczko, A. and Lange, H. (2001) Carbon Nanotubes: Experimental Evidence for a Null Risk of Skin Irritation and Allergy. *Fullerene Sci. Technol.*, **9**, 247–250.

114 Huczko, A. *et al.* (2001) Physio Logical Testing of Carbon Nanotubes: Are they Asbestos-like? *Fullerene Sci. Technol.*, **9**, 251–254.

115 Porter, A.E., Gass, M., Muller, K,. Skepper, J.N., Midgley, P.A., and Well, M. (2007) Direct Imaging of Single-walled Carbon Nanotubes in Cells, *Nat. Nanotechnol.* **2**, 713–717.

116 Kolosnjaj, J., Szwarc, H., and Moussa, F. (2007) Toxicity studies of carbon nanotubes. *Adv. Exp. Med. Biol.*, **620**, 181–204.

117 Nel, A., Xia, T., Madler, L., and Li, N. (2006) Toxic potential of materials at the nanolevel. *Science*, **311** (5761), 622–627.

118 http://nanotechweb.org/cws/article/tech/26825.

119 Glass, J. (1998) Asian Shipbreakers face US crackdown: Pentagon review to focus on PCB disposal risk, *Lloyd's List*, Informa, London, UK.

120 Hall, J.R. (1997) *The Total Cost of Fire in the United States Through 1994*, National Fire Protection Agency, Quincy, MA.

121 Morikawa, T., Okada, T., Kajiwara, M., Sato, Y., Tsuda, Y., (1995) Toxicity of Gases from Full-scale Room Fires involving Fire Retardant Content, *J. Fire Sci.*, **13**, 1, 23–42.

122 Kutz, M. (ed.) (2005) Handbook of Environmental Degradation of Materials, William Andrew Pub. ISBN: 0-8155-1500-6 (William Andrew Inc.).

18
Conductive and Magnetic Fillers

Theodore Davidson

18.1
Introduction

Polymers are often the favored materials for the manufacture of devices and components. They combine ready formability and molding characteristics with good insulating and dielectric properties. These attributes are sufficient to make plastics and elastomers functional for many purposes. But in some instances, it would be desirable to have polymers that could dissipate static charges or shield their contents from electromagnetic fields. In other applications, the control of static charge (as in xerography) or control of conductivity as a function of current density (self-limiting switches and heaters) is attractive. All of these functions have been achieved by the addition of conductive fillers to polymers. In this chapter we describe some of these applications and the roles played by functional fillers in controlling polymer conductivity. In a similar manner, we will discuss how to confer the properties of permanent magnetization on polymers through composite materials consisting of a magnetic solid dispersed in a polymer matrix or binder. These range from the familiar refrigerator magnets and magnetic recording tape to data disks that can store information at very high densities.

It is widely observed when adding carbon black (CB) to an insulating polymer that there is a threshold concentration, V_c, at which the composite changes from insulating to conductive. Conductivity (and resistivity) change by several orders of magnitude for a small increment in carbon black concentration. This threshold, which typically occurs from 3 to 15% CB, has been the subject of many investigations. Unfortunately, in these systems there are not only many independent variables such as the surface chemistry and dispersibility of various blacks but, to complicate these studies, there are also significant effects of polymer interfacial chemistry, various modes of compounding and segregation to phase boundaries. We attempt to elucidate several of these effects by examples from the literature. Our selections are not all-inclusive and do not attempt to address the very large patent literature in this field. Previously, the field has been surveyed by Sichel [1] and Rupprecht [2] and reviewed by Huang [3].

Functional Fillers for Plastics: Second, updated and enlarged edition. Edited by Marino Xanthos

ISBN: 978-3-527-32361-6

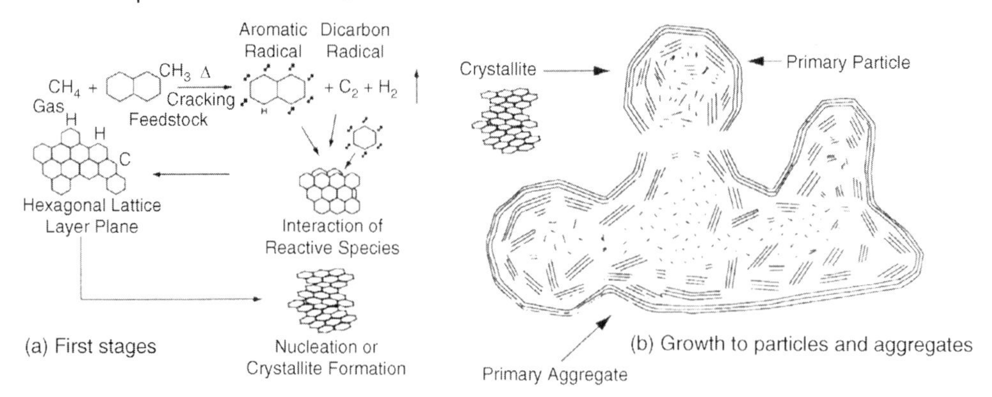

Figure 18.1 Schematic diagram of carbon black formation. (After Ref. [5], reprinted with permission from Cabot Corporation.)

18.2 Carbon Black

18.2.1 General

Carbon black (often referred to as "black" or "CB" in this chapter) consists of agglomerates of small assemblies of carbon particles called aggregates, which are formed in the processes by which solid carbon condenses from the vapor. An agglomerate consists of a number of aggregates held together physically as distinguished from the continuous pseudo-graphitic structure of the aggregates. Figure 18.1 shows a schematic of CB formation. Blacks are variously described as "high structure" or "low structure" that correlates with their spatial extent, the former having larger dimensions than the latter. Their structure is further characterized by the adsorption of liquid dibutyl phthalate (DBP or DBPA) with those blacks that take up more DBP being termed "high structure." They typically have a more reticulated form whereas the low structure blacks are more compact [4].

18.2.2 Varieties

Carbon blacks are made by the combustion of natural gas, or acetylene, or various hydrocarbon oil feedstocks under reducing conditions. In earlier times, "channel blacks" were made by thermal decomposition of natural gas in an open system with the black being collected on lengths of channel iron in the shape of long inverted vees of angle iron supported over a line of gas flames.

Increased efficiency and reduced loss to the surroundings are achieved in closed furnace systems yielding "furnace black" or "lampblack." Thermal decomposition of natural gas produces "thermal blacks" with large primary particles ($\leq$500 nm) and

Table 18.1 Representative grades of carbon black used in plastics.

Type	Surface area by nitrogen adsorption (m^2/g)	DBPA ($cm^3/100\,g$)	Volatiles (%)
High color	240	50	2.0
Medium color	200	117	1.5
Medium color	210	74	1.5
Regular color	140	114	1.5
Regular color	84	102	1.0
Low color	42	120	1.0
Aftertreated	138	55	5.0
Conductive	254	178	1.5
Conductive	67	260	0.3
Conductive	950	360	1.0

Adapted from Ref. [7].

surface areas of 6–8 m^2/g. Details of these various types are given by Kühner and Voll [6] and Medalia [7]. Currently manufactured grades cover a range from approximately 9–300 m^2/g in external surface area; 30–180 $cm^3/100\,g$ in DBPA absorption; and 0.5–15% "volatiles" [7]. There is a continuous exothermic process fed by acetylene that yields "acetylene blacks," which are more crystalline and have high structures. These are difficult to densify but are useful for their enhanced electrical conductivity and antistatic properties when compounded into rubber and plastics (Table 18.1).

When dry, CB forms aggregates (Figure 18.2). The proportions of these groupings have been measured for 19 blacks and may be compared to their "structure" as evidenced by DBPA uptake (Table 18.2) [8].

Ninety percent of CB production goes to the rubber industry. Out of the nonrubber consumption, 36% is for additives to plastics while 30% goes into printing inks [6]. The process technology for making each type of carbon black is a large subject in itself. Surveys can be found in [6] and [9]. The mechanism of formation, involving nucleation, aggregation, and agglomeration of the aggregate particles is dealt with by Bansal and Donnet [10].

18.2.3 Commercial Sources

There are numerous manufacturers of basic carbon black. A partial list would include Akzo Nobel, Cabot Corp., Columbian Chemicals, Chevron-Phillips, and Degussa Corp. Most manufacturers describe the various grades they produce on their web sites.

As with many fillers, colorants, and additives, CB is frequently compounded with a thermoplastic to form a concentrate or masterbatch that is then "let down" and dispersed in a chosen polymer to the final desired concentration. Masterbatches are available from a variety of compounders including Americhem, Inc., Ampacet Corp.,

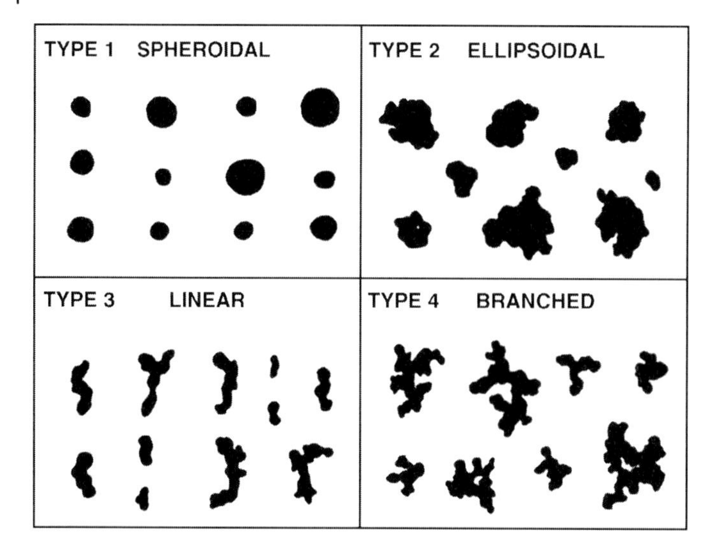

Figure 18.2 Rendering of shape categories for carbon black aggregates. Compare to data in Table 18.2. (After Ref. [8], reprinted with permission from Marcel Dekker, Inc.).

Table 18.2 Weight percent of aggregates in four shape categories for various CB grades in the dry state.

		Weight % of each shape category			
Carbon black	**DBPA (cm^3/100 g)**	**1**	**2**	**3**	**4**
CD-2005	174	0.1	4.8	17.8	77.3
N358	155	0.1	7.9	34.9	57.1
HV-3396	138	0.2	4.2	33.4	62.2
N121	134	0.4	8.8	28.7	62.1
N650	129	0.2	9.2	47.0	43.6
N234	124	0.3	9.0	32.5	58.3
N299	124	0.4	10.0	33.2	56.4
N351	120	0.1	9.2	46.9	43.8
N550	120	0.6	13.8	45.3	40.3
N339	118	0.2	9.5	36.6	53.7
N110	115	0.3	8.7	31.1	59.9
N220	115	0.6	11.9	34.0	53.5
N330	100	0.2	10.2	44.1	45.5
N660	91	0.4	15.4	52.5	31.7
N630	78	0.4	21.4	49.0	29.2
N774	77	1.3	20.8	46.3	31.6
N326	72	1.6	23.4	35.2	39.8
N762	67	2.5	22.4	47.7	27.4
N990	35	44.9	34.8	14.4	5.9

After Ref. [8], reprinted with permission from Marcel Dekker, Inc.
Note: N designates furnace blacks with "normal" curing in rubber compounds. The first digit after N relates to particle size or surface area. The second and third numbers are arbitrary grade designators.

Cabot Corp., Colloids Ltd. (UK), Hubron Manufacturing Div., Ltd. (UK), Polychem USA, A. Schulman, Singapore Polymer Corp. (PTE.) Ltd, and others.

18.2.4
Safety and Toxicity

Carbon black is reported to have occupational exposure limits (OEL) of 3.5 mg/m^3 according to the American Conference of Governmental Industrial Hygienists (ACGIH) regarding threshold limit value (TLV); National Institute for Occupational Safety and Health (NIOSH) regarding recommended exposure limits (REL); and the Occupational Safety and Health Administration (OSHA) regarding permissible exposure limits (PEL). A detailed discussion of CB safety, health, and environmental considerations can be found on the web site of the Cabot Corporation [11].

18.2.5
Surface Chemistry and Physics

In general, the quenching step of manufacture presents an opportunity for oxidation of the fine CB powder. Surface groups can be further enhanced intentionally by oxidation with acid or ozone. Surface groups formed in this way are primarily carboxylates and phenolics. An untreated black typically has $pH > 6$ and volatile content of 0.5–1.5%. Oxidized blacks have $pH < 6$ and volatile content of 3–10%.

Pantea *et al.* [12] examined 10 different CBs having a range of surface areas from 150 to 1635 m^2/g and DBP absorption as a measure of structure from 115 to 400 $cm^3/100\,g$. In compacted powder specimens of these highly conductive CBs, there was no correlation between the measured conductivity and the surface concentrations of oxygen or sulfur. The measured ranges were from "not detectable" (n.d.) to 0.8 at.% oxygen and n.d. to 0.3 at.% sulfur as determined by X-ray photoelectron spectroscopy (XPS). However, upon examining the C_{1s} spectra at high resolution – and particularly in regard to the C_{1s} peak at 284.5 eV associated with graphitic character – it was found that the full width at half maximum (FWHM) of this peak correlated well with the observed conductivities. Static secondary ion mass spectroscopy (SIMS) spectra and low-pressure nitrogen adsorption isotherms confirm the effect that the graphitic surface structure has in influencing "dry" CB conductivity [12].

18.3
Phenomena of Conductivity in Carbon Black-Filled Polymers

18.3.1
General

When the concentration of CB in a polymer is progressively increased, there is a critical volume fraction, V_c, at which the electrical conductivity increases by several

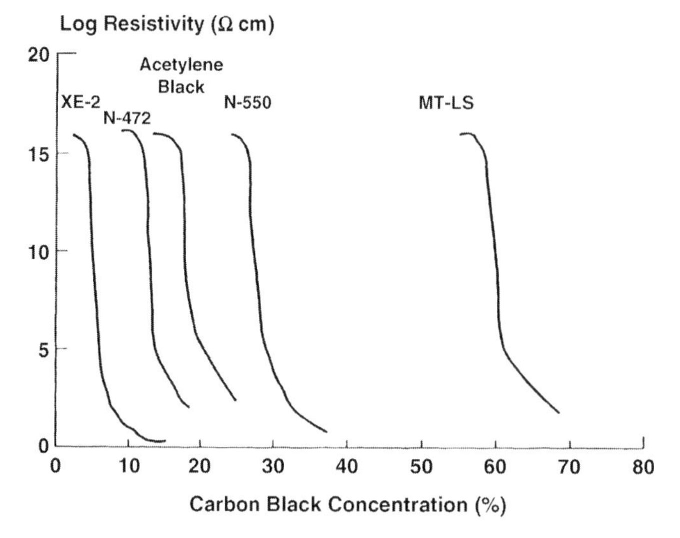

Figure 18.3 Log resistivity versus concentration curves for various carbon blacks in HDPE. (After Ref. [18], reprinted with permission from Marcel Dekker, Inc.)

orders of magnitude [13, 14]. This critical volume fraction differs for various polymers and different blacks (see Figure 18.3 and Table 18.3). As expected, it is also dependent upon mixing and on the resulting mesoscopic distribution of CB in the insulating matrix polymer. It has been assumed that the carbon particles need not make physical contact everywhere but within some small distance, say 2 nm,

Table 18.3 Threshold concentrations for XE-2, a highly conductive carbon black, in various polymers.

Polymer	Threshold concentration (wt%)
High-density polyethylene (MI = 20)	4.0
High-density polyethylene (MI = 2)	2.6
Cross-linked HDPE	1.8
Low-density polyethylene (LDPE)	1.6
Polypropylene	1.6
Poly(ethylene-*co*-14% vinyl acetate), EVA	1.2
Poly(ethylene-*co*-28% vinyl acetate), EVA	3.2
High impact polystyrene, HIPS	2.8
Transparent styrene-butadiene copolymer, SB	2.0
Polycarbonate	4.4
Polyphenylene oxide, PPO	1.6
Polyphenylene sulfide, PPS	1.2
Polyacetal, POM	1.0
Polyamide 66, PA-66	1.4
Oil-extended thermoplastic elastomer, TPE	2.4

After Ref. [18], reprinted with permission from Marcel Dekker, Inc.

electrons can tunnel through the insulating polymer and provide pathways for conduction. These observations have led to explanations of conductivity based on percolation theory that is formulated to account for the movement of charge in a disordered medium. Of course, there are some agglomerates of CB in closer contact and within aggregates a different "graphitic" mechanism of conduction may be expected [8, 12].

18.3.2 Percolation Theories

Systems involving granular metal particles and polymers have been modeled in terms of mixtures of random voids and conductive particles. There is also an inverted random void model [15]. For CB in polymers, percolation is the process by which charge is transported through a system of interpenetrating phases. In these systems the first phase consists of conductive particle agglomerates and the second is an insulating polymer. Since charge transport is at least in part by tunneling through thin insulating regions, a combined theory is called the tunneling-percolation model (TPM) [16].

As Balberg notes in a review: "The electrical data were explained for many years within the framework of interparticle tunneling conduction and/or the framework of classical percolation theory. However, these two basic ingredients for the understanding of the system are not compatible with each other conceptually, and their simple combination does not provide an explanation for the diversity of experimental results [17]." He proposes a model to explain the apparent dependence of percolation threshold critical resistivity exponent on "structure" of various carbon black composites. This model is testable against predictions of electrical noise spectra for various formulations of CB in polymers and gives a satisfactory fit [16].

18.3.3 Effects of Carbon Black Type

The conductivity of a polymer–CB composite depends upon such factors as

- carbon black loading,
- physical and chemical properties of the chosen carbon black,
- chemistry of the polymer and its morphology in the solid state,
- mixing and finishing processes used to create the composite.

For various CBs dispersed in the same matrix, for example, high-density polyethylene (HDPE), the critical concentration, V_c, ranges from 8 to 62 wt% (see Figure 18.3). Printex XE-2 shows the lowest critical volume fraction while MT-LS shows the highest. When the XE-2 blend has reached its asymptotic minimum resistivity (at 14% CB), blends of acetylene black, N550 and MT-LS at this loading in HDPE are still insulating. Comparable results for various CBs in polypropylene (PP) have also been reported [19].

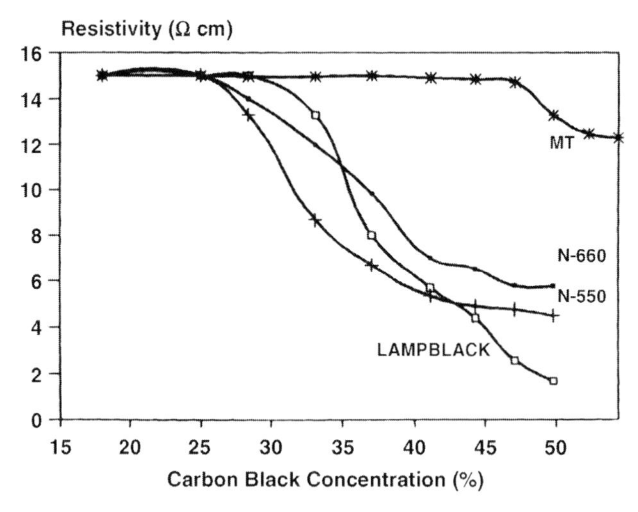

Figure 18.4 Log resistivities versus concentrations for various carbon blacks dispersed in SBR compounds. (After Ref. [18], reprinted with permission from Marcel Dekker, Inc.)

Figure 18.4 shows that the profile of resistivity versus concentration differs markedly for the four blacks studied in SBR. A possible explanation was thought to lie in differences in the surface chemistry of the blacks, particularly in the amount of oxygen-containing groups present on their surfaces as suggested by Sichel *et al.* [14]. Foster [5] reports resistivity measurements in an amine cured epoxy matrix and in a thermosetting acrylic polymer with various CBs. The data show that "aftertreated" grades (containing more surface functionality) have a higher surface resistivity than untreated furnace blacks. Also, low surface area, low structure CBs, dispersed in these polymers show consistently higher resistivity than high structure, high surface area blacks. On the other hand, XPS and SIMS analyses by Pantea *et al.* [12] suggest that "the concentration of noncarbon elements on the carbon black surface is not a determining factor for the electrical conductivity." To them, the amount of graphitic surface character seems to be the determinative. However, the ultrahigh vacuum environment of XPS and SIMS may promote desorption of oxygenated species and other organic compounds. It would be of great interest to measure blacks of higher surface functionality by these techniques.

18.3.4
Effects of Polymer Matrix

When comparing the same CB in different matrices, the threshold concentrations of XE-2 (a high structure CB) range from 1.0 in polyacetal to 4.4 wt% in polycarbonate (PC) (see Table 18.3). There are explicable variations in V_c among the two types of polyethylene of different MI cited. But it is notable that all the values of V_c are within

Table 18.4 Influence of type of polypropylene (homopolymer versus copolymer) and conductive filler on surface resistivity and impact strength of filled composites.

	Homopolymer	Copolymer				
Carbon black (wt%)	1.2	1.2	3.7	4.6	4.8	5.1
Glass fibers (wt%)	15.4	15.4	10	15.4	10	10
Surface resistivity (Ω/sq)	10^6	$>10^{12}$	10^8	10^6	10^5	10^4
Notched Izod impact strength (J/m)	57	117	140	115	140	135

After Ref. [19], reprinted with permission from Marcel Dekker, Inc.
Note: CB is Ketjenblack EC 600 JD (Akzo, The Netherlands). The glass fibers are 10 μm in diameter and 3 mm long (Vetrotex, Owens-Corning).

a fairly limited range of 1–5 wt%. This suggests dominance of the CB properties over those of the matrix [18].

Differences are observed for both surface and bulk conductivities of one CB in various polymers as for example when Narkis *et al.* [19] compare PP, PS, HIPS, PE, Noryl®. There is even a subtle effect of composition on resistivity in the case of PP homopolymer versus PP copolymer because the latter contains a rubbery component (Table 18.4). The CB apparently segregates into the rubbery phase and a higher loading of CB is needed to reach levels of conductivity comparable to the homopolymer [19]. Effects of titanate surface treatments on the conductivity of CB compounds are included in Chapter 5.

18.3.5 Other Applications

A variety of electronic devices based upon dispersed particles in a polymer matrix have been fabricated. Composites of CB in certain thermoplastics exhibit the property of a step change from low to high resistance when the device temperature exceeds a certain value. This positive temperature coefficient of resistance (PTCR) behavior arises from Joulean heating that causes thermal expansion and, in some cases, melting of the polymer matrix, thereby disrupting the continuity of the CB filler pathways. When the device cools, continuity is restored [20]. This effect has been used to create temperature-limited heating tapes and self-resetting fuses. The microphysics of this switching have been analyzed recently [21].

A device consisting of 15 vol% nickel powder and 40 vol% SiC powder dispersed in a silicone resin binder has been fabricated. Thermal curing is performed via a peroxide initiator. The resulting device switches from a high resistance state ($>10^6$ Ω) to a conductive state (1–10 Ω) when subjected to electrostatic discharge circuit transients. This response effectively shunts transients to ground in times of <25 ns [22].

One of the major applications of CB-loaded polymers is for materials that will dissipate an electrostatic charge. These are referred to as ESD compounds. According to the Electronics Industry Association (EIA) "conductive" materials have surface resistivity of $<10^5$ Ω/sq; dissipative materials 10^5–10^{12} Ω/sq; insulators $>10^{12}$ Ω/sq.

In practice, electrostatic dissipative materials have preferred resistivity in the range of 10^6–10^9 Ω/sq [19].

As an example, a commercial ESD compound uses as little as 1–2 wt% of high structure Ketjenblack EC 600 (Akzo, The Netherlands) to achieve stable resistivities of 10^6–10^9 Ω/sq in combination with glass fibers at loadings of 10–25 wt%, to create an inorganic interface where the CB presumably concentrates [19]. The glass fibers are ~3 mm long and 10 μm in diameter, giving an aspect ratio of 300. The resulting compound is ideal for electrostatic dissipation requirements and uses a much lower loading of CB than would be required without the glass fiber. Without glass fibers to promote CB clustering, a loading of 25 vol% would be typical to achieve resistivity in the range of 10^8 Ω/sq.

18.4 Distribution and Dispersion of Carbon Black in Polymers

18.4.1 Microscopy and Morphology [8]

Carbon black was an oft-examined substance during the evolution of the transmission electron microscope. It was a tour-de-force to resolve the graphitic planes in carbon black [23]. With the scanning electron microscope (SEM), more fully dimensional views became available. In recent years, with the advent of the atomic force microscope (AFM) there have been attempts to correlate fine-scale microstructure with electrical conductivity.

18.4.2 Percolation Networks

It has been possible to directly image the percolation network at the surface of a CB–polymer composite. An early report is that of Viswanathan and Heaney [24] on CB in HDPE in which it was shown that there are three regions of conductivity as a function of the length L, used as a metric for the image analysis. Below $L = 0.6$ μm, the fractal dimension D of the CB aggregates is 1.9 ± 0.1. Between 0.8 and 2 μm, the data exhibit $D = 2.6 \pm 0.1$ while above 3 μm, $D = 3$ corresponding to homogeneous behavior. Theory predicts $D = 2.53$. “It is not obvious that the carbon black–polymer system should be explainable in terms of standard percolation theory, or that it should be in the same universality class as three-dimensional lattice percolation problems [24].” Subsequent experiments of this kind were made by Carmona [25, 26].

It has been correctly realized that most microscopies – including AFM and electrical probes – render only a two-dimensional section of a three-dimensional conducting network. Gubbels [27, 28] applied methods of image analysis to blends consisting of two polymers plus CB. The oft-observed segregation of CB to the phase boundaries or interphase was confirmed.

18.4.3
CB in Multiphase Blends

It is known that in some polymers, addition of CB to levels above V_c causes a decrease in the mechanical properties. This is a large effect for some polymers, but is minor for others [3] Tensile elongation varies with CB loading for polycarbonate and polypropylene but is little affected in ethylene vinyl acetate (EVA) and an ethylene–octene copolymer (Engage®) [3].

Both for reasons of economy and to minimize unwanted effects on mechanical behavior, it is desirable to use the minimum concentration of CB to achieve the required electrical properties. By employing polymer blends, it is possible to create morphologies in which the CB additive concentrates in one phase or, better, in the interphase. There the CB aggregates approach each other closely and the percolation threshold is low. Examples of systems [27–29] where such phase segregation can occur are

- polyolefins with other crystallizable or noncrystallizable polymers [30–33],
- phase-separating block copolymers such as SBS, SEBS, and so on [34],
- other incompatible polymer blends [35, 36].

In a noncrystallizable polymer such as atactic polystyrene, V_c is close to 8 wt%. In semicrystalline PE it is 5 wt%, presumably due to segregation of the CB to the noncrystalline phase or the phase boundaries in PE. This segregation is further enhanced in PE/PS blends when the composition allows for continuity of the PE phase and "double percolation" of the phases and conductive regions [27, 37]. This has been observed at a 45/55 ratio by weight of PE/PS in a melt-blended composition with more than 0.4 wt% (0.2 vol%) carbon black [27]. The effect seems to depend upon the relative interfacial tensions of the polymers and the CB in a manner consistent with the independent observations of Miyasaka *et al.* [38].

18.4.4
Process Effects on Dispersion

All the above situations reiterate the fact that the observed conductivities are highly dependent on the morphology or microstructure of the composite and on the mixing process that gives rise to it. Regarding mixing, see Chapter 3 in this book and also Refs [39–42]. Manas-Zloczower *et al.* considered that CB agglomerates would break up into particles roughly half the size of the original under the action of mixing stresses [43]. Another viewpoint is that agglomerates "erode" as small pieces break off from their surface [44]. Subsequent studies in the laboratory of Manas-Zloczower using a transparent cone-and-plate viscometer with fluids and polymer melts covering a wide viscosity range showed that both mechanisms operate. The early stages are predominately agglomerate breakup followed later by erosion [45]. Still, it is well known that in a process like injection molding there are segregation effects that inherently lead to skin-core differences in CB distribution and to more complex spatial distributions. Even with the simplest flows and part shapes there can be measurable variations of resistivity between different locations in a single part [46].

A different approach for controlling the dispersion of CB is based on a collective motion of fluid elements called chaotic advection. The theory behind this mode of mixing is set out by Aref [47, 48]. Danescu and Zumbrunnen [49, 50] have built equipment incorporating an eccentric cavity formed between two offset cylinders. In such a device, the high structure CB Printex XE-2 (Degussa Corp.) precompounded into atactic polystyrene was let down to loadings ranging from 0.4 to 2.5 wt%. A drop of two orders of magnitude in resistivity was observed in the concentration range from 0.8 to 1.0 wt%. The conductivity threshold value – 0.8 wt% CB – is 70% lower than the percolation threshold of this system compounded by conventional means. Further studies have been published including an apparatus for continuously extruding CB-polymer films with chaotic advection [51].

18.5 Other Carbon-Based Conductive Fillers

For *carbon nanotubes*, discussed in detail in Chapter 10, conductivity is achieved at lower loadings (by weight) but these materials are difficult to disperse in molten polymers. Methods of surface functionalization and lower cost manufacturing must be developed before carbon nanotubes will find wider use as conductive fillers [52, 53]. As an alternative to nanotubes, Fukushima and Drzal [54] have observed conductivity thresholds of less than 3 vol% in composites containing acid-etched or otherwise functionalized exfoliated graphite. These composites retain or improve upon their mechanical properties compared to other carbon-filled polymers.

For carbon fibers (see also Chapter 10) increasing the aspect ratio of the filler should, predictably, lower the resistivity compared to an equiaxed filler. It is interesting to observe that for one such carbon fiber compounded into five thermoplastics, the resistivity versus loading curves are identical; namely, the critical volume concentration of fiber is $\sim$7 vol%. This is the concentration at which the volume resistivity drops from 10^{12} to 10^{1} Ω cm [55]. Inevitably, for all thermoplastic matrices, attrition of the carbon fiber (as measured by a diminished L/D) increases the resistivity of the composite (Table 18.5). King and coworkers [56] have shown that still lower resistivities are produced when carbon fiber is combined with other fillers such as graphite powder or CB. It seems that an effect from increased internal interfacial area is dominant. Narkis *et al.* [19, 57] showed that even glass fibers create interfacial regions where added CB at low concentrations can significantly increase conductivity.

18.6 Intrinsically Conductive Polymers (ICPs)

Shortly after the discovery of intrinsic conductivity in polymers experiments were conducted to disperse these organic materials into host polymers to improve

Table 18.5 Effects of compounding technique and screw type on the fiber length and resistivity of compounds filled with carbon fibers.

Matrix	Extruder compounder	Concentration (vol%)	Resistivity (Ω cm)	Fiber length (mm)	Fiber aspect ratio	Polymer form
ABS	Single screw	0.20	0.60	0.44	55	Pellets
ABS	Twin screw	0.20	1.70	0.24	30	Pellets
PPO, Noryl®	Single screw	0.25	0.630	0.18	22	Pellets
PPO, Noryl®	Twin screw	0.25	3.30	0.11	14	Pellets
Polyamide-6,6	Single screw	0.19	0.56	0.23	29	Pellets
Polyamide-6,6	Twin screw	0.19	1.04	0.13	16	Pellets
PPS, Ryton®	Single screw	0.21	3.40	0.15	19	Powder
PPS, Ryton®	Twin screw	0.21	1.25	0.22	27	Powder

After Ref. [55], reprinted with permission from John Wiley & Sons, Inc.

processability and increase stability. The quest is still on, although some ICP-polymer blends are being sold commercially, for example, by RTP Company, Winona, MN and Eeonyx Inc., Pinole, CA. For example, polyaniline (PANI) shows improved processability and stability when conjugated to a protonic acid such as poly(styrene sulfonic acid) or to the small molecule dodecylbenzenesulfonic acid [58, 59]. In this state it is more compatible with certain polymers. Such blends, as interpenetrating polymer networks, show $V_c < 1\%$ [60]. ICPs exhibit a volume resistivity of 10^5 compared to 10^3–10^9 Ω cm for a CB dispersion. This comparison is for each additive in a matrix of polypropylene [61]. In PANI/poly(styrene sulfonic acid) dispersions, the particle sizes are <1 μm and thin coatings with resistivities of 1–10 Ω cm can be cast as transparent films [60].

18.7 Metal Particle Composites

Most polymers can be filled with metal particles to render them conductive. However, consideration must be given to specifics of the possible combinations [62]. For example, ABS and PC are widely used for instruments, TV, and similar housings but copper is known to promote degradation of PC so another filler is often preferable [63].

Polymers that phase-separate upon solidification may contribute to segregation of conductive filler to the interphase or to noncrystalline regions. This can produce favorably high conductivities with less metal because of the local concentration of metal particles. This is illustrated in Figure 18.5 [64]. Note two cases: when polymer and metal particles are of comparable size (panels a and b) versus small metal particles (panels c and d). Also in Figure 18.5 note the microstructures for $V < V_c$ versus $V > V_c$.

This effect of metal segregation is borne out by measurements of resistivity versus volume percent metal filler as shown in Figure 18.6. A threshold for

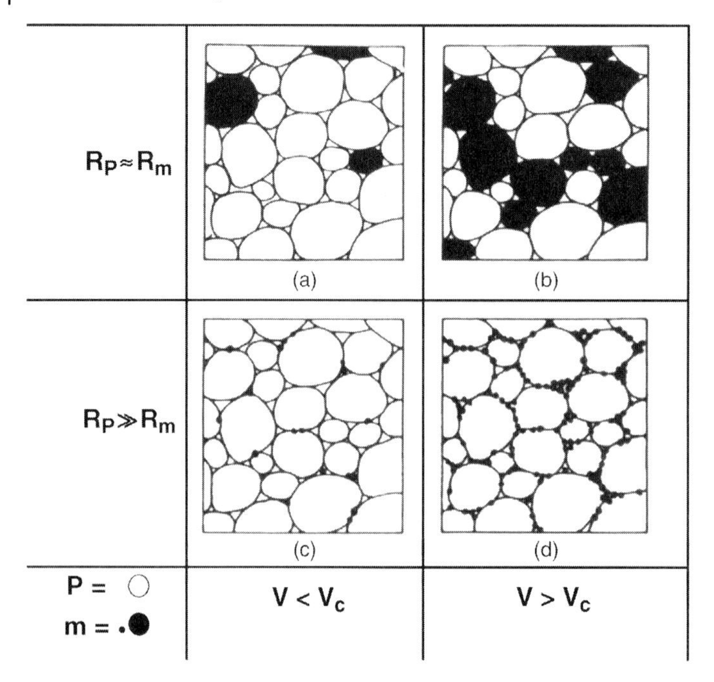

Figure 18.5 Schematic illustrations of "random" (a and b) versus "segregated" (c and d) particle distributions. (After Ref. [63], reprinted with permission from Marcel Dekker, Inc.)

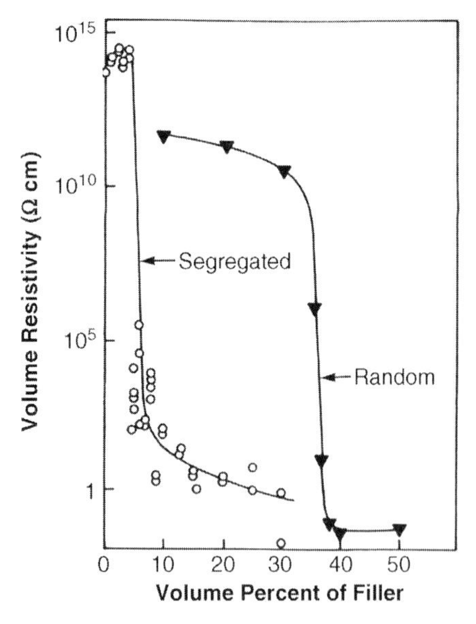

Figure 18.6 Influence of specimen microstructure and filler content on electrical resistivity for a random versus a segregated distribution of metal particles. The "random distribution" ▼ is 50% silver in phenol–formaldehyde resin while the segregated case ○ is 7% silver in PVC. (After Ref. [63], reprinted with permission from Marcel Dekker, Inc.)

conductivity in metal-filled polymers was first observed by Gurland [65] in a composite of silver spheres in phenol–formaldehyde resin (Bakelite®). The threshold or critical volume concentration, V_c observed was 0.38.

V_c depends on the particle shape but, in general, for equiaxed particles V_c equals 0.4 for broad and 0.2 for narrow particle size distributions (PSD), respectively [66]. Generally, when equiaxed metal particles are used, a narrow PSD creates conductive pathways at lower loadings than a broad PSD [66]. Among metal fibers, aluminum fibers with an aspect ratio 12.5 : 1 showed a threshold at ~12 vol% in either thermosetting polyester or polypropylene. As the aspect ratio of metal fibers is increased, the threshold loading for conductivity decreases. For example, stainless steel fibers 6 mm long and 8 μm in diameter in ABS show resistivity of 0.70 Ω cm at 1 vol% loading [66, 67]. Metal fibers may be coated or sized to increase adhesion to the polymer matrix.

The shielding effectiveness of various fillers versus frequency is shown in Figure 18.7 [66]. Adopting the guideline that 30–40 dB of attenuation will meet the majority of requirements, all materials in Figure 18.7 meet this criterion up to at least 50 MHz. Polymers filled with aluminum flakes or metal fibers are effective in terms of cost and performance and may offer significant economic advantages over the laborious methods used for painting or coating parts made of unfilled polymers. In some instances, metal-filled polymers may be pigmented to give colored products. This may be used as a design advantage since carbon black fillers give only black parts.

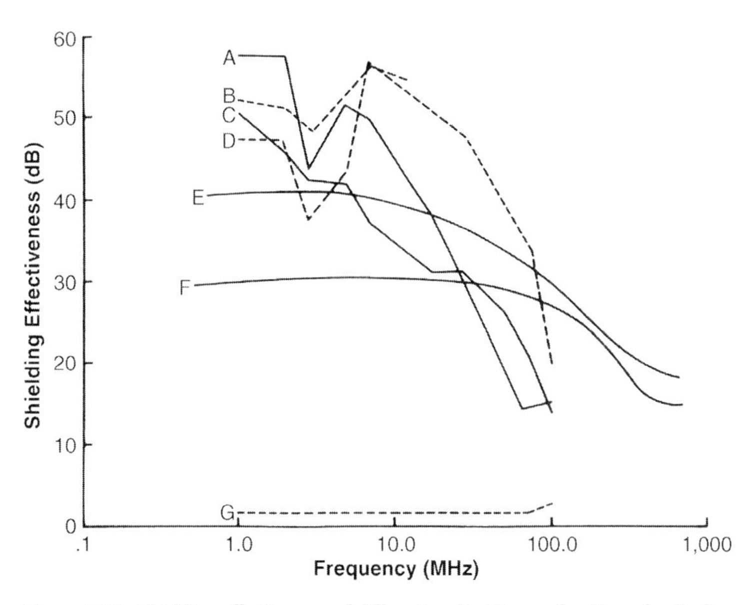

Figure 18.7 Shielding effectiveness of different materials as a function of radio frequency. (a): Nickel-coated glass fiber composite; (b): copper plate; (c): carbon black composite; (d): steel; (e): 20% aluminum fiber in polyester; (f): 30% aluminum flake in polyester; (g): unfilled polymer. (After Ref. [63], reprinted with permission from Marcel Dekker, Inc.)

Table 18.6 Static charge decay rate of selected polycarbonate-based plastics.

Material	Applied voltage (%)	Decay rate (s)	
		At +5 kV	At −5 kV
Polycarbonate	50	>100	>100
	10	>100	>100
	0	>100	>100
Polycarbonate coated with nickel paint	50	0.02	0.02
	10	0.03	0.04
	0	0.05	0.05
Polycarbonate filled with 30% metal	50	0.02	0.02
	10	0.04	0.03
	0	0.06	0.05
Polycarbonate filled with 40% metal	50	0.02	0.02
	10	0.04	0.03
	0	0.05	0.05

After Ref. [63], reprinted with permission from Marcel Dekker, Inc.

Static decay rates of carbon or metal-filled versus metal-painted or neat polycarbonate are shown in Table 18.6. All the filled polycarbonates performed similarly: a charge of ±5 kV decayed in less than 0.06 s compared to more than 100 s for neat PC [63]. Metal-filled composites may show ohmic behavior in the interior but appear nonohmic overall due to surface depletion of the conductive species [68].

Highly conducting composites can be produced with low concentrations of metal fillers provided their aspect ratio is high. Fiber L/D is a significant consideration in processing such materials [66]. A summary of the effects of processing on carbon fibers in various polymer melts is presented in Table 18.5 [55]. Metal fibers are better able to withstand processing than metallized glass or carbon fiber. Stainless steel fibers at a loading of 1 vol% produce a composite with RF signal attenuation of up to 50 dB. Such compounds can be successfully processed in screw extruders and may, if desired, be colored [69].

18.8 Magnetic Fillers

In polymers filled with magnetic materials, the magnetic moment is just proportional to the volume loading of magnetic particles. As an example, consider the case of magnetite (Fe_3O_4), a ferrite, in polypropylene or polyamides. In practice, to make resin-bonded magnets, relatively high filler levels are used, for example 60–80 wt% or 25–45 vol% [70]. A significant drop in electrical resistivity of this composite is noted at 44 vol% magnetite. In terms of process conditions, the maximum practical loading of magnetite was 80 wt% when operating at 90% maximum torque and 300 rpm using a downstream side feeder on a twin-screw extruder with $L/D = 39$. In these experiments, production rate of the magnetic compound was 85 kg/h [70].

Other magnetic particles compounded into plastics include barium ferrite, alnico, samarium cobalt (SmCo) and rare earth iron borides. A sampling of magnetics terminology and properties of the representative permanent magnets: ferrite, alnico, SmCo, and NdFeB are given by Trout [71]. Magnetic powders are generally pre-compounded with the matrix polymer and then injection molded to give near net-shape parts or stock shapes. The matrix can provide added value by decreasing the corrosion effects on the magnetic fillers. A recommended general treatment of magnetic materials and their physics may be found in the book by Cullity [72]. Neodymium–iron–boron powder is a preferred material for the bonded magnet producer. Manufacturers prepare these composites by various processes: injection molding of ferrites such as barium ferrite; sintering of SmCo; and processing NdFeB by sintering, hot pressing, and molding.

While making magnetic tape and “floppy disks” the magnetic filler is dispersed in a concentrated polymer solution that is coated onto a substrate and oriented by an externally applied magnetic field before the composite coating “sets” by drying or cross-linking. Orientation is very important for recording media in order to achieve high magnetic remanence that aids in avoiding inadvertent demagnetization. Some of the technology for preparing magnetic recording media has been described [73] but much is kept as trade secrets.

18.9 Concluding Remarks

Often, real composite materials are far from homogeneous in their properties or microstructure. In some instances, an immediate technical problem may be dealt with by adjusting one parameter or another, but no discriminating fundamental measurements can be made on such systems. Therefore, it is to be hoped that future studies will describe the polymers and fillers in considerable detail, the mixing and processing variables will be annotated, and the microstructure of the resulting composites will be characterized. Only then will meaningful measurements and comparisons be possible among these various filled plastics that have such interesting electrical and magnetic properties.

18.10 Appendix: Measurements of Resistivity

One distinguishes “bulk” or volume resistivity (ohm-cm) in which the dimensions of the conductor are considered, from surface resistivity (ohms per square). Definitions and methods can be found in ASTM D 4496-87 and BS 2044 for volume or “bulk” resistivity and in IEC 167 and AFNOR C26-215 for surface resistivity. Figure 18.8

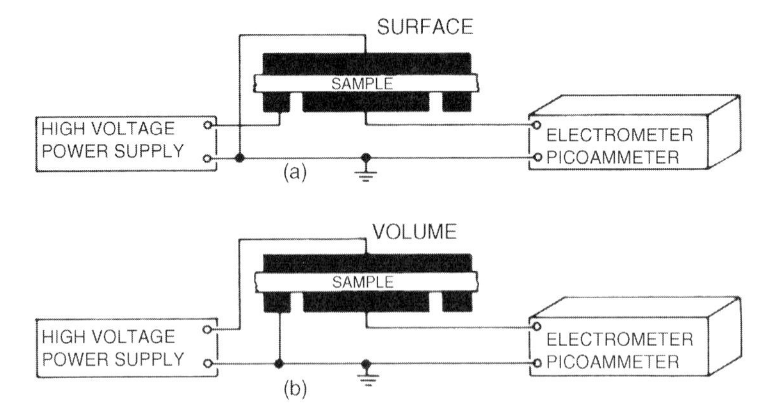

Figure 18.8 Setups for measuring surface and bulk resistivity. (After Ref. [74].)

Table 18.7 Resistivity classification of conductive thermoplastics.

	Bulk resistivity (Ω cm)	Surface resistivity (Ω/sq)
Undoped	10^{14}–10^{16}	10^{12}–10^{14}
Antistatic	10^{9}–10^{14}	
Dissipative	10^{5}–10^{9}	10^{6}–10^{9}
Conductive	10^{0}–10^{5}	
EMI shielding	<1	$<10^{6}$

After Refs [3, 19, 69].

shows measurement setups and illustrates the concepts involved [74]. Table 18.7 classifies plastic compounds according to their resistivity level.

Acknowledgments

The author appreciates the bibliographical assistance from Mr. Bruce Slutsky of the Van Houten Library at NJIT and Ms. K. Fitzgerald.

References

1 Sichel, E.K. (ed.) (1982) *Carbon Black-Polymer Composites: The Physics of Electrically Conducting Composites*, Marcel Dekker, New York.
2 Rupprecht, L. (ed.) (1999) *Conductive Polymers and Plastics in Industrial Applications*, Plastics Design Library, Norwich, NY.
3 Huang, J.-C. (2002) Carbon black filled conducting polymers and polymer blends. *Adv. Polym. Technol.*, **21** (4), 299–313.
4 Donnet, J.-B., Bansal, R.C., and Wang, M.-J. (eds) (1993) *Carbon Black: Science and Technology*, 2nd edn, Marcel Dekker, Inc., New York.

5 Foster, J.K. (1991) Effects of carbon black properties on conductive coatings. Presented at the 2nd International Exhibition of Paint Industry Suppliers, San Paulo, Brazil, On the Web at http://www.cabotcorp.com/cws/businesses.nsf/8969ddd26dc8427385256c2c004dad01/9d77475ac031436285256c7a005021dd/$FILE/CB-002.pdf.
6 Kühner, G. and Voll, M. (1993) Manufacture of carbon black, Chapter 1, in *Carbon Black: Science and Technology*, 2nd edn (eds J.-B. Donnet, R.C. Bansal and M.-J. Wang), Marcel Dekker, Inc., New York.
7 Medalia, A.I. (1994) Chapter 15, in *Mixing and Compounding of Polymers: Theory and Practice* (eds I. Manas-Zloczower and Z. Tadmor), Hanser Publishers, Munich and Cincinnati.
8 Hess, W.M. and Herd, C.R. (1993) Microstructure, morphology, and general physical properties, Chapter 3, in *Carbon Black: Science and Technology*, 2nd edn (eds J.-B. Donnet, R.C. Bansal and M.-J. Wang), Marcel Dekker, Inc., New York.
9 Dannenberg, E.M. (1978) Carbon Black, in *Kirk-Othmer Encyclopedia of Chemical Technology*, 3rd edn, vol. 4, John Wiley & Sons, Inc., New York, pp. 631–666.
10 Bansal, R.C. and Donnet, J.-B. (1993) Mechanism of carbon black formation, Chapter 2, in *Carbon Black: Science and Technology*, 2nd edn (eds J.-B. Donnet, R.C. Bansal and M.-J. Wang), Marcel Dekker, Inc., New York.
11 Cabot Corporation, Carbon Black User's Guide: Safety, Health, and Environmental Information, http://www.cabot-corp.com/wcm/download/en-us/unknown/carbonblackuserguide.pdf; accessed Nov. 2009.
12 Pantea, D. *et al.* (2003) Electrical conductivity of conductive carbon blacks: influence of surface chemistry and topology. *Appl. Surf. Sci.*, **217**, 181–193.
13 Janzen, J. (1975) On the critical conductive filler loading in antistatic composites. *J. Appl. Phys.*, **46** (2), 966–969.
14 Sichel, E., Gittleman, J.I., and Sheng, P. (1978) Transport properties of the composite material carbon-poly(vinyl chloride). *Phys. Rev.*, **B18** (10), 5712–5716.
15 Balberg, I. (1998) Limits on the continuum-percolation transport exponents. *Phys. Rev.*, **B57**, 13351.
16 Rubin, Z. *et al.* (1999) Critical behavior of the electrical transport properties in a tunneling percolation system. *Phys. Rev.*, **B59** (19), 12196–12199.
17 Balberg, I. (2002) A comprehensive picture of the electrical phenomena in carbon black-polymer composites. *Carbon*, **40**, 139–143.
18 Probst, N. (1993) Conducting carbon black, Chapter 8, in *Carbon Black: Science and Technology*, 2nd edn (eds J.-B. Donnet, R.C. Bansal and M.-J. Wang), Marcel Dekker, Inc., New York.
19 Narkis, M. *et al.* (1999) Novel electrically conductive injection moldable thermoplastic composites for ESD applications, in *Conductive Polymers and Plastics in Industrial Applications* (ed. L. Rupprecht), Plastics Design Library, Norwich, NY.
20 Doljack, F.A. (1981) PolySwitch PTC devices-a new low-resistance conductive polymer-based PTC device for overcurrent protection. *IEEE Trans. Components, Hybrids, Mfg Tech.*, **CHMT-4** (4), 372–378.
21 Azulay, D. *et al.* (2003) Electrical-thermal switching in carbon black-polymer composites as a local effect. *Phys. Rev. Lett.*, **90**, 236601.
22 Rector, L. and Hyatt, H. (1998) Polymer composite varistor materials. Proceedings of the 56th SPE ANTEC, pp. 1381–1385.
23 Heidenreich, R.D., Hess, W.M., and Ban, L.L. (1968) Test object and criteria for high resolution electron microscopy. *J. Appl. Cryst.*, **1**, 1.
24 Viswanathan, R. and Heaney, M.B. (1995) Direct imaging of the percolation network in a three-dimensional disordered conductor-insulator composite. *Phys Rev. Lett.*, **75** (24), 4433–4436, and errata in (1996) *Phys Rev. Lett*, **76** (19), 3661.
25 Carmona, F. and Ravier, J. (2002) Electrical properties and mesostructure of carbon black-filled polymers. *Carbon*, **40**, 151–156.
26 Carmona, F. and Ravier, J. (2003) To what extent is the structure of a random composite compatible with a percolation model? *Physica B*, **338**, 247–251.

27 Gubbels, F., Gubbels, F., Jerome, R., Teyssie, Ph., Vanlathem, E., Deltour, R., Calderone, A., Parente, V., and Bredas, J.L. (1994) Selective localization of carbon black in immiscible polymer blends: a useful tool to design electrical conductive composites. *Macromolecules*, **27**, 1972–1974.

28 Gubbels, F., Blacher, S., Vanlathem, E., Jerome, R., Deltour, R., Brouers, F., and Teyssie, Ph. (1995) Design of electrical conductive composites: key role of the morphology on the electrical properties of carbon black filled polymer blends. *Macromolecules*, **28**, 1559–1566.

29 Knackstedt, M.A. and Roberts, A.P. (1996) Morphology and macroscopic properties of conducting polymer blends. *Macromolecules*, **29**, 1369–1371.

30 Tchoudakov, R., Breuer, O., Narkis, M., and Siegmann, A. (1996) Conductive polymer blends with low carbon black loading: polypropylene/polyamide. *Polym. Eng. Sci.*, **36**, 1336.

31 Thongruang, W., Spontak, R.J., and Balik, C.M. (2002) Correlated electrical conductivity and mechanical property analysis of high-density polyethylene filled with graphite and carbon fiber. *Polymer*, **43**, 2279.

32 Thongruang, W., Spontak, R.J., and Balik, C.M. (2002) Bridged double percolation in conductive polymer composites: an electrical conductivity, morphology, and mechanical property study. *Polymer*, **43**, 3717.

33 Thongruang, W., Balik, C.M., and Spontak, R.J. (2002) Volume-exclusion effects in polyethylene blends filled with carbon black, graphite, or carbon fiber. *J. Polym. Sci.: Part B: Polym. Phys.*, **40**, 1013.

34 Tchoudakov, R., Breuer, O., Narkis, M., and Siegmann, A. (1999) Conductivity/morphology relationships in immiscible polymer blends: HIPS/SIS/carbon black, in *Conductive Polymers and Plastics in Industrial Applications* (ed. L. Rupprecht), Plastics Design Library, Norwich, NY, pp. 51–56.

35 Zhang, M.Q., Yu, G., Zeng, H.M., Zhang, H.B., and Hou, Y.H. (1998) Two-step percolation in polymer blends filled with carbon black. *Macromolecules*, **31**, 6724–6726.

36 Feng, J. *et al.* (2003) A method to control the dispersion of carbon black in an immiscible polymer blend. *Polym. Eng. Sci.*, **43**, 1058–1063.

37 Levon, K., Margolina, A., and Patashinsky, A.Z. (1993) Multiple percolation in conducting polymer blends. *Macromolecules*, **26**, 4061–4063.

38 Miyasaka, K., Watanabe, K., Jojima, E., Aida, H., Sumita, M., and Ishikawa, K. (1982) Electrical conductivity of carbon-polymer composites as a function of carbon content. *J. Mater. Sci.*, **17**, 1610.

39 Todd, D.B. (ed.) (1998) *Plastics Compounding: Equipment and Processing*, Hanser Publishers, Munich.

40 Hornsby, P.R. (1999) Rheology, compounding and processing of filled thermoplastics. *Adv. Polym. Sci.*, **139**, 155.

41 Manas-Zloczower, I. and Tadmor, Z. (eds) (1994) *Mixing and Compounding of Polymers: Theory and Practice,* Hanser Publishers, Munich and Cincinnati.

42 White, J.L.*et al.* (eds) (2001) *Polymer Mixing: Technology and Engineering*, Hanser Publishers, Munich and Cincinnati.

43 Manas-Zloczower, I., Nir, A., and Tadmor, Z. (1982) Dispersive mixing in internal mixers – a theoretical model based on agglomerate rupture. *Rubber Chem. Technol.*, **55**, 1250.

44 Shiga, S. and Furuta, M. (1985) Processability of EPR in an internal mixer (II) morphological changes of carbon black agglomerates during mixing. *Rubber Chem. Technol.*, **58**, 1.

45 Hong, C.-M., Kim, J., and Jana, S.C. (2003) The effect of shear-induced migration of conductive fillers on conductivity of injection molded articles. Proceedings of the 61st SPE ANTEC, pp. 1625–1629.

46 Jana, S.C. (2003) Loss of surface and volume electrical conductivities in polymer compounds due to shear-induced migration of conductive particle. *Polym. Eng. Sci.*, **43**, 570.

47 Aref, H. (1984) Stirring by chaotic advection. *J. Fluid Mech.*, **143**, 1.

48 Aref, H. (2002) The development of chaotic advection. *Phys. Fluids*, **14**, 1315.

49 Danescu, R.I. and Zumbrunnen, D.A. (1999) in *Conductive Polymers and Plastics in Industrial Applications* (ed. L. Rupprecht), Plastics Design Library, Norwich, NY, pp. 77–83.

50 Danescu, R.I. and Zumbrunnen, D.A. (1999) Production of electrically conducting plastics at reduced carbon black concentrations by three-dimensional chaotic mixing, in *Conductive Polymers and Plastics in Industrial Applications* (ed. L. Rupprecht), Plastics Design Library, Norwich, NY, pp. 85–91.

51 Zumbrunnen, D.A. *et al.* (2006) Smart blending technology enabled by chaotic advection. *Adv. Polym. Technol.*, **25** (3), 152.

52 Potschke, P., Bhattacharyya, A.R., and Janke, A. (2003) Morphology and electrical resistivity of melt mixed blends of polyethylene and carbon nanotube filled polycarbonate. *Polymer*, **44**, 8061.

53 Sandler, J.K.W., Kirk, J.E., Kinloch, I.A., Shaffer, M.S.P., and Windle, A.H. (2003) Ultra-low electrical percolation threshold in carbon-nanotube-epoxy composites. *Polymer*, **44**, 5893.

54 Fukushima, H. and Drzal, L.T. (2003) A carbon nanotube alternative: graphite nanoplatelets as reinforcements for polymers. Proceedings of the 61st SPE ANTEC, pp. 2230–2234.

55 Bigg, D.M. (1984) The effect of compounding on the conductive properties of EMI shielding compounds. *Adv. Polym. Technol.*, **4** (3–4), 255.

56 Heiser, J.A., King, J.A., Konell, J.P., and Sutter, L.L. (2004) Electrical conductivity of carbon filled nylon 6,6. *Adv. Polym. Technol.*, **23** (2), 135–146.

57 Narkis, M., Lidor, G., Vaxman, A., and Zuri, L. (1998) Novel Electrically Conductive Injection Moldable Thermoplastic Composites for ESD Applications. Proceedings of the 56th SPE ANTEC, pp. 1375–1380.

58 Haba, Y., Segal, E., Narkis, M., Titelman, G.I., and Siegmann, A. (2000) Polyaniline–DBSA/polymer blends prepared via aqueous dispersions. *Synth. Met.*, **110**, 189–193.

59 Segal, E., Haba, Y., Narkis, M., and Siegmann, A. (2001) On the structure and electrical conductivity of polyaniline/polystyrene blends prepared by an aqueous-dispersion blending method. *J. Polym. Sci.: Part B: Poly. Phys.*, **39**, 611–621.

60 Heeger, A.J. (2002) Semiconducting and metallic polymers: the fourth generation of polymeric materials. *Synth. Met.*, **125**, 23–42.

61 Dahman, S. (2003) All polymeric compounds: conductive and dissipative polymers in ESD control materials. Proceedings of the EOS/ESD Symposium, Las Vegas.

62 Bhattacharya, S.K. (ed.) (1986) *Metal-Filled Polymers: Properties and Applications*, Marcel Dekker, Inc., New York.

63 Kusy, R.P. (1986) Applications, Chapter 1, in *Metal-Filled Polymers: Properties and Applications* (ed. S.K. Bhattacharya), Marcel Dekker, Inc., New York.

64 Kusy, R.P. (1977) Influence of particle size ratio on the continuity of aggregates, *J. Appl. Phys.*, **48**, 5301.

65 Gurland, J. (1966) An estimate of contact and continuity of dispersions in opaque samples, *Trans. Metal. Soc. AIME*, **236**, 642.

66 Bigg, D.M. (1986) Electrical properties of metal-filled polymer composites, Chapter 3, in *Metal-Filled Polymers: Properties and Applications* (ed. S.K. Bhattacharya), Marcel Dekker, Inc., New York.

67 Bigg, D.M. and Stutz, D.E. (1983) Plastic composites for electromagnetic interference shielding applications. *Polym. Compos.*, **4**, 40.

68 Reboul, J.P. (1986) Nonmetallic fillers, Chapter 6, in *Metal-Filled Polymers: Properties and Applications* (ed. S.K. Bhattacharya), Marcel Dekker, Inc., New York.

69 Murphy, J. (2000) Fillers for electrical/electronics. *Plast. Addit. Compound.*, **2** (9), 22–27.

70 Duifhuis, P.L. and Janssen, J.M.H. (2001) Magnetite functional filler: a compounding study in polypropylene and polyamide. *Plast. Addit. Compound.*, **3** (11), 14–17.

71 Trout, S.R. Understanding permanent magnet materials: an attempt at universal magnetic literacy. Paper

presented at Coil Winding 2000 Conference, Available on the Internet at http://spontaneousmaterials.com/Papers/CoilWinding2000.pdf.

72 Cullity, B.D. (1972) *Introduction to Magnetic Materials*, Addison-Wesley, Reading, MA.

73 Mee, C. and Daniel, E. (1988) *Magnetic Recording*, McGraw-Hill, New York.

74 Keithley Instruments Co . (1977) *Electrometer Measurements*, 2nd edn, Keithley Instruments Co., Cleveland, OH.

19
Surface Property Modifiers

Subhash H. Patel

19.1
Introduction

A variety of organic/inorganic additives, in either solid or liquid form, are incorporated into plastic articles to achieve desired surface property modification. Depending upon the specific function they perform in modifying the surface property, they could be categorized as follows:

1)	Lubricants	Prevent sticking to processing equipment
2)	Antiblocking and slip agents	Prevent sheet and film sticking
3)	Antifogging agents	Disperse moisture droplets on films
4)	Coupling agents	Enhance interfacial bonding between the filler and the polymer matrix
5)	Antistatic agents	Prevent static charge buildup on surfaces
6)	Wetting agents	Stabilize filler dispersions

In the paints and coatings industry, surface modifiers are used to modify the appearance or to improve the performance characteristics of a cured film. Typical performance features include antiblocking, slip, abrasion resistance, matting, and scratch/mar resistance [1].

It would be beyond the scope of this chapter to discuss all the organic/inorganic additives used by the plastics industry for surface property modification. Thus, the discussion would be limited only to two types of additives meeting the definition of functional fillers as used in this book. They are (a) inorganic or organic lubricants/ tribological modifiers and (b) antiblocking inorganic fillers. The emphasis of this chapter is placed on the first type of fillers. Fillers of the second type, the antiblocking function of which is not covered in other chapters of the book, are briefly presented. Fillers improving scratch/mar resistance as a secondary function are included in other chapters of the book.

Functional Fillers for Plastics: Second, updated and enlarged edition. Edited by Marino Xanthos

ISBN: 978-3-527-32361-6

Figure 19.1 Schematic of solid lubrication mechanism. Reproduced with permission from Ref. [3].

19.2 Solid Lubricants/Tribological Additives

Different types of solid materials most widely used at present include [2, 3]

1) **Specific fillers (layer-lattice solids)**: molybdenite or molysulfide (MoS_2) and graphite.
2) **Polymers**: PTFE (polytetrafluoroethylene), polychlorofluoroethylene, silicones.
3) **Less frequently used materials**: ceramics, for example, BN (boron nitride), aramid, or carbon fibers; miscellaneous, for example, calcium fluoride, cerium fluoride, tungsten disulfide (WS_2), mica, borax, silver sulfate, cadmium iodide, lead iodide, and talc.

Among the above listed materials, MoS_2 and graphite are the predominant solid lubricants. In dry powder form, these are effective lubricants due to their lamellar structures. The lamellae orient parallel to the surface in the direction of motion as shown in Figure 19.1 [3].

Such lamellar structures are even able to prevent contact between highly loaded stationary surfaces. In the direction of motion, the lamellae easily shear over each other resulting in low friction. While larger particles perform best on relatively rough surfaces at lower speeds, finer particles perform best on relatively smoother surfaces and at higher speeds. A comparison of various solid lubricants with respect to their coefficients of friction is shown in Figure 19.2 [4].

Solid/dry lubricants are useful under conditions when conventional liquid lubricants are inadequate [3], namely high temperatures, conditions of reciprocating motion, and extreme contact pressures.

19.2.1 General

19.2.1.1 Molybdenite

Molybdenite or molybdenum disulfide (MoS_2) or molysulfide is a mineral [5] found in granites, syenites, gneisses, and crystalline limestones. A NIOSH (National Institute for Occupational Safety and Health) web site [6] has listed a total of 31 synonyms for molybdenite. The use of molybdenite as a lubricant was apparently recorded in the early seventeenth century by John Andrew Cramer [7]. Chrysler, an automotive manufacturer, was the first to widely use MoS_2 grease in the 1960s.

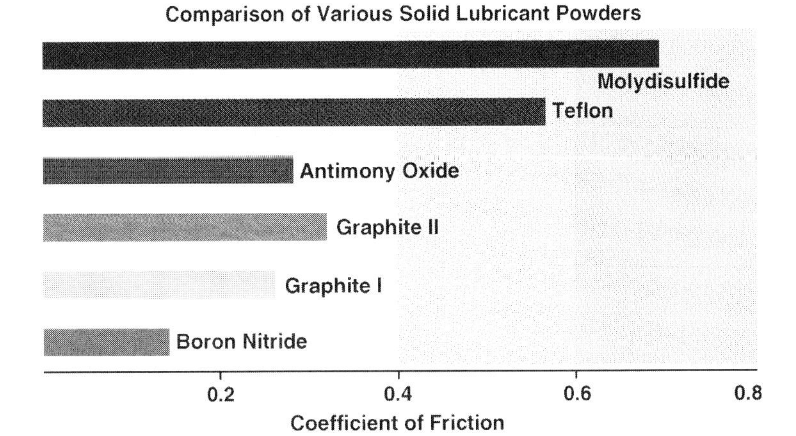

Figure 19.2 Comparison of coefficient of friction of various solid lubricant powders. Reproduced with permission from Ref. [4].

19.2.1.2 Graphite

Graphite is one of the oldest and the most widely used (due to its lower price compared to MoS_2 or BN) solid lubricants. It is obtained both as a soft mineral and as a man-made (synthetic) product. Graphite powders are used as solid lubricants in three ways: (1) in dry films, (2) as an additive in liquids (oils) or semisolids (greases), and (3) as a component of self-lubricating (internally lubricated) composites. A NIOSH web site [8] has listed a total of 55 synonyms for graphite.

19.2.1.3 Polytetrafluoroethylene

PTFE is a perfluorinated, straight chain, high molecular weight synthetic polymer [9]. In contrast to most inorganic functional fillers, PTFE is an organic filler having a unique combination of high heat and chemical resistance together with the lowest friction coefficient of any known internal lubricant, high purity, and dielectric properties. The features and benefits of PTFE include excellent slip, antiblocking, improved stability against polishing, and improved abrasion, scratch, mar, and scuff resistances [1]. A variety of synonyms and trade names exist for PTFE, with Teflon™, being the most well known.

19.2.1.4 Boron Nitride

Boron nitride is a synthetic, high temperature, white solid lubricant, used for parts that are highly resistant to wear. Although it was discovered in the early nineteenth century, it was not developed as a commercial material until the latter half of the twentieth century [10]. BN is often referred to as "white graphite" because it is a lubricious material with the same plate-like hexagonal crystal structure as black graphite [11]. In the same way that carbon exists as graphite and diamond, BN can be synthesized in hexagonal (soft like graphite) and cubic (hard like diamond) crystal forms. A NIOSH web site [12] has listed a total of 24 synonyms for BN.

19.2.2 Production

19.2.2.1 Molybdenite

By far the majority of the world production of MoS_2 comes from the USA, Chile, Canada, and China. Smaller quantities are mined in the rest of the world including Mexico, Australia, South Korea, Namibia, and several European countries.

There are three types of mines/ore bodies from where molybdenite can be recovered [13]:

1) Primary mines, from which solely molybdenite is recovered.
2) By-product mines, from which molybdenite is recovered during copper recovery.
3) Coproduct mines, from which both molybdenite and copper-bearing minerals are recovered.

The raw ore is pulverized using a series of crushers and rotating ball and/or rod mills to fine particles. This liberates the molybdenite from its host rock. The product is then beneficiated by flotation separation, subsequent regrinding, and reflotation to increase the molybdenite content of the new concentrate stream by steadily removing the unwanted material. The final concentrate may contain 70–90% molybdenite. An acidic leach may be employed to dissolve copper and lead impurities, if required. A schematic of the production of molybdenum compounds including MoS_2 can be found in Ref. [13].

In addition to its natural occurrence, MoS_2 can be prepared synthetically by several ways including direct combination of its elements under pure nitrogen at 800 °C, thermal decomposition of ammonium tetrathiomolybdate or molybdenum trisulfide, and by the reaction of MoO_3 with H_2S or H_2S/H_2 mixtures at 500 °C. These techniques result in the formation of hexagonal crystalline MoS_2, by far the most common form; however, the rhombohedral form has been prepared synthetically, but it is also found naturally. Natural and synthetic MoS_2 of both crystalline types possess lubricant properties, but the natural hexagonal material is preferred when cost and overall performance are considered [14].

19.2.2.2 Graphite

Graphite is considered as an "archaic" industrial mineral since it has been mined for its useful properties (lubrication, pigmentation, writing, etc.) for thousands of years [15]. There are two types of graphite used in industry: natural and synthetic.

Natural graphite is obtained in three commercially used varieties: *flake*, *crystalline vein*, and *amorphous* graphite [15–18]. Most *flake* graphite is formed in a metamorphic geological environment by the heat and pressure induced transformation of dispersed organic material. Flake graphite is removed from its enclosing "ore" rock by crushing the rock and separating the graphite flakes by froth flotation. "Run of mine" graphite is available in 80–98% carbon purity ranges. However, most processors are also capable of supplying 99% carbon flake graphite through various postflotation purification methods. The impurities in flake graphite are virtually identical in composition to the enclosing rock.

Crystalline vein graphite is unique, as it is believed to be naturally occurring pyrolytic (deposited from a fluid phase) graphite. Vein graphite gets its name from the fact that it is found in veins and fissures in the enclosing "ore" rock. This variety is formed from the direct deposition of solid, graphitic carbon from subterranean, high-temperature pegmatitic fluids. It typically shows needle-like macromorphology and flake-like micromorphology. Due to the natural fluid-to-solid deposition process, vein graphite deposits are typically above 90% pure, with some actually reaching 99.5% graphitic carbon in the "as-found" state. Vein graphite is mined using conventional shaft or surface methods. Although several small deposits of this type of graphite are known to exist worldwide, Sri Lanka is the only area presently producing commercially viable quantities of this unique mineral.

Most commercial-grade *amorphous graphite* is formed from the contact or regional metamorphism of anthracite coal. It is considered a seam mineral and is extracted using conventional, coal-type mining techniques. *Synthetic graphite*, also known as "artificial graphite," is a man-made product. Synthetic graphite is manufactured by heat-treating amorphous carbons, that is, calcined petroleum coke, pitch coke, and so on, in a reducing atmosphere to temperatures above 2500 °C. At high temperatures, the "pregraphitic" structures present in these "graphitizable carbons" become aligned in three dimensions. The result is the transformation of a two-dimensionally ordered amorphous carbon into a three-dimensionally ordered crystalline carbon. Feedstocks for synthetic graphite production are chosen from product streams that have a high concentration of polynuclear aromatics structures that can coalesce to graphene layers under the influence of heat.

19.2.2.3 **Polytetrafluoroethylene**

Commercially, PTFE is produced from the monomer tetrafluoroethylene by two different polymerization techniques, namely, suspension and emulsion polymerization. These processes give two vastly different physical forms of chemically identical PTFE. While suspension polymerization produces granular PTFE resin, emulsion polymerization produces an aqueous PTFE dispersion and PTFE fine powders (after coagulating the dispersion).

Fine powder PTFE resins, which are relevant to this chapter, are made by variations in the emulsion polymerization technique. It is extremely important that the dispersion remains stable enough during polymerization, but is, subsequently, sufficiently unstable to be able to coagulate into a fine powder [19]. Emulsion polymerization latex has an average particle size of about 150–300 nm. However, by using low reaction conversion it is possible to obtain 100 nm particles [20]. In addition, by the perfluorinated microemulsion polymerization technology [21, 22], it is possible to obtain particles in the size range 10–100 nm. Suspension polymer is obtained as reactor beads with dimensions of a few millimeters, which, after posttreatment and milling, may be ground to micron particle size [23].

19.2.2.4 **Boron Nitride**

BN has at least four crystal structures, namely, hexagonal (h-BN), cubic or zinc blende or sphalerite (c-BN), wurtzite (w-BN), and rhombohedral (r-BN). Among these, the first two are commercially important, and only h-BN is used as a solid lubricant. Three major ways used today for the production of h-BN are

$$B_2O_3 + 2NH_3 \rightarrow 2BN + 3H_2O \quad (T = 900\,^{\circ}C)$$

$$B_2O_3 + CO(NH_2)_2 \rightarrow 2BN + CO_2 + 2H_2O \quad (T > 1000\,^{\circ}C)$$

$$B_2O_3 + 3CaB_6 + 10N_2 \rightarrow 20BN + 3CaO \quad (T > 1500\,^{\circ}C)$$

These processes yield refractory grades with 92–95% BN and 5–7% B_2O_3. The B_2O_3 is removed by evaporation in a second step by reheating to >1500 °C to obtain ceramic grade with >98.5% BN. The c-BN is usually prepared from the hexagonal form at high pressures (4–6 GPa or 40–60 kbar) and temperatures (1400–1700 °C) in the presence of lithium or magnesium nitride catalysts.

19.2.3
Structure/Properties

19.2.3.1 Molybdenite

Molybdenite or "Moly Ore" as it is sometimes called is a very soft, very high luster, metallic mineral, which could be easily confused with graphite [24]. Whereas graphite has a darker black-silver color and a black-gray or brown-gray streak, molybdenite has a bluish-silver color and streak. Color pictures of molybdenite from various parts of the world can be seen at various web sites [24–33].

Figure 19.3 [3] shows a comparison of the crystal structures of MoS_2 and graphite. Molybdenite's hexagonal crystal structure is composed of molybdenum ions sandwiched between layers of sulfur ions. The sulfur layers are strongly bonded to the molybdenum, but are not strongly bonded to other sulfur layers, rendering it softness, easy shear, and perfect cleavage [24]. Thus, the principle of action of molybdenite as a dry lubricant is based on the formation of bonds between the metal and sulfur. These bonds slip under shear forces and are continuously reformed, thereby holding the lubricating film on the surface [34].

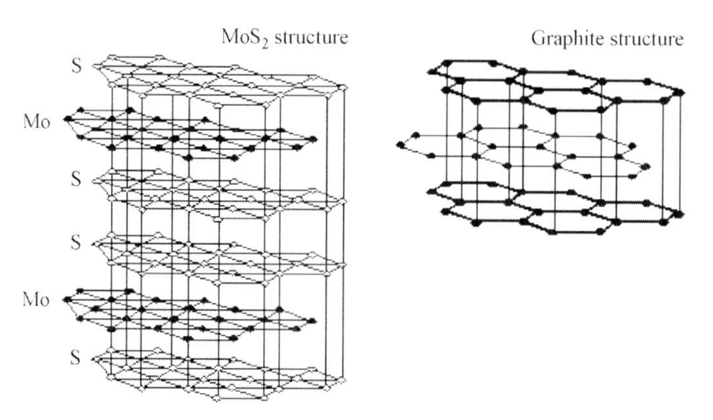

Figure 19.3 Comparison of crystal structures of MoS_2 and graphite. Reproduced with permission from Ref. [3].

Salient physical and chemical properties of molybdenite with regard to its use as filler are summarized in Table 19.1. Although a melting point of 1185 °C has been reported in the older literature, it is most likely incorrect. No melting was observed when MoS_2 was heated under high vacuum at 1800 °C for 10 min, although it is doubtful that it remains intact at that temperature. MoS_2 has been shown to dissociate in two stages, first to $Mo_2S_3 + S_2$ and then to its elements at 1100 °C. Similar more recent work has shown that at 1.3×10^{-7} Pa, the onset of the thermal dissociation of MoS_2 is detected at 927 °C (by TGA) and at 1093 °C (by mass spectrometry) [14].

MoS_2 is diamagnetic. Magnetic susceptibility in the basal plane is smaller and decreases less rapidly with increasing temperature than in the plane perpendicular to it. The average magnetic susceptibility of several commercial samples of MoS_2 was found to be 0.25×10^{-6} emu/g as received and 0.32 emu/g after purification with HCl [14].

The lubrication performance of MoS_2 often exceeds that of graphite, and it is also effective in vacuum or an inert atmosphere up to 1200 °C, temperatures at which graphite cannot be used. The temperature limit of 400 °C for the use of molybdenite in air is imposed by oxidation. It begins to sublime at 450 °C. It is insoluble in water, dilute acids, and most organic solvents. It reacts with azides of alkali and alkaline earth metals to rapidly generate nitrogen gas in large volumes at relatively low temperature. This reaction is utilized for the inflation of passive restraint "air bags" for passenger cars and light trucks [14, 35]. When used as a lubricant, the particle size should be matched to the surface roughness of the plastic or metal substrate. Large particles may result in excessive wear by abrasion caused by impurities, while small particles may promote accelerated oxidation [3].

The lubricating behavior of MoS_2 is not affected by exposure to nuclear radiation. Most of this work has been conducted on compounded solid film lubricants containing MoS_2, other solid lubricants and binders (phenolic resins, sodium silicate, etc.). The limiting factor in these bonded film evaluations appears to be the binder rather than MoS_2. Graphite has been reported to suffer lattice distortion when exposed to a neutron dose of 3.66×1020 nvt in a mixed reactor flux, while similarly exposed MoS_2 was unaffected [14].

Another member of the same chemical family, namely tungsten disulfide (WS_2) is one of the most lubricious materials known to science and is used extensively by NASA, and the military, aerospace, and automotive industries [36]. With coefficient of friction at 0.03, it offers excellent dry lubricity and can also be used at high-temperature and high-pressure applications. It offers temperature resistance from −270 to 650 °C in normal atmosphere and from −188 to 1316 °C in vacuum. WS_2 has a thermal stability advantage of 93 °C over MoS_2. Coefficient of friction of WS_2 actually decreases at higher loads [36].

19.2.3.2 **Graphite**

Graphite and diamond, polymorphs of carbon, share the same chemistry, but have very different structures (cubic for diamond versus hexagonal for graphite) and properties. While diamond is the hardest mineral known to man, graphite is one of the softest. Diamond is the ultimate abrasive and an excellent electrical insulator,

Table 19.1 Comparison of physical/chemical properties of MoS_2, graphite, PTFE, and BN [3, 7, 9, 11, 14, 24, 27, 28, 32, 33, 36, 37, 39, 42, 45, 46, 58, 77–84].

Property	Molybdenite	Graphite	PTFE	Boron nitride
Chemical formula	MoS_2	C	$-(-CF_2-CF_2-)-$	BN
CAS #	1317-33-5	7782-42-5	9002-84-0	10043-11-5
Molecular weight	160.07	12.01	Up to 10^7	24.82
Color	Lead gray/bluish-gray	Black-silver	White-to-translucent	White
Luster	Metallic	Metallic to dull	—	—
Streak	Green to bluish-gray	Black-gray to brownish-gray	—	—
Magnetism	Diamagnetic [14]; nonmagnetic [36]	Strongly diamagnetic	Nonmagnetic	Nonmagnetic
Crystal system	Hexagonal	Hexagonal	Various forms, for example, trigonal, hexagonal	Cubic (abrasive); hexagonal (lubricant)
Cleavage	Perfect basal, easy to remove small flakes	Perfect in one direction	—	—
Crystal features	Crystals are flexible, but not elastic, greasy feel	Thin flakes are flexible but inelastic	From triclinic to disordered hexagonal at 19 °C	Layered or zinc blende
Water absorption, % at room temperature	—	0.5–3.0	Nil (<0.01%)	0.0–1.0
Density (g/cm^3)	4.7–5.0	1.4–2.4	2.14–2.20	2.27 (hexagonal); 3.48 (cubic)
Mohs hardness	1–1.5	1–2	50–60 (Shore-D)	2.0
Melting point (°C)	1185, sublimes at 450	Withstands temperature up to 2820	327, at >400 appreciable decomposition	2700–3000 (melting); 3000 (sublimation); 2700 (dissociation in vacuum)
Onset of oxidation in air (°C)	360	450	>400	>2000

Maximum operating temperature (°C) (without oxygen ingress)	420	550	260	1200 (oxidizing atm.); 3000 (inert atm.)
Minimum operating temperature (°C)	−180	−20	−200	—
Products of oxidation, decomposition	MoO_2, MoO_3	CO, CO_2	Mainly monomer, trifluoroacetate, hexafluoropropene, mono- and difluoroacetic acid, and so on	At >2200 °C, boron oxides, nitrogen, fluoride fumes. With strong oxidizer, may produce ammonia
Solubility	Soluble in hot H_2SO_4, aqua regia, HNO_3 Insoluble in water, dilute acids, and most solvents	Soluble in molten iron	No solvent at room temperature	Degrades in hot conc. alkali Not wetted by molten metals
Resistance to chemicals	Good	Very good	Excellent	Good
Resistance to corrosion	Poor	Good	Good	Good
Thermal conductivity (W/(m K))	0.13–0.19	At 273 K, 160 (natural), 80 (parallel to *c*-axis), 250 (perpendicular to *c*-axis)	0.20	20 (at room temperature)
Coefficient of thermal expansion $\times 10^{-6}$ (1/C)	10.7	Overall: 7.8 (293 K); 8.9 (500 K)	160 (from 25 to 100 °C)	0.46 (perpendicular); 0.6 (parallel)
Electrical resistivity (Ω m)	—	1.2×10^{-6} (natural)	—	10^{11} (at room temperature)
Friction coefficient	0.03–0.06	0.08–0.10	0.04 (dynamic); 0.09 (static)	0.12

Figure 19.4 Scanning electron micrograph of flake graphite pinacoid structure. Reproduced with permission from Ref. [16].

whereas graphite is a very good lubricant and a good conductor of electricity [37]. Table 19.1 contains representative properties of graphite of relevance to its use as filler.

Due to its impervious laminar structure, *flake* graphite is an effective coating additive and barrier filler in plastics. When properly dispersed, overlapping graphite lamellae form a tough, impervious coating, which is not only lubricious but also inert and both electrically and thermally conductive. Also, flake graphite is nonphotoreactive and will not be bleached or affected by ultraviolet light. Figure 19.4 shows a flake graphite pinacoid surface [16]. The morphology of flake graphite is consistently laminar regardless of particle size.

Also known as "expandable graphite," *intumescent flake* graphite is a synthesized intercalation compound of graphite that expands or exfoliates when heated. This material is manufactured by treating flake graphite with various types of intercalation reagents, which migrate between the graphene layers. If exposed to a rapid increase in temperature, these intercalation compounds decompose into gaseous products, which results in high inter-graphene layer pressure that pushes apart the graphite basal planes. The result is an increase in the volume of the graphite up to 300 times, a lowering of bulk density, and an approximately 10-fold increase in surface area. *Intumescent flake* graphite may be used as a fire-suppressant additive. Its fire-suppressant function may be affected by mechanisms associated with (a) the formation of a char layer (see also Chapter 17), (b) the endothermic exfoliation, which effectively removes heat from the source, and (c) out-gassing from decomposing intercalation reagents that displace oxygen in an advancing flame front.

In polymer applications, *crystalline vein* graphite, in addition to lubricity, may offer superior performance since it has slightly higher thermal and electrical conductivity, which result from its high degree of crystalline perfection and good oxidation resistance. *Amorphous* graphite is the least "graphitic" of the natural graphites. However, the term "amorphous" is a misnomer since this material is truly crystalline.

This graphite variety is "massive" with a microcrystalline structure (anhedral), as opposed to *flake* and *vein*, both of which have relatively large, visible crystals (euhedral). *Amorphous* graphite is typically lower in purity than other natural graphites due to the intimate contact between the graphite "microcrystals" and the mineral ash with which it is associated. *Amorphous* graphite tends to be much less reflective in both large and small particle sizes. Therefore, it has a darker color, bordering on black, while other natural graphites have a color closer to "silver-gray."

The morphology of *synthetic* graphite is generally a function of particle size. For particles larger than about 20 μm, the macroscopic morphology is very similar to that of the coke feed used to manufacture the graphite. However, as size is reduced below about 20 μm, the basic flaky structure common to all graphites becomes apparent in the primary particle.

19.2.3.3 Polytetrafluoroethylene

PTFE is generally considered as a thermoplastic polymer, retaining a very high viscosity at 327 °C. It can be employed at any temperature from −200 to +260 °C [9]. The addition of micronized PTFE powder to unfilled polymeric resins and to polymeric resins containing glass fibers provides greatly increased resistance to surface wear and abrasion.

PTFE has a very low coefficient of friction (0.04) and is therefore useful as an internal lubricant processing aid. As the PTFE particles are inert, processing and physical properties of the thermoplastic resin are not adversely affected. The optimum loading is typically 15–20 wt% PTFE. As a result of the improved resistance to surface wear and slip, thermoplastic products containing PTFE retain longer their surface appearance while in use [38].

The following properties and those shown in Table 19.1 are important if PTFE is to be used as a functional filler.

Thermal properties: PTFE is one of the most thermally stable plastic materials. At 260 °C, it shows minimum decomposition and maintains most of its properties; appreciable decomposition begins only at over 400 °C. The arrangement of the PTFE molecules (crystalline structure) varies with the temperature. There are different transition points, with the most important ones being that at 19 °C, corresponding to crystal disordering relaxation, and that at 327 °C, which corresponds to the disappearance of the crystalline structure; PTFE then assumes an amorphous aspect conserving its own geometric form. The linear thermal expansion coefficient varies with temperature. By contrast, thermal conductivity of PTFE does not vary with temperature and is relatively high, such that the material can be considered to be a good insulating material.

Surface properties: The molecular configuration of PTFE imparts a high degree of antiadhesiveness to its surfaces, and for the same reason these surfaces are hardly wettable. PTFE possesses the lowest friction coefficients of all solid materials, between 0.05 and 0.09. The static and dynamic friction coefficients are almost equal, so that there is no seizure or stick-slip action. Wear depends upon the condition and type of the other sliding surface and obviously depends upon the speed and loads.

Mechanical properties: PTFE maintains its tensile, compressive, and impact properties over a broad temperature range. Hence, it can be used continuously at temperatures up to 260 °C, while still possessing a certain compressive plasticity at temperatures near absolute zero. PTFE is quite flexible and does not break when subjected to stresses of 0.7 MPa according to ASTM D 790. Flexural modulus is about 350–650 MPa at room temperature, about 2000 MPa at −80 °C, and about 45 MPa at 260 °C. The Shore D hardness, measured as per ASTM D 2240, has values between D50 and D60. PTFE exhibits "plastic memory," that is, if subjected to tensile or compression stresses below the yield point, part of the resulting deformations remain after the discontinuance of the stresses. If the piece is reheated, the induced strains tend to release themselves within the piece, which resumes its original form.

Environmental resistance: PTFE is practically inert against known elements and compounds. It is attacked only by the alkaline metals in their elementary state, and by chlorine trifluoride and elementary fluorine at high temperatures and pressures. PTFE is insoluble in almost all solvents at temperatures up to about 300 °C. Fluorinated hydrocarbons cause a certain swelling that is, however, reversible; some highly fluorinated oils, at temperatures over 300 °C, exercise a certain dissolving effect. Resistance to high-energy radiation is rather poor.

Electrical properties: The dielectric strength of PTFE varies with the thickness and decreases with increasing frequency. It remains practically constant up to 300 °C and does not vary even after a prolonged treatment at high temperatures (6 months at 300 °C). PTFE has very low dielectric constant and dissipation factor values that remain unchanged up to 300 °C in a frequency field of up to 10 GHz, even after a prolonged thermal treatment.

19.2.3.4 **Boron Nitride**

Hexagonal BN powder exhibits the same characteristics of solid lubricants as seen in graphite and molybdenum disulfide. These include crystalline structure, low shear strength, adherence of the solid lubricant film, low abrasivity, and thermochemical stability [3]. In many instances, (h)BN exceeds the performance levels of these conventional solid lubricant characteristics, particularly adherence and thermochemical stability. Figure 19.5 [39] is a typical scanning electron micrograph (SEM) of a commercial BN powder.

Table 19.2 shows the effect of temperature of synthesis of BN on various properties, namely, surface area, crystallinity, coefficient of friction, and oxygen content [40]. It may be noted that with increase in synthesis temperature, the coefficient of friction decreases. Typical properties of (h)BN relevant to its use as a filler are summarized in Table 19.1.

Figure 19.6 [39, 40] compares the coefficient of friction at various temperatures for graphite, MoS_2, talc, and (h)BN. In contrast to (h)BN, graphite and MoS_2 undergo major increases in coefficient of friction (lose their lubricity) between 400 and 500 °C; talc shows increases in coefficient of friction at much lower temperature. The ability to retain lubricity at elevated temperatures is an important characteristic of (h)BN. Inorganic ceramic materials such as (h)BN have inherent advantages over polymers like PTFE and other low melting point materials. BN has an oxidation threshold of approximately 850 °C, and the rate of reaction is negligible even up to 1000 °C.

Figure 19.5 Scanning electron micrograph of BN powder (GE Advanced Ceramics-Grade: AC 6004). Reproduced with permission from Ref. [39].

Table 19.2 Effect of synthesis temperature on properties of (h)BN [39].

Temperature of synthesis (°C)	800	1400	1900	2000
Surface area (m^2/g)	50–100	20–50	10–20	<10
Crystallinity	Turbostatic	Quasiturbostatic	Mesographitic	Graphitic
Coefficient of friction	0.6	0.4	0.2–0.3	0.15
Oxygen content (%)	>5	1.5–5	0.5–1.5	<0.5

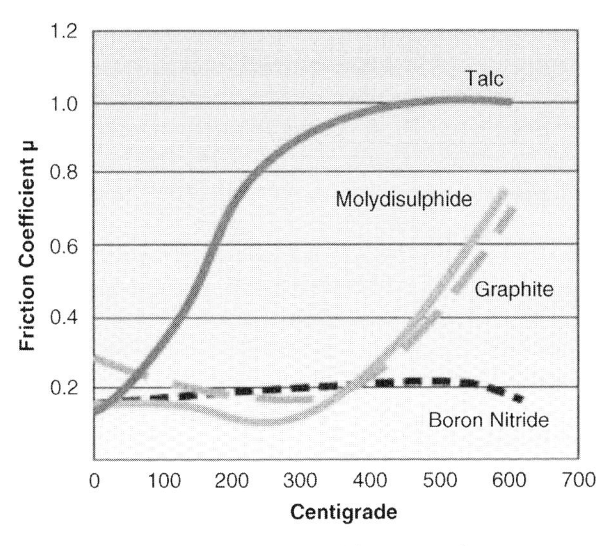

Figure 19.6 Changes in coefficient of friction as a function of temperature for different solid lubricants. Reproduced with permission from Ref. [39].

19.2.4
Suppliers/Manufacturers

The web site of the International Molybdenum Association (IMOA) [41] includes a list of the IMOA member companies that offer "unroasted" Mo concentrates that contain molybdenite. Table 19.3 lists suppliers/manufacturers of MoS_2. Table 19.4 lists suppliers/manufacturers of different types and grades of graphite. Tables 19.5 and 19.6 lists major suppliers/manufacturers of BN and PTFE powders, respectively.

19.2.5
Cost/Availability

Prices of MoS_2 powders depend on degree of fineness and order size, with prices in the range of $105/kg, to $1750/25 kg. The costs of various grades of PTFE powders for lots of 2000–3000 kg were (2004–2005) in the $5.5–6.5/kg range depending upon the composition/property of the product. The costs of BN powders could vary, depending upon quantity, from $230–335/kg for low density, off-white grades to $250–370/kg for higher density grades. High-purity grades can be purchased at $200–400/kg, depending upon quality and size of the order.

With respect to graphite, the following bulk price ranges per ton have been reported depending upon type, grade (particle size), purity, and quantity ordered: Flake, $650–2250; amorphous, $550–1000; vein, $2000–4000; synthetic, $650–2500; expandable, $2000–4000.

Flake graphite is available in sizes ranging from 0.5 mm flakes to 3 μm powder. The morphology of flake graphite is consistently laminar regardless of particle size. *Intumescent* graphite is available in purity ranging from 80 to 99% carbon. Both coarse and fine grades are available. The degree of intumescence, also known as "expandability," generally ranges from 80 to 300 times volume increase. Products can be specified as low (acidic), neutral, and high (alkaline) pH to allow compatibility with a variety of aqueous and nonaqueous coating/paint systems [15, 16, 18].

Crystalline vein graphite commercial grades are available from 85 to 99% carbon. Sized materials from 2.5 cm to 3 μm are available. Most of the current supply of *amorphous* graphite available in the United States is imported from Mexico and China. Amorphous graphite is typically lower in purity than other natural graphites. Commercial grades of *amorphous* graphite are available with purities in the range 75–85% and in sizes from 10 cm lumps to 3 μm powder. This graphite variety is typically lower in cost than other types, but is still lubricious, conductive, and chemically stable.

Synthetic graphite is typically available in purities above 99%. High purity is the rule, rather than the exception with this material because the feedstocks used to make it are typically petroleum-based materials that are inherently low in mineral impurities, and the manufacturing method tends to expel impurities, which are vaporized at the high-process temperatures. This variety of graphite is available in sizes from 1.2 cm down to 3 μm [15, 16, 18].

Table 19.3 Major suppliers/manufacturers of MoS_2.

Supplier/manufacturer	Grades
AML Industries, Inc. Warren, OH, USA www.amlube.com Fax: 1-330-399-5005	Amlube 510 (technical grade) Amlube 511 (fine technical grade)
Climax Molybdenum Company (A subsidiary of Freeport-McMoRan Copper & Gold Inc.) Phoenix, AZ, USA Tel: 1-800-255-7684 United Kingdom, The Netherlands, Japan, China www.climaxmolybdenum.com	Technical Technical fine Super fine (suspension)
Dow Corning Corporation Midland, MI, USA www.dowcorning.com Fax: +1 989 496 6731	Molykote® Z Powder (3–4 μm) Molykote 7365, 7495, 7604 Molykote D-29, D-55, D-78, D-79
Everlube Products Peachtree City, GA, USA (China, UK, France, India, Israel, Italy, Japan, S. Korea, Taiwan) www.everlubeproducts.com Tel: 1-800-428-7802	Dry/bonded film lubricant coatings with MoS_2, graphite, BN, PTFE: Everlube, Lube-Lok, Lubri-Bond, Ecoa-lube, Esnalube, Perma-Slik, Kal-Gard
Jinduicheng Molybdenum Mining Corp. Germany, Japan, P.R. China, USA e-mail: jck@jdcmmc.com e-mail: jdcjap@jdcmmc.com e-mail: jdceu@jdcmmc.com	Super fine grade Grade 1 Reagent grade
Strategic Metals Ltd Amersham, Buckinghamshire, UK www.strategicmetals.co.uk/ Fax: +44 (0) 1494 581159	MoS_2 powder in various grades
Langeloth Metallurgical Company, LLC (An Affiliate of Thompson Creek Metals) Langeloth, PA, USA www.langeloth.com Fax: 724-947-2240	High performance moly (HPM): Large particle grade 40.0 μm Technical grade 10.0 μm Technical fine grade 3.0 μm Super fine grade 1.6 μm
Thompson Creek Metals Company Englewood, CO, USA Fax: 303-761-7420	
M.K. Impex Canada Mississauga, Ontario, Canada www.lowerfriction.com Fax: 905-461-9238	MKN-MoS_2-050 (Nano-powder 50 nm)
McGee Industries/McLube Aston, PA, USA www.Mclube.com Tel: 1-800-262-5823	McLube MoS_2-98, 99, 100

Table 19.4 Major suppliers/manufacturers of graphite.

Producer/supplier	Grades/types
AML Industries, Inc. Warren, OH, USA www.amlube.com Fax: (330) 399 5005	Natural graphite powder: AmLube 611 High-purity graphite powder: AmLube 610, 613
Asbury Carbons Asbury, NJ, USA www.asbury.com Fax: 1-908-537-2908	Flake graphite, amorphous graphite, vein graphite, synthetic graphite, intumescent flake graphite (expandable graphite)
Superior Graphite Co. Chicago, IL, USA Fax +1 312 559 9064 Fax +1 800 542 0200 www.superiorgraphite.com Superior Graphite Europe Ltd Sundsvall, Sweden Fax + 46 60 13 41 28	*ThermoPURE products:* Purified crystalline flake and crystalline vein graphite: carbon (LOI): 99.7–99.9% Purified synthetic graphite: carbon (LOI): 99.7–99.9% *Signature products*: Natural crystalline vein and flake graphite: carbon (LOI): 80–99% Amorphous graphite: carbon (LOI): 60–90% (microcrystalline); synthetic graphite: carbon (LOI): 98.0–99.7%
Timcal Ltd Bodio, Switzerland Fax: 41 91 873 2019 Westlake, OH, USA Fax: 1-440-871-6026 www.timcal.com	*Synthetic graphite (Timrex)*: grades KS, T, SFG, MX/MB, HSAG, KB/KL, SLP *Natural flake graphite (Timrex)*: grades BNB, GA/GB
M.K. Impex Canada Mississauga, Ontario, Canada www.lowerfriction.com Fax: 905-461-9238	MKN-CG-400 (nanopowder 400 nm)

19.2.6 Environmental/Toxicity Considerations

19.2.6.1 Molybdenite

According to IMOA [6, 42], MoS_2 has been shown not to be harmful to rats by inhalation and by ingestion (MoS_2 in 1% w/v aqueous methylcellulose at a dose level of 2000 mg/kg bodyweight).

The 4 h LC_{50} (lethal concentration for 50% kill) in air has been reported to be more than 2.82 mg/l. The inhalation hazard associated with acute exposure to MoS_2 is low. MoS_2 has been shown not to be harmful to rats, guinea-pigs, and rabbits in contact with skin (acute lethal dermal dose >2000 mg/kg body weight), not to cause sensitization by skin contact, and not to be irritating to skin or eyes.

Table 19.5 Major suppliers/manufacturers of BN.

Supplier/manufacturer	Grades/types
Momentive Performance Materials, Inc. (formerly GE Advanced Ceramics) Albany, NY, USA www.momentive.com Fax: 1-440-878-5928 (USA) 49 4152 938 303 (Germany) 81 3 5114 3779 (Japan) 86 21 5079 3736 (China)	BN powder grades: HCP, HCPH, HCPL, AC6004, NX, HCV, AC6003, AC6097, AC6069, AC6028, AC6103, AC6110, AC6091, HCR48, HCJ48, HCM
Industrial Supply, Inc. Loveland, CO, USA Fax: 1-970-461-8429	(h)BN grades: PG (Premium), SG (Standard), CG (Custom)
National Nitride Technologies Co., Ltd Taichung Hsien, Taiwan Fax: 886-4-2276-6810 Pomona, CA, USA www.nntbn.com	BN powder grades: N, NA, NW, S, SW, SA,
Saint-Gobain Advanced Ceramics (formerly Carborundum Co.) Amherst, NY, USA Fax: 1-716-691-2090 www.bn.saint-gobain.com	Combat BN powder grades: MCFP, PHPP325A, PHPP325B, PCPS3005, PCPS302, PCPS308, PCPS3012, PCPS330, PSHP605, PSHP325, PSHP040 CarboTherm™ BN platelets grades: CTP-05, 2, 5, 8, 12, 30 CarboTherm Agglomerates grades: CTL7MHF, CTL20MHF, CTH7MHF, CTH10MHF
San Jose Delta Associates, Inc. Santa Clara, CA, USA www.sanjosedelta.com Fax: 1-408-727-6019	BN grades: HBN, HBR, HBC, HBT
M.K. Impex Canada Mississauga, Ontario, Canada www.lowerfriction.com Fax: 905-461-9238	MK-hBN powders: 70 nm, 150 nm, 1.5 μm, 5–30 μm
Zibo ShineSo Chemical New Material Co., Ltd Zibo, China www.shineso.com.cn Fax: +86-533-6280067	Hexagonal BN (hBN) grades: HE108, HS05, XK02, N70, HS05New Cubic BN (cBN) grades: SCBN110, 115, 120, 210, 230, 280
Denka Group Japan www.denka.co.jp Fax: +81-3-5290-5078	hBN powder grades: SP-2, HGP, GP, SGP

(Continued)

Table 19.5 (Continued)

Supplier/manufacturer	Grades/types
Denka Corporation New York, NY, USA Fax: 1-212-688-8727	
PlasmaChem GmbH Berlin, Germany Fax: +49 30 6392 6314 www.plasmachem.com	BN nanopowder: PL-IS-CBN

Table 19.6 Major suppliers/manufacturers of PTFE.

Producer/supplier	Grade
Asahi Glass Co., Ltd Tokyo, Japan www.agc.co.jp/english	Fluon™ PTFE powders: G100 series: fine particle powders G200 series: presintered extrusion powders G300 series: agglomerated (free flowing) Fluon® PTFE coagulated dispersion powders: CD1 series: low reduction ratio polymers CD0 series: high reduction ratio copolymers
Daikin America, Inc. Orangeburg NY, USA Tel: 1-800-365-9570 www.daikin.cc Daikin Industries, Ltd Osaka, Japan Fax: 81-6-6373-4390 Czech Republic, Belgium, Saudi Arabia, Germany, France, Netherlands, China, Hong Kong, Taiwan, Thailand, Singapore	DAIKIN-POLYFLON™ PTFE fine powders: F-104, F-104U, F-107, F-201, F-201L, F-205, F-207, F-208, F-301, F-303
DuPont Corp. Wilmington, DE, USA www.dupont.com/zonyl Tel: 1-866-828-7009 Fax: 1-302-992-2695	Zonyl: MP 1000, 1100, 1150, 1200, 1300, 1400, 1500, 1600N Zonyl: TE 3808, 3887, 6519
Dyneon LLC (a 3M Company) Oakdale, MN, USA www.dyneon.com Fax: +1 800 635 8061 Dyneon GmbH & Co. KG (Germany) Fax: +49 6107 772 517 Sumitomo 3M Limited (Japan) Fax: +81 33709 8743	Dyneon PTFE fine powders: TF 2021, 2025, 2029, 2053, 2071, 2072, 2073, TFX 2035, TFM 2001 Dyneon PTFE micropowders: TF 9201, 9205, 9207 J14, J24

Table 19.6 (*Continued*)

Producer/supplier	Grade
The Lubrizol Corporation Wickliffe, OH, USA www.lubrizol.com Tel: 1-440-943-4200	Micronized powder grades: Pinnacle 6001, 6003, 6005, 6007, 6020, 9000, 9001, 9002, 9003, 9007, 9008, 9500, 9600 Lanco 1792, 1793, 1795 NV, 1797, 1799, 1890 Lanco TF 1725, 1725 EF, 1778, 1780, 1780 EF, 1830 Lanco SM 2001, TFW 1765
Shamrock Technologies, Inc. Newark, NJ, USA www.shamrocktechnologies.com Fax: 973-242-8074	Fluoro-T 802, 803, 807, 811, 815 Fluoro-T 702, 707, 715 Fluoro-T 602, 603, 607, 611, 615 Fluoro MF 1253, 1437, 1437FG, 1433FG Micronized powders: SST series PTFE, FluoroSLIP series PTFE-PE Fluoro M290, E, HP, FG, Raven 5372
Solvay Solexis, S.p.A. Bollate (MI), Italy Fax: 39 02 3835 2129 Solvay Solexis, Inc. West Deptford, NJ, USA Fax: 1-856-853-6405 Japan, France, Brazil, China, India, Korea, Singapore, Taiwan, Thailand www.solvaysolexis.com	Micronized powder grades: Polymist: F-5, 5A, 5A EX, 510, XPP 500 series Algoflon: D 1200 series, D3200 series, DF 100, 200, 300 series, E, F & G series
Micro Powders, Inc. Tarrytown, NY, USA Fax: 1-914-472-7098 www.micropowders.com	Micronized PTFE: Fluo: HT, HTI-2, HTG, HT-LS, HTG-LS, 300, 300XF, 600 series, 750TX, 850TX Polyfluo (PTFE/PE combinations): 100, 200, 300, 400, 500 series

Up to 1998, there had been no reports of dermatitis in exposed workers. The following occupational exposure limits are applicable: OSHA (Occupational Safety and Health Administration) TWA (time-weighted average) (total dust), 10 mg/m^3; DFG (Deutsche Forschungsgemeinschaft) MAK (maximum concentration in the workplace) TWA (total dust), 15 mg/m^3; ACGIH (American Conference of Governmental Industrial Hygienists, Inc.) TLV (threshold limit values) TWA, 10 mg/m^3.

19.2.6.2 **Graphite**

The following exposure limits have been reported [43]: TLV, 2.0 mg/m^3 as respirable dust (ACGIH 1996–1997); OSHA PEL (permissible exposure limits), TWA: 15 mppcf (million particles per cubic foot); NIOSH REL (recommended exposure limits), TWA: 2.5 mg/m^3; NIOSH IDLH (immediately dangerous to life or health concentration): 1250 mg/m^3.

With regard to effects of long term or repeated exposure to dust, lungs may be affected resulting in graphite pneumoconiosis. OSHAs proposed 8 h TWA PEL for synthetic graphite was 10 mg/m^3 (total particulate), and this limit is established by the final rule; the 5 mg/m^3 limit for the respirable fraction is retained. The ACGIH also has a TLV–TWA limit of 10 mg/m^3 for graphite as total dust [44].

19.2.6.3 **Polytetrafluoroethylene**

PTFE is not classified as dangerous according to European Commission directives. It should be handled in accordance with good industrial hygiene and safety practices. This material has not been tested for environmental effects and as per the International Agency for Research on Cancer (IARC), it belongs to Group 3, unclassifiable as to carcinogenicity to humans. For PTFE thermal decomposition products, air concentration should be controlled since they are likely to be toxic monomer, mono- and difluoroacetic acid, trifluoroacetate, hexafluoropropene, and so on [45].

19.2.6.4 **Boron Nitride**

BN is a nonflammable, nonreactive solid material. It is supplied in the form of an odorless, white powder and is considered as a nuisance dust [46]. Exposure to BN has not been shown to result in direct poisoning. Under normal operating conditions or thermal decomposition, BN has not been shown to liberate free boron. According to MSDS-102 [47], BN powders may contain about 0.1–8.0% boron oxide (CAS # 1303-86-2) that may cause irritation, alter kidney function, and produce changes in the blood as a result of occupational exposure. Boron oxide may cause reversible effects that are generally not life threatening. The exposure limits for boron oxide are as follows: ACGIH, 10 mg/m^3 TWA; OSHA, total dust 10 mg/m^3 TWA; NIOSH, 10 mg/m^3 TWA.

19.2.7
Applications

Some typical applications of MoS_2, graphite, PTFE, and BN in various plastics are listed in Table 19.7. Specific recent examples include

- A process for preparing high-strength UHMWHDPE (ultrahigh molecular weight high-density polyethylene) composite plastics, useful for manufacturing mechanical drives and rotation parts, contains 5–15% MoS_2 dry powder [48].
- The outer edges of the blades of the rotor used in air motor are made of polyamide or acetal copolymer containing MoS_2 dry lubricant [49].
- A helical blade used in fluid compressors made of plastics, such as PTFE, PFA (perfluoroalkoxy resin), PEEK (polyetheretherketone), PES (polyether sulfone),

Table 19.7 Various applications of MoS_2, Graphite, PTFE and BN in different polymers.

Polymer(s)	MoS_2	Graphite	PTFE	BN	References
Polyolefins	X	X	X	X	[85–87]
Acrylonitrile rubber	X				[88]
Polyurethane	X				[89]
PTFE copolymer	X	X			[90]
Epoxy	X	X			[91]
Acetal, HDPE	X		X		[89, 92, 93]
EPR, NBR, rubber products	X	X	X		[94–96]
Nylon 12				X	[97]
PTFE, PA, PAI, PI, PPS	X	X	X	X	[98–101]
Epoxy			X		[102]
Thermoplastics			X		[103]
PPQ (poly(phenyl quinoxaline))		X			[104]
Thermoset (Kerimid)		X			[105]
PC, polyester, phenolic	X				[106]
PE, FEP			X	X	[107]
PCO (polycyclooctene)				X	[108]

PEI (polyether imide), PAI (polyamide imide), PPS (polyphenylene sulfide), or LCP (liquid crystal polymer), contains solid lubricants, including MoS_2, graphite, BN, and so on [50].

- The screens for waste water filtration are made of fiber-reinforced plastic containing solid lubricants such as MoS_2 [51].
- Flame and smoke suppression activity in PVC and PVDC compositions has been reported for MoS_2, although MoO_3 and ammonium octamolybdate are almost exclusively used in these applications [14, 52].
- MoS_2 nanostructures are of great interest for a wide variety of nanotechnological applications ranging from the potential use of inorganic nanotubes in electronics to the active use of nanoparticles in heterogeneous catalysis. By analyzing the atomic-scale structure of clusters, the origin of the structural transitions occurring at unique cluster sizes has been identified. The novel findings suggest that good size control during the synthesis of MoS_2 nanostructures may be used for the production of chemically or optically active MoS_2 nanomaterials with superior performance [53].
- Combinations of MoS_2 and graphite have generally been found to exhibit a synergistic effect in extreme pressure and antiwear characteristics, with the level of synergism depending on the ratio of two components [7].
- With the addition of 30% graphite, the friction coefficient of polyamide-6 was reduced by 30% with only a small increase in wear, while for polystyrene, both friction coefficient and wear were decreased. A further increase in the graphite concentration increased both the wear and the friction coefficient [34]. Yan *et al.* [54] reported similar observations for a PTFE/graphite system and ratio-

nalized the increased wear rate in terms of increased porosity with an increase in graphite concentration.

- Various micronized polytetrafluoroethylene powders were compounded with silicone rubber (MQ) and the mechanical properties of the composites were evaluated. At a PTFE level of only 5 wt%, the fractured surface of the composites showed layered structure morphology. This structure effectively improved the tear strength of the MQ but it also lowered the tensile properties of the composites. The addition of fluorosilicone rubber (FMQ) as a compatibilizer, improved considerably the tensile and tear strength of the composites. Extrusion of the MQ/PTFE/FMQ composites on an electric wire indicated that the spherical PTFE powder was suitable for the extrusion process [55].
- Electron beam modification of PTFE nanopowder resulted in increasing concentration of radicals and carboxylic groups (–COOH) with increase in irradiation dose. Low-temperature reactive mixing of the modified PTFE with ethylene-propylene-diene-monomer (EPDM) rubber produced PTFE coupled EPDM rubber compounds with the desired physical properties due to the formation of a compatible interphase; this was confirmed by transmission electron microscopy (TEM) and differential scanning calorimetry (DSC) [56].
- For PES lubricated with 20% PTFE, the dynamic coefficient of friction decreases from 0.37 to 0.11 and the wear factor drops from 1500 to 3 [57].
- Optimum PTFE loading of 15% in amorphous and elastomeric base resins and 20% for crystalline base resins provide the lowest wear rates. Higher PTFE loadings have minimal effects in terms of further reduction in wear rate, although the coefficient of friction will continue to decrease [56]. The effects of PTFE on the wear characteristics of various engineering resins are strongly dependent on the type of resin [57].
- In addition to its primary function as a lubricating filler and wear-resistant barrier (Table 19.7) for applications such as sliding or rotating motion [39, 58], secondary functions of the nonabrasive hexagonal BN powder are (a) enhancement of thermal conductivity [59] (see also Chapter 5) due to excellent particle-to-particle contact that provides a superior thermal path, (b) rheology modification during processing of PE films, (c) enhancement of dielectric strength, (d) high-temperature composite stability due to its chemical inertness and low coefficient of thermal expansion, (e) improvement of crystal nucleation efficiency during processing, and (f) enhanced mold release characteristics.

19.3 Antiblocking Fillers

19.3.1 General

For plastic sheets/films, "blocking" is generally defined as a condition that occurs when two or more sheets/films stick together when stacked on top of one another.

For coatings, blocking is a measure of the ability of a given coating to resist adhesion to itself (or another freshly coated surface) or adhesion to another substrate [18]. Thus, an antiblocking agent can be defined as an additive that prevents the undesirable sticking together or adhesion of coated surfaces under moderate pressure, or under specified conditions of temperature, pressure, and humidity. The antiblocking agents function by producing invisible, microsurface imperfections on the film/coating surface, which entrap air and voids on a microscopic scale, thereby reducing the adhesion/sticking of the film layers to each other [60]. In other words, antiblocking additives in particulate form simply act as "spacer-bars" between the two film layers/surfaces [61], lowering the interlayer blocking (adhesive) force. Two parameters, namely the particle size and the number of antiblock particles on the film surface, dominate the antiblocking effect, besides other factors such as type of film material. It has been shown that the higher the surface roughness (i.e., the more particles that there are on the surface) the better the antiblocking performance. In general, small particle size antiblocks are preferred for thin films (20–30 μm), while coarse particles are employed for thick films (>30 μm). Agglomerates of antiblocks reduce the antiblocking performance [61]. The selection of the appropriate antiblock depends on the polymer type and the quality requirements of the final film product.

Several inorganic fillers/organic additives, such as silica, talc, kaolin, $CaCO_3$, titania, zeolites, cross-linked acrylic copolymers, spherical silicon beads, and so on, are employed in the plastics/coatings industry to attain the desired blocking performance. Some of these fillers are discussed elsewhere in this book in terms of their primary function; only amorphous silica forms (natural and synthetic), used for antiblocking, will therefore be discussed in this chapter.

Silicas, which are in competition with carbon blacks as functional fillers for plastics and rubbers, have one significant advantage: their white color [62]. The most important role of silicas is as elastomer reinforcements, inducing an increase in the mechanical properties. Other functions, in addition to their use as antiblocks for PE, PP, and other films, are (a) to promote adhesion of rubber to brass-coated wires and textiles, (b) to enhance the thermal and electrical properties of plastics, (c) in accumulator separators, and (d) as rubber chemical carriers.

As fumed silica enhances several different properties in formulation, such as thickening liquids or improving flowability of powders, it is used in a variety of applications [63], including plastics, rubbers, coatings, cosmetics, adhesives, sealants, inks, and toners.

19.3.2 Silica as Antiblocking Filler

19.3.2.1 Production

The term silica is used for the compound silicon dioxide, SiO_2, which has several crystalline forms as well as amorphous forms, and may be hydrated or hydroxylated [64]. It is the chemical inertness and durability of silica that has made it very popular in many applications [34].

Figure 19.7 An example of a DE deposit with a single principal Diatom present, 1000×. Courtesy of World Minerals Inc.

Natural silicas can be divided into crystalline and amorphous. Crystalline varieties include sands, ground silica (silica flour), and a form of quartz – Tripoli. The amorphous types used as antiblocks include diatomaceous earth (DE) or diatomite. DE is a chalky sedimentary rock composed of the skeletons of single-cell aquatic organisms, the diatomites, grown in a wide variety of shapes and varying in size from 10 μm to 2 mm. Figure 19.7 shows an example of a DE deposit. The skeletons are composed of opal-like, amorphous silica ($SiO_2(H_2O)_x$) having a wide range of porous, fine structures. The natural grades are uncalcinated powders classified according to particle size distribution. During the calcination process, the moisture (~40% in DE) is also removed due to high-process temperature, which may also cause sintering of DE particles to clusters.

Amorphous synthetic silicas, used as antiblocks, are produced by two different processes: pyrogenic or thermal (generally referred to as fumed silica grades) and wet process (known as precipitated or particulated silica). The ingredients used in the manufacture of fumed silica are chlorosilanes, hydrogen, and oxygen, and the process involves the vapor phase hydrolysis of silicon tetrachloride in a hydrogen oxygen flame [64]. The reactions are

$$2H_2 + O_2 \rightarrow 2H_2O$$
$$SiCl_4 + 2H_2O \rightarrow SiO_2\downarrow + 4HCl\uparrow$$

(Overall reaction)

$$SiCl_4 + 2H_2 + O_2 \rightarrow SiO_2\downarrow + 4HCl\uparrow$$

Silica can be precipitated from a sodium silicate solution by acidifying (by sulfuric or hydrochloric acid) to a pH less than 10 or 11, usually using a lower concentration than in the silica gel preparation described below. Typical reactions are

$$3SiO_2 + Na_2CO_3 \rightarrow 3SiO_2 \cdot Na_2O + CO_2$$

$$(SiO_2 \cdot Na_2O)_{aq.} + 2H^+ + SO_4^{2-} \rightarrow SiO_2 + Na_2SO_4 + H_2O$$

The product is separated by filtration, washed, dried, and milled. The final product properties, such as porosity, specific surface area, size and shape of particles and agglomerates, density, hardness, and so on, are dependent on process variables such as reactant concentration, rates of addition, temperature, and fraction of theoretical silicate concentration in the reaction [34].

Synthetic silica gel, $SiO_2(H_2O)_x$, is a solid, amorphous form of hydrated silicon dioxide distinguished by its microporosity and hydroxylated surface [61]. It is produced according to the following reaction [34]:

$$Na_2O(SiO_2)_x + H_2SO_4 \rightarrow xSiO_2 + Na_2SO_4 + H_2O$$

The product, containing about 75% water, is dried in a rotary kiln, washed with hot alkaline water (which reinforces the matrix, decreases shrinkage, and produces larger pores), and milled to produce xerogels. Replacing the water with methanol before drying or supercritical drying produces aerogels with up to 94% air space.

In a recent US patent [65], a new process for producing ultrafine silica particles has been described. It involves directing a plasma jet onto a silicon-containing compound so as to form silica vapor, and condensing this vapor on a collection surface. The silica particles have high porosity and surface area and may be used as catalyst supports or as antiblocking or antislipping agents in plastic films.

19.3.2.2 **Structure/Properties**

Synthetic silica gel is a major antiblocking filler used for high-quality film applications. The structure of silica gel is an interconnected random array of polymerized spheroidal silicate particles with 2–10 nm diameter and 300–1000 m^2/g surface area [61]. Its refractive index of 1.46 is very close to those of PE (~1.50) and PP (~1.49), which helps in retaining the high transparency and clarity of the films. Because of the highly porous structure, synthetic silica gel provides many particles per unit weight, thereby directly enhancing the antiblocking performance versus a nonporous material. Its high purity, >99.5 wt%, and low level of deleterious impurities further minimize the likelihood of any major quality deterioration of polymer films [61]. Amorphous diatomite, DE, used in commercial applications contains 86–94 wt% SiO_2, with a skeletal structure that contains 80–90% voids. Various properties of diatomite, fumed silica, and precipitated silica are compared in Table 19.8 [61, 64]. Further information on fumed silica may be found in Chapter 20.

19.3.2.3 **Suppliers/Manufacturers**

The Unites States is the world's largest producer and consumer of diatomite. Other major producers include Russia, France, Germany, Mexico, Spain, Italy, Brazil, Peru, and so on.

Table 19.9 lists various suppliers/manufacturers of various types/grades of silica antiblocks.

19.3.2.4 **Environmental/Toxicity Considerations**

In evaluating the effects of exposure to silica, it is important to differentiate between amorphous and crystalline forms. According to NIOSH [66], at least 1.7 million U.S. workers are exposed to respirable crystalline silica in a variety of industries and

Table 19.8 Comparison of properties of fumed silica versus diatomite and precipitated silica [61, 64, 67].

Property	Diatomite, DE	Fumed silica	Precipitated silica
CAS #	61790-53-2	112945-52-5 69012-64-2	112926-00-8 7699-41-4
SiO_2 (%)	85.5–92.0	96.0–99.9	97.5–99.4
CaO (%)	0.3–0.6	trace	0.5
Na_2O (%)	0.5–3.6	Trace	0–1.5
Loss on ignition (%)	0.1–0.5	1.0–2.5	3–18
Decomposition temperature (°C)	2000	2000	2000
Thermal conductivity (W/(m K))	0.015	0.015	0.015
Thermal expansion coefficient (K^{-1})	0.5×10^{-6}	0.5×10^{-6}	0.5×10^{-6}
Specific heat (J/(kg K))	794	794	794
Surface area (m^2/g)	0.7–3.5	15–400	45–700
pH, aq. suspension	9–10	3.5–8	4–9
Water solubility (g/100 ml)	None	0.015	0.015
Acid solubility	None, exp. HF	None, exp. HF	None, exp. HF
True density (g/cm^3)	2.0–2.5	2.16	2.0–2.1
Hardness (Mohs)	NA	6.5	1.0
Refractive index	1.42–1.48	1.45	1.45
Volume resistivity (Ω cm)	NA	10^{13}	10^{11}–10^{14}
Surface resistivity (Ω cm)	5×10^9	5×10^9	5×10^9
Dielectric constant (10^4 Hz)	1.9–2.8	1.9–2.8	1.9–2.8

Table 19.9 Major suppliers/manufacturers of antiblock silicas.

Supplier/manufacturer	Grades/types
Cabot Corporation World Headquarters: Boston, MA, USA www.cabot-corp.com Fax: 1-617-342-6103	Cab-O-Sil (hydrophilic fumed silica): LM-150, M-5, M-5P, MS-75, H-5, HS-5, EH-5 HP-60, PTG Cab-O-Sil (hydrophobic fumed silica): TS-500, TS-530, TS-610, TS-720 Cab-O-Sil (densified silica grades): LM-150D, M-5DP, M-7D, M-75D
Evonik Degussa GmbH Frankfurt, Germany www.evonik.com Fax: +49-69-218-2533 Evonik Degussa Corp. Parsippany, NJ, USA Fax: 1-973-541-8502	Sipernat 101M, 120, 160, 200, 35, 310, 320, 50, 500LS, 880, D10, D17, and so on Sident 8, 9, 10, 22S, and so on Ultrasil 360, 880, 7000GR, AS7, VN2, 866, and so on Hydrophilic fumed silica: AEROSIL 90, 130, 150, 200, 300, 380, OX50, EG50, TT600 Hydrophobic fumed silica: AEROSIL R 972, 104, 202, 711, 812, 7200, 8200, 9200, and so on Special hydrophobic silica: AEROSIL RY50, RY200, RX300, DT4, LE1, and so on

Table 19.9 (*Continued*)

Supplier/manufacturer	Grades/types
Fuji Silysia Chemical, Ltd Aichi-ken, Japan www.fuji-silysia.co.jp Fax: + 81-568-51-8557 Res. Triangle Park, NC, USA Fax: 1-919-544-5090	Sylysia: (micronized, synthetic, amorphous, pure, colloidal silica) Sylophobic Sylosphere Fuji Baloon
Dicalite, Dicaperl Minerals, Inc., Bala Cynwyd, PA, USA Fax: 1-610-822-9119 www.dicalite.com	Diatomaceous earth fillers: Dicalite
Harwick Standard Distribution Corp. Akron, OH, USA Fax: 1-330-798-0214 www.harwickstandard.com	Silica S (fumed silica) HI-SIL 132, 135, 900, 915 (synthetic silicas)
INEOS Silicas Americas Joliet, IL, USA Fax: 1-815-727-5312 www.ineossilicas.com	Synthetic silica grades: Gasil AB705, 710, 720, 725, 905, 920, 23D, 200DF, 114, EBN
Minerals Technologies Inc. New York, NY, USA Tel: 1-212-878-1800 www.mineralstech.com	Optibloc – 8, 10, 25, and so on (clarity antiblock) Clear-Bloc 80 Sylobloc 45 (synthetic silica) ABT – 1000, 2500, 2501 Microbloc Polybloc
OCI International, Inc. Subsidiary of OCI (Hong Kong) Ltd Houston, TX, USA Fax: 1-832-379-0002 http://ocicorp.co.kr/e_index.asp	Micloid (micronized silica)
PPG Industries, Inc. Pittsburgh, PA, USA **PPG Silica Products** Monroeville, PA, USA Tel: 1-800-243-6745 www.ppg.com	Precipitated silicas: Hi-Sil: 132, 134G,135, 190G, 190G-M, 210, 233, 233D, 255C-D, 532EP, 900, 915, Lo-Vel: 27, 275, 29, 6200, 2003, 4000, 6000, HSF, 39A (nontreated) Lo-Vel: 66, 8100, 2023, 2033, 2010, 2018 (wax-treated) Inhibisil: 33, 73, 75 Ciptane: I, LP
Sibelco Dessel, Belgium Fax: + 32 14 83 72 12 www.sibelco.be	Sibelite M72 Sibelite Cristobalite flour

(*Continued*)

Table 19.9 (*Continued*)

Supplier/manufacturer	Grades/types
Unimin Corp. New Canaan, CT, USA Fax: 1-203-966-3453 www.unimin.com	Microcrystalline silica
W.R. GRACE & Co. Columbia, MD, USA www.gracedavison.com **Grace Silica GmbH** Düren, Germany W.R. Grace Specialty Chemicals (Malaysia) Grace Brasil Ltda	LUDOX (colloidal silica for antiblocking) PERKASIL KS & SM (precipitated silica) SYLOWHITE SYLOBLOC SYLOJET SYLOID (synthetic amorphous silica gel) DURAFILL
Wacker-Chemie AG Munchen, Germany Fax: +49 89 6279-1770 www.wacker.com	Hydrophilic amorphous silica: HDK C10, C10P, N20, N20P, N20ST, S13, T30, T30P, T40, T40P, V15, V15P Hydrophobic amorphous silica: HDK H13L, H15, H20, H30,
World Minerals Inc. Santa Barbara, CA, USA Fax: 1-805-735-7981 www.worldminerals.com World Minerals Europe S.A., France Fax: 33 1 49 55 66 57 World Minerals Asia Pacific, China Fax: 86 10 6 567 3168 DC Chemical Seoul, South Korea www.dcchem.co.kr	ActivBlock (for polyolefins) Diatomite (natural silica) products: Celite 209, 350, 350MC, Celpure, Micro-Ken 801, Kenite, Diactiv 17, Primisil Diafil 525, 570, 590 CelTiX SuperFloss, A, E, MX, SuperFineSuperFloss, Snow Floss, White Mist PF Synthetic silicates brands: Micri-Cel, Celkate, Calflo KONASIL K-90, K-150, K-200, K-300 (fumed silica)
Guangzhou GBS High-Tech & Industry Co., Ltd Guangzhou, China www.gzgbs.com	HL-150, 200, 200B, 300, 380 (hydrophilic) HB-215, 220, 615, 620, 630, 720 (hydrophobic)

occupations. Silicosis, an irreversible but preventable disease, is an illness most closely associated with occupational exposure to the material, which also known as silica dust. Occupational exposures to respirable crystalline silica are associated with the development of silicosis, lung cancer, pulmonary tuberculosis, and diseases of the respiratory tract. Crystalline silica has been classified by IARC as carcinogenic to humans when inhaled.

In general, fumed or pyrogenic silica, which is X-ray amorphous, is nontoxic, does not cause silicosis, and is safe to work with. However, excessive inhalation should be avoided by appropriate ventilation or the wearing of protective masks [67].

OSHAs current limit for amorphous silica is 20 mppcf, which is equivalent to 6 mg/m^3 TWA (ACGIH 1984), measured as total dust. The ACGIH has established a limit for this dust (measured as total dust containing <1% quartz) of 10 mg/m^3 (8 h TLV–TWA). In contrast, TLVs for quartz, cristobalite, and tridymite dusts are only 0.05 mg/m^3. Diatomaceous earth is largely noncrystalline; the presence of varying amounts of crystalline quartz has led, in the opinion of the ACGIH (1986/Ex. 1–3, p. 520), to conflicting results in studies of the pulmonary effects of exposure to this colorless to gray, odorless powder. Acute toxicity LD_{50}/LC_{50}-values for amorphous silica relevant to classification are oral > 5000 mg/kg rat, dermal > 5000 mg/kg rabbit, inhalation > 0.139 mg/l/4 h rat. In general, according to FDA regulations, amorphous silica is generally recognized as safe (GRAS).

19.3.2.5 Applications of Silicas

Polyolefin (PE, PP) and polyester (PET) films are inherently prone to "blocking." "Blocking" is particularly common in films produced by casting, blown film extrusion, biaxial orientation, tubular water quench, and calendering. In general, typical concentration of 2500–4000 ppm of natural silica is employed for LDPE, while 1000–2000 ppm of synthetic silica is employed for PP, LDPE, LLDPE, and PET. The benefits of antiblocking silicas are summarized in Ref. 68. Some recent applications of silica antiblocks for plastic films are given below.

Davidson and Swartz [60] have studied the antiblocking performance of various synthetic silicas, talc, and diatomaceous earth fillers for high-clarity polyethylene film applications and reported antiblocking performance versus haze, yellowness index (YI), coefficient of friction (COF), and gloss for LDPE, LLDPE, and mLDPE. The antagonism between five commercial grades of erucamide and seven commercial grades of inorganic antiblock agents used in LDPE film formulations has been investigated by Peloso *et al.* [69]. Use of silica from rice husk ash (RHA) as an antiblocking agent in LDPE film has been studied by Chuayjuljit *et al.* [70]. It was found that LDPE film with 2000–3000 ppm RHA silica showed similar properties to commercial LDPE films filled with 500–1000 ppm silica in terms of blocking behavior, mechanical strength, and film clarity. It should be noted that diatomaceous earth, in addition to its traditional antiblocking applications in films and sheets, has found other uses such as nucleation site for polyurethane foams, reinforcement in siloxane (RTV) with extension of the pyrogenic silica, and as an additive in various thermoset formulations [71].

In a review article, Shiro [72] has compiled the applications of antiblocking fillers, such as silica, kaolin, talc, and so on, in plastics, transparent films, magnetic recording materials, electrical insulating films, and so on. Polyester films and coatings for magnetic recording materials with improved antiblocking property, adhesion and solvent resistance are prepared using silica as the antiblocking filler [73]. Similarly, pressure-sensitive adhesives, with good antiblocking property, heat resistance, and abrasion resistance, containing silica gel have been developed [74]. Thermoplastic films for use in stretch/cling wrapping are reported to contain a silica-based antiblocking polyolefin layer on the opposite side from the cling layer [75]. Coating compositions for imparting a suede finish to leather substitutes contain polyurethanes, polybutylene, and silica [76].

Synthetic silicas are used preferentially as antiblocks in PP and PVC films compared to diatomite and other mineral fillers. For thicker films, coarser silicas with an average particle size of about 11 μm are used. The advantages of using synthetic silicas are their high efficiency and adsorption capacity (due to high porosity) for PVC plasticizers. As a secondary function, antiblocking silicas not only create a microrough surface, but also improve the printability of PVC films. Antiblocks are also used in the skin layer of multilayer films containing polymers such as PA, EVOH, ABS, and so on, and in various applications of polyester films, such as magnetic tapes, video tapes, packaging films, capacitors, and so on. For magnetic films, very fine silica (<1–2 μm) or submicron-sized calcium carbonate or China clay are used besides other minerals such as barites and aluminum oxides [61].

References

1 Rohr, F E. (2003) *Paint Coat. Ind.*, Defining and Predicting Performance of Surface Modifiers in Coatings, **19** (10), 110, accessed at www.pcimag.com.

2 SpecialChem. S.A . (2003) Additives for tribological polymers used in bearing applications, article accessed at www.specialchem4polymers.com/resources/articles.

3 TribolCogy-abc, Solid lubricants/dry lubrication at www.tribology-abc.com/abc/solidlub.htm.

4 GE Advanced Ceramics, Lubrication at www.advceramics.com/geac/applications/lubrication/.

5 Molybdenite at www.mindat.org/min-2746.html.

6 National Institute for Occupational Safety and Health (US NIOSH), Molybdenum sulfide at www.cdc.gov/niosh/rtecs/qa47aba8.html.

7 Klofer, A. (1997) A study of synergistic effects of solid lubricants found in lubrication packages of constant velocity joints, Thesis, Imperial College of Science Technology and Medicine, London, March 14, accessed at www.mitglied.lycos.de/akloefer/Technik2/Technik2.html.

8 National Institute for Occupational Safety and Health (US NIOSH), Graphite at www.cdc.gov/niosh/rtecs/md9364d0.html.

9 Gapi Group, Virgin PTFE at www.gapigroup.com/ptfe2.HTM.

10 Ceram Research, Boron nitride at www.azom.com/details.asp?ArticleID=78.

11 Accuratus, Boron nitride at www.accuratus.com/boron.html.

12 National Institute for Occupational Safety and Health (US NIOSH), Boron nitride at www.cdc.gov/niosh/rtecs/ed7704c0.html.

13 International Molybdenum Association UK (IMOA) at www.imoa.org.uk/.

14 Risdon, T.J., Properties of molybdenum disulfide at http://www.climaxmolybdenum.com/NR/rdonlyres/793808C9-200E-4E1F-B2AE-A816899D1519/0/PropertiesMoSulfide_1107.pdf.

15 Asbury Carbons, Graphite at www.asbury.com.

16 Tamashausky, A.V. (2003) Graphite – a multifunctional additive for paint and coatings. *Paint Coat. Ind.*, **19** (10), 64, accessed at www.pcimag.com.

17 Graphtek LLC, Graphite at www.graphtekllc.com/graphite.htm.

18 Koleske, J.V. *et al.* (2008) Two thousand eight – additives guide. *Paint Coat. Ind.*, **24** (6), 38.

19 Gangal, S.V. (1996) Fluorine compounds, organic, in *Kirk-Othmer Encyclopedia of Chemical Technology*, 3rd edn, vol. 11, John Wiley & Sons, Inc., New York, p. 4.

20 Bladel, H., Felix, B., Hintzer, K. Lohr, G., and Mitterberger, W.D. (1994) US Patent 5,576,381, Hoechst Aktiengesellschaft, Germany
21 Giannetti, E. and Visca, M. (1987) US Patent 4,864,006, Ausimont S.p.A., Italy
22 Visca, M. and Chittofrati, A. (1988) US Patent 4,990,283, Ausimont, S.p.A., Italy.
23 Marchese, E. and Kapeliouchko, V., High solids-content PTFE nanoemulsions, Ausimont S.p.A., Italy, accessed at www.fondazione-elba.org/73.htm.
24 The mineral molybdenite at http://mineral.galleries.com/minerals/sulfides/molybden/molybden.htm.
25 Molybdenite gallery at www.mindat.org/gallery.php?min=2746.
26 Sulphides at www.geocities.com/ijkuk/ik_sulphide.htm.
27 Molybdenite at www.theimage.com/mineral/molybdenite/index.htm.
28 Molybdenite at www.webmineral.com/data/Molybdenite.shtml.
29 Molybdenite at www.gwydir.demon.co.uk/jo/minerals/molybdenite.htm.
30 Systematic mineralogy – sulfides at http://geology.csupomona.edu/drjessey/class/gsc215/minnotes17.htm.
31 Molybdenite at www.chem4kids.com/misc/photos/molybdenite1.jpg.
32 Molybdenite at www.mineralgallery.co.za/molybdenite.htm.
33 Mineral description: molybdenite at www.geology.neab.net/minerals/molybden.htm.
34 Wypych, G. (2000) *Handbook of Filllers*, ChemTec Publishing, Toronto, Ont., Canada.
35 Hendrickson, R.R. *et al.* (1973) US Patent 3,741,585.
36 Dry/solid lubricants at www.lowerfriction.com.
37 Graphite at http://mineral.galleries.com/minerals/elements/graphite/graphite.htm.
38 Thermoplastic grade PTFE additives at www.shamrocktechnologies.com.
39 Lelonis, D.A. *et al.*, Boron nitride powder – a high performance alternative for solid lubrication, article accessed at www.advceramics.com.
40 Deacon, R. (1957) *Proc. Roy. Soc.*, **243A**, 464.
41 International Molybdenum Association UK (IMOA), Concentrates – unroasted, at www.imoa.info/Default.asp?Page=197.
42 International Molybdenum Association UK (IMOA), Molybdenum disulfide at www.imoa.org.uk/Default.asp? Page=155.
43 National Institute for Occupational Safety and Health (US NIOSH), Graphite (natural) at www.cdc.gov/niosh/ipcsneng/neng0893.html.
44 National Institute for Occupational Safety and Health (US NIOSH), Graphite (synthetic) at www.cdc.gov/niosh/pel88/syngraph.html.
45 Ellis, D.A. *et al.* (2001), Thermolysis of fluoropolymers as a potential source of halogenated organic acids in the environment, *Nature*, **412**, 321.
46 Electronic Space Products International, MSDS of BN at www.espi-metals.com/msds's/Boron%20Nitride.htm.
47 Saint-Gobain Advanced Ceramics, MSDS 102 for Combat BN powders, Revision Date: 02/09/2004, Amherst, NY.
48 Yichuan, J. (2002) Faming Zhuanli Shenqing Gongkai Shoumingshu, China, CN 1338484, AN: 2003:204208
49 Cooper Power Tools GmbH & Co . (2002) DE 20021980, Germany, AN: 2002:446050
50 Fujiwara, T. *et al.* (2002) US Patent 6,354,825, Kabushiki Kaisha Toshiba, Japan.
51 Hiroshi, K. and Koji, S. (2004) JP 2004042010, Yamayichi Technos K.K., Japan, AN: 2004:117492
52 Kroenke, W.J. (1979) US Patent **4**, 153, 792.
53 Lauritsen, J.V. *et al.* (2007) Size-dependent structure of MoS_2 nanocrystals. *Nat. Nanotechnol.*, **2**, 53–58, at www.nature.com/nnano.
54 Yan, F. *et al.* (1996), The Correlation of Wear Behaviors and Microstructures of Graphite–PTFE Composites Studied by Position Annihilation, *J. Appl. Polym. Sci.*, **61** (7), 1231–1236.
55 Park, E.-S. (2008) Processibility and mechanical properties of micronized polytetrafluoroethylene reinforced silicone rubber composites. *J. Appl. Polym. Sci.*, **107**, 372–381.
56 Khan, M.S. *et al.* (2008) Modification of PTFE nanopowder by controlled electron

beam irradiation: useful approach for the development of PTFE coupled EPDM compounds. *eXPRESS Polym. Lett.*, **2** (4), 284–293.
57 LNP engineering plastics, A guide to plastic gearing at www.lnp.com/LNP/Products/AvailableBrochures/WearResistant.html.
58 Saint-Gobain Advanced Ceramics Corp., Boron nitride at www.bn.saint-gobain.com.
59 Zhuo, Q. *et al.* (2000) WO 00/42098, Ferro Corporation.
60 Davidson, D.L. and Swartz, D. (2004) Proceedings of the 62nd SPE ANTEC, vol. 50, p. 1111.
61 Zweifel, H. (ed.) (2001) Chapters 7 and 17, in *Plastics Additives Handbook*, Hanser Publishers, Munich.
62 Biron, M. Silicas as polymer additives, article accessed at www.specialchem4polymers.com/resources/articles/article.aspx?id=1295.
63 Cabot Corp., Untreated fumed silica: general application guide, Technical Bulletin at www.cabot-corp.com.
64 Willey, J.D. (1985) Amorphous silica, in *Kirk-Othmer Concise Encyclopedia of Chemical Technology*, 3rd edn, John Wiley & Sons, Inc., p. 1054.
65 Debras, G. (2002) US Patent 6,495,114, Fina Research, S.A., Feluy (BE).
66 National Institute for Occupational Safety and Health (US NIOSH), Silica at www.cdc.gov/niosh/topics/silica/default.html.
67 Katz, H.S. and Milewski, J.V. (eds) (1978) Chapter 8, in *Handbook of Fillers and Reinforcements for Plastics*, Van Nostrand Reinhold Co., New York.
68 Gasil – antiblocking silicas at http://www.ineossilicas.com/index2.asp.
69 Peloso, C.W. *et al.* (1998), Characterising the degradation of the polymer slip additive erucamid in the presence of inorganic antiblock agents, *Polym. Degrad. Stab.*, **62** (2), 285–290.
70 Chuayjuljit, S. *et al.* (2003), Use of Silica from Rice Husk Ash as an Antiblocking Agent in Low-Density Polyethylene Film, *J. Appl. Polym. Sci.*, **88** (3), 848–852.
71 Product literature at www.worldminerals.com.
72 Shiro, M. (1995) *Purasuchikkusu*, Function and application of fillers, Antiblocking function, **46** (10), 38, AN: 1995:882737.
73 Juzo, S. *et al.* (1994) JP 06056979, Toray Industries Japan, AN: 1994:458262
74 Tsutomu, S. (1989) JP 01168779, Toppan Moore Co. Ltd., Japan, AN: 1990: 8634
75 Masten, P. (1989) EP 317166, Exxon Chemicals, AN: 1989:555491
76 Takamitsu, D. (1986) JP 61285268, Honny Chemicals, Japan, AN: 1987:41861.
77 Rodriguez, F. (1989) *Principles of Polymer Systems*, 3rd edn, Hemisphere Publ. Corp.
78 Asahi Glass Company, Fluon PTFE at www.agc.co.jp/english/chemicals/jushi/ptfe/PTFE3.htm.
79 Jaszczak, J.A.The Graphite Page at www.phy.mtu.edu/faculty/info/jaszczak/graphite.html.
80 Gapi Group, Virgin PTFE datasheet at www.gapigroup.com/virgin2.HTM.
81 Ferro-Ceramic Grinding, Inc., Ceramic properties at www.ferroceramic.com.
82 Boron nitride at www.a-m.de/englisch/lexikon/bornitrid.htm.
83 BN – boron nitride at www.ioffe.rssi.ru/SVA/NSM/Semicond/BN/basic.html.
84 Ranan, B. (1986–87) *Handbook of Chemistry and Physics*, 67th edn, CRC Press, p. B-109.
85 Allod GmbH & Co. K.-G . (2002) DE 20118862, Germany, AN: 2002:240365
86 Yongwei, H. (1992) CN 1064906, Nanning Automobile Flexible Axel Factory, China, AN: 1993:519870.
87 Muliawan, E. and Hatzikiriakos, S.G. (2004) Proceedings of the 62nd SPE ANTEC, vol. 50, p. 256.
88 Kazuhiko, K. and Kimihiro, N. (1997) JP 09040807, Mitsubishi Cable Ind., Japan, AN: 1997:264610
89 Kazumasa, H. and Hiroshi, Y. (1996) JP 08252874, Mitsubishi Belting Ltd., Japan, AN: 1996:733587
90 Antonio, C. and Ewald, M. (1995) EP 658611, Ringsdorff Sinter GmbH, Germany, AN: 1995:890103
91 Jaroslav, S. and Ladislav, M. (1992) CS 276646, Tos Hulin, Czechoslovakia, AN: 1994:558821

92 Daisuke, S., Takatoshi, A., and Tyoichi, S. (1996) US Patent 5,508,581, Nikon Corp., Japan, AN: 1996:298576

93 Sunao, I. *et al.* (1993) JP 05247351, Matsushita Electric Co., Japan, AN: 1994:78661

94 Ludomir, S. *et al.* (1996) PL 168619, Politechnika Lodzka, Poland, AN: 1996:485617

95 Ludomir, S. *et al.* (1988) PL 144270, Politechnika Lodzka, Poland, AN: 1990:120439.

96 Haberstroh, E. *et al.* (2004) Proceedings of the 62nd SPE ANTEC, vol. 50, p. 355.

97 Itaru, O. and Masaaki, S. (1987) JP 62264601, Seiko Epson Corp., Japan, AN: 1989:106850

98 Heinz, K.W. and Erich, H. (1993) DE 4200385, Glyco-Metall-Werke Glyco & Co. KG, Germany, AN: 1994:109032.

99 Bharat, B. (1987) *ASTM Special Tech. Publ.*, **947**, 289–309.

100 Vinogradova, O.V. *et al.* (1996), Tribochemical processes in poly(phenylene sulfide) filled with molybdenum disulfide and graphite, *Trenie I Iznos*, **17** (4), 544–549, AN: 1997:615201.

101 Behncke, H. and Kaehne, H.H. (1971) DE 1937390, AN: 1971: 112839

102 Ciora, P. *et al.* (1986) RO 89228, Intreprinderea de Utilaj Chimic, Romania, AN: 1987:555861.

103 Bezard, D. (1988) *Oesterreichische Kunststoff-Zeitschrift*, **19** (1–2), **18** (20–1), 24, AN: 1988:438801.

104 Korshak, V.V. *et al.* (1986), Effect of friction temperature on the surface structure and wear resistance of antifriction self-lubricating plastics based on poly(phenylquinoxaline), *Trenie I Iznos*, **7** (1), 16–20, AN: 1986:208216.

105 Toray Industries, Inc . (1985) JP 60020958, Japan, AN: 1985: 204960.

106 Avaliani, D.I. and Arveladze, I.S. (1979) *Soobshcheniya Akademii Nauk Gruzinskoi SSR*, **96** (1), 149–152, AN: 1980:77260.

107 Hatzikiriakos, S. and Rathod, N. (2003), Boron nitride processing aids, *Korea-Australia Rheo. J.*, **15** (4), 173.

108 Liu, C. and Mather, P.T. (2004) Proceedings of the 62nd SPE ANTEC, vol. 50, p. 3080.

20
Processing Aids

Subhash H. Patel

20.1
Introduction

It is widely accepted by the plastics community that "Processing Aids" is a term given to additives that are employed to overcome processing problems of various plastics, predominantly polyvinyl chloride (PVC). Typically, processing aids are combinations of high molecular weight (MW) polymers/copolymers (such as polymethylmethacrylate (PMMA), styrene-acrylonitrile (SAN) copolymer), oligomers, or other resins, which, when added to PVC, enhance fusion and improve melt properties. Thus, processing aids, when added at low concentrations, contribute to processing improvements, higher productivity, and better product quality.

Addition of small amounts of a processing aid to a polymer may lead to a major change in the rheology/viscosity and/or morphology of the material thereby improving processability. Rheology modifiers are compounds that alter the deformation and flow characteristics of matter under the influence of stress. They are used extensively in paints, coatings, plastisols, and liquid thermosetting agents prior to cross-linking, to modify viscosity/shear rate characteristics or to impart thixotropy. A thixotropic coating exhibits a time-dependent decrease in viscosity with increased shear rate up to a limiting value (due to a loss of structure in the coating system). This loss is usually temporary and the system regains its original state when given enough time. The generated curve is commonly known as a "thixotropic" or "hysteresis" loop [1].

Rheological additives can be roughly divided according to their chemical nature into inorganic and organic thickeners, with a subsequent distinction being made between thickeners for solvent-borne systems and thickeners for water-borne systems [2]. A variety of inorganic fillers and organic compounds may be used in cross-linkable liquid systems. Inorganic thickeners/viscosity modifiers are unmodified or organically modified fillers such as bentonites, hectorites, silicas, and organoclays. Alkaline earth oxides or hydroxides are a special class of reactive fillers acting as thickeners/viscosity modifiers used in sheet- and bulk-molding compounds (SMC/BMC) based on unsaturated polyesters (UP).

Functional Fillers for Plastics: Second, updated and enlarged edition. Edited by Marino Xanthos

ISBN: 978-3-527-32361-6

The majority of the processing aids employed by the thermoplastics industry are organic compounds, and these usually lower melt viscosity. In other words, processing aids are also rheology modifiers. However, viscosity stabilizers that act by minimizing polymer degradation may also be considered as processing aids in thermally sensitive systems such as PVC. However, in view of the scope and title of this book, one can expand the definition of processing aids to inorganic fillers that act as rheology modifiers or rheology stabilizers, for a variety of thermosets, coatings/paints, thermoplastics, and elastomers. Three representative fillers whose primary functionality is to assist processing will be discussed in this chapter:

- Magnesium oxide (MgO or magnesia) as a thickening agent/viscosity modifier in SMC/BMC, along with some applications in other polymer systems.
- Fumed silica as rheology modifier in liquid cross-linkable coatings.
- Hydrotalcites as process stabilizers (acid scavengers) for PVC and as acid neutralizers in polyethylene (PE) and polypropylene (PP), as well as applications in other polymer systems.

20.2 Production

20.2.1 Magnesium Oxide

The majority of magnesium oxide produced today is obtained from the processing of naturally occurring magnesium salts present in magnesite ore, magnesium chloride-rich brine, and seawater [3–5]. Worldwide, about two thirds of the magnesium oxide (some 11 million tons per annum) produced is from seawater (coproduced with salt production) and the rest comes from the natural mineral deposits: magnesite, dolomite, and salt domes [5]. Large mineral deposits of magnesite are located in Australia, Brazil, Canada, China, the Russian Federation, Europe, Turkey, and the United States. When heated from 700 to 1000 °C, magnesium carbonate thermally decomposes to produce magnesium oxide and carbon dioxide:

$$MgCO_3 \rightarrow MgO + CO_2$$

Other sources of MgO are underground deposits of brine, which are essentially saturated salt solutions containing magnesium chloride and calcium chloride. Deposits of brine located approximately 750 m below ground are used in the Martin Marietta's process [3]. The process involves the conversion of $MgCl_2$ into $Mg(OH)_2$ through a series of steps involving the reaction of brine with a calcined dolomitic limestone ($CaMg(CO_3)_2$), followed by filtration and calcination of the obtained $Mg(OH)_2$ to produce magnesium oxide:

$$2Mg(OH)_2 \rightarrow 2MgO + 2H_2O$$

Several types of kilns can be used in the calcination step. Calcination not only converts magnesium hydroxide to magnesium oxide, but is also the most important step for determining the activity of the product as related to its end use. The process of extraction of magnesia from seawater may involve reaction of lime slurry with $MgCl_2$ and $MgSO_4$ dissolved in seawater to form $Mg(OH)_2$ that is further calcined as in the Premier Periclase process [4] or thermal decomposition of a concentrated $MgCl_2$ brine in a special reactor as in the Dead Sea Periclase (DSP) process, also known as the "Aman Process" [6].

In the reactor, a fine spray of brine droplets is contacted with hot gases obtained from the combustion of clean fuels. Under these conditions the magnesium chloride is thermally decomposed to raw magnesium oxide

$$MgCl_2 + H_2O \rightarrow MgO + 2HCl$$

At this stage the reactor product contains residual salts (sodium, potassium, and calcium chlorides) from the brine that were not affected by the thermal regime of the spray reactor. The raw magnesium oxide is mixed with water in settling tanks and converted to insoluble magnesium hydroxide,

$$2MgO + 2H_2O \rightarrow 2Mg(OH)_2$$

while the residual salts are dissolved and are readily separated from the magnesium hydroxide by filtration and washing procedures.

Subsequent drying or calcining of the high-purity magnesium hydroxide yields magnesium hydroxide, or oxide, respectively of very high purity (99.7% as MgO). These materials undergo further processing to yield a wide range of products that are differentiated by properties such as particle size, bulk density, reactivity, and surface area.

20.2.2 Fumed Silica

Fumed silica, or fumed silicon dioxide, is produced by the vapor phase hydrolysis of silicon tetrachloride in an H_2/O_2 flame. The reactions are shown in Chapter 19. Hydrophilic fumed silica bearing hydroxyl groups on its surface is produced by this process. Hydrophobic fumed silica is made by processing fumed hydrophilic silica through in-line hydrophobic treatments such as with silanes, siloxanes, silazanes, and so on [1]. Examples of different types of hydrophobic fumed silica coatings include DMDS (dimethyldichlorosilane), TMOS (trimethoxyoctylsilane), HMDS (hexamethyldisilazane).

20.2.3 Hydrotalcites

Hydrotalcite is a natural mineral with a white color and pearl-like luster. It is mined in small quantities in Norway and the Ural area of Russia. Other mines of hydrotalcite are located in Austria, Canada, Czech Republic, Germany, Sweden, South Africa, and United States [7]. Natural hydrotalcite is a hydrated magnesium-, aluminum- and carbonate-containing mineral with a layered structure that is represented alternatively as $6MgO{\cdot}Al_2O_3{\cdot}CO_2{\cdot}12H_2O$ or $Mg_6Al_2(OH)_{16}CO_3{\cdot}4H_2O$.

Natural hydrotalcite deposits are generally found intermeshed with spinel and other materials due to the existence of nonequilibrium conditions during the formation of the deposits. Other minerals, such as penninite and muscovite, as well as heavy metals, are also found in natural hydrotalcite deposits. There are as yet not known the techniques for separating these materials and purifying the natural hydrotalcite.

Kyowa Chemicals Company, Japan, was the first in the world to succeed in the industrial synthesis of hydrotalcite in 1966 [8]. Synthetically produced hydrotalcite can be made to have the same composition as natural hydrotalcite, or, because of flexibility in the synthesis, it can be made to have a different composition by replacing the carbonate anion with other anions, such as phosphate, nitrate, or chloride [9]. In general, the hydrotalcite compositions are prepared from aqueous solutions of soluble magnesium and aluminum salts, which are mixed in a molar ratio of about 2.5 : 1 to 4 : 1, together with a basic solution containing at least a twofold excess of carbonate and a sufficient amount of a base to maintain a pH of the reaction mixture in the range of 8.5–9.5. The product particle size is determined by controlling process parameters including reaction temperature, reaction stirring speed, reaction pH, reactant addition rates and reactant concentrations [9]. Synthetic hydrotalcites can have a plate-like, a needle-like, or a spheroidal morphology through control of the synthesis parameters. The highly pure nanocrystal Mg,Al-hydrotalcite with titania doping was synthesized from $MgC1_2$, NaOH, $A1_2(CO_3)$, and TiO_2 by a one-step aqueous reaction method for 16 h at about 90 °C and at atmospheric pressure. The reagents were weighed in a proper molar ratio according to the nominal composition $Mg_{6-x}Ti_xAl_2(OH)_{16}CO_3{\cdot}4H_2O$ [10]. Hydrotalcites can be modified through anion exchange and intercalation, or through calcination to mixed oxides, to a variety of products including catalysts, drugs, stabilizers and fireproof agents for plastics, and adsorbents [11, 12].

The empirical formulae for some synthetic hydrotalcites are as follows [13, 14]:

- DHT-4A® (Kyowa Chemical Ind., Japan)
  - $Mg_{4.5}Al_2(OH)_{13}(CO_3){\cdot}3.5H_2O$
- Baeropol MC 6280® (Baerlocher GmbH, Germany)
  - $Mg_4Al_2(OH)_{12}(CO_3){\cdot}2.85H_2O$
- SORBACID® 911 (SUD-CHEMIE)/Hycite® 713 (Ciba Specialty Chemicals)
  - $[Mg_{1-x}Al_x(OH)_2](CO_3)_{x/2}{\cdot}nH_2O$
  - $(0.25 < x < 0.33)$
- HI-TAL (Shin Woun Chemical Co., South Korea) & CLC-120 (Doobon, Inc., South Korea)
  - $Mg_4Al_2(OH)_{12}CO_3{\cdot}3H_2O$
- Polylizer – 121 (Doobon, Inc. South Korea)
  - $Mg_3ZnAl_2(OH)_{12}CO_3{\cdot}3H_2O$

20.3
Structure/Properties

20.3.1
Magnesium Oxide

Magnesium oxide is a white powder broadly similar to calcium oxide and is rarely found in Nature as such but more commonly as the carbonate form, including the less common mineral complex with calcium carbonate (carnallite) [5].

Three basic types or grades of "burned" magnesium oxide can be obtained from the calcination of magnesium hydroxide, with the differences between each grade related to the degree of reactivity remaining after being exposed to a range of extremely high temperatures. Temperatures used during calcination to produce refractory grade magnesia are in the range 1500–2000 °C and the magnesium oxide is referred to as "dead-burned" since most, if not all, of the reactivity is eliminated. A second type of MgO produced by calcining at temperatures in the range 1000–1500 °C is termed "hard-burned." The third grade of MgO is produced by calcining in the range 700–1000 °C and is termed "light-burned" or "caustic" magnesia. This material has a wide reactivity range and is used in plastics, rubber, paper, and other industrial applications. A typical scanning electron micrograph of "light-burned" MgO is shown in Figure 20.1 [3].

Figure 20.1 A typical SEM of "light-burned" MgO submicron particles with surface area 0.1–1.0 m^2/g. Reproduced with permission from Ref. [3].

Table 20.1 Typical properties of MgO [15, 16].

Property	Value
Density (g/cm^3)	3.6 (varies), AluChem (3.36–3.4)
Melting point (°C)	2800
Boiling point (°C)	3600
Refractive index	1.7085, AluChem (1.73)
Thermal conductivity (W/(m K)) at 273 K	42
Thermal expansion (K^{-1}) at 273 K	10.8×10^{-6}
Dielectric constant at 1 MHz	9.65
Young's modulus (GPa)	249
Solubility in water (%)	0.00062
Hardness (Mohs)	6.0–6.5 (AluChem)

Some typical properties of MgO are given in Table 20.1 [15, 16].

20.3.2 Fumed Silica

Fumed silica (silicon dioxide) is generally regarded as a unique material because of its unusual particle characteristics, enormous surface area and high purity. It is a fine, white and extremely fluffy powder. When added to liquids and polymers with which its refractive index (1.46) is a close match, it appears colorless or clear. Unlike crystalline silicas, amorphous fumed silica is safe to handle, thus eliminating the serious health problems associated with crystalline silica dust. Its high surface area, ranging from 130 to 380 m^2/g, affects dispersibility, rheology control, thixotropic behavior, and reinforcement efficiency [16]. The higher the surface area, the more the rheological control and thixotropic behavior increases, and the greater the potential for reinforcement; however, dispersion becomes more difficult. Various properties of fumed silica are compared with those of precipitated silica in Table 19.8.

20.3.3 Hydrotalcites

Natural hydrotalcites, with $Mg_6Al_2(OH)_{16}CO_3 \cdot 4H_2O$ structure, which crystallize in a trigonal–hexagonal scalenohedral system, are characterized by a lamellar structure, and have a density of 2.06 g/cm^3, and a Mohs hardness of 2 [7, 17]. Figure 20.2 shows a three-dimensional double-layered structure of a typical hydrotalcite [8] consisting of magnesium and aluminum hydroxide octahedra interconnected via the edges. Additional interstitial anions between the layers compensate the charge of the crystal and determine the size of the interlayer distance (basal spacing). While hydrotalcites are accessible through the corresponding metal salts [18], the metal alcoholate route, may provide certain advantages over other ways of synthesis such as variation of the Al/Mg ratio over a wide range, high purity and controlled anion content [19]. Compared to aluminum trihydrate (pH 8–9), hydrotalcites are even more alkaline

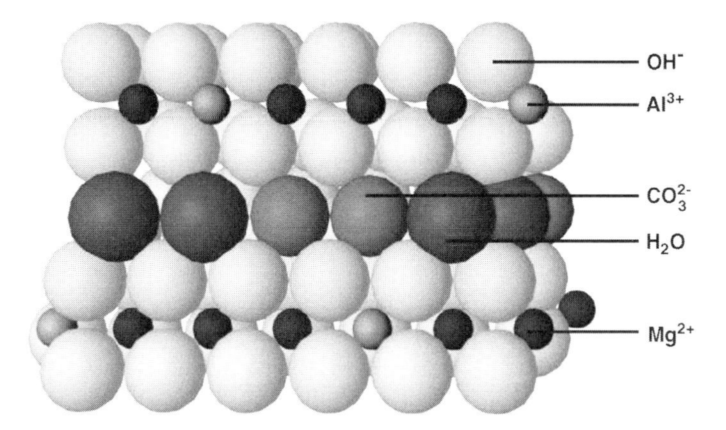

Figure 20.2 Schematic of the double-layered metal hydroxide structure of a typical hydrotalcite. Reproduced with permission from Ref. [8].

by nature. Basicity is adjustable by increasing the Mg/Al ratio and/or incorporating anions other than hydroxyl.

Synthetic hydrotalcites employed in plastics' applications are finely divided, free-flowing, odorless, amorphous powders. For polyolefin applications, hydrotalcites with an average particle size of 0.5 μm and a specific surface area of less than 20 m^2/g (BET method) are recommended. Their density varies around 2.1 g/cm^3 depending on the composition. Sodium, zinc, and calcium stearate-coated hydrotalcites are available, which show enhanced compatibility with polyolefins [13]. Figure 20.3 is a scanning electron micrograph of a typical submicron-size synthetic hydrotalcite [20].

A useful and unique feature of hydrotalcite lies in its special acid adsorbing mechanism and its inherent anion exchanging property. For example, in the case of

Figure 20.3 Scanning electron micrograph of synthetic hydrotalcite submicron particles [20].

hydrochloric acid, the CO_3^{2-} in the hydrotalcite structure is easily replaced by Cl^-, and these chlorine ions are incorporated into the crystal structure to produce chlorinated compounds such as $Mg_{4.5}Al_2(OH)_{13}Cl_2 \cdot 3.5H_2O$, which is insoluble in both water and oil. Also, the Cl^- is not desorbed from the crystal structure up to a temperature of approximately 450 °C [8, 13]. These reactions allow hydrotalcite to be used as an HCl scavenger in PVC stabilization and as acid neutralizer of acidic catalyst remnants in polyolefins.

20.4 Suppliers/Manufacturers

Major suppliers/manufacturers of MgO are given in Table 20.2. Major suppliers/manufacturers of hydrotalcites are given in Table 20.3. Major suppliers/manufacturers of fumed silica are covered along with antiblock silicas in Table 19.9.

Table 20.2 Suppliers/manufacturers of magnesium oxide.

Supplier/manufacturer	Grades/types
AluChem, Inc. Cincinnati, OH, USA www.aluchem.com	Dead-burned MgO grades: ADM-98H ADM-99P
Baymag Calgary, Alberta, Canada www.baymag.com	Baymag 30, 40, 96 (natural MgO)
Dead Sea Periclase (ICL Industrial Products) Mishor Rotem D.N. Arava, Israel, 86805 www.periclase.com	Active MgO grades: RA-150, 150A, 110, 40, 25
Martin Marietta Magnesia Specialties, LLC Baltimore, MD 21220, USA www.magspecialties.com	MagChem HSA-10, HSA-30 MagChem 40, 50, 60 (light-burned grades) MagChem 50M (micronized, high and moderate activity grades) Elastomag 100, 170, 170 MM, FE Marinco FCC (food grade)
Premier Chemicals, LLC W. Conshohocken, PA 19428, USA www.premierchemicals.com	MAGOX – Super Premium MAGOX – Premium MAGOX – 98 HR
Premier Periclase Ltd (PPL) Drogheda, County Louth, Ireland www.premierpericlase.ie/	Calcined MgO: TechMag Ground MgO: MagBase
Scora S. A. Caffiers, France www.scora.com	SCORA MgO (for plastics, rubbers and elastomers, paints and inks)

Table 20.3 Suppliers/manufacturers of hydrotalcites.

Supplier/manufacturer	Grades/types
Baerlocher GmbH Unterschleissheim, Germany Baerlocher USA LLC Dover, OH 44622, USA www.baerlocher.com	Baeropol MC 6280 (customer-specified additive blends)
DOOBON, Inc. Choongchungbuk-Do, South Korea www.doobon.co.kr	CLC-120 Polylizer-120 Polylizer-121
Kyowa Chemical Industry Co., Ltd Kagawa, Japan www.kyowa-chem.co.jp/indexe.html	DHT-4A Alcamizer
Sasol North America, Inc. Houston, 77079, USA www.sasolnorthamerica.com	PURAL MG 61 HT PURAL MG 30, MG 50, MG 70
Sud-Chemie, Inc. Louisville, KY 40232, USA	SORBACID 911 SYNTAL HYCITE 713 (distributed by Ciba Specialty Chemical)
Shin Woun Chemical Co., Ltd Shiheung, South Korea www.swchem.co.kr	HI-TAL
FCC, Inc. (Division of Union Co.) Zhejiang, China www.nanoclay.net	SUPLITE RB

20.5 Environmental/Toxicological Considerations

20.5.1 Magnesium Oxide

A NIOSH (National Institute for Occupational Safety and Health) web site [22] has listed 24 synonyms for MgO. According to the Material Safety Data Sheet for a grade recommended for use in plastics (MagChem HAS-10) [3] MgO does not fit the definition of hazardous material and no LD_{50} (lethal dose for 50% kill) or LC_{50} (lethal concentration for 50% kill) data are available for MgO dust. MgO is stable under ambient temperatures and pressures. Exposure to water may cause this product to slowly hydrate through an exothermic reaction. MgO products may present a nuisance dust hazard but are not flammable or combustible. MgO is not considered a teratogen, mutagen, sensitizer, toxin, or carcinogen by NTP

(National Toxicology Program), IARC (International Agency for Research on Cancer), and OSHA (Occupational Safety and Health Administration). However, if magnesium oxide is heated in a reducing atmosphere to the point of volatilization (i.e., $>1700\,^{\circ}$C), magnesium oxide *fume* may be generated. Exposure limits for MgO *fume* are reported as follows: ACGIH (American Conference of Governmental Industrial Hygienists, Inc.): 10 mg/m^{3}TWA (time-weighted average); OSHA: final PELs (permissible exposure limits), total particulate 15 mg/m^3 TWA.

20.5.2 Fumed Silica

It is important to differentiate between amorphous and crystalline forms of silica when evaluating its effects upon exposure. In general, fumed or pyrogenic silica, which is X-ray amorphous, is not toxic, does not cause silicosis, and is safe to work with. However, excessive inhalation should be avoided by proper ventilation or protective masks [16]. According to MSDS for hydrophilic fumed silica (CAS# 112945-52-5), no acute toxic effects are expected upon eye or skin contact or by inhalation [23]. This material does not contain any reportable carcinogenic ingredients or any reproductive toxins at OSHA or WHMIS (Workplace Hazardous Materials Information System) reportable levels. It is not mutagenic in different *in vitro* and *in vivo* test systems. A long-term exposure exceeding TLV can lead to damaging effects as a result of mechanical overloading of the respiratory tract. Animal tests have shown no indication of carcinogenic or reproduction effects. The ACGIH recommends a TLV–TWA of 10 mg/m^3 measured as total dust containing less than 1% quartz. Acute toxicity LD_{50}/LC_{50}-values relevant to classification are oral >5000 mg/kg rat, dermal >5000 mg/kg rabbit, and inhalation >0.139 mg/l/4 h rat.

20.5.3 Hydrotalcites

Hydrotalcites are, in general, environmentally safe, nontoxic, noncorrosive, and nonvolatile. According to MSDS for hydrotalcites with the general name magnesium aluminum hydroxycarbonate (CAS# 11097-59-9), OSHA–PEL values are 15 mg/m^3 dust and 5 mg/m^3 respirable dust and the ACGIH–TLV is 10 mg/m^3 total dust [24]. The product itself is noncombustible. Acute toxicity values of LD_{50} oral (rat) >10 000 mg/kg have been reported. Dust inhalation may cause respiratory tract irritation.

20.6 Applications

20.6.1 Magnesium Oxide

In plastics/rubber adhesives manufacture/compounding operations, MgO has different functions depending upon its type/grade and the specific application. It may be used as a filler, anticaking agent, pigment extender, acid neutralizer, or thickening agent [3]. For example, viscosity development in unsaturated polyester SMC and BMC may be controlled using magnesium oxide. Magnesia is a low cost acid acceptor for general purpose use where water resistance is not a critical property. Because it is nondiscoloring, magnesia can be used in both black and nonblack rubber compounds. Certain high activity grades of MgO regulate the rate of vulcanization and provide excellent scorch protection in various rubbers such as neoprene, hypalon, chlorobutyl rubber, and so on.

In chlorosulfonated polyethylene it acts as a cross-linker, acid acceptor, and scavenger to neutralize acidic products and prevent ordinary mold steel corrosion [25]. In silicone rubber it neutralizes peroxide generated acids during curing and promotes oil resistance. In all these and other acid-acceptor systems (e.g., chlorinated polyethylene), fine particle size (usually less than 325 mesh) and closely controlled surface area (as high as 160–200 m^2/g) or reactivity are vital [3, 6, 25]. As another example of its acid scavenging function, a PVC-containing plastic waste was mixed with MgO to give a mixture, which during injection-molding showed good suppression of HCl generation [26]. In polychloroprene–phenol resin adhesives, magnesia acts as an acid acceptor and is an integral part of the bonding system. Magnesium oxide is used regularly in the emulsion polymerization process of ABS acting as an acid scavenger and "fixing" the excess emulsifier that remains in the resin [6]. General applications of MgO in specific polymers are summarized in Ref. [27].

The primary role of MgO in unsaturated polyesters is as a reactive filler/thickening agent in SMC/BMC manufacturing. Incorporation of up to 5% MgO in the formulation allows the production of a tack-free sheet within a few days at room temperature. The sheets can then be easily handled prior to the high-temperature molding step. The increased viscosity assists dispersion uniformity and improves the flow characteristics of the compound prior to the onset of the cross-linking reaction.

Two mechanisms have been proposed in the literature to account for the interaction of polyester and magnesium oxide in the reactive monomer (e.g., styrene) medium. One is a chain extension mechanism and the second is the formation of a coordinate complex, also known as two-stage thickening mechanism [28]. The common starting point for these two mechanisms is the formation of basic and neutral salts with the polyester carboxylic acid (–COOH) end groups according to the following reactions:

$$-\mathrm{COOH} + \mathrm{MgO} \rightarrow -\mathrm{COOMgOH}$$

$$-\mathrm{COOH} + \mathrm{HOMgOOC}- \rightarrow -\mathrm{COOMgOOC}- + \mathrm{H_2O}$$

$$\sim\sim \mathrm{UP} \sim \mathrm{COO^{-}\,^{+}Mg^{+}\,^{-}OOC} \sim \mathrm{UP} \sim \mathrm{COO^{-}\,^{+}Mg^{+}\,^{-}OOC} \sim \mathrm{UP} \sim\sim\sim$$

In the chain extension mechanism, it is postulated that dicarboxylic acid groups on the UP chains react with MgO to produce a very high molecular weight species (via condensation polymerization) and, thus, give rise to a large increase in viscosity. However, this theory only applies to those UP molecules terminated by carboxylic groups with the structure HOOC∼∼∼COOH. For other possible polyester structures, such as HOOC∼∼∼OH and HO∼∼∼OH, the MW of the polyester will only increase twofold or not at all if the chain-extension mechanism is followed. Accordingly, this mechanism cannot explain the large increase of viscosity in thickening of UPs with structures having terminal OH functional groups, as reported in other publications [29].

In the two-stage mechanism, it is postulated that a high MW salt is formed initially, and then a complex is formed between the salt and carbonyl groups of the ester linkages, as shown below. The second stage of this theory is considered to be responsible for the large increase in viscosity. Several publications support this mechanism, for example, Refs [29, 30].

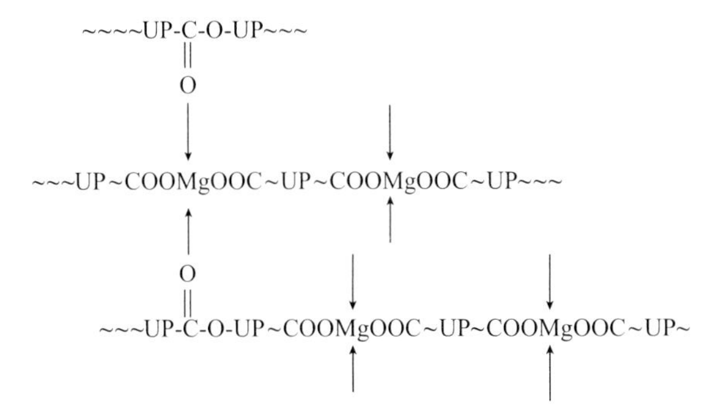

In recent work [31, 32], attempts were made to produce similar magnesium salts by melt processing mixtures of an unsaturated polyester oligomer (in the absence of a reactive monomer) with various amounts of MgO. As an example, Figure 20.4 shows the effects of increasing MgO concentration on the reaction between UP and MgO in a batch mixer at 220 °C. As the torque data indicate, the reaction rate

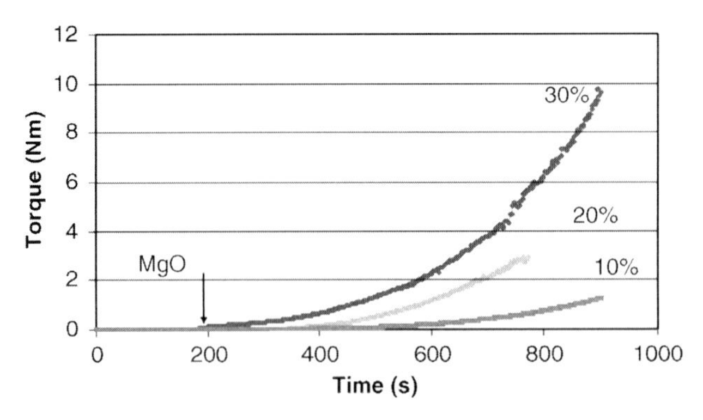

Figure 20.4 Changes in torque of unsaturated polyester oligomer upon addition of various amounts of MgO in a batch mixer at 220 °C [31, 32].

increases significantly with increasing MgO concentration. Thus, depending on concentrations and processing conditions, different products with different rheological characteristics and different glass transition temperature are possible. Such products may be considered as thermoplastic ionomers containing unsaturation that would be useful for further reactions.

20.6.2
Fumed Silica

Both hydrophilic and hydrophobic silicas are used in solvent-borne coatings to improve rheological properties, and as flow control agents and antisettling additives for pigments [1]. Fumed silica is a weak acid, bearing hydroxyl groups on its surface. The thickening mechanism of liquid coating systems is explained by hydrogen-bond formation between neighboring aggregates of silica, leading to the formation of a regular network. Some of these hydrogen bonds could be broken under shear forces, resulting in reduced viscosity. Usually, fumed silica contains 0.5–2.5% moisture, which not only aids the thickening process, but also facilitates curing of some polyurethane prepolymer systems [25]. Fumed silica is a thixotropic additive, which, when dispersed into epoxy or polyester resins, increases viscosity, imparts thixotropic behavior, and adds antisag and antisettling characteristics during the potlife of the resins.

Besides its principal use mainly in coating/paint systems, fumed silica may also be used in thermoplastics. Melt viscosity of poly(ethylene 2,6-naphthalate) (PEN) was reported to decrease with the employment of small amounts of fumed silica nanoparticles. Additional effects on mechanical properties and crystallization behavior by using untreated and surface-treated silica are reported in Ref. [33].

In addition to their function as rheology modifiers, silica, colloidal or fumed, and clays are among the most widely studied inorganic fillers for improving the scratch/abrasion resistance of transparent coatings. These fillers are attractive from the standpoint that they do not adversely impact the transparency of coatings due to the fact that the refractive indices of these particles closely match those of most resin-based coatings. The drawback of silica-based fillers is that high concentrations of their particles are generally required to show a significant improvement in the scratch/abrasion resistance of a coating, and these high loadings can lead to various other formulation problems associated with viscosity, thixotropy, and film formation [34, 35].

Recent publications describe the effects of silica on the conductivity and mechanical properties of a polyethylene oxide/ammonium bifluoride complex containing propylene carbonate [36], as a foam stabilizer in polyester polyurethane foams, and on the properties of polylactic acid nanocomposites prepared by the sol–gel technique [37] (see also Chapter 24), on the mechanical properties and permeability of i-PP composites [38], on the surface hardness of polymers for biomedical devices [39], on enhanced properties of polymer interlayers that are used in multiple layer glazing panels [40].

20.6.3 Hydrotalcites

Synthetic hydrotalcites are mainly used as process stabilizers for PVC, acting as scavengers for the HCl evolved during processing and hence minimizing degradation. They are mostly used as costabilizers with metal soap stabilizers to improve color and heat stability [13] at concentrations from 0.5 to 3.5 phr, and may be used as acid scavengers in commingled recyclable waste plastics containing PVC [41]. The thermal stability of melt mixed PVC composites, containing dodecyl sulfonate-pillared hydrotalcite prepared by ion-exchange reaction, improved with increasing modified hydrotalcite contents [42].

Hydrotalcites are also used as acid neutralizers in polyolefins, particularly those containing residual acidic products from Ziegler-Natta catalysts used for polymerization. In the latter case they are part of the normal stabilizer packages at typical concentrations in PP, LLDPE, and HDPE, not exceeding 500 ppm [13]. They function as processing aids minimizing corrosion of metallic surfaces during processing and also lead to less discoloration (yellowing) of the molded products. Studies on model hydrotalcite/phenolic antioxidant compositions, in the absence of polyolefins containing residual acidity, confirmed that color ("pink") development at processing temperatures is due to enhanced hydrotalcite/antioxidant interactions, and occurs irrespective of the hydrotalcite used [43].

Hydrotalcites may exhibit a series of additional functions. For example, due to their ability to release water and carbon dioxide at relatively low temperature they have been extensively evaluated as flame retardants. Data with a mass-loss calorimeter indicated that an EVA polymer filled with 50 wt% of hydrotalcite had the slowest heat release rate and the lowest evolved gas temperature as compared with aluminum hydroxide or magnesium hydroxide. XRD data, combined with thermal analysis results, indicated that the layered structure of hydrotalcite may play a role in the degradation mechanism. The improved fire resistance of EVA filled with hydrotalcite also results from its intumescing behavior [44]. Fire-resistant polycarbonate compositions containing 0.1–1.5% hydrotalcite had a UL 94 fire resistance rating V-2 [45]. Hydrotalcites can also be part of the polycondensation catalyst package for producing flame retardant thermoplastic polyesters [46], or incorporated, swollen with a glycol, at any stage in the course of polycondensation to produce polyester resins with improved strength, rigidity, and gas barrier properties [47].

Hydrotalcite-containing plastic films for agricultural applications showed good filler dispersion combined with adequate transparency, controlled permeation, synergistic light stability, and satisfactory mechanical properties [21]. A hydrotalcite compound containing specific interlayer ions showed excellent ability to absorb infrared rays and had excellent light transmission when used in an agricultural film [48]. In other applications, hydrotalcite can inhibit water blushing in agricultural polyolefin films, blooming in vinyl chloride resin films, and the decrease in electrical resistance of PVC electric wires [49]. Hydrotalcite has also been shown as a suitable replacement for lead containing stabilizers in heat and water resistant chlorosulfonated polyethylene formulations [50]. Hydrotalcite additions (up to 10 phr) can stabilize unplasticized polyvinyl chloride against TiO_2 photocatalyzed degradation by removing chloride through an anion exchange mechanism, thus modifying the autocatalytic pathway [51]. Mg–Zn–Al-based hydrotalcite-type particles comprising of core particles composed of a Mg–Al-based hydrotalcite and a Mg–Zn–Al-based hydrotalcite layer formed on the surface of the core particle, have been produced with an average plate surface diameter 0.1–1.0 μm and an adjustable refractive index to a required value 1–1.56 by using platelets with appropriate thickness. Resin compositions show not only higher resin stability and functional properties, but also excellent transparency, compared to using conventional hydrotalcite-type particles [52].

Finally, recent research on bioactive composites (see also Chapter 22) suggests that hydrotalcite promotes both thermal and hydrolytic degradation of poly(L-lactic acid) and may also promote bioactivity. In addition to its reinforcing characteristics, the inherent acid neutralizing capacity of the hydrotalcites can provide pH control during polymer biodegradation, which is usually accompanied by the formation of acidic low molecular weight fragments. This could be of significant importance in *in vivo* tissue engineering applications [20, 53] where significant pH stability is required.

References

1 Koleske, J.V. *et al.* (2008) 2008 Additives Guide. Paint & Coatings Industry Magazine, 24, 6, 38; also available at www.pcimag.com.

2 Manshausen, P. Role and function of rheological additives in modern emulsion and industrial coatings, **available at** http://www.pcimag.com/Articles/Feature_Article/dc51fd89f66a7010VgnVCM100000f932a8c0_.

3 Martin Marietta Magnesia Specialties LLC technical information at www.magspecialties.com/students.htm.

4 Premier Periclase (subsidiary of CRH plc (www.crh.com)), Ireland at www.premierpericlase.ie/.

5 Chemlink Australia Consultants at www.chemlink.com.au/mag&oxide.htm.

6 ICL-Industrial Products/Dead Sea Periclase at www.periclase.com.

7 Mindat., Hydrotalcite at www.mindat.org/min-1987.html.

8 Kyowa Chemical Industry Co. Ltd., Technical information at www.kyowa-chem.co.jp/english/.

9 Cox, S.D. and Wise, K.J. (1994) US Patent 5,364,828, Minerals Technologies, USA.

10 Qing-li, R.E.N. and Qiang, L.U.O. (2006) Preparation and thermal decomposition mechanism of Mg,Al-hydrotalcite nano-crystals with titania doping. *Trans. Nonferrous Met. Soc. China*, **16** (1), 402–405.

11 Hibino, T. (2006) Synthesis and applications of hydrotalcites. *Nendo Kagaku*, **45** (2), 102–109.

12 Tang, Y., Wu, M., and Luo, Y. (2006) The structure and properties of hydrotalcites and its application as additives of plastics. *Huagong Keji*, **14** (2), 65–69.

13 Zweifel, H. (ed.) (2001) *Plastics Additives Handbook*, Hanser Publishers, Munich.

14 Ashton, H. (2003) Proceedings of the Functional Fillers for Plastics 2003, Intertech Corp. Atlanta, GA, October 2003.

15 Crystran Co. Magnesium oxide – data sheet at www.crystran.co.uk/mgodata.htm.

16 Katz, H.S. and Milewski, J.V. (eds) (1978) Chapters 8 and 10, in *Handbook of Fillers and Reinforcements for Plastics*, Van Nostrand Reinhold Co., New York.

17 Mineralogy database "Hydrotalcite – mineral data" at http://webmineral.com/data/Hydrotalcite.shtml.

18 Miyata, H. *et al.* (1977), Physiochemical proporties of synthetic hydrotalcites in relation to composition, *Clay Miner.*, **25**, 14–18.
19 Sasol North America, Product information at www.sasoltechdata.com/.
20 Chouzouri, G. *et al.* (2004) Proceedings of the 62nd SPE ANTEC, vol. 50, p. 3366.
21 Chen, Z. and Kang, J. (2006) Application of novel additive of hydrotalcite in agriculture-films. *Zhongguo Suliao*, **20** (2), 13–15.
22 NIOSH (National Institute for Occupational Safety and Health). "Magnesium oxide" at www.cdc.gov/niosh/rtecs/om3abf10.html
23 Cabot Corp . (August 2000) MSDS for CAB-O-SIL® untreated fumed silica, Revised.
24 J.M. Huber Corp . (2003) MSDS for Hysafe® 539, Revised December 17.
25 Wypych, G. (2000) *Handbook of Fillers*, ChemTec Publishing, Toronto, Ont., Canada, pp. 132–137, 651–652.
26 Shioya, K. *et al.* (2005) Suppression of hydrochloric acid generated in recycling of waste plastics, JP 2005067196, Jpn. Kokai Tokkyo Koho.
27 Magnesium oxide applications in specialty polymers at www.premierchemicals.com/magox/rubber.htm
28 Judas, D. *et al.* (1984), Mechanism of the thickening reaction of polyester resins: Study on models, *J. Polym. Sci.: Polym. Chem. Ed.*, **22**, 3309.
29 Vancsó-Szmercsányi, I. and Szilágyi, Á. (1974), Coordination polymers from polycondensates and metal oxides. II. Effect of water molecules on the reactions of polyesters with MgO and ZnO, *J. Polym. Sci.: Polym. Chem. Ed.*, **12**, 2155.
30 Rao, K.B. and Gandhi, K.S. (1985) *J. Polym. Sci.: Polym. Chem. Ed.*, **23**, 2135.
31 Wan, C. (2004) Reactive modification of polyesters and their blends, Ph.D. Thesis, New Jersey Institute of Technology, Newark, NJ.
32 Wan, C. and Xanthos, M. (2003) Proceedings of the 61st SPE ANTEC, vol. 49, p. 1503.
33 Seong, H.K. *et al.* (2004) Proceedings of the 62nd SPE ANTEC, vol. 50, p. 1357.
34 Nanoparticle composites for coating applications (May 2004) Paint & Coatings Industry Magazine, accessed at www.pcimag.com/cda/articleinformation/.
35 Degussa Corp. Technical Library Document GP-89, accessed at www.epoxyproducts.com/silica.html.
36 Sekhon, S.S., Sharma, J.P., and Park, J.S. (2007) Conductivity behavior of nano-composite polymer electrolytes: role of fumed silica and plasticizer. *Macromol. Symp.*, **249/250**, 209–215.
37 Yan, S., Yin, J., Yang, J., and Chen, X. (2007) Structural characteristics and thermal properties of plasticized poly(L-lactide)-silica nanocomposites synthesized by sol–gel method. *Mater. Lett.*, **61** (13), 2683–2686.
38 Vassiliou, A., Bikiaris, D., and Pavlidou, E. (2007) Optimizing melt-processing conditions for the preparation of iPP/fumed silica nanocomposites: morphology, mechanical and gas permeability properties. *Macromol. React. Eng.*, **1** (4), 488–501.
39 Sullivan, M.H. *et al.* (2007) A method for hardening surface of plastics with ceramics for medical devices, PCT Int. Appl. WO 2007138323.
40 Yuan, P. (2007) Polymeric interlayers comprising modified fumed silica, PCT Int. Appl. WO 2007143746.
41 Shioya, K. *et al.* (2006) Method for suppressing hydrochloric acid in recycling of waste plastics, JP 2006070237, Jpn. Kokai Tokkyo Koho.
42 Wang, H., Bao, Y.-Z., Huang, Z.-M., and Weng, Z.-X. (2004) Structure and properties of poly(vinyl chloride)/dodecyl sulfonate-pillared hydrotalcite composites. *Zhongguo Suliao*, **18** (11), 54–58.
43 Patel, S.H. *et al.* (1995), Mechanism and performance of hydrotalcide acid neutralizers in thermoplastics, *J. Vinyl Addit. Tech.*, **1** (3), 201.
44 Camino, G. (2001), Effect of hydroxides and hydroxycarbonate structure on fire retardant effectiveness and mechanical properties in ethylene – vinyl acetate copolymer, *Polym. Degrad. Stab.*, **74** (3), 457.

45 Chung, J.Y.J. and Paul, W.G. (2003) Fire-resistant polycarbonate compositions containing hydrotalcite, PCT Int. Appl., WO 2003046067.

46 Nakajima, T. and Tsukamoto, K. (2005) Polycondensation catalysts with high catalytic activity for preparation of heat-resistant polyesters with decreased foreign matter content, JP 2005023160, Jpn. Kokai Tokkyo Koho.

47 Yatsuka, T. *et al.* (2003) Polyester resin compositions and producing process thereof, PCT Int. Appl., WO 2003066734.

48 Takahashi, H. and Okada, A. (2002) US Patent 6,418,661, Kyowa Chemical Industry Co Ltd.

49 Tsujimoto, H., Suzuki, M., and Kurato, M. (2006) Hydrotalcite as additives for synthetic resin compositions, PCT Int. Appl. WO 2006043352.

50 Fuller, R.E. and Macturk, K.S. (2000) *KGK-Kautschuk und Gummi Kunststoffe*, **53** (9), 506.

51 Martin, G. P., Robinson, A. J., and Worsley, D.A. (2006) Hydrotalcite mineral stabilization of titanium dioxide photocatalyzed degradation of un-plasticized polyvinyl chloride, *ECS Trans.*, **1** (18), 13–19.

52 Kobayashi, N. and Honmyo, T. (2004) Mg–Zn–Al based hydrotalcite-type particles, their production, and resin composition, Eur. Pat. Appl., EP 1462475.

53 Chouzouri, G. and Xanthos, M. (2003) Modification of biodegradable polyesters with inorganic fillers. Proceedings of the 61st SPE ANTEC, vol. 49, pp. 2561–2565.

21
Spherical Fillers

Anna Kron and Marino Xanthos

21.1
Organic Spherical Fillers

21.1.1
General

Organic spheres are predominantly polymeric, consisting of synthetic or natural polymers. The field of polymeric nano- and microparticles is vast, comprising, for instance, latex particles for coatings, hollow particles for syntactic foams, and microcapsules for foaming and additive release. In addition, there are core-shell microbeads and coated polymeric particles, where the particles can exhibit multiple functionalities, thanks to the individual features of their different layers [1]. As fillers in thermosets and thermoplastics, hollow microspheres and expandable microcapsules are among the most frequently used in commercial applications.

21.1.2
Production and Properties

In their function as fillers, the organic spheres share the performances and benefits of the spherical form, similar to glass and ceramic spheres. However, their effect on a polymer matrix is normally not the enhancement of mechanical strength, such as tensile strength and abrasion resistance. Instead, they can impart new features to thermoplastics and thermosets, such as reduced density, improved resilience and ductility, mechanical and thermal stress absorption, or enhanced thermal and electrical insulating properties. When added to binders and plastisols for coatings, the function of the spheres can be surface modification of the coated surface; this may include the creation of a visual effect or antislip properties, or to make a protective coating [2, 3].

The production of polymeric particles follows several possible routes. Heterogeneous polymerizations are common manufacturing processes, normally starting from O/W-emulsions and forming a latex or a particle dispersion, either through

Functional Fillers for Plastics: Second, updated and enlarged edition. Edited by Marino Xanthos

ISBN: 978-3-527-32361-6

emulsion, suspension, or dispersion polymerizations [4]. Solid thermoplastic particles such as polyacrylate or polystyrene microspheres are often produced this way, as well as core-shell microbeads and microcapsules such as thermoexpandable microspheres [4, 5]. As a result of the dynamic state of droplet formation in emulsions, the produced particles are polydisperse. However, monosized/narrow size distributed acrylic and styrenic polymer microbeads may be produced, as for example, through seeded polymerization. Monodisperse particles are commercially available as a development from the findings of Ugelstad and co-workers [6]. Another technique for producing narrow size particles is by membrane emulsification polymerization, where the organic phase is forced through a mesh into the aqueous phase at some state of the emulsification [7].

Thermoexpandable microspheres are a special type of microcapsules, developed not only as a filler in polymers, but also for paper and board, inks and emulsion explosives. They are, essentially, microballoons where the interior is a volatile hydrocarbon acting as a blowing agent, surrounded by a thermoplastic copolymer shell. The copolymers are based on high cohesive energy density monomers, such as acrylonitrile and vinylidene chloride, with their corresponding polymer units designed to soften at a specific temperature (corresponding to their glass transition temperatures, T_g). Thanks to the simultaneous increase in pressure from the trapped blowing agent and the symmetric spherical form, the volume change upon expansion can be dramatic, as illustrated in Figure 21.1 [2, 3, 5, 8]. As a result, bulk densities can be as low as 5 kg/m^3 in the case of free expansion; in a surrounding binder or plastics,

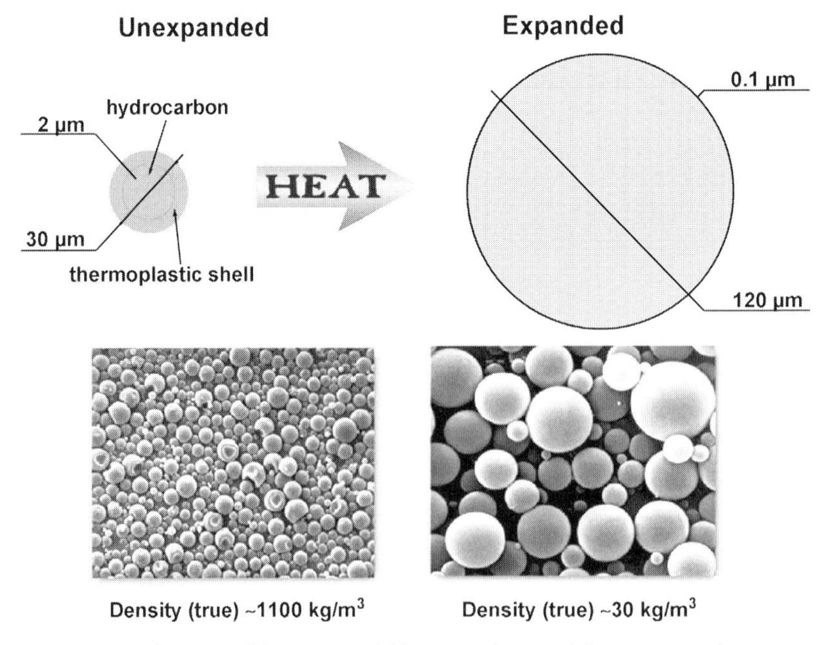

Figure 21.1 Schematics of thermoexpandable microspheres and their corresponding scanning electron micrographs.

however, they are somewhat higher, that is, 30–50 kg/m^3. Upon continued heating the gradual leakage through the thinned shell reduces the internal pressure so that the microspheres stop their expansion and finally start to shrink. Thermoexpandable microspheres are today produced by few manufacturers, where the two major ones are Akzo Nobel and Matsumoto Yushi-Seiyaku.

There is a wide range of morphologies and compositions of polymeric microspheres. Core-shell microbeads can be created by sequentially polymerizing vinyl monomers, typically methacrylates and styrene. Each layer may consist of a homopolymer or a copolymer; in the copolymers multifunctional vinyl monomers (e.g., diacrylates and divinylbenzene) may be used to create a cross-linked material. An optional final step can be grafting of functional groups to the surface in order to enhance compatibility to the surrounding medium.

By adjusting reaction parameters such as interfacial tension, initiator and inhibitor types, diluent content, monomer composition, and degree of cross-linking, several morphologies of the spheres can be tailor-made such as porous spheres, spheres with porous outer shells, smooth beads, closed hollow spheres, one-hole particles, or solid swellable spheres [4, 9, 10]. A common product is expanded polystyrene (EPS) beads where the porosity of the beads is obtained through high temperature, high pressure impregnation with a blowing agent, typically a low molecular weight hydrocarbon [11].

Emulsion and suspension polymerizations are best suited for chain polymerizations of vinyl group monomers. Alternative techniques are employed for condensation polymerizations; phenolic hollow spheres are produced by spray-drying in a process developed by Sohio, performed today by Asia Pacific Microspheres. In this process, phenolic resole resins are dissolved in water and a blowing agent is added, followed by spray-drying the mixture to obtain uniform particles with size of 5–50 μm [12]. Interfacial polycondensation, using emulsions where the reaction between lipophilic and hydrophilic monomers occurs at the interface, can be employed for making melamine–formaldehyde, urea–formaldehyde, and polyurethane resin spheres. Microencapsulation can be performed through interfacial polycondensation [13].

21.1.3 General Functions

The organic spheres benefit from the uniformity of the spherical shape and the resulting low surface area of the particle, as described later for their inorganic counterparts. One implication is enhanced processability with a minimum increase in viscosity, with the spheres being thought to act as microscopic ball-bearing components. The low surface area allows higher solids loading with less impact on the flow of the melt or of the resin components during shaping of the plastic product. Process improvements such as easier machinability and faster cycle times are beneficial. When introducing hollow spheres, process parameters such as clamping forces, cycle times, and injection pressures can be reduced in comparison to the processing of the unfoamed, solid material. In that sense, foaming of the polymeric material can be a cost saver. In addition, mold filling is improved resulting in reduced shrinkage problems. Thanks to their uniformity, the spheres give inherently isotropic properties to the plastic or composite, and can even obstruct directional orientation of other more high aspect ratio fillers. Stresses are, thereby, more evenly distributed, enhancing dimensional stability and reducing warpage [3].

The solid organic microspheres make use of their size uniformity in certain applications. For general purpose polystyrene, the addition of monosized polystyrene particles can be used as a dispersion agent for other additives or as a melt flow modifier without any changes of the matrix's physical properties [14]. As additives in coatings such as paints or polymer films, monosized polyacrylic and polystyrene particles can affect the flow characteristics of fluids on surfaces, and improve the coatings cleanability, abrasion, and corrosion resistance. Microcapsules for slow release are mostly produced for medical or cosmetic applications, for example, controlled drug release. In plastics, however, the same function can be used for additive release, where monosized particles of styrene, acrylic, or resorcinol formaldehyde polymers slowly release additives claimed to result in a prolonged effect compared to traditional additive loadings [14].

In comparison with inorganic spheres, hollow organic microspheres have, in some respects, a somewhat different performance. The elastic nature of the polymer shell and the cushioning effect resulting from the compressible interior, is the key factor. Organic microspheres are, thereby, much more likely to withstand the pressure and shear forces during common processing steps such as pumping, calendering, injection molding, and extrusion, than glass spheres. While the latter balance the density reduction against minimum breakage during processing by shell thickness and particle size changes, the organic spheres do not have to make this compromise. As a result, the lower required amounts of organic microspheres make foaming economically feasible for many thermoplastics, in some cases even for commodity materials such as polyethylene, polypropylene, and PVC. A further benefit generated by the absence of breakage problems is that the wear of the processing equipment is minimized with the use of organic microspheres. The heat expandable microspheres can have an even more pronounced shear resistance when mixed into the matrix in their unexpanded form, and then expanded *in situ*. They can resist the high shear operations and thereafter, in forming the end-product, effectively expand to give the requested volume.

Figure 21.2a shows cross sections of a range of unexpanded high temperature microspheres with particles exhibiting a considerable relative wall thickness and also a rough architecture. As a comparison, the same grade after open air expansion is shown in Figure 21.2b. The perfect spherical shape of the expanded microspheres resulting in large volume increase is evident, despite the initial rough texture of the particle. For compounding followed by molding, a well defined expansion temperature of the microspheres is a must, in order to avoid unwanted pre-expansion. The continuous change in volume upon prolonged times at high temperatures also restricted the freedom in selecting polymer and microsphere grades [2, 3]. Glass spheres possess the clear advantage of withstanding both chemically harsh environments and the high temperatures needed to process some plastic matrices, while for organic microspheres there is most often a limitation on the possible plastics to be used. For thermoplastic materials, such as PEEK and polyamides, glass spheres are still the dominating lightweight filler.

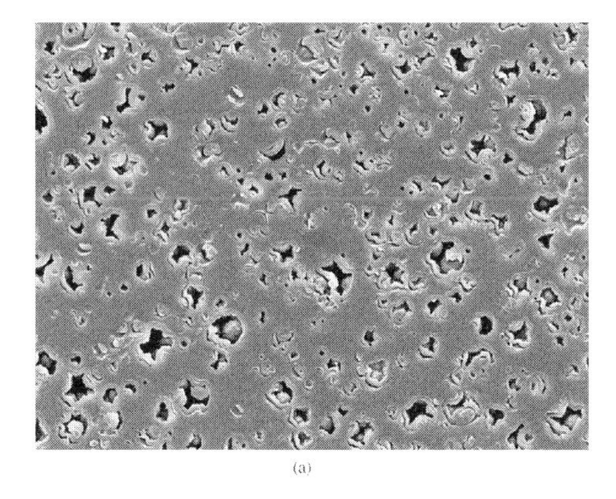

(a)

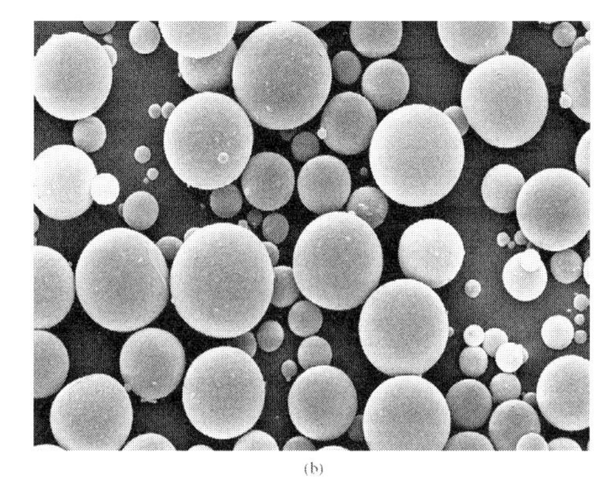

(b)

Figure 21.2 (a) Cross-sectional scanning electron micrograph of unexpanded, high expansion temperature microspheres (Expancel 092 DU 40). Magnification 500×. (b) Scanning electron micrographs of expanded, high expansion temperature microspheres (Expancel 092 DU 40). Magnification 500×.

Hollow organic spheres, with their high level of elastic response to stresses, act as microcushions enhancing ductility and resilience in many plastic parts [2, 3, 15]. Impact strength, crack resistance, and elastic response to pressure and shear forces can all be increased. However, elongation at break and tensile strength are lowered due to the presence of voids in the matrix. The same voids tend to enhance electrical and thermal insulating properties of the material. The thermal insulation enhancement is illustrated in Table 21.1 for polypropylene foamed with heat expandable spheres.

21.1.4 Applications in Thermoplastics

Typical product applications for hollow organic spheres are shoe soles of TPU, SBS or PVC, synthetic wine stoppers, cables, extruded profiles, hoses, and interior auto-

Table 21.1 The effect of foaming PP with heat expandable microspheres on its thermal conductivity[a].

Density (kg/m^3)	Thermal conductivity (W/(m K))
900 (unfoamed reference)	0.20
530	0.099
305	0.066

a) Specimens are 5 mm thick injection-molded plaques of polypropylene BF 330 MO (Borealis, Sweden).

motive details. Organic hollow microspheres play the role of a lightweight filler with the purpose of density reduction. Thermoexpandable microspheres, expanded *in situ*, have been found to foam thermoplastics like polyethylene (HD/MD/LD), polypropylene, PP/EPDM copolymers, PVC, polystyrene, ABS, and EVA, as well as thermoplastic elastomers such as TPOs, SEBS/SBS, and TPUs. It is realistic to obtain density reductions of 40–60%; examples are shown in Table 21.2. Processes such as extrusion, injection molding, film blowing, rotational molding, and thermoforming have been found to work well for certain sets of parameters, for example, a low melt viscosity most often results in better foaming [2, 16, 17]. This is illustrated in Figure 21.3 for polypropylene resins with six different melt flow rates where the resulting density reduction is plotted versus foaming temperature. Thermoexpandable microspheres have successfully been used in wood plastic composites (WPC) to reduce composite weight by 5–30% and also enhance the material's ability to accept nailing, drilling, and cutting in a manner more like natural wood. Common polymer materials for WPC are polypropylene, polyethylene, PVC, and ABS [17, 18].

Expandable microspheres are among the lightest fillers available, and compete with chemical blowing agents (CBA) as foaming agents in thermoplastics. Although a more costly additive, the benefits possessed by the microspheres are their lower sensitivity to processing instabilities, uniform cells, less dependence on melt strength, and more easily obtained uniformity in foaming. Thanks to the encapsulated blowing agent, there are more opportunities for spray-up applications and open

Table 21.2 Density of five common thermoplastics upon injection molding into 5 mm thick plaques with simultaneous foaming using different heat expandable microspheres.

Material	Expancel microsphere grade[a]	Addition (%)	Temperature (°C)	Density (kg/m^3)	Density reduction (%)
SBS	092 DU 120	3	165	585	40
HDPE	093 DU 120	3	180	525	40
PVC-P	092 DU 120	3	180	750	37
TPU	093 DU 120	3.25	170	690	40
ABS	098 DU 120	3	205	770	27

a) Expancel is a registered trademark of Akzo Nobel in a number of world territories.

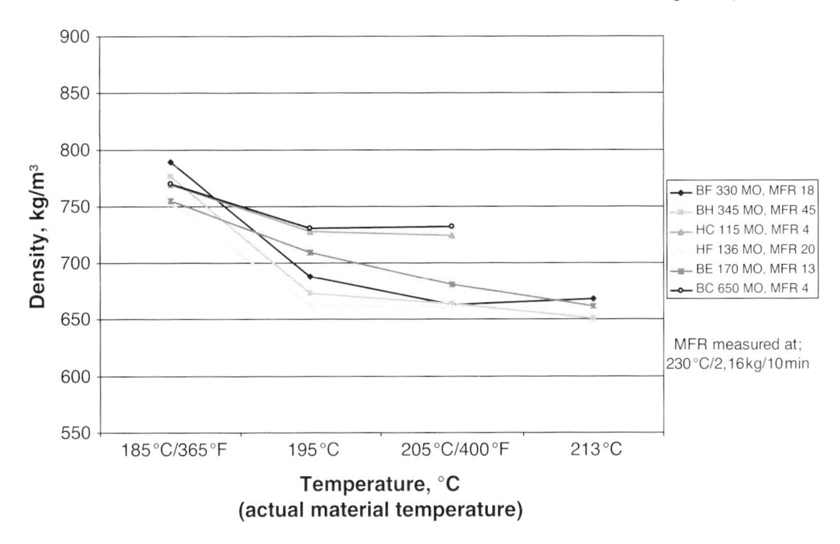

Figure 21.3 Densities of six polypropylene resins with different melt flow rates, foamed with 3% Expancel 098 DUX 120 and injection molded at four different temperatures. Initial density 900 kg/m^3.

mold processing such as production of sheets, films, blankets, and coatings [3]. The closed structure has also been reported to improve flexural strength, compared to CBA foams. Measurements and calculations on foamed and unfoamed WPC show that even though there is a reduction in flexural modulus upon foaming, shown in Figure 21.4, a comparison of the data on a weight basis (i.e., the moduli divided by the corresponding densities) results in approximately a 20% increase in the foamed materials [18]. Within this context, the hollow organic spheres do exhibit a reinforcing effect. The presence of closed cells is assumed to be the reason, in combination with the adhesion of the walls to the matrix. This has also been observed for phenolic resin microspheres [19].

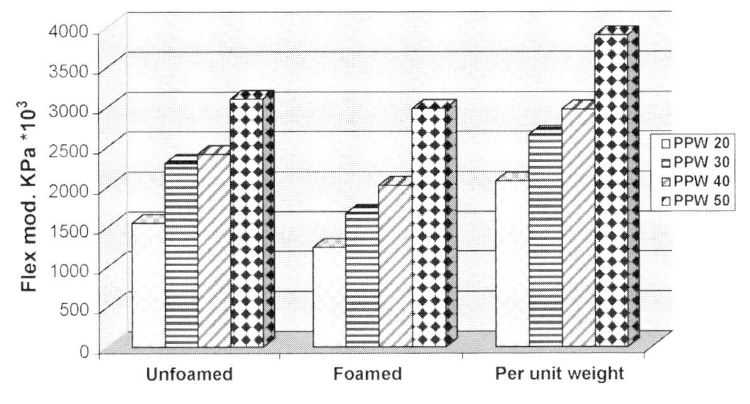

Figure 21.4 Flexural modulus comparison for foamed and unfoamed polypropylene filled with wood flour (PPW 20, 30, 40, and 50%). Foaming with 3.25% Expancel 092 MB 120.

21.1.5
Applications in Thermosets and Rubbers

Thermosets also benefit from the foam structure, as evidenced by improved thermal insulation, sound dampening and mechanical stress absorption responses to temperature changes or impact. Hollow spheres with an already set volume are normally used, that is, pre-expanded microspheres. The reason is that the curing reactions often interfere with any expansion before a sufficient volume increase has been obtained. Hollow organic spheres are found in products such as sealants, adhesives, putties, pipes, cultured marble, body fillers, model-making materials, and pastes [2, 3, 19]. Common suitable matrix materials are epoxies, PUR, and polyesters.

In different types of rubbers hollow spheres are used either as lightweight fillers or as foaming agents. The processing technique and curing temperature determine which role is most suitable. In natural rubber latex and room temperature curing silicones, the pre-expanded spheres are used as lightweight fillers, while in EPDM rubber and high temperature curing silicones the heat expandable spheres are acting like foaming agents. The performance of heat expandable microspheres added at 4% to EPDM is shown in Figure 21.5. In their role as foaming agent, the spheres were expanded to very low true densities in all three experiments; 14 kg/m^3 in a heat cabinet (unrestricted expansion) compared to 30 kg/m^3 in a closed mold (similar setup as in compression molding). However, the results show that a restriction in the allowed volume for the product, clearly affects the degree of expansion.

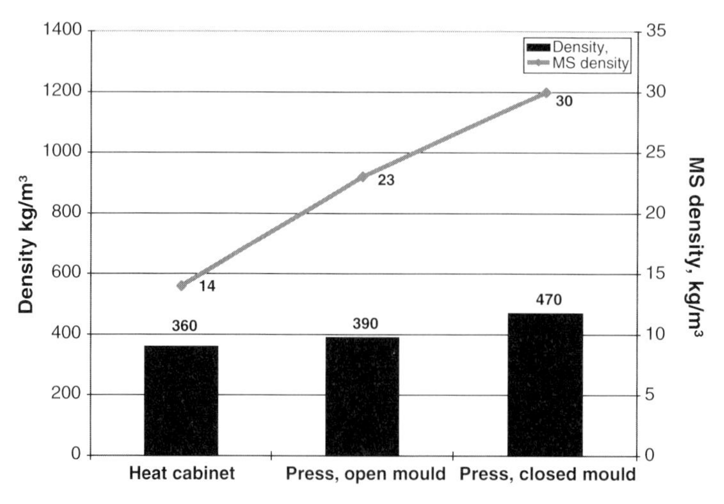

Figure 21.5 EPDM rubber cured at 175 °C for 10 min in three parallel experiments, illustrating the effect of different process pressures on foaming. Left axis: density of the EPDM specimen. Right axis: density of the microspheres in the specimen, that is, MS-density.

21.1.6
Applications in Coatings

Phenolic microspheres are an example of flame resistant hollow spheres, exploited in protective coatings where the spheres char and act either as sacrificial media or as insulators to protect structures exposed to high temperatures [19]. They are also used in abrasion applications like abrasive papers and grinding wheels. Protective underbody coatings with hollow organic fillers in automotive applications provide sound reduction as well as improved stone-chip resistance. Hollow spheres may also be of use in decorative paints and varnishes, added to give visual effects in various applications, for instance a smooth surface, a "velvet effect," a matte look, antislip properties, or 3D-prints [2, 3, 20]. Hollow spheres may also be used as a partial replacement of solvent. The latter function may imply shorter drying times, enhanced applicability, and possibility of increased pigment volume concentration.

Microcapsules containing a small amount of additives that will be released by crack propagation or other triggering mechanisms have been incorporated into polymeric coatings [21]. Urea/formaldehyde capsules containing silicone oil are shown in Figure 21.6.

This process has been used for self-healing in polymer composites via release of a polymerizable "healing agent" that would bridge cracks after reaction in the presence of appropriate catalysts.

21.1.7
Surface Functionalization

Incompatibility problems with organic and inorganic spherical fillers in plastics may be overcome by grafting reactive groups on their surface or adding coupling agents.

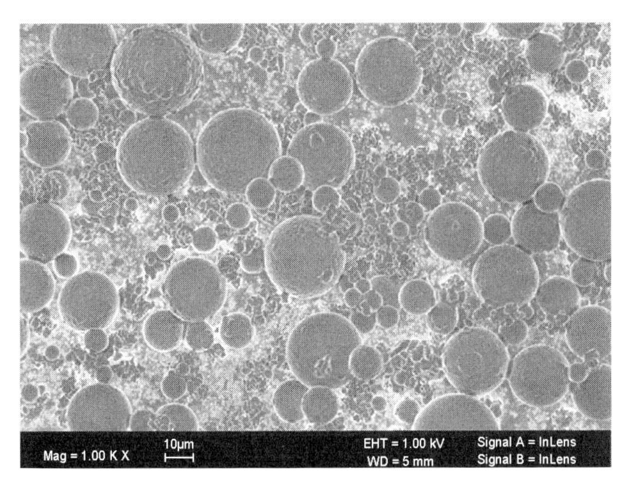

Figure 21.6 Urea–formaldehyde microcapsules containing silicon oil. (Courtesy of Dr K. Park, NJ Institute of Technology, Newark, NJ.)

Functionalized organic microspheres can, thus, create a composite material where the walls of the filler are bonded to the surrounding matrix. Expected benefits are elimination of delamination for an overall stronger composite material and prevention of crack propagation. Furthermore, the polymer particle can be functionalized by incorporating monomers with functionalities such as hydroxyl, oxirane, and amine. When immersed in reactive resins such as epoxies, polyurethanes, and phenolics, the functionalized filler is bonded to the matrix simultaneously with the curing reactions. The manufacture of metal coated organic spheres is another example of surface functionality. Such a filler combines metal-related properties such as high electrical conductivity with the polymer's lower weight [14].

21.1.8 Cost

Organic spheres, being relatively high cost additives, are seldom used with the sole purpose to minimize cost, even if this implies a reduction in the volume of the matrix material. Processing costs can, however, be reduced as described above, but normally the enhancement of one or more product performance parameters are the reason for adding organic microspheres. The cost of hollow spheres, especially heat expandable microspheres, are best evaluated and compared on a per volume basis, rather than on per product weight basis. When doing so their price level is quite comparable with that of commodity inorganic fillers.

21.1.9 Suppliers

Suppliers of thermoexpandable microspheres are Expancel® (Eka Chemicals AB, Akzo Nobel), Sweden; Matsumoto Yushi-Seiyaku Co. Ltd, Japan; Sekisui Chemical Company Ltd, Japan, and Kureha Chemical Industry Co. Ltd, Japan. Suppliers of solid microspheres are Microbeads AS, Norway; Matsumoto Yushi-Seiyaku Co. Ltd, Japan, and Sekisui Chemical Company Ltd, Japan. Supplier of microcapsules for release purposes is Microbeads AS, Norway. Supplier of phenolic microspheres is Asia Pacific Microspheres, Malaysia. Supplier of EPS particles: Kaucuk a,s-Unipetrol Group, Czech Republic; Dow Chemical Company, Michigan, USA; BASF SE, Germany. Suppliers of metal coated microspheres: Microbeads AS, Norway and Sekisui Chemical Company Ltd, Japan.

21.2 Glass and Ceramic Spheres

21.2.1 Introduction

Glass and ceramic spheres are used extensively in thermoplastics and thermosets. They may be solid or hollow with densities varying from 2.5 to 0.1 g/cm^3. The spheres aspect ratio of unity may provide only moderate positive effects for their use as mechanical property modifiers (see Chapter 2). More important functions for both solid and hollow spheres are associated with their spherical form; they include enhanced processability and dimensional stability. Overall reduced composite density is an additional, and the most important function of hollow spheres.

21.2.2
Production and Properties

Solid glass spheres are produced by firing crushed glass with subsequent collection and cooling of the spheroid product, or by melting the formulated glass batch and subsequent break up of the free-falling molten stream to form small droplets [22]. The spheres are mostly based on the A-glass composition (see Chapter 7) although E-glass spheres are also available. A-glass is recommended for all polymers except for alkali sensitive resins such as polycarbonate, acetal, and PTFE, for which the use of E-glass is recommended [23]. Commonly used mean particle sizes may range from 200 to 35 μm. Essentially all glass beads are coated with appropriate coupling agents depending on the intended use. Metal-coated, for example, silver-coated spheres, are also available for shielding applications. They are not considered as hazardous materials and nuisance dust OSHA exposure limits are applicable.

Hollow glass spheres are produced by heating crushed glass containing a blowing agent. During the process of liquefaction, the surface tension causes the particles to assume a spherical shape. The gas formed from the blowing agent expands to form hollow spheres. Other methods involve passing spray-dried sodium borosilicate solutions through a flame or spray-drying modified sodium silicate solutions [22, 24]. The mean particle size may range from 20 to 200 μm depending on the manufacturer, and as in the case of the solid beads they are also coated with a variety of coupling agents. The crush strength of hollow spheres is determined by the thickness of the walls and, as expected, the higher the sphere density, the higher the crush strength. Available effective densities are much lower than those of all polymers, ranging from 0.15 to 0.8 g/cm^3; these correspond to crush strength values ranging from 2 to >150 MPa. Low density spheres are shear sensitive and broken fragments can cause excessive wear in processing equipment. This largely restricted their first uses to liquid or low-pressure processes such as casting or in open-mold reinforced thermosets (see Chapter 1) and selected thermoplastic operations such as plastisol processing. Development of higher strength spheres led to uses in higher shear, thermoset closed mold applications and eventually in extrusion and injection molding of thermoplastics. Denser hollow spheres with 0.6 g/cm^3 density are reported to have crush strength sufficient for surviving injection molding [25]. Some physical properties of solid and hollow glass spheres are compared in Table 21.3 [24, 26]. It should be noted that the chemical properties of both solid and hollow glass spheres are basically those of their precursor silicate glasses.

Ceramic hollow spheres are aluminosilicates produced from a variety of minerals or reclaimed from fly ash waste. Ceramic spheres have higher densities than glass beads, but are less expensive, more rigid, and mechanically more resistant,

Table 21.3 Comparison of A-glass-based solid and hollow glass spheres.

Property	Solid	Hollow
Softening temperature (°C)	700	700
Density (g/cm^3)	2.3–2.5	0.1–1.1
Hardness (Mohs)	5.5–6	5
Modulus (GPa)	60–70	200
Thermal conductivity (W/(m K))	0.7	0.0084
Thermal expansion coefficient (K^{-1})	8.6×10^{-6}	8.8×10^{-6}
Dielectric constant (10^4 Hz)	5	1.5

apparently due to their thicker walls. Their true densities may vary from 0.3 to 0.8 g/cm^3 with mean particle size from 30 to 125 μm. Figures 21.7 and 21.8 show commercial ceramic microspheres with a broad particle size distribution; wall thickness is estimated at about 1 μm. Detailed properties of available ceramic spheres are given in Refs [22, 24].

21.2.3
Functions

The spherical form of these fillers gives rise to distinct functions and properties compared with other directional fillers. These include

- high packing fractions, little effect on viscosity, and high attainable loading levels (up to 50 vol%);
- better flow characteristics than high aspect ratio fillers;

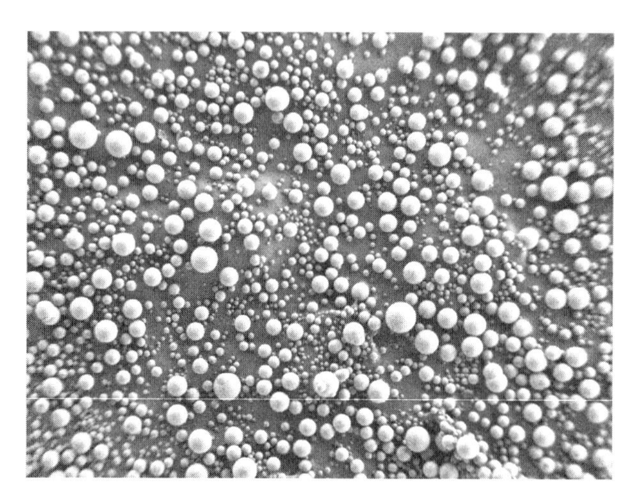

Figure 21.7 Photomicrograph of commercial ceramic microspheres with a top size of about 50 μm and broad particle size distribution, 300×. (Courtesy of Dr S. Kim, Polymer Processing Institute, Newark, NJ.)

Figure 21.8 Photomicrograph of a broken ceramic microsphere with a wall thickness of about 1 μm, 7430×. (Courtesy of Dr S. Kim, Polymer Processing Institute, Newark, NJ.)

- more uniform stress distribution around the spherical inclusions and better dimensional stability;
- no orientation effects and enhanced isotropy;
- reduced uniform and predictable shrinkage and less warpage in injection-molded parts;
- smoother surfaces than with directional fillers.

Effects on mechanical properties depend on the particle size, volume fraction, and surface treatment (see Chapter 2). With solid spheres, mechanical properties that are positively affected are usually modulus, compressive strength, and, in certain cases, tensile strength. Properties that are adversely affected are ductility, manifested as elongation at break, and often impact strength. A variety of thermoplastics such as PS, Nylon, SAN, ABS, PC, and PVC containing solid glass spheres or glass fiber/sphere combinations, are used in automotive, appliances, and connector applications. An increase in extrusion throughput with increasing glass bead loadings has been reported for nylon compounds [27], and increased scratch and abrasion resistance in nylon and HDPE have been documented [27, 28]. Thermoset applications include incorporation in epoxies, polyesters, polyurethanes, and silicones, and as partial replacement of the fibers in fiberglass-reinforced processes.

For hollow spheres, in addition to the effects imparted by their spherical shape, the most important function is density reduction. The effects on mechanical properties are strongly dependent on loadings and wall thickness of the spheres, but a certain modulus increase is usually accompanied by lower tensile and impact strengths. High dielectric strength, reduced dielectric constant, and good thermal insulation are additional attributes of hollow spheres. A significant application of hollow spheres is in thermoset syntactic foams based on liquid epoxy, polyurethane, and polyesters, as well as in vinyl plastisols and foams and in other materials processed at low pressures. Applications are found in cultured marble, the automotive industry, recreational items, sports goods, the electronics industry, ablative composites, and in flotation and buoyancy [29]. In injection moldable thermoplastics such as Nylon or PP, high crush strength hollow spheres with a density of 0.6 g/cm^3 have produced composite density reductions very close to the theoretically predicted on the basis of Eq. (1.4) [25]. This is an indication of the very small degree of breakage during processing. Considering the fairly high cost of the high strength glass bubbles (about US$ 6.00/kg, 2003 prices), overall cost reductions calculated from Eq. (1.5) are only possible for some higher priced engineering thermoplastics such as PEEK.

21.2.4
Suppliers

Major suppliers of solid glass spheres are Potters Industries, Inc., Valley Forge, PA; Sovitec France, Florange, France. Major suppliers of hollow glass and ceramic spheres are 3M Specialty Additives, St Paul, MN; PQ Corp. Valley Forge, PA; Grefco Minerals, Inc., Torrance, CA; Potters Industries, Inc., Valley Forge, PA; Envirospheres Pty Ltd, Australia; Trelleborg Fillite Ltd, UK; Advanced Minerals Ltd, UK, Cenosphere Co., TN; Zeelan Industries, St Paul, MN. Additional suppliers/distributors are listed in Refs [24, 30, 31].

Acknowledgment

Ms. Lena Jonsson, Eka Chemicals AB, is gratefully acknowledged by A.K. for sharing her expertise in processing and foaming of thermoplastics.

References

1 Arshady, R. (ed.) (1999) *Microspheres, Microcapsules and Liposomes, Vol. 1: Preparation and Chemical Applications*, Citus Books, London, UK.

2 Jönsson, L. (2005) Expandable microspheres as foaming agents in thermoplastics, thermosets and elastomers. Proceedings of the Advances in Plastics Technology Conference, Inst. for Plastics Processing, November 2005, Katowice, Poland.

3 Wood, K. (2008) Microspheres – fillers filled with possibilities, Composites World, April, www.compositesworld.com/articles/microspheres-fillers-filled-with-possibilities, website accessed on March 12, 2009.

4 Arshady, R. (1992) Suspension, emulsion and dispersion polymerization: a methodological survey. *Colloid Polym. Sci.*, **270**, 717–732.

5 Jonsson, M., Nordin, O., Malmstrom, E., and Hammer, C. (2006) Suspension polymerization of thermally expandable core/shell particles. *Polymer*, **47**, 3315–3324.

6 Ugelstad, J., Mork, P.C., Kaggerud, K.H., Ellingsen, T., and Berge, A. (1980) Swelling of oligomer-polymer particles – new methods of preparation of emulsions and polymer dispersions. *Adv. Colloid Interface Sci.*, **13**, 101–140.

7 Wang, R., Zhang, Y., Ma, G., and Su, Z. (2006) Preparation of uniform poly (glycidyl methacylate) porous microspheres by membrane emulsion polymerisation. *J. Appl. Polym. Sci.*, **102**, 5018–5027.

8 Eka Chemicals AB, Akzo Nobel, Sweden, accessed on August 25, 2008 www.expancel.com.
9 Ma, G. and Li, J. (2004) Compromise between dominant polymerization mechanisms in preparation of polymer microspheres. *Chem. Eng. Sci.*, **59**, 1711–1721.
10 Dowding, P.J. and Vincent, B. (2000) Suspension polymerization to form beads. *Colloids Surf. A*, **161**, 259–269.
11 Villalobos, M.A., Hamielec, A.E., and Wood, P.E. (1993) Bulk and suspension polymerization of styrene in the presence of *n*-pentane. An evaluation of monofunctional and bifunctional initiation. *J. Appl. Polym. Sci.*, **50**, 327–343.
12 Kopf, P.W. (1987) Phenolic fillers, in *Encyclopedia of Polymer Science and Technology*, vol. 7, John Wiley & Sons, Inc., New York, pp. 322–367.
13 Su, J.-F., Wang, L.-X., Ren, L., Huang, Z., and Meng, X.-W. (2006) Preparation and characterization of polyurethane microcapsules containing *n*-octadecane with styrene-maleic anhydride as a surfactant by interfacial polycondensation. *J. Appl. Polym. Sci.*, **102**, 4996–5006.
14 Microbeads AS, Norway, website accessed on August 25, 2008 www.micro-beads.com.
15 Ahmad, M. (2001) Flexible vinyl resiliency property enhancement with hollow thermoplastic microspheres. *J. Vinyl Addit. Technol.*, **7** (3), 156–161.
16 Elfving, K. (2003) New developments with expandable microspheres. Proceedings of the RAPRA Conference, Blowing Agent and Foaming Processes, Munich, Germany.
17 Ahmad, M. (2004) Thermoplastic microspheres as foaming agents for wood plastic composites. Proceedings of the WPC 2004 Conference, Wood Plastics Composite Org., September 2004, Vienna, Austria.
18 Kron, A. (2005) Low weight filler – thermally expandable hollow polymer microspheres. Proceedings of the High Performance Fillers RAPRA Conference, March 2005, Cologne, Germany.
19 Asia Pacific Microspheres, Malayan Adhesives & Chemicals Sdn Bhd, Malaysia, website accessed on August 25, 2008, www.phenoset.com.
20 Kawaguchi, Y. and Oishi, T. (2004) Synthesis and properties of thermoplastic expandable microspheres: the relation between crosslinking density and expandable property. *J. Appl. Polym. Sci.*, **93**, 505–512.
21 Feng, W., Patel, S.H., Young, M.-Y., Zunino, J.L., III, and Xanthos, M. (2007) Smart polymeric coatings – recent advances. *Adv. Polym. Technol.*, **26** (1), 1–13.
22 Katz, H.S. and Milewski, J.V. (eds) (1978) Chapters 18 and 19 in *Handbook of Fillers and Reinforcements for Plastics*, Van Nostrand Reinhold Co., New York.
23 Potters Industries Inc., http://www.pottersbeads.com/markets/polyspheriglass.asp, accessed March 28, 2004.
24 Wypych, G. (2000) *Handbook of Fillers*, ChemTec Publishing, Toronto, Ont., Canada, pp. 72–75, 87–91.
25 Israelson, R. Proceedings of the Functional Fillers for Plastics 2003, Intertech Corp., October 2003, Atlanta, GA.
26 Hohenberger, W. (2001) Chapter 17, in *Plastics Additives Handbook* (ed. H. Zweifel), Hanser Publishers, Munich.
27 Anonymous (2002) Solid glass beads offer major benefits for polyamide compounds. *Plast. Addit. Compound.*, **4** (6), 32–33.
28 Beatty, C.L. and Elrahman, M.A. (1999) Fillers (glass bead reinforcement), in *Concise Polymeric Materials Encyclopedia* (ed. J.C. Salamone), CRC Press, Boca Raton, FL, pp. 475–476.
29 Muck, D.L and Ritter, J.R. (1979) Plastics Compounding, January–February, pp. 12–28.
30 World Buyers' Guide 2004 (2004) Plastics Additives & Compounding, vol. 5, p. 7.
31 Advanstar Communications, Inc . (1999) *Plastics Compounding Redbook Directory*, Advanstar Communications, Inc., Cleveland, OH.

22
Bioactive Fillers

Georgia Chouzouri and Marino Xanthos

22.1
Introduction

During the last few decades, the need for biomaterials in dental, craniofacial, and orthopedic applications has increased and so has the necessity for further development of new engineering composite materials. These types of materials are required to provide distinctive mechanical performance in bone growth applications, as well as biocompatibility and biological active response, known as bioactivity. According to Hench [1], bioactivity is the ability of a material to elicit a specific biological response at its interface with a living tissue, which results in the formation of a bond between the tissue and the material. It is essential to understand that no single biomaterial is appropriate for all tissue engineering applications, and also that the mechanical and biological behavior can be "tailored" for a given application. The so-called biomedical composites that are classified as bioinert, bioactive, and bioresorbable consist of a matrix, which can be metallic (e.g., titanium), inorganic (e.g., glass), or polymeric (e.g., high-density polyethylene, HDPE), in combination with miscellaneous fillers [2]. Usually, these functional fillers are responsible for the *in vivo* bioactive response, although the matrix may also exhibit a biological response in certain bioresorbable compositions containing degradable synthetic (e.g., polylactic acid, PLLA) or natural (e.g., collagen, polysaccharides) polymeric matrices. Certain types of glasses, ceramics, and minerals have been reported in the literature to act as bioactive fillers. They fall under the generic term of bioceramics [3], which encompasses all inorganic, nonmetallic materials to be used in the human body. Such materials have been used on their own as implants, have been dispersed in matrices such as polymers and inorganics, or have been applied as coatings on metallic implants.

22.2
Bone as a Biocomposite

Bone is a natural composite material with major components type I collagen, calcium phosphate minerals (hydroxyapatite (HA) is the predominant one), carbonate

Functional Fillers for Plastics: Second, updated and enlarged edition. Edited by Marino Xanthos

ISBN: 978-3-527-32361-6

substituted apatite, and water [4]. Bone is a brittle anisotropic material with low elongation at fracture (3–4%) with properties that may vary broadly. Tensile modulus and strength for a long human bone are reported to be 17.4 GPa and 135 MPa, respectively, in the axial direction, and much lower in the radial direction: 11.7 GPa and 61.8 MPa, respectively [5]. Compressive strength, a property more relevant to the actual use of the bone, is higher, approaching 196 MPa and 135 MPa in the axial and transverse directions, respectively. In biomedical polymer composites, attempts are made to reach these high modulus/strength levels through the introduction of high volume loadings (as high as 45%) of bioactive fillers.

Bone has a complex structure with several levels of organization. In developing bone substitutes, two structure levels are considered. The first is a bone apatite-reinforced collagen that forms lamellae at the nanometer to micrometer scale, and the second is the osteon-reinforced interstitial bone at the micrometer to millimeter scale [2]. The apatite-collagen composite triggered researchers to investigate composites of bioactive ceramics in polymer matrices as alternatives for bone replacement [2]. Usually, these polymers can be either biostable or biodegradable, depending on the intended application. In composites containing bioactive ceramics in biodegradable polymers, the rate of degradation needs to be controlled so that mechanical integrity is retained in the early stages of bone healing. Table 22.1 lists some polymers that have been reported in the literature to be widely used for the production of biocomposites for tissue engineering applications.

22.3
Synthetic Biomedical Composites and Their Bioactivity

22.3.1
Bioceramics for Tissue Engineering Applications

Bioceramics is a generic term that covers all inorganic, nonmetallic materials that have been used in the human body as implants or prostheses. According to the type of tissue attachment, there are three types of bioceramics: bioinert, bioactive, and bioresorbable [1, 6]. Inert bioceramics are biologically inactive, with a characteristic lack of

Table 22.1 Examples of polymers used in tissue engineering applications.

Biostable polymers	Biodegradable polymers
Polyethylene (PE, HDPE)	Polylactic acid (PLLA)
Polyetheretherketone (PEEK)	Polyglycolic acid (PGA)
Polysulfone (PSU)	Poly-ε-caprolactone (PCL)
Polyurethane (PU)	Poly-β-hydroxybutyrate (PHB)
Polymethylmethacrylate (PMMA)	Polyorthoesters
Bisphenol-α-glycidyl methacrylate (bis-GMA)	Poly-δ-valerolactone
	Blends of starch with ethylene vinyl alcohol (SEVA)
	Chitosan

Table 22.2 Bioactive fillers used in tissue engineering applications.

Chemical name	Chemical formula	Calcium/phosphorous ratio[a)]
Hydroxyapatite	$Ca_{10}(PO_4)_6(OH)_2$	1.67
Dicalcium phosphate	$CaHPO_4 \cdot 2H_2O$	1
β-Tricalcium phosphate	$Ca_3(PO_4)_2$	1.5
Tetracalcium phosphate	$Ca_4P_2O_9$	2
Calcium carbonate	$CaCO_3$	NA
Wollastonite	$CaSiO_3$	NA
Bioactive glasses	(see Table 22.3)	
A–W glass ceramics	$Ca_{10}(PO_4)_6(OH,F)_2 - CaSiO_3$ in $MgO–CaO–SiO_2$ matrix	1.67

a) From Ref. [7].

interaction between the tissue and the bioceramic and vice versa. Common examples are alumina and zirconia. Bioactive ceramics are surface reactive ceramics that form a bonding layer with the tissue in order to accelerate the bone growth. Typical examples are hydroxyapatite, bioactive glasses, and bioactive glass ceramics. Bioresorbable active ceramics are designed to gradually degrade and be slowly replaced by the host tissue. Tricalcium phosphates, calcium phosphate salts, and calcium carbonate minerals are common bioresorbable ceramics. Table 22.2 summarizes the most significant ceramics that have been reported to show *in vitro* bioactivity and in most cases *in vivo* bioactivity. Table 22.3 contains information on the major suppliers of bioactive fillers and related products.

Similarly to other functional fillers, shape, size, size distribution, pH, and volume percentage of the bioactive filler, as well as type and level of bioactivity and filler distribution in the matrix play important roles on the properties of the composites. In addition, the properties of the matrix, the state of the filler-matrix interfacial region, as well as manufacturing conditions are of great importance in the performance of the final biomaterial [2]. In the following sections, some examples of composites containing specific bioactive and bioresorbable active ceramics are presented. Matrices to be considered are mostly polymers. The majority of such composites are prepared by conventional melt processing methods (extrusion compounding followed by injection or compression molding), although some composites are prepared by solution casting techniques. Attempts have been made to simulate the bone structure and its properties through specialized forming technologies, including shear controlled orientation injection molding (SCORIM®) [8] and hydrostatic extrusion [2]. The state of the art and recent developments in bioinert, biodegradable, and injectable polymer composites for hard tissue replacement have been recently reviewed by Mano *et al.* [8].

22.3.2
Mechanisms and Procedure for *In Vitro* Evaluation of Bioactivity

Bone regeneration through bioactive fillers is accelerated by the formation of a bonding layer with apatite structure that is reinforced with collagen. This apatite

Table 22.3 List of major bioactive filler suppliers.

Supplier's name – web site	Supplier's location, product trade name (filler type)
ApaTech – www.apatech.com	UK, Actifuse® (silicate substituted calcium phosphate), ApaPore™ (HA)
Berkeley Advanced Biomaterials – www.hydroxyapatite.com	USA, Cem-Ostetic® & Bi-Ostetic™ (HA) (calcium phosphates)
Biomet – www.biometmicrofixation.com	USA, Mimix® & Mimix® QS (calcium phosphates mixtures)
Cam Implants – www.camimplants.nl	Netherlands, CAMCERAM® (HA/βTCP)
Ceraver – www.ceraver.fr	France, CERAPATITE® (HA), CALCIRESORB® (βTCP), CALCIRESORB® 35 (HA/βTCP)
Curasan – www.curasan.com	Germany, Cerasorb® (βTCP)
DePuy – www.depuyusa.com	USA, α-BCM (calcium phosphate)
Isotis – www.isotis.com	USA, OsSatura TCP™ (βTCP)
Mo-Sci – www.mosci.com	USA, Bioactive glasses
NovaBone® – www.novabone.com	USA, NovaBone®, PerioGlass®, NovaBone C/M® (bioactive glass), NovaBone Putty®
Orthovita – www.orthovita.com	USA, Vitoss® (βTCP), Cortoss®
Schott – www.us.schott.com	USA, Bioactive glasses
Stryker – www.stryker.com	USA, BoneSource® BVF (calcium phosphate powders)
Synthes – www.synthes.com	USA, ChronOS (βTCP)
Geistlich – www.geistlich.com	Switzerland, Orthoss®

structure can be formed when the material comes into contact with human blood plasma. Nucleation of the apatite layer may be promoted by a variety of functional groups (carboxyl, hydroxyl) that are present on the filler surface or formed through contact with physiological fluids. In order to screen the *in vitro* bioactivity of a material, researchers have used a simulated body fluid (SBF), which contains similar inorganic ion concentrations and has similar pH to human blood plasma [9]. Bioactivity is investigated by analyzing for elements like calcium and phosphorus that could be part of the apatite type layer formed at the surface. The Ca/P needs to be calculated and compared with the ratio of 1.67 that is assumed to be equivalent to the one needed for bone ingrowth. For certain systems, the concentrations of calcium and phosphorous in the biological or simulated body fluid can support the formation of the apatite layer needed for bone ingrowth, making the presence of calcium or phosphorous elements in the filler structure unnecessary.

22.4 Bioceramics as Functional Fillers

22.4.1 Hydroxyapatite

Hydroxyapatite is considered to be a biocompatible and osteoconductive material, exhibiting only an extracellular response leading to bone growth at the bone–filler interface [2]. By contrast, osteoproductive fillers such as bioactive glasses (see Section 22.4.4) elicit both an extra- and an intracellular response at the interface [3]. One of the main reasons for investigating HA as bioactive filler is its similarity to the biological hydroxyapatite in impure calcium phosphate form found in the human bone and teeth. HA has a Ca:P ratio of 10 : 6 and its chemical formula is $Ca_{10}(PO_4)_6(OH)_2$. The biological HA contains, in addition, magnesium, sodium, potassium, and a poorly crystallized carbonate containing apatite phase, as well as a second amorphous calcium phosphate phase [10].

Several biocomposites containing HA have been described in the literature, although not all of them have achieved clinical success. The matrix in the most widely known and investigated HA composites is high-density polyethylene, a biocompatible and biostable polymer broadly used in orthopedics. The composite known as HAPEX™, firstly introduced by Smith and Nephew Richards in 1995 [2], was the result of pilot studies, laboratory testing, clinical trials, and pilot plant production efforts spanning a period of about 15 years until regulatory approval was attained [3]. A range of 0.2–0.4 volume fraction HA was determined to be the optimum. HAPEX was the first composite designed to mimic the structure and retain the properties of bone, and is mainly used for middle ear implants. It has mechanical properties similar to those of the bone and it is easy to trim, which allows surgeons to precisely fit it at the time of implantation. A further goal of the inventors was to produce similar composite materials that can carry greater loads for other parts of the body. Both commercially available and "in house" synthesized hydroxyapatites have been evaluated along with different polyethylene types. By varying the amount and particle size of HA a range of mechanical properties approaching those of bone and different degrees of bioactivity can be obtained depending on the application [2]. The *in vitro* and *in vivo* responses have also been studied extensively. In human osteoblast cell primary cultures used for *in vitro* experiments, the osteoblast cells appeared to attach to HA; cell proliferation followed, thus confirming the bioactivity of the composites. In *in vivo* experiments with adult rabbits, the composite implant surface was covered by newly formed bone.

Sousa *et al.* [11] investigated HDPE filled with 25 wt% commercially available HA having average particle size of 10 μm; the composites were produced by melt mixing, followed by shear controlled orientation injection molding (SCORIM®) to simulate bone structure. Sousa *et al.* [12] also produced composites of blends of starch with ethylene vinyl alcohol (SEVA-C) with 10, 30, and 50 wt% hydroxyapatite by twin screw extrusion compounding followed by SCORIM®, as well as by conventional injection molding. SCORIM® processing appeared to improve the stiffness of the composites, giving a better control in their mechanical properties compared to conventional injection molding. No data were reported regarding the bioactivity of these composites. Similarly, SEVA-C filled with 30 wt% commercially available HA was produced by Leonor *et al.* [13] by melt mixing followed by injection molding; circular samples were used to study the formation of a calcium phosphate layer when immersed in SBF. Figures 22.1 and 22.2 are scanning electron micrographs (SEMs) of the unfilled and filled matrix before and after immersion in the SBF; they clearly show, in the case of the filled polymer, the formation of a bond to the tissue apatite layer.

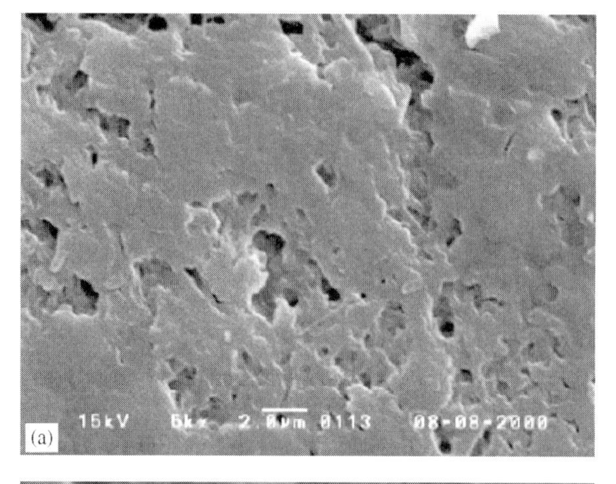

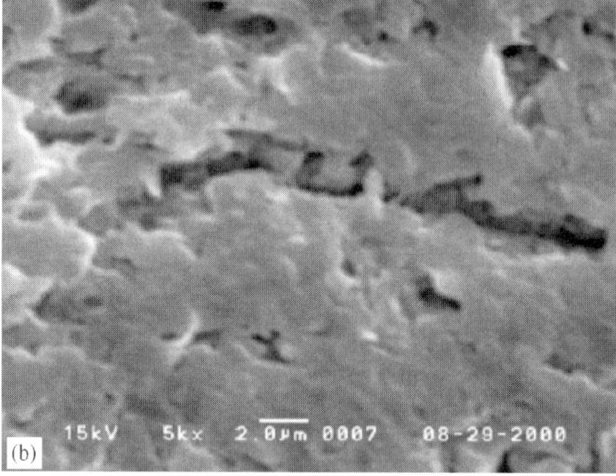

Figure 22.1 Scanning electron micrographs of an unfilled SEVA-C surface before (a) and after 7 days (b) immersion in the SBF. (Reproduced with permission from Elsevier, Ref. [13].)

Biocomposites of polysulfone (PSU) filled with 40 vol% HA have also been produced for hard tissue replacement [2]. PSU is a better matrix candidate than HDPE due to its higher strength and modulus, and, hence, better performance in load bearing applications. The PSU/HA composites were produced similarly to HDPE/HA composites by using conventional compounding methods, followed by compression or injection molding to the desired shape. By increasing the HA content, the stiffness of the composite was increased to levels approaching the lower limit for human bone. Of particular importance in this and other composites containing HA and bioactive glass is the ability to control the polymer/filler interfacial strength, a complex problem as bioactivity is a surface-related phenomenon.

Yu *et al.* [14] produced HA-reinforced polyetheretherketone (PEEK) composites by mixing PEEK and HA (10–40 vol%) powders, followed by compaction, pressureless sintering, and evaluation for bioactivity in SBF. The surface of the 40 vol% composite

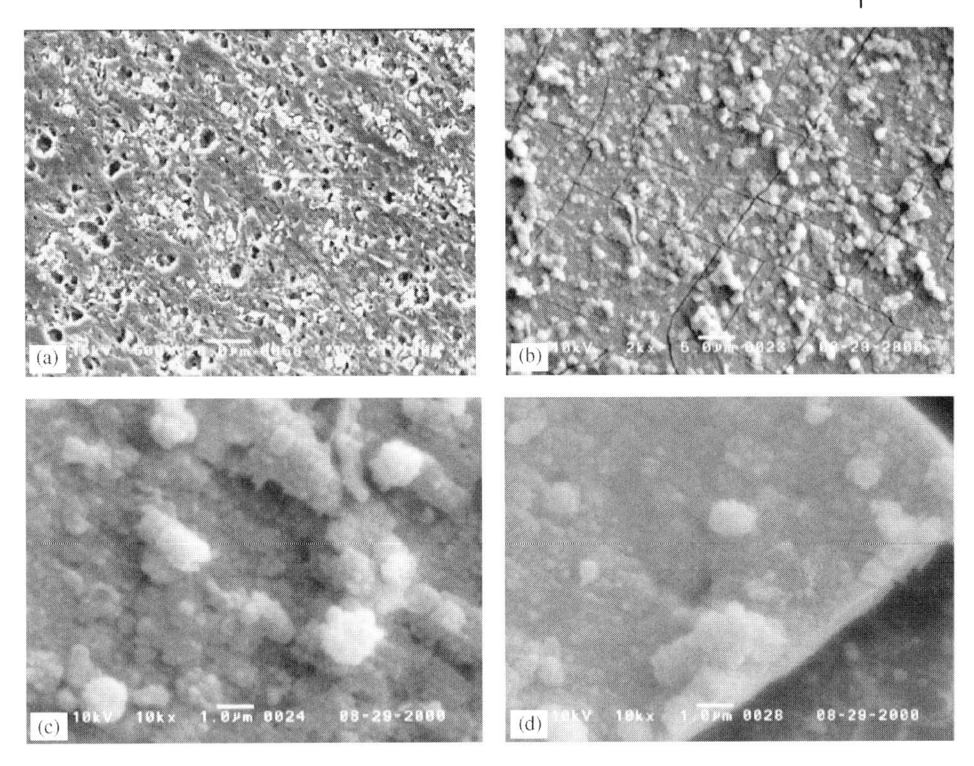

Figure 22.2 Scanning electron micrographs of SEVA-C + 30% HA composite before (a) and after 7 days (b) immersion in the SBF and at different magnification (c) and at a different cross section (d). (Reproduced with permission from Elsevier, Ref. [13].)

was covered by an apatite layer in a short immersion period of 3 days, whereas the surface of the 10 vol% composite required 28 days to be fully covered with apatite. Thus, constant growth rate and subsequent bioactivity of the composite increased with increasing HA volume fraction.

In another study, Ni and Wang [15] introduced different loadings (10, 20, and 30 vol %) of HA particles into polyhydroxybutyrate (PHB) matrix and conducted *in vitro* studies. After a short period of time (within 1 day in SBF) formation of apatite was observed. The number of nucleation sites of apatite crystals was proportional to the HA content, and the composite with the higher loading had, as expected, a faster apatite layer growth. Dynamic mechanical analysis (DMA) showed that the storage modulus of the composite increased initially, due to apatite formation, and after prolonged immersion periods eventually decreased due to polymer degradation.

In an *in vivo* study that lasted 5–7 years, Hasegawa *et al.* [16] investigated the biocompatibility and biodegradation of HA/PLLA composite bone rods using non-calcined HA (n-HA) and calcined HA (c-HA) that were implanted into the distal femurs of 25 rabbits. The n-HA/PLLA composites showed excellent biodegradability and osteoconductivity. Newly formed bone surrounded the residual material and trabecular bone, bonded to the rod, was observed toward the center of the implant.

22.4.2 Calcium Phosphate Ceramics

Different phases of calcium phosphate ceramics have been used depending on the application. The stability of these ceramics is subject to temperature and the presence of water. In the body ($T = 37\,^\circ C$ and pH = 7.2–7.4), calcium phosphates are converted to HA. At lower pH (<4.2) dicalcium phosphate ($CaHPO_4{\cdot}2H_2O$) is the stable phase. At higher temperatures, other phases of phosphate minerals exist, such as β-tricalcium phosphate ($Ca_3(PO_4)_2$), which is chemically similar to HA with Ca:P ratio of 3 : 2, and tetracalcium phosphate ($Ca_4P_2O_9$). Tricalcium phosphate (TCP) is not a natural bone mineral component, although it can be partly converted to HA in the body according to the following reaction [1]:

$$4Ca_3(PO_4)_2\,(\text{solid}) + 2H_2O \rightarrow Ca_{10}(PO_4)_6(OH)_2\,(\text{surface}) + 2Ca^{2+} + 2HPO_4{}^{2-}$$

TCP is an osteoconductive and resorbable material, with a resorption rate dependent on its chemical structure, porosity, and particle size [10].

Composites of polyhydroxybutyrate, a natural biodegradable thermoplastic β-hydroxy acid, with TCP have been prepared by conventional melt processing technologies (extrusion, injection, or compression molding) [2, 17, 18]. *In vitro* experiments in SBF produced an apatite like structure on the composite surface suggesting bioactivity. When immersion in SBF was extended to 2 months or more, the onset of matrix degradation could be followed by the decrease in storage modulus.

Composites of chitosan and β-TCP with improved compressive modulus and strength have been prepared by a solid–liquid phase separation of the polymer solution and evaporation of the solvent [8]. The composites exhibited bioactivity when immersed in SBF. Variation of polymer/filler ratio and development of different macroporous structures resulted in products with potential applications in tissue engineering.

22.4.3 Calcium Carbonate

Calcium carbonate ($CaCO_3$) minerals can exist in the forms of vaterite, aragonite, and calcite (see also Chapter 16). All forms have the same chemistry, but different crystal structures and symmetries. Aragonite is orthorhombic, vaterite is hexagonal, and calcite is trigonal. Natural coral is calcium carbonate in the aragonite form (>98% $CaCO_3$). It is a porous, slowly resorbing material with an average pore size of 150 μm and very good interconnectivity. For use in periodontal osseous defects, it can be supplied with an average particle size of 300–400 μm. The major advantage of calcium carbonate is that when other bioactive materials such as HA have to go through the formation of carbonate containing structures, calcium carbonate can pass over this step; consequently, this can result to a more rapid bone ingrowth [10].

An application of calcium carbonate as a bioactive filler was discussed by Kasuga *et al.* [19], who incorporated vaterite powders prepared by a carbonation process in methanol into a polylactic acid matrix. Composites containing 20–30 wt% vaterite

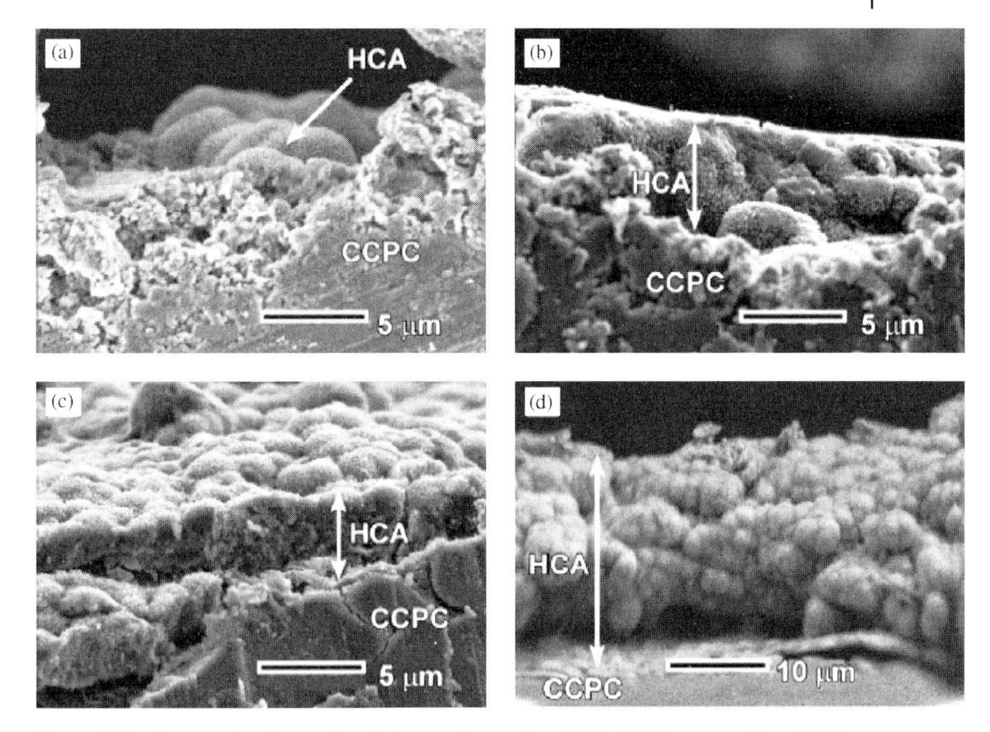

Figure 22.3 Cross-sectional scanning electron micrographs of PLA–$CaCO_3$ composites (CCPC) upon exposure to SBF; (a and b) 20% filler; (c and d) 30% filler; (a,c) 1 days; (b,d) 3 days of immersion. (Reproduced with permission from Elsevier, Ref. [19].)

were prepared by solution mixing and immersed in SBF at 37 °C. Scanning electron micrographs (Figure 22.3) show the formation of a thick apatite layer even after a short period of 1–3 days.

22.4.4
Bioactive Glasses

Special compositions of glass appear to have the ability to develop a mechanically strong bond to bone. The so-called bioactive glasses contain SiO_2, Na_2O, CaO, and P_2O_5 at specific ratios [1, 10, 20]. Figure 22.4 represents a scanning electron micrograph of bioactive glass in powder form that was used in a recent study on PCL bioactive composites [7, 21]. Bioactive glasses differ from the traditional soda-lime-silica glasses (see also Chapter 7) as they contain less than 60 mol% SiO_2, have high Na_2O and CaO contents, and a high CaO/P_2O_5 ratio. As a result, when these glasses are exposed to physiological liquids they can become highly reactive. This feature distinguishes the bioactive glasses from bioactive ceramics such as HA. When the latter comes into contact with physiological fluids, both its composition and physical state remain unchanged, in contrast to the bioactive glass, which undergoes a chemical transformation. A slow exchange of ions between the glass

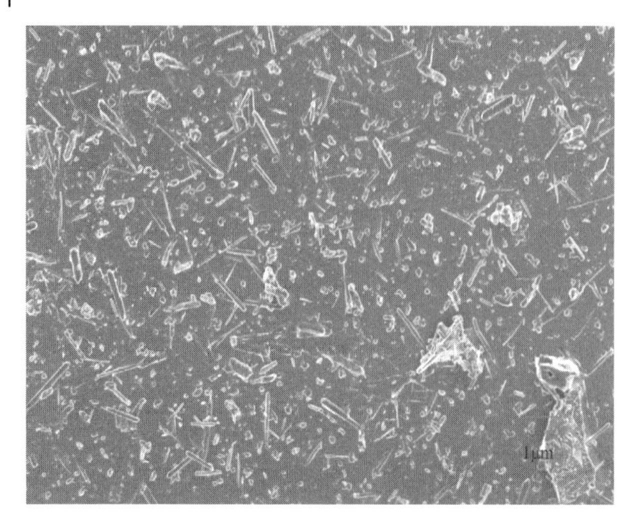

Figure 22.4 Scanning electron micrograph of bioactive glass powder [7, 21].

and the fluid takes place [6], resulting in the formation of a biologically active carbonated HA layer that provides bonding to the bone and also to soft connective tissues. Silicon and calcium that are slowly dissolved from the glasses activate families of genes in old bone cells, which then form new bone cells [3].

Most of the bioactive glasses are based on bioglass designated as 45S5, which implies 45 wt% SiO_2 and CaO/P_2O_5 molar ratio of 5 : 1. Glasses with a lower CaO/P_2O_5 ratio do not bond to the bone. Nevertheless, based on modifications of the 45S5 bioglass, a series of other bioactive glasses have been investigated by substituting, for instance, 5–15 wt% SiO_2 with B_2O_3 or 12.5 wt% CaO with CaF_2 [1, 6, 10, 22, 23]. Table 22.4 provides typical compositions of bioactive glasses.

A large variety of bioactive glass polymer composites has been investigated. Rich *et al.* [24] and Jaakkola *et al.* [25], synthesized a copolymer of poly(ε-caprolactone-co-DL-lactide) (96/4 molar ratio) and produced composites in a batch mixer with bioactive glass having two different ranges of particle size (<45 and 90–315 μm) at concentrations ranging from 40 to 70 wt%. They concluded that the higher the glass content and the glass surface/volume ratio in the matrix, the faster the apatite formation.

Table 22.4 Bioactive glasses and their composition in weight percent [1].

Glass designation	SiO_2	P_2O_5	CaO	CaF_2	Na_2O	B_2O_3	MgO
45S5[a)]	45	6	24.5	—	24.5	—	—
45S5F	45	6	12.25	12.25	24.5	—	—
45S5.4F	45	6	14.7	9.8	24.5	—	—
40S5B5	40	6	24.5	—	24.5	5	—
45S5.OP	45	—	24.5	—	30.5	—	—
45S5.M	48.3	6.4	—	—	26.4	—	18.5

a) Also known as bioglass.

Similarly, Närhi *et al.* [26], explored the biological behavior of a composite based on a copolymer of degradable poly(ε-caprolactone-co-DL-lactide) filled with glass S53P4 in experimental bone defects in rabbits. Bone ingrowth was mainly observed in the superficial layers of the composites that contained higher concentrations of fillers of larger particle size. In another example, Bioglass®-reinforced polyethylene containing 30 vol% filler prepared by melt processing exhibited excellent biocompatibility and enhanced osteoproductive properties compared with the HAPEX™ material containing HA. Microscopic examination of the interface between human osteoblast-like cells and the composite indicated direct bonding between the hydroxyl carbonate apatite layers formed on the filler particles *in vitro* and the cells [27–29]. Bioglass®/polysulfone composites have been shown to provide a closer match to the modulus of cortical bone, with an equivalent strain to failure [3].

In another study, Yao *et al.* [30], reported on the optimal synthesis parameters and the kinetics of formation of calcium phosphate layer at the surface of PLGA/30 wt% bioactive glass porous composites. The porous structure supported marrow stromal cells (MSC) proliferation and promoted MSC differentiation into osteoblast phenotype cells. The porous composite was found to be bioactive and demonstrated a significant potential as a bone substitute.

22.4.5 Apatite–Wollastonite Glass Ceramic

Apatite–wollastonite glass ceramics (AWGC) consist of crystalline apatite $[Ca_{10}(PO_4)_6(OH)F_2)]$ and wollastonite ($CaSiO_3$) (see also Chapter 14) in a MgO–CaO–SiO_2 glassy matrix. The nominal composition by weight is MgO, 4.6; CaO, 44.7; SiO_2, 34.0; P_2O_5, 16.2; CaF_2, 0.5 [31]. This composition has been used as a bone replacement material due to its high bioactivity and its ability to instantaneously bond to living tissue without forming a fibrous layer. The mechanical properties of AWGCs are better than those of both bioactive glass and HA [1, 31–33]. In addition, AWGCs appear to have long-term mechanical stability *in vivo*, as they chemically bond to living bone 8–12 weeks after implantation [32, 33]. According to Hench [1], additions of Al_2O_3 or TiO_2 to the AWGC may inhibit bone bonding.

Shinzato *et al.* [31] evaluated AWGCs as fillers in bisphenol-α-glycidyl methacrylate (bis-GMA) composites. An AWGC filler with an average particle size of 4 μm was synthesized and incorporated in the polymer at 70 wt% loading. The composite had a cured surface on one side and an uncured surface on the other in order to differentiate between their bone bonding abilities. Such composites were implanted into the tibiae of male white rabbits. Direct bone formation through a Ca–P rich layer was observed histologically only for the uncured surfaces, presumably due to enhanced diffusion in the non-cross-linked state and faster exposure to the filler surface [31]. In another study [34], Juhasz *et al.* investigated composites of HDPE filled with AWGC of average particle size in the range 4.4–6.7 μm at filler contents ranging from 10 to 50 vol%. With an increase in AWGC volume fraction, increases in Young's modulus, yield strength, and bending strength were achieved while the fracture strain decreased. Specifically, a transition from ductile to brittle fracture was observed at certain filler concentrations. Based on mechanical and bioactivity test data, composites with 50 vol% AWGC appear to have potential as implants for maxillofacial applications.

22.4.6 Other Bioactive Fillers

Within the class of bioceramics, wollastonite ($CaSiO_3$) was also shown to be bioactive and biocompatible. The structure of wollastonite is similar to that of bioactive glasses and glass-ceramics, and consequently, the formation of an apatite layer when in contact with biological fluids, can be similarily explained. Figure 22.5 shows a scanning electron micrograph of calcium silicate in powder form [7, 21]. Liu *et al.* [35] used commercial available wollastonite of particle size in the range 10–60 μm for coating Ti–6Al–4V substrates through a plasma spayed method. The obtained specimens were further soaked in a lactic acid solution to activate surface functional groups, and then rinsed and immersed in SBF. As expected, an apatite layer was formed through surface reactions.

In the family of glass ceramics, Ceravital™, a low-alkali, bioactive, silica, glass ceramic, shows similar surface activity toward biological fluids as bioactive glass [36]. Gross and Strunz [1] observed, however, that even small additions of Al_2O_3, Ta_2O_5, TiO_2, Sb_2O_3, or ZrO_2 could inhibit bone bonding. Another bioactive filler produced by combining carbonate-containing amorphous calcium phosphate, as a basic ceramic material, with crystalline carbonate (calcite) was used by Schiller *et al.* [37] to formulate composites through solution mixing with polylactides and their copolymers. The filler was shown to increase bioactivity and at the same time maintained the pH in the physiological range for long-term applications. Based on preliminary bone growth results, such composites appear to have future potential as skull implants with specific geometries.

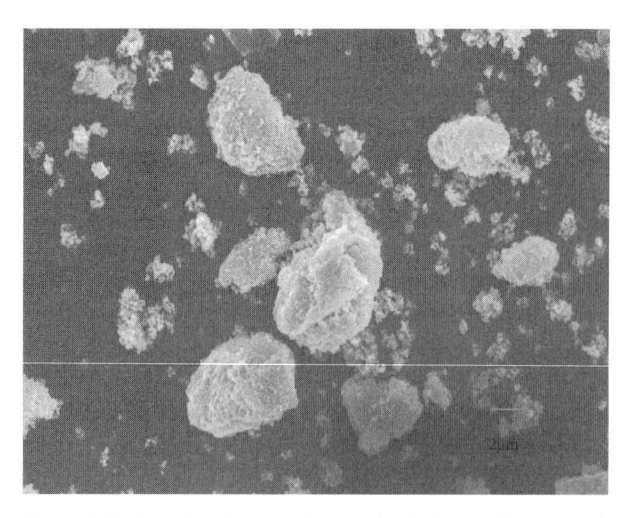

Figure 22.5 Scanning electron micrograph of calcium silicate powder [7, 21].

22.5
Modification of Bioceramic Fillers

In order to further improve their properties, bioactive ceramics have been modified by incorporating various elements. Yamasaki *et al.* [38] added magnesium, an important element controlling biological functions, during the synthesis of functionally graded carbonate apatite crystals. The composites prepared with magnesium containing filler appeared to promote higher bone density than those without it. Similarly, Blaker *et al.* [39] incorporated Ag_2O into bioactive glasses. The silver-doped glasses were used as coatings for surgical sutures and appeared to minimize the risk of microbial contamination without in any way compromising their bioactivity. In another study, Ito *et al.* [40] used zinc, a trace element found in the human bone, to modify TCP ceramics. TCP is an appropriate zinc carrier, since its crystal structure has an atomic site that can accommodate divalent cations with ionic radius similar to that of zinc. Zinc was found to stimulate bone formation *in vitro* as well as *in vivo* when its concentration was within noncytotoxic levels. Porter *et al.* [41], incorporated silicate ions into hydroxyapatite; this lead to an increase of the number of defects related to the specific sites within the ceramic (Si-HA) that were most likely to dissolve. Thus, an increase in the number of defects led to an increased HA solubility and, consequently, to an increased rate of osseointegration. At the surface of the Si-HA, larger needle-like crystallites in the deeper regions of the implant were observed, whereas smaller plate-like apatite crystallites were observed at the bone–HA interface. This suggests that two different biological processes took place. The needle-like crystallites were generated by a loss of material from the grains of Si-HA and were not due to the heterogeneous nucleation of the biological apatite. More recent developments deal with the incorporation of strontium into bioactive fillers. Specifically, Renaudin *et al.* [42] prepared Sr-doped calcium phosphates by a sol–gel process. The samples were porous and had a multiphase morphology due to the addition of strontium. Upon interaction with the physiological fluids, a partial release of Sr^{2+} in solution showed promising results combining good bioactivity, antiosteoporotic, and antiinflammatory properties. In another study, Wu *et al.* [43] incorporated strontium ions into $CaSiO_3$ ceramics to improve their physical and biological properties. A phase transition from β to α-$CaSiO_3$, an enhanced densification of the $CaSiO_3$, a decrease in its ionic dissolution rate and also a decrease in the pH value of the SBF were demonstrated. In addition, the bioactivity mechanism of $CaSiO_3$ and its ability to form apatite in the SBF did not change. Proliferation of human bone derived cells was also stimulated by the Sr incorporation.

22.6
Fillers Formed *In Situ*

Surface area, pore volume, and pore size distribution are very significant factors controlling the surface reactivity of bioceramic materials. *In situ* filler formation by sol–gel methods (see also Chapter 24) has been used to prepare bioactive glasses with high surface area and porosity; as a result, for a given glass composition, an increase in growth rate of the interfacial apatite layer can be obtained.

Pérez-Pariente *et al.* [44] investigated the effect of composition and textural properties on the bioactivity of glasses prepared by the sol–gel method by hydrolysis and polycondensation of proper amounts of tetraethyl orthosilicate (TEOS), triethyl phosphate (TEP), calcium nitrate ($Ca(NO_3)_2{\cdot}4H_2O$), and magnesium nitrate ($Mg(NO_3)_2{\cdot}4H_2O$) with 1 M HNO_3 as a catalyst. After immersion of the formed gel in the SBF solution, glasses with higher CaO content developed higher porosity, which led to apatite nucleation on the surface from the very first stages; in contrast, glasses with higher SiO_2 content had increased surface area and as a result showed an increased growth rate of the Ca–P layer on their surface. Similar results were obtained by Peltola *et al.* [45], who prepared sol–gel derived SiO_2 and $CaO–P_2O_5–SiO_2$ compositions and examined their bioactivity in SBF. In contrast to a higher phosphorous concentration, which was ineffective, a higher calcium concentration appeared to favor apatite nucleation.

In another study by Rhee [46], silanol groups appeared to provide nucleation sites to favor the formation of apatite crystals in organic polymer/silica hybrids of low and high molecular weight polycaprolactone (PCL) prepared through the sol–gel method. When immersed in SBF, fast and uniform nucleation and growth occurred in the case of the low molecular weight hybrid due to an increase in the number of interaction points with the silica and the decreased size of the silica domain. Additionally, the lower molecular weight of PCL means faster degradation and faster exposure of the silica phase to the SBF.

22.7 Recent Developments – Nanostructured Bioactive Fillers

In order to improve the properties of the tissue engineering biomaterials nanostructured bioactive fillers have been investigated. Nanoparticles can disperse more uniformly in the polymer matrix, thus enhancing the coating characteristics of the apatite layer formed and also result in a better cell attachment and proliferation.

Some examples include the work of Deng *et al.* [47] who incorporated 20 wt% nano-HA into PDLLA using solvent coblending and hot pressing techniques. The filler particles had a width of about 7–50 nm and a length of 70–350 nm. SEM, EDX, and XRD analysis showed the formation of a layer of nonstoichiometric apatite after 7 days immersion in SBF, demonstrating moderate *in vitro* bioactivity. In another study, Xianmiao *et al.* [48] developed nano-HA/chitosan composite membranes by solvent casting and evaporation methods. Their results indicated that the membranes were rough and microporous and could enable adhesion and growth of cells. It was also shown that the composite membranes had no negative effects on the morphology, viability and proliferation of cells making them suitable for tissue engineering applications. Sundaram *et al.* [49] used gelatin–starch and nano-HA to produce a composite scaffold via a novel microwave vacuum drying and cross-linking process. The scaffold appeared to become stronger and more rigid upon the nano-HA addition and could be a promising material in tissue engineering research. Lee *et al.* [50] added nanofibrous bioactive glass, with a mean diameter of 240 nm, to PCL in order to produce a composite in a thin membrane form. The nanocomposite enhanced the rapid formation of an apatite layer when immersed in SBF. In addition, murine-derived osteoblastic cells adhered and grew over the nanocomposite with improved cell viability compared to the pure PCL membrane. Fujihara *et al.* [51] designed a new type of guided bone regeneration membrane (GBR) using PCL/$CaCO_3$ composite nanofibers produced by electrospinning. The GBR membranes showed good cell attachment and proliferation when observed under SEM.

22.8 Concluding Remarks

Considering the wide range of fillers and matrices, and the great variety in the available methods for biocomposite preparation, it is clear why research in the field of tissue engineering is showing significant growth. Although there are many methods for producing biocomposites, each one is unique with a very specific application in its field. The major challenge in R&D efforts is the ability to "tailor" the properties of the resulting bioactive and bioresorbable composites in order to enhance the process known as osseointegration. Particularly important is the need to balance mechanical properties (modulus, strength, fracture toughness) and, in the case of erodable matrices, degradation rate with biological response. Times for bone growth, the possibility of tissue inflammation, and compensation for pH decrease due to formation of acidic degradation components are important parameters that need to be taken into account. Continuing research aimed at identifying novel, bioactive functional fillers (including nanomaterials) will undoubtedly address these issues. It is recognized, however, that regulatory, long-term performance and liability concerns may extend the time interval between successful clinical trials and market introduction of new biocomposites.

References

1 Hench, L.L. (1996) Classes of materials used in medicine, Chapter 2, in *Biomaterials Science. An Introduction to Materials in Medicine* (eds B.D. Ratner, A.S. Hoffman, F.J. Schoen, and J.E. Lemons), Academic Press, San Diego, pp. 73–84.

2 Wang, M. (2003) Developing bioactive composite materials for tissue replacement. *Biomaterials*, **24** (13), 2133–2151.

3 Hench, L.L.in http://www.in-cites.com/papers/ProfLarryHench.html and http://www.bg.ic.ac.uk/Lectures/Hench/BioComp/Chap3.shtml.

4 Lutton, C., Read, J., and Trau, M. (2001) Current chemistry: nanostructured biomaterials: a novel approach to artificial bone implants. *Aust. J. Chem.*, **54** (10), 621–623.

5 Callister, W.D., Jr (2003) *Materials Science and Engineering, An Introduction*, 6th edn, John Wiley & Sons, Inc., Hoboken, NJ, p. 598.

6 Krajewski, A. and Ravaglioli, A. (2002) Bioceramics and biological glasses,

Chapter 5, in *Integrated Biomaterials Science* (eds R. Barbucci *et al.*), Kluwer Academic/Plenum Publishers, New York, pp. 208–254.
7 Chouzouri, G. (2007) Ph.D. Dissertation, New Jersey Institute of Technology, Newark, NJ.
8 Mano, J.F., Sousa, R.A., Boesel, L.F., Neves, N.M., and Reis, R.L. (2004) Bioinert, biodegradable and injectable matrix composites for hard tissue replacement: state of the art and recent developments. *Compos. Sci. Technol.*, **64** (6), 789–817.
9 Kokubo, T., Kushitani, H., Sakka, S., Kitsugi, T., and Yamamuro, T. (1990) Solutions able to reproduce *in vivo* surface structure changes in bioactive glass-ceramic A–W. *J. Biomed. Mater. Res.*, **24** (6), 721–734.
10 Ashman, A. and Gross, J.S. (2000) Synthetic osseous grafting, Chapter 8, in *Biomaterials Engineering and Devices: Human Applications* (eds D.L. Wise *et al.*), Humana Press, New Jersey, pp. 140–154.
11 Sousa, R.A., Mano, J.F., Reis, R.L., and Cunha, A.M. (2001) Interfacial interactions and structure development in injection molded HDPE/hydroxyapatite composites. Proceedings of the 59th SPE ANTEC, 2001, vol. 47.
12 Sousa, R.A., Mano, J.F., Reis, R.L., Cunha, A.M., and Bevis, M.J. (2002) Mechanical performance of starch based bioactive composite biomaterials molded with preferred orientation. *Polym. Eng. Sci.*, **42** (5), 1032–1045.
13 Leonor, I.B., Ito, A., Onuma, K., Kanzaki, N., and Reis, R.L. (2003) *In vitro* bioactivity of starch thermoplastic/hydroxyapatite composite materials: an *in situ* study using atomic force microscopy. *Biomaterials*, **24** (4), 579–585.
14 Yu, S., Hariram, K.P., Kumar, R., Cheang, P., and Ajk, K.K. (2005) *In vitro* apatite formation and its growth linetics on hydroxyapatite/polyetheretherketone Biocomposites. *Biomaterials*, **26** (15), 2343–2352.
15 Ni, J. and Wang, M. (2002) *In vitro* evaluation of hydroxyapatite reinforced polyhydroxybutyrate composite. *Mater. Sci. Eng. C*, **20** (1), 101–109.
16 Hasegawa, S., Ishii, S., Tamura, J., Furukawa, T., Neo, M., Matsusue, Y., Shikinami, Y., Okuno, M., and Namakura, T. (2006) A 5–7 years *in vivo* study of high-strength hydroxyapatite/poly(L-lactide) composite rods for the internal fixation of bone fractures. *Biomaterials*, **27** (8), 1327–1332.
17 Wang, M., Ni, J., Goh, C.H., and Wang, C.X. (2000) Developing tricalcium phosphate/polyhydroxybutyrate composite as a new biodegradable material for clinical applications *Key Eng. Mater. Bioceramics*, **13**, 741–744.
18 Wang, M., Ni, J. and Wang, J. (2001) In vitro bioactivity and mechanical performance of tricalcium phosphate/ polyhydroxybutyrate composites *Key Eng. Mater. Bioceramics*, **14**, 429–432.
19 Kasuga, T., Maeda, H., Kato, K., Nogami, M., Hata, K.I., and Ueda, M. (2003) Preparation of poly(lactic acid) composites containing calcium carbonate (vaterite). *Biomaterials*, **24** (19), 3247–3253.
20 Hench, L.L. (1988) Bioactive ceramics, in Part II of *Bioceramics: Material Characteristics Versus In Vivo Behavior* (eds P. Ducheyene and J.E. Lemons), The New York Academy of Sciences, New York, pp. 54–71.
21 Chouzouri, G. and Xanthos, M. (2007) *In vitro* bioactivity and degradation of polycaprolactone composites containing silicate fillers. *Acta Biomater.*, **3** (5), 745–756.
22 Fujibayashi, S., Neo, M., Kim, H.M., Kokubo, T., and Nakamura, T. (2003) A comparative study between *in vivo* bone ingrowth and *in vitro* apatite formation on Na_2O–CaO–SiO_2 glasses. *Biomaterials*, **24** (8), 1349–1356.
23 Brink, M., Turunen, T., Happonen, R.P., and Yli-Urpo, A. (1997) Compositional dependence of bioactivity of glasses in the system Na_2O–K_2O–MgO–CaO–B_2O_3–P_2O_5–SiO_2. *J. Biomed. Mater. Res.*, **37** (1), 114–121.
24 Rich, J., Jaakkola, T., Tirri, T., Närhi, T., Yli-Urpo, A., and Seppälä, J. (2002) *In vitro* evaluation of poly(ε-caprolactone-co-DL-lactide)/bioactive glass composites. *Biomaterials*, **23** (10), 2143–2150.
25 Jaakkola, T., Rich, J., Tirri, T., Närhi, T., Jokinen, M., Seppälä, J., and Yli-Urpo, A.

(2004) *In vitro* Ca–P precipitation on biodegradable thermoplastic composite of poly(ε-caprolactone-co-DL-lactide) and bioactive glass (S53P4). *Biomaterials*, **25** (4), 575–581.

26 Närhi, T.O., Jansen, J.A., Jaakkola, T., Ruijter, A., Rich, J., Seppala, J., and Urpo, A.Y. (2003) Bone response to degradable thermoplastic composite in rabbits. *Biomaterials*, **24** (10), 1697–1704.

27 Huang, J., Di Silvio, L., Wang, M., Tanner, K.E., and Bonfield, W. (1997) In vitro assessment of hydroxyapatite and Bioglass®–reinforced polyethylene composites. *Key Eng. Mater. Bioceramics*, **10**, 519–522.

28 Huang, J., Di Silvio, L., Wang, M., Rehman, I., Ohtsuki, C., and Bonfield, W. (1997) Evaluation of *in vitro* bioactivity and biocompatibility of Bioglass®-reinforced polyethylene composite. *J. Mater. Sci.: Mater. Med.*, **8** (12), 809–813.

29 Huang, J., Di Silvio, L., Kayser, M., and Bonfield, W. (2000) TEM examination of the interface between Bioglass®/ polyethylene composites and human osteoblasts cells in vitro. *Key Eng. Mater. Bioceramics*, **13**, 649–652.

30 Yao, J., Radin, S., Phoebe, S., Leboy, P.S., and Dusheyne, P. (2005) The effect of bioactive glass content on synthesis and bioactivity of composite poly (lactic-co-glycolic acid)/bioactive glass substrate for tissue engineering. *Biomaterials*, **26** (14), 1935–1943.

31 Shinzato, S., Kobayashi, M., Mousa, W.F., Kamimura, M., Neo, M., Choju, K., Kokubo, T., and Nakamura, T. (2000) Bioactive bone cement: effect of surface curing properties on bone-bonding strength. *J. Biomed. Mater. Res.*, **53** (1), 51–61.

32 Yamamuro, T. *et al.* (1988) Novel methods for clinical applications of bioactive ceramics, in Part II of *Bioceramics: Material Characteristics Versus In Vivo Behavior* (eds P. Ducheyene and J.E. Lemons), The New York Academy of Sciences, New York, pp. 107–114.

33 Juhasz, J.A., Best, S.M., Kawashita, M., Miyata, N., Kokubo, T., Nakamura, T., and Bonfield, W. (2003) Bonding strength of the apatite layer formed on glass-ceramic apatite-wollastonite-polyethylene composites. *J. Biomed. Mater. Res.*, **67A** (3), 952–959.

34 Juhasz, J.A., Best, S.M., Brooks, R., Kawashita, M., Miyata, N., Kokubo, T., Nakamura, T., and Bonfield, W. (2004) Mechanical properties of glass-ceramic A–W-polyethylene composites: effect of filler content and particle size. *Biomaterials*, **25** (6), 949–955.

35 Liu, X., Ding, C., and Chu, P.K. (2004) Mechanism of apatite formation on wollastonite coatings in simulated body fluids. *Biomaterials*, **25** (10), 1755–1761.

36 Reck, R. *et al.* (1988) Bioactive glass-ceramics in middle ear surgery, in Part II of *Bioceramics: Material Characteristics Versus In Vivo Behavior* (eds P. Ducheyene and J.E. Lemons), The New York Academy of Sciences, New York, pp. 100–106.

37 Schiller, C., Rasche, C., Wehmöller, M., Beckmannc, F., Eufinger, H., Epple, M., and Weihe, S. (2004) Geometrically structured implants for cranial reconstruction made of biodegradable polyesters and calcium phosphate/ calcium carbonate. *Biomaterials*, **25** (7), 1239–1247.

38 Yamasaki, Y., Yoshida, Y., Okazaki, M., Shimazu, A., Kubo, T., Akagawa, Y., and Uchida, T. (2003) Action of FGMgCO_3Ap-collagen composite in promoting bone formation. *Biomaterials*, **24** (27), 4913–4920.

39 Blaker, J.J., Nazhat, S.N., and Boccaccini, A.R. (2004) Development and characterisation of silver-doped bioactive glass-coated sutures for tissue engineering and wound healing applications. *Biomaterials*, **25** (7), 1319–1329.

40 Ito, A., Kawamura, H., Otsuka, M., Ikeutsi, M., Ohgushi, H., Ishikawa, K., Onuma, K., Kanzaki, N., Sogo, Y., and Ichinose, N. (2002) Zinc-releasing calcium phosphate for stimulating bone formation. *Mater. Sci. Eng.*, **22** (1), 21–25.

41 Porter, A.E., Patel, N., Skepper, J.N., Best, S.M., and Bonfield, W. (2003) Effect of glass composition on the degradation properties and ion release characteristics of phosphate glass-polycaprolactone composites. *Biomaterials*, **24** (25), 4609–4620.

42 Renaudin, G., Laquerrière, P., Filinchuk, Y., Jallot, E., and Nedelec, J.M. (2008) Structural characterization of sol–gel derived Sr-substituted calcium phosphates with anti-osteoporotic and anti-inflammatory properties. *J. Mater. Chem.*, **18** (30), 3593–3600.

43 Wu, C., Ramaswamy, Y., Kwik, D., and Zreiqat, H. (2007) The effect of strontium incorporation into $CaSiO_3$ ceramics on their physical and biological properties. *Biomaterials*, **28** (21), 3171–3181.

44 Pérez-Pariente, J., Balas, F., Román, J., Salinas, A.J., and Vallet-Regí, M. (1999) Influence of composition and surface characteristics on the *in vitro* bioactivity of SiO_2–CaO–P_2O_5–MgO sol–gel glasses. *J. Biomed. Mater. Res. A*, **47** (2), 170–175.

45 Peltola, T., Jokinen, M., Rahiala, H., Levänen, E., Rosenholm, J.B., Kangasniemi, I., and Yli-Urpo, A. (1999) Calcium phosphate formation on porous sol–gel-derived SiO_2 and CaO–P_2O_5–SiO_2 substrates *in vitro*. *J. Biomed. Mater. Res. A*, **44** (1), 12–21.

46 Rhee, S. (2003) Effect of the molecular weight of poly(ε-caprolactone) on interpenetrating network structure, apatite-forming ability, and degradability of poly(ε-caprolactone)/silica nano-hybrid materials. *Biomaterials*, **24** (10), 1721–1727.

47 Deng, C., Weng, J., Lu, X., Zhou, S.B., Wan, J.X., Qu, S.X., Feng, B., and Li, X.H. (2008) Preparation and *in vitro* bioactivity of poly(D,L-lactide) composite containing hydroxyapatite nanocrystals. *Mater. Sci. Eng. C*, **28** (8), 1304–1310.

48 Xianmiao, C., Yubao, L., Yi, Z., Li, Z., Jidong, L., and Huanan, W. (2008) Properties and *in vitro* biological evaluation of nano-hydroxyapatite/chitosan membranes for bone guided regeneration. *Mater. Sci. Eng. C*, **29** (1), 29–35.

49 Sundaram, J., Durance, T.D., and Wang, R. (2008) Porous scaffold of gelatin-starch with nanohydroxyapatite composite via novel microwave vacuum drying. *Acta Biomater.*, **4** (4), 932–942.

50 Lee, H.H., Yu, H.S., Jang, J.H., and Kim, H.W. (2008) Bioactivity improvement of poly(ε-caprolactone) membrane with the addition of nanofibrous bioactive glass. *Acta Biomater.*, **4** (3), 622–629.

51 Fujihara, K., Kotaki, M., and Ramakrishna, S. (2005) Guided bone regeneration membrane made of polycaprolactone/calcium carbonate composite nano-fibers. *Biomaterials*, **26** (19), 4139–4147.

23
Polyhedral Oligomeric Silsesquioxanes

Chris DeArmitt

23.1
Introduction

Polyhedral oligomeric silsesquioxanes, known as POSS®, are a unique family of molecular fillers comprised of a central, silica-like core surrounded by covalently attached organic groups [1–5] (Figure 23.1). These molecules can be viewed as inorganic–organic hybrids where the silica core imparts rigidity and high temperature resistance and the organic moieties provide compatibility and functionality. As hybrid materials, POSS provide a spectrum of new properties. As an indicator of the global interest in POSS, there are now over 2500 articles and 900 patents dealing with this topic. POSS research continues to flourish and commercialization has taken off in recent years.

23.2
Production

POSS is produced by the condensation reaction of organosilanes [1, 3]. Interestingly, these are the same organosilanes [6] that are commonly used for the surface treatment of mineral fillers to improve dispersion in, and/or adhesion to, the polymer matrix (see Chapter 4). Either trichlorosilanes or trialkoxysilanes may be used to make POSS [1, 3]. Normally, such organosilanes self-condense into an amorphous network and POSS cage formation does not occur to any significant extent. Initial POSS syntheses suffered from extremely long reaction times (several months), coupled with very low yields of imperfect cage structures and resin formation necessitating arduous separation methods to obtain a pure product [7]. These problems were overcome and now short reaction times and good yields have enabled rapid commercialization of POSS in thermoplastics, thermosets, elastomers, and coatings [8].

Functional Fillers for Plastics: Second, updated and enlarged edition. Edited by Marino Xanthos

ISBN: 978-3-527-32361-6

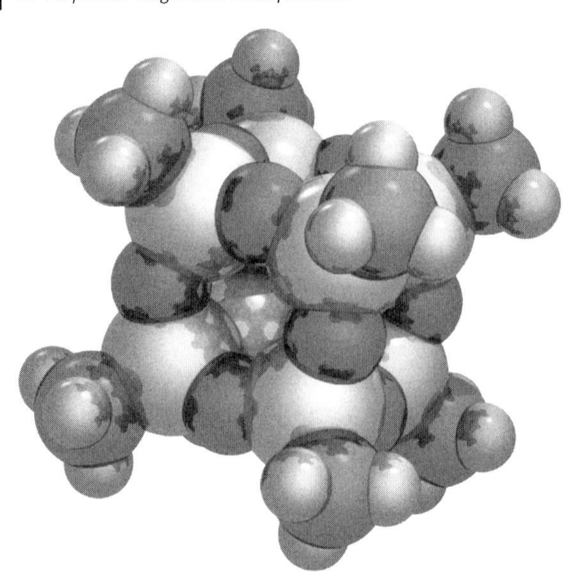

Figure 23.1 POSS molecule showing the central rigid cage and surrounding methyl groups.

23.3 Structure and Properties

For conventional mineral fillers such as calcium carbonate, dolomite, mica, and wollastonite, the chemistry of each is constant across all grades of a given material and the grades are differentiated by their particle size, size distribution, particle shape, and presence (or otherwise) of a surface treatment. In the case of POSS, every grade is chemically distinct and so properties such as density, polarity, refractive index, and so on, are different for each POSS [8] (Table 23.1).

Table 23.1 Overview of typical POSS properties.

Density range	0.9–1.3 g/cm^3 typical (up to 1.82 g/cm^3)
Refractive index range	1.40–1.65
Molecular size	1–5 nm
Form	Colorless, odorless crystalline solids, some waxes, and liquids
Polarity	Very low (fluoroalkyl), low (alkyl), phenyl (medium) to polyionic (high)
Chemical and pH stability	Molecular silicas (closed cage) very stable, trisilanols good stability
Thermal stability	250–350 °C typical (>400 °C for some types)
Safety	All testing performed has shown POSS to be safe
Purity	Standard purity >97% (higher purity and electronics grades are available)

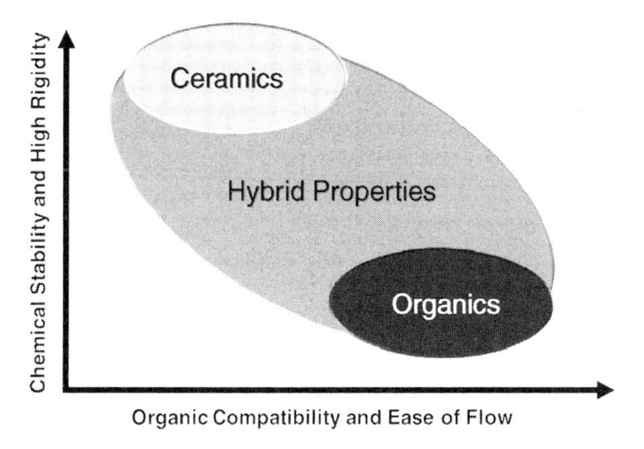

Figure 23.2 Organic–inorganic hybrid properties of POSS.

As the size of the organic groups on the POSS is increased, so the properties shift progressively from those resembling silica toward those of organic compounds. Thus, it is possible to access properties in between those of organic and inorganic materials (Figure 23.2).

As an example, the density of a POSS cage with hydrogen at the corners of the cage is 1.82 g/cm^3, which is much higher than the density of common organic compounds. For POSS with progressively longer alkyl substituents, the silica-like core dominates less and so the density decreases to lower values, ~1.0 g/cm^3, more common for organic molecules (Table 23.2). This change occurs for other properties such as thermal conductivity, modulus, dielectric constant, and so on. For instance, the Young's modulus of common organic polymers such as PE and PP is in the range 1–1.5 GPa, whereas the value for silica is ~70 GPa. The Young's modulus of octacyclopentyl POSS has been calculated to be ~12 GPa [9]. POSS with smaller organic groups are expected to have commensurately higher moduli, whereas larger organic groups such as *iso*-octyl chains lead to low modulus, liquid POSS types [8].

Table 23.2 Comparison of POSS densities with those of inorganic silica and organic molecules.

Material Type	Density (g/cm^3)
Quartz	2.60
Amorphous silica	2.18
Octa hydrido POSS	1.82
Octamethyl POSS	1.50
Octaethyl POSS	1.33
Octaiso-butyl POSS	1.13
Octaiso-octyl POSS	1.01
iso-Octane	0.69

The most prevalent POSS types may be divided into four categories:

1) **Molecular silicas**: a closed cage with an inert organic group at each vertex (e.g., alkyl, fluoroalkyl, phenyl, PEG). Usually all the eight groups are the same but alternatives exist with seven identical groups plus one different one.
2) **Functionalized POSS**: a closed cage with a reactive organic group at each corner (e.g., epoxy, amine, acrylate, methacrylate, alcohol, isocyanate, sulfonate, carboxylic acid, thiol, imides, silane, or nitrile.) It is also possible to have seven reactive groups and one unreactive or vice versa.
3) **POSS trisilanols**: where there are seven identical groups but one silicon atom is missing to leave an open corner with three reactive silanols.
4) **POMS**: where a metal atom has been reacted into the corner of a POSS trisilanol.

Conceptually, the inert POSS (type 1) can be viewed as molecular fillers. It can be shown that POSS are clearly molecules since

1) POSS are spontaneously soluble in solvents;
2) POSS can be analyzed by methods such as HPLC, GPC, and solution NMR;
3) POSS exhibit distinct molecular weights;
4) POSS can be crystallized;
5) pure POSS can be a solid, liquid, or gas.

Although POSS are molecules, they do look reminiscent of particles. This molecule–particle duality or "molicle" appearance leads to some interesting observations. For example, all particulate fillers tend to agglomerate [10] and, as the particle size decreases, the problem is greatly exacerbated such that it becomes very difficult to make nanocomposites with a proper level of particle dispersion [11]. That is one of the great challenges for nanocomposites. In stark contrast, POSS molecular fillers are able to dissolve into a solvent or polymer with perfect dispersion and no tendency to agglomerate (Figure 23.3). If the POSS polarity is chosen to match that of the solvent then the Gibbs free energy of mixing is negative and dissolution is spontaneous with no need for stirring.

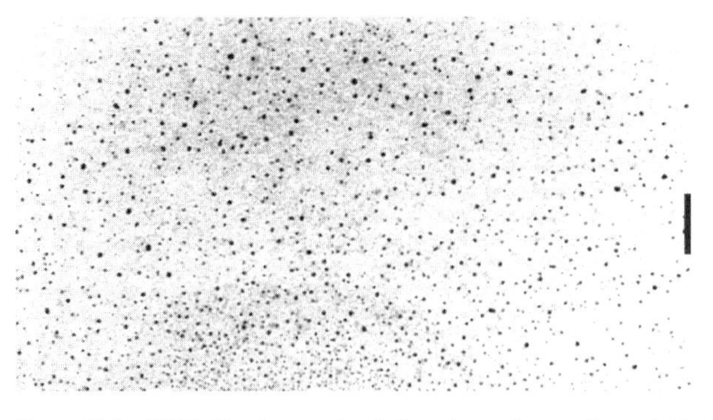

Figure 23.3 POSS dissolves molecularly and spontaneously; each black dot is a 1.5 nm POSS molecule (scale bar is 50 nm).

Traditional particulate fillers tend to agglomerate already during manufacture, and once agglomerated, it is difficult to subsequently disperse them. Dispersion requires addition of a dispersant additive (e.g., stearic acid or alkysilanes) and the input of energy to overcome the attractive forces between particles [6]. The correct selection of dispersant type and application method is crucial. In contrast, POSS as synthesized, already possesses its own intrinsic dispersant in the form of covalently bonded organic groups and dispersion is easy.

In order to ensure good compatibility between POSS and the matrix, one must match the polarity or solubility parameter of the POSS and the solvent, polymer or coating in which it is to be dissolved. POSS polarities span the whole gamut from extremely hydrophobic fluoroalkyl POSS used to make ultrahydrophobic surfaces [12], through progressively more polar variants such as alkyl POSS (low polarity), phenyl POSS (medium polarity), and water soluble types including PEG POSS, POSS trisulfonic acids, and octaammonium POSS.

POSS is usually found to be soluble up to around 5 vol% in thermoplastics such that up to that concentration, no light scattering occurs and the material remains optically clear. In thermoset resins and solvents, POSS can be mixed at all ratios if the polarities of POSS and medium are properly matched.

Instead of matching the polarity of the POSS to the matrix, one can also imbue compatibility through reaction of POSS with the matrix by choosing appropriate chemistry. Virtually all chemistries are available to facilitate reactions such as alcohol, carboxylic acid, sulfonic acid, epoxy, chloroalkyl, acrylate, methacrylate, isocyanate, amine, thiol, silanol, and several others.

23.4 Suppliers/Cost

Initially, extremely long reaction times (months) and very low yields meant that cost of the POSS was rather high, approximately \$3000/lb (€4000/kg) and, therefore, the commercial appeal was limited at that time. However, the potential of POSS was such that effort was made to lower reaction times and to improve yields. Hybrid Plastics has worked to scale up POSS production and now has a capacity of several hundred tons per year (as of 2009) that is increasing steadily in line with demand. Correspondingly, the improved synthesis methods and scale-up have resulted in a 1000-fold reduction in POSS selling price to just \$30–40/lb (€50/kg) for larger amounts of some of the most popular POSS variants [8]. Over 250 types of POSS have been synthesized, 80 types are presently available and of those, several have large-scale commercial availability.

Major commercial suppliers of POSS are Hybrid Plastics Inc., Hattiesburg, MS, USA. Distributors of POSS are Sigma-Aldrich Inc., Gelest and Toyotsu Chemiplas Corporation. Some POSS types are available in R&D amounts from Mayaterials Inc.

Suppliers of POMS are Hybrid Plastics Inc., Hattiesburg, MS, USA, Hybrid Catalysis BV.

23.5 Environmental/Safety Considerations

POSS are chemical compounds and should be handled with the normal precautions used for all other chemicals. Most POSS types are in the form of a crystalline white powder (typical size range 5–100 μm). Some types are colorless, viscous liquids.

Oral toxicity testing both in the United States and in Europe has shown that all three POSS tested are in the safest possible category [8] and do not require the risk phrase R22 "Harmful if Swallowed."

Octaisobutyl POSS	US	Category IV Oral $LD_{50} > 5000$ mg/kg (highest US method dose)
Octamethyl POSS	EU	Oral $LD_{50} > 2000$ mg/kg (highest EU method dose)
Dodecaphenyl POSS	EU	Oral $LD_{50} > 2000$ mg/kg (highest EU method dose)

Several of the larger production volume POSS types are TSCA listed. These include octaiso-octyl POSS, dodecaphenyl POSS, octamethyl POSS, octaiso-butyl POSS, *iso*-butyl trisilanol POSS, octaglycidyl POSS, aminopropyl *iso*-butyl POSS, and aminoethylaminopropyl *iso*-butyl POSS [8]. TSCA listing and FDA approval of other POSS types is ongoing.

23.6 Functions

23.6.1 Primary Function

The primary function of POSS depends upon the type of POSS chosen and the matrix. In thermoplastics, the most significant benefit of POSS is improved melt flow. In particular, improved flow for high temperature polymers, such as PEEK, PEI, COC, PA6, PPS, and PPO enables one to fill complicated, thin-walled parts. Flow aids for polymers are well established, however, conventional flow aids are not designed to withstand the high processing temperatures needed for these polymers. Also noteworthy is the finding that POSS does not degrade the mechanical properties, so for example, PEEK, PEI, PPS, PA6, and COC all retain full modulus and yield strength while melt flow is dramatically enhanced [13, 14]. The retention of yield strength indicates that the flow enhancement is achieved without molecular weight degradation, which has been further proven by gel permeation chromatography. Addition of chemically analogous hydrolyzed organosilane resin worsened melt flow so there is some particular attribute of the POSS structure responsible for its utility in melt flow enhancement (Figure 23.4).

POSS trisilanols act as highly effective dispersants to lower viscosity and improve mechanical properties. The POSS trisilanols have proven effective for many fillers

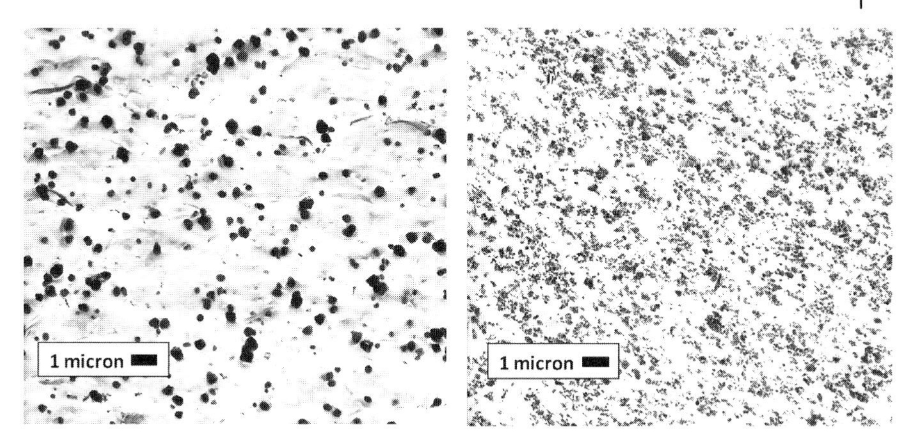

Figure 23.4 TiO_2 in PP with no dispersant (left) and the same TiO_2 in PP with POSS trisilanol dispersant added in the extruder (right) [15].

and pigments [14] including silica, titania [15], mica, alumina, and boron nitride. It comes as no surprise that the POSS trisilanols are effective on those materials because it is well known that organosilanes work well on the same materials [6, 10]. More surprising is that the POSS trisilanols are able to disperse fillers such as calcium carbonate, other carbonates, and barium sulfate, which are generally recognized as unsusceptible to organosilanes. POSS dispersants offer some advantages over traditional dispersants such as stearic acid and organosilanes. In particular, the reaction of POSS trisilanols with fillers leads only to the evolution of innocuous water. In contrast, organosilanes produce VOCs (usually methanol or ethanol) when reacted with fillers and pigments. Furthermore, the POSS has high thermal stability and binds strongly making POSS the obvious choice for pigments and fillers that need to be dispersed in polymers that have high processing temperatures [14]. For highly filled systems, the high flow effect in the matrix can be combined with the dispersant effect to provide a dual action flow improvement.

Addition of inert POSS molecular silica (MS0825, octaiso-butyl POSS) results in no significant change in viscosity because that POSS type cannot adsorb and cannot, therefore, act as a dispersant. The POSS trisilanols (SO1458, a phenyl trisilanol and SO1455, an isooctyl trisilanol) both adsorb strongly and act as efficient dispersants [14]. The POSS trisilanols coat the filler and reduce particle–particle interactions, and this accounts for the large drop in viscosity.

In thermosets, where POSS is reacted into the material, the primary advantage is excellent high temperature modulus retention beyond 300 °C [8]. For example, octa functional POSS epoxies or POSS amines can be added to conventional epoxy or BMI resins to provide very high temperature resins. While traditional resins lose rigidity at and above the glass transition temperature, the POSS modified materials retain modulus even under extreme temperature conditions. Such materials are in demand for oil well applications where the drive to deeper, hotter wells pushes existing

Table 23.3 Effect of POSS on swelling of silicone rubber.

Material type	Weight increase (%)
Silicone + 0% POSS	25
Silicone + 20% POSS	19
Silicone + 40% POSS	14
Silicone + 60% POSS	10

materials beyond their limits. The advantage of the POSS is attributed to two factors. First, the rigidity of the POSS cage [9] and second, the high cross-link density attainable due to the extraordinarily high functionality of the POSS.

In elastomers, the primary benefit of reacting POSS into the polymer is the improvement in solvent resistance. For example, while silicone elastomers are readily swollen when exposed to acetone, adding POSS reduces the solvent swelling dramatically (see Table 23.3). Two mechanisms are believed to be at work. The POSS allows a high cross-link density, which is well known to reduce swelling (in fact, solvent swelling can be used to estimate cross-link densities). The other mechanism is due to the structure of the POSS cage. The cage itself cannot swell, it is rigid with an immutable configuration. Thus, the POSS cage provides a volume of unswellable material analogous to conventional inorganic fillers.

23.6.2 Secondary Function

POSS exhibits several secondary effects which, when combined with primary or other secondary effects, can lead to commercial adoption. Thus, one may decide to use POSS for a primary benefit such as flow enhancement in a high temperature thermoplastic and see additional benefits such as improved mold-release, lower part friction [16, 17], and better surface finish. Another known benefit is flame retardance. POSS does not provide stand-alone flame retardance but it has been reported to be an effective synergist when used in conjunction with primary flame retardants. Under combustion conditions, the POSS vitrifies [18] to form a char that provides some intumescent flame inhibition (Chapter 17).

The vitrification of POSS has proven useful in other instances besides flame retardance. POSS has been used to protect polyimide used on satellites [19]. The aggressive radiation and atomic oxygen present in Low Earth Orbit combine to destroy standard polyimide. During testing, the regular polyimide was destroyed and only small traces of it remained. When POSS was added to the polyimide, the POSS formed a glassy protective coating upon exposure in space. This latter sample retained its mechanical integrity at the end of the test. The results were so encouraging that NASA is performing further space testing of POSS modified polymers. Similarly, it has been reported that, in the presence of oxygen plasma, POSS can vitrify and protect polymers from erosion [20]. Further work on POSS in polyimides resulted in a colorless polyimide sold as film by NeXolve Corp. under the trade name Corin XLS. The product won an R&D 100 Award in 2008.

References

1 Pielichowski, K., Njuguna, J., Janowski, B., and Pielichowski, J. (2006) *Adv. Polym. Sci.*, **201**, 225–296.

2 Joshi, M. and Butola, B.S. (2004) *J. Macromol. Sci., Part C – Polym. Rev.*, **44** (4), 389–410.

3 Pan, G. (2007) *Physical Properties of Polymers Handbook, Part VI* (ed. J.E. Mark), Springer, New York, USA.

4 Phillips, S.H., Haddad, T.S., and Tomczak, S.J. (2004) *Curr. Opin. Solid State Mater. Sci.*, **8**, 21.

5 Guizhi, L., Lichang, W., Hanli, N., and Pittman, C.U., Jr (2001) *J. Inorg. Organometal. Polym.*, **11** (3), 123.

6 Rothon, R.N. (2003) *Particulate-Filled Polymer Composites*, 2nd edn (ed. R.N. Rothon), RAPRA Technology Ltd., Shawbury, Shrewsbury, Shropshire, UK, p. 152.

7 Feyer, F.J., Newman, D.A., and Walzer, J.F. (1989) *J. Am. Chem. Soc.*, **111**, 1741–1748.

8 Hybrid Plastics Inc. technical information; http://www.hybridplastics.com/pdf/user-v2.06.pdf.

9 Capaldi, F.M., Boyce, M.C., and Rutledge, G.C. (2006) *J. Chem. Phys.*, **124**, 214709.

10 DeArmitt, C. and Rothon, R. (2002) *Plast. Addit. Compound.*, **4** (5), 12–14.

11 Rothon, R.N. and DeArmitt, C. (2003) *Particulate-Filled Polymer Composites*, 2nd edn (ed. R.N. Rothon), RAPRA Technology Ltd., Shawbury, Shrewsbury, Shropshire, UK, p. 489.

12 Tuteja, A. *et al.* (2007) *Science*, **318**, 1618.

13 DeArmitt, C. and Wheeler, P. (2008) *Plast. Addit. Compound.*, **10** (4), 36–39.

14 Hybrid Plastics Inc. technical information; http://www.hybridplastics.com/pdf/flowdisperse-v1.00.pdf.

15 Wheeler, P.A., Misra, R., Cook, R.D., and Morgan, S.E. (2008) *J Appl. Polym. Sci.*, **108** (4), 2503–2508.

16 Misra, R., Morgan, S., and Fu, B. (2007) *Proc. SPE Ann. Tech. Conf.*, **1**, 62–66.

17 Lichtenhan, J.D. *et al.* (2007) US Patent 2007/0225434. POSS Nanostructured Chemicals as Dispersion Aids and Friction Reducing Agents.

18 Vannier, A., Duquesne, S., Bourbigot, S., Castrovinci, A., Camino, G., and Delobel, R. (2008) *Polym. Degrad. Stab.*, **93**, 818–826.

19 Poe, G.D. and Farmer, S.F. (2009) US Patent 2009/0069508. Polyimide polymer with oligomeric silsesquioxane.

20 Eon, D. (2006) *J. Vac. Sci. Technol.*, **B24** (6), 2678–2688.

24
In Situ-Generated Fillers: Bicontinuous Phase Nanocomposites

Leno Mascia

24.1
Introduction

Over the past 10 years or so, there has been a considerable research interest in producing new materials from combinations of supermolecular organic compounds and an inorganic component capable of producing segregated domains with dimensions in the region of 5–100 nm. These materials are generally known as *organic–inorganic hybrids, ceramers*, and *bicontinuous nanocomposites* [1–8].

The concept of *in situ*-generated fillers derives from the fact that the inorganic phase of the nanocomposite is produced during processing through sol–gel reactions, unlike the case of conventional fillers that are added as preformed particles, even though these may also have been produced by the sol–gel method. The diagram in Figure 24.1, adapted from Ref. [3], illustrates the mechanism leading to the formation of silica by the sol–gel process starting from a precursor obtained by the prehydrolysis of a tetraalkoxysilane derivative.

From an examination of the gelation mechanism depicted in Figure 24.1, the possibility of producing a nanostructured composite material by entrapping the organic matrix between the gel particles can be envisaged. Indeed, the addition of an organic component to an inorganic oxide precursor solution was originally used simply to prevent cracking during drying of the gel. In this case, the organic matter is driven off in the subsequent desiccation and sintering steps to produce a pure inorganic oxide glass or ceramic, whereas for the production of polymer nanocomposites containing an *in situ*-generated filler, the entrapped organic matrix remains in place by omitting the sintering step. This approach has now been adapted to produce nonsintered silica or other ceramic-like products known as *ormosils* and *ormocers* (terms derived from *or*ganic *mo*dified *sil*icas and *or*ganic *mo*dified *cer*amic). In these materials the organic component forms part of a heterogeneous network within the silica domains to provide the necessary macromolecular structure for the development of mechanical strength. Gross phase separation is prevented by chemically binding the two

Functional Fillers for Plastics: Second, updated and enlarged edition. Edited by Marino Xanthos

ISBN: 978-3-527-32361-6

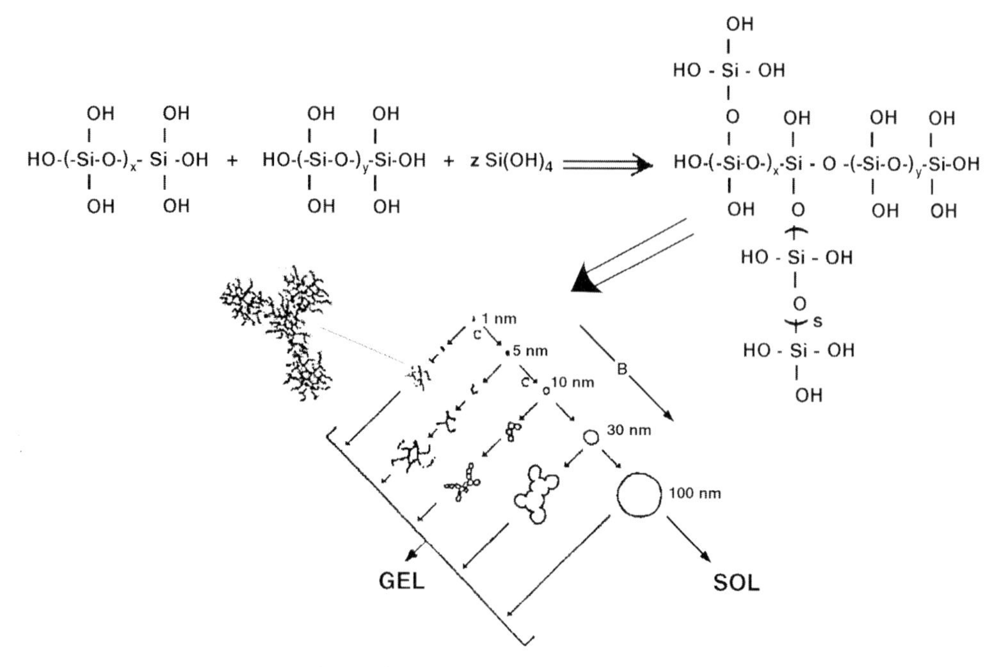

Figure 24.1 Mechanism for the formation of silica from hydrolyzed tetraethoxysilane by the sol–gel process.

phases so that the finished material contains domains of the two components as cocontinuous phases similar to block copolymers [4–6].

The materials discussed in this chapter substantially differ from the above-mentioned hybrid materials, in so far as the organic component of the systems considered here constitutes the predominant phase, which is bound to the inorganic phase through an interphase consisting of *ormosil* nanostructured domains [7–9]. The typical morphological structure of bicontinuous nanocomposites is shown in Figure 24.2, which contains a pictorial interpretation of their morphology (left) and a TEM micrograph taken on an epoxy/silica system at 15% nominal silica content (right) [9].

This type of morphology creates the conditions to achieve the most efficient mechanism for the transfer of external excitations through the two main phases, thereby maximizing the contribution of the inorganic component to the overall properties of the nanocomposite.

24.2 Methodology for the Production of Bicontinuous Nanocomposites

There are generally two approaches used for the production of bicontinuous nanocomposites.

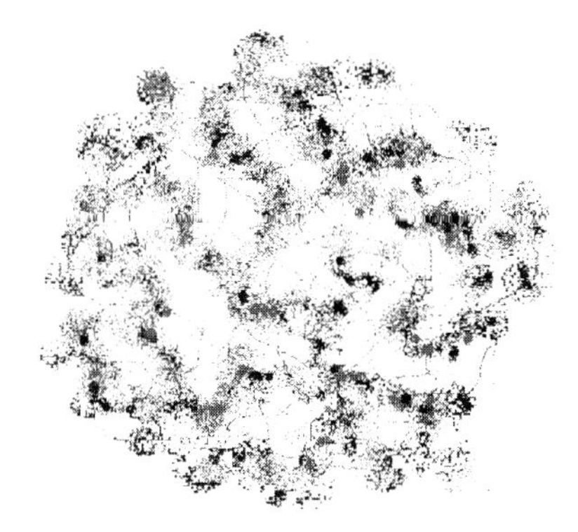

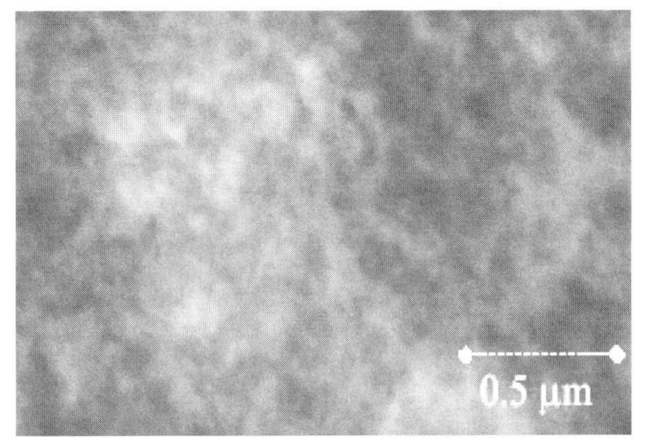

Figure 24.2 Nanostructure of bicontinuous nanocomposites. Top: Pictorial description of interconnected organic and inorganic domains with gradient density interphase. Bottom: TEM micrograph of an epoxy/silica bicontinuous nanocomposite.

In one case (Method I), the organic component forms a phase-separated macromolecular network, which is chemically bound to the surrounding inorganic domains (e.g., epoxy-based systems).

In the second case (Method II), the organic component consists of a linear polymer interdispersed within nanostructured inorganic domains.

Method I. A suitably functionalized organic oligomer, containing alkoxysilane terminal groups, is mixed with a solution of a "suitable" metal alkoxide, usually in alcohol and water, although many aprotic polar solvents are also suitable. The organic and inorganic precursors can react both intramolecularly and intermolecularly to form two chemically bonded nanostructured cocontinuous domains, one containing the metal oxide network and the other containing the organic component. This morphology is achieved by controlling the rate of the competing hydrolysis and condensation reactions. In the early days of the development of these materials the morphology was frequently controlled by allowing gelation to take place very slowly, in a closed environment, to prevent the solvent from escaping [1, 2, 10].

The very early systems were obtained from hydroxyl-terminated low molecular weight (MW) polydimethyl siloxanes, which were reacted with tetraethoxysilane (TEOS) in the presence of water and an acid catalyst to form separate macromolecular aggregates similar to *ormosils*.

Other oligomers that have also been used for this purpose are polytetramethylene oxide and polycaprolactones, both containing alkoxysilane functional groups at the chain ends [1, 10, 11]. More recently the organic oligomers that have been used for the production of bicontinuous nanocomposites are commercial thermosetting resins, particularly epoxy resins.

Method II. This method consists of mixing solutions of high molecular weight polymers and metal alkoxide, often in combination with conventional functionalized alkoxysilane-coupling agents. This is the preferred method for producing materials with a higher level of ductility than can be achieved with Method I. However, it is also the most difficult system to produce due to the low miscibility of high MW polymers and the related propensity of the inorganic component to segregate into large particulate domains [10–12].

The types of reactions involved in the production of silica-based bicontinuous nanocomposites are as follows:

Step 1. Hydrolysis

$$\mathrm{Si(OEt)_4} + n\mathrm{H_2O} \rightarrow \mathrm{(HO)}_n\mathrm{Si(OEt)}_{4-n} + n\mathrm{EtOH}$$

$$\text{Oligomer } [\mathrm{Si(OEt)_3}]_x + m\mathrm{H_2O} \rightarrow \text{Oligomer } (\mathrm{HO})_m\ [\mathrm{Si(OEt)}_{x-m}]_x + m\mathrm{EtOH}$$

Step 2. Condensation

$$\equiv\mathrm{SiOH} + \mathrm{HOSi}\equiv\ \rightarrow\ \equiv\mathrm{Si}-\mathrm{O}-\mathrm{Si}\equiv + \mathrm{H_2O}$$

$$\equiv\mathrm{SiOEt} + \mathrm{HOSiO}\equiv\ \rightarrow\ \equiv\mathrm{Si}-O\equiv\mathrm{Si}\equiv + \mathrm{EtOH}$$

For nanocomposites produced by Method II, the oligomer unit of the silane-coupling agent is functionalized to react with the components of the polymer matrix within which the siloxane domains are to be generated.

24.3
General Properties of Bicontinuous Phase Nanocomposites

The most important characteristics of nanocomposites, which arise from the cocontinuity of the two phases, are as follows:

(a) Large increase in modulus, strength and hardness, particularly around the glass transition temperature (T_g) of the organic phase.
(b) Considerable reduction in thermal expansion coefficient.
(c) Enhanced barrier properties.
(d) Notable improvement in thermal oxidative stability.

For the case of properties listed in (a) and (b) one notes that the phase cocontinuity produces conditions for which the strain in the two phases, resulting from an externally applied stress, is constant across the two phases and uniform through any cross section of the composites.

In view of the continuity of the reinforcing inorganic phase, the global stress acting on the "composite" across this section is fairly close to the sum of the stresses on the two phases, weighted by the respective volume fractions, that is,

$$\sigma_h = K[V_i\sigma_i + (1-V_i)\sigma_o], \quad (24.1)$$

where σ is the stress, V is the volume fraction, K is the stress transfer efficiency factor, which is closest to 1 than any other systems in all three directions; the subscripts i and o stand for the inorganic phase and the organic phase, respectively.

For conditions in which the strains in the two phases are equal, Eq. (24.1) (law of mixtures) can also be written in terms of Young's modulus, E (see Chapter 2). The modulus obtained from the law of mixtures corresponds to the maximum value achievable in composite materials.

Since the maximum stress that a material will withstand is synonymous with its strength, that is, the value of the stress at which fracture occurs, it can be stipulated that due to the isostrain conditions within the two phases, fracture in bicontinuous nanocomposites takes place when the most brittle phase, in this case the inorganic phase, reaches its ultimate strain value, $\varepsilon_{\text{ifracture}}$. When this condition is reached, fracture will rapidly propagate through the organic phase without any further increase in strain. Expressed in terms of fracture conditions Eq. (24.1), therefore, becomes

$$\sigma_{\text{bnc(fracture)}} = \varepsilon_{\text{i(fracture)}} K[V_i E_i + (1-V_i)E_o], \quad (24.2)$$

where $\sigma_{\text{bnc(fracture)}}$ is the strength of nanocomposite, E_i is the modulus of inorganic phase, and E_o is the modulus of organic phase. In practice, the strength of bicontinuous nanocomposites can be substantially higher than the value predicted by Eq. (24.2) if the organic matrix is very ductile. In any case, the strength of bicontinuous nanocomposites is always limited by the low strain to fracture exhibited by the inorganic component. The strength of these types of nanocomposites can be enhanced by incorporating organic segments in the network of the inorganic phase to produce structures similar to *ormosils*, which brings about an increase in its strain at break. Although this would, obviously, result in a reduction in the modulus of the inorganic component and a corresponding reduction in the modulus of the nanocomposites, the approach offers the possibility of "tailoring" the properties according to requirements.

Using similar arguments for the prediction of hardness, it can be deduced that the level of enhancement achievable is expected to be between those achievable for the Young's modulus and those for strength. The way these concepts can be exploited to enhance the properties of conventional fiber composites will be illustrated later.

The reasons for the improvements of diffusion-related properties, quoted earlier can be related to the tortuosity of the two cocontinuous domains. This is due to the fact that diffusion of gases primarily occurs through the organic phase as a result of the high density of the inorganic phase.

For permeation phenomena involving the absorption of diffusing species through swelling, the very large reduction in absorption of liquids achievable with bicontinuous nanocomposites can be related to the "quasi" isostrain conditions within the two phases. Since the inorganic component is impervious to the diffusing species, and has a much higher Young's modulus than the organic phase, it will severely restrict the swelling of the surrounding organic phase, thereby considerably suppressing the total amount of the solvent being absorbed. The improved thermal oxidative stability of the organic polymer, within the inorganic domains of a bicontinuous nanocomposite material, can also be related to the enhancement in barrier properties provided by the inorganic phase, which reduces the rate of inward diffusion of oxygen and outward diffusion of volatiles formed as a result of the degradation reactions.

The main systems that have been investigated by the author's research group over the past several years will be used to illustrate the above principles and to provide additional insight into the properties achievable with organic–inorganic hybrids and their potential applications.

24.4 Potential Applications of Bicontinuous Nanocomposites

24.4.1 Epoxy/Silica Systems for Coating Applications

Epoxy resins are widely used in coatings in applications requiring high resistance to abrasion and chemical resistance. Different types of resin and hardener combinations are used to tailor the properties to specific end uses. The increasingly more stringent requirements in many applications makes the enhancement of properties by the production of bicontinuous nanocomposites described previously a viable route for further developments in epoxy resin coatings.

In this context it must be borne in mind that, although the oligomeric products resulting from hydrolysis and condensation reactions of TEOS can react with the components of epoxy resin mixtures, particularly the hardeners, the reactions are difficult to control and can cause severe embrittlement of the cured products. For this reason some authors have used aliphatic resins and/or long chain aliphatic hardeners so that the final cured products are rubbery in nature and can tolerate more than glassy systems the embrittlement effect of the above reactions [12].

With aromatic resins, such as those based on bisphenol-A, cured with aromatic or cyclic aliphatic hardeners, it is difficult to prevent gross phase separation without introducing an adequate level of alkoxysilane functionality in the epoxy resin. Many commercially available silane-coupling agents can be used for this purpose,

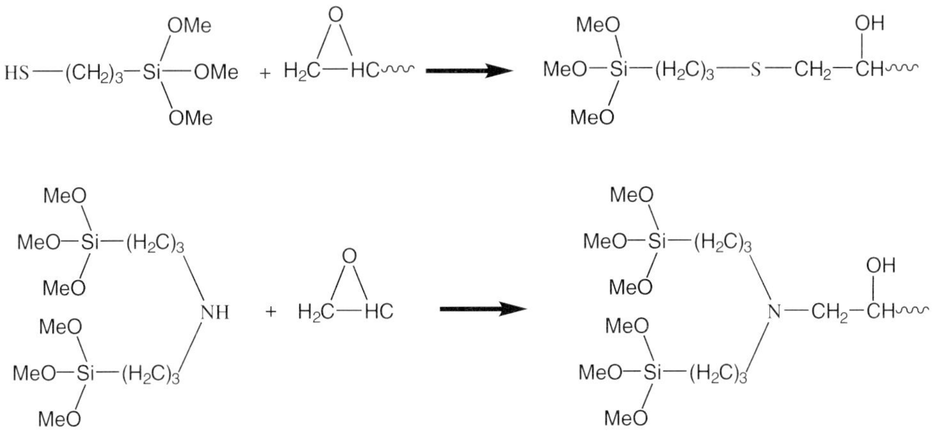

Figure 24.3 Silane functionalization of epoxy resins. Top: Functionalization with γ-mercapto-propyltrimethoxysilane. Bottom: Functionalization with amino-bis-(γ-propyltrimethoxysilane).

particularly γ-mercaptopropyltrimethoxysilane (MPTS) and amino-bis-(γ-propyl-trimethoxysilane) (APTMS, commercially known as A-1170).

The functionalization reactions with these two coupling agents are shown in Figure 24.3. An insufficient level of compatibilization to achieve the required nanostructure was found with the sole use of γ-glycidyloxytrimethoxysilane (GOTMS) added to the TEOS in the preparation of the silica precursor solution. With such systems it was expected that the reaction of the epoxy groups in GOTMS with the epoxy resin hardener would produce the required compatibilizing species for the epoxy resin and alkoxysilane component to prevent gross phase separation.

To achieve a high level of miscibility between the two precursor components a mixture of low MW (350) and high MW (5000) bisphenol-A type resins at a weight ratio of 9 : 1 was used to produce the organic network. The resulting miscibility was attributed to the presence of a large number of hydroxyl groups in the high MW epoxy component, which strongly interact with the silanol groups of the alkoxysilane solution, possibly through a series of equilibrium ester exchange reactions [13].

A comparison was made between the functionalization with MPTS and the functionalization with APTMS at 10% (molar) with respect to the epoxy groups present in the resin mixture. The functionalized epoxy resin was dissolved in a butanol/xylene mixture, subsequently mixed with a prehydrolyzed alkoxysilane solution containing 12% molar GOTMS, followed by the addition of the *p*-dicyclohexane aminomethane (PACM) hardener. The solutions were cast in PTFE molds to produce thin, about 0.75 mm thick, plaques and cured first for 24 h at room temperature and then for 1 h at 120 °C [13].

In order to demonstrate the much higher property-enhancement efficacy of cocontinuous domains over particulate domains, systems similar to the above were

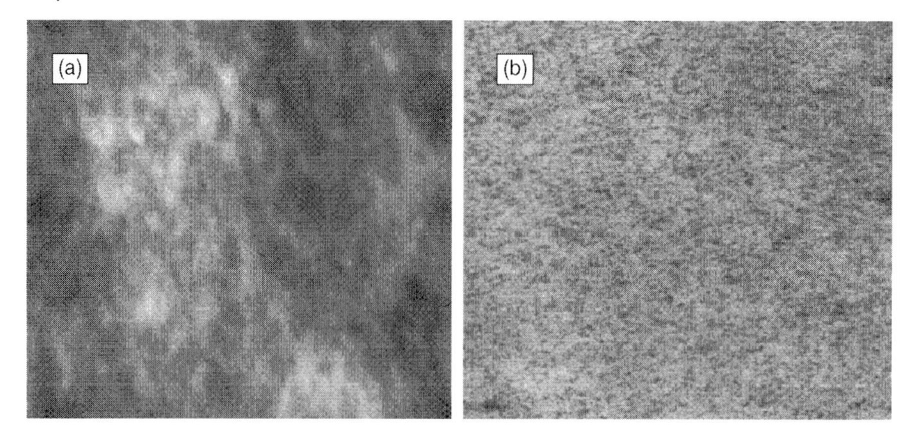

Figure 24.4 TEM micrographs of systems containing 15 wt% SiO_2, both produced from a bisphenol-A epoxy resin grafted with an amine alkoxysilane-coupling agent: (a) epoxy/silica hybrid (cocontinuous domains) and (b) epoxy/silica nanocomposite (particulate silica domains).

produced using dispersions of silica sols in isopropanol (particle diameter 7–9 nm). The difference in morphology of the two systems is illustrated by the TEM micrographs in Figure 24.4 [9], from which it is clear that the use of TEOS produces nanostructured cocontinuous domains whereas the silica sol dispersion produces the expected particulate structure.

The linear expansion resulting from an increase in temperature, the solvent absorption behavior in tetrahydrofuran (THF), and the microhardness are shown in Figures 24.5–24.7 for samples containing 15% silica prepared by different methods [9]. Not only these demonstrate the superiority of the systems containing cocontinuous domains but these also illustrate the very high efficiency of the latter type of morphology in enhancing both mechanical and barrier properties of the base macromolecular component of the hybrid, particularly useful in applications such as surface coatings.

Additional constituents have been introduced to epoxy–silica hybrids to achieve specific properties. For instance, the incorporation of small amounts of a silane-functionalized perfluoroether oligomer has been found to considerably reduce the surface energy of membranes. This has been attributed to the migration of the fluoroligomer-derived species from the bulk to the surface, which continues even after curing, through a slow-release mechanism [14].

In conventional coatings, the concept of slow release of active species, such as corrosion inhibitors, is well established and examples have also widely reported in the literature [15–17]. Owing to the toxicity and associated environmental hazards of chromates, and all hexavalent chromium compounds, there has been considerable interest in studies related to the use of molybdates as corrosion inhibitors in surface treatments and coatings [18–22]. Although some examples of organic–inorganic hybrids containing SiO_2–MoO_3 have been reported in the literature [23], the possibility of introducing molybdate species in polymer–silica hybrids for coatings

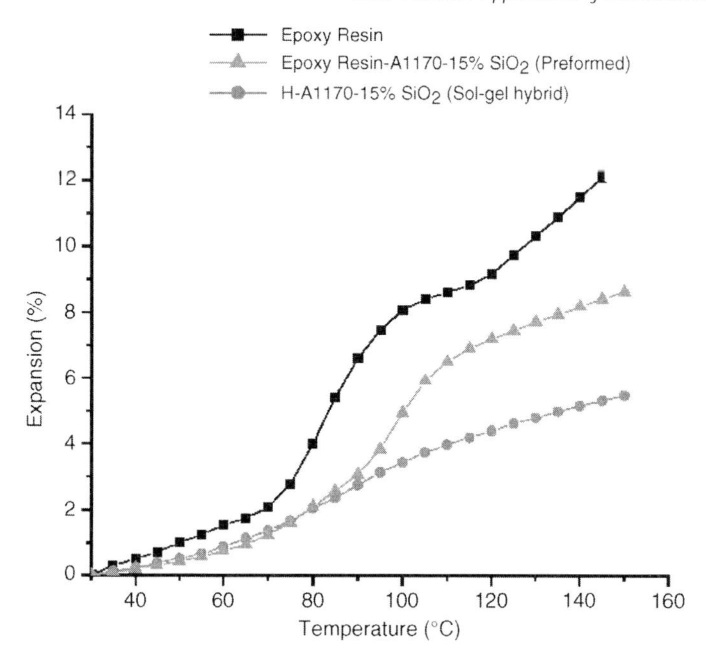

Figure 24.5 Comparison of linear expansion as function of temperature of an epoxy resin control with an epoxy/silica nanocomposite and an epoxy/silica hybrid.

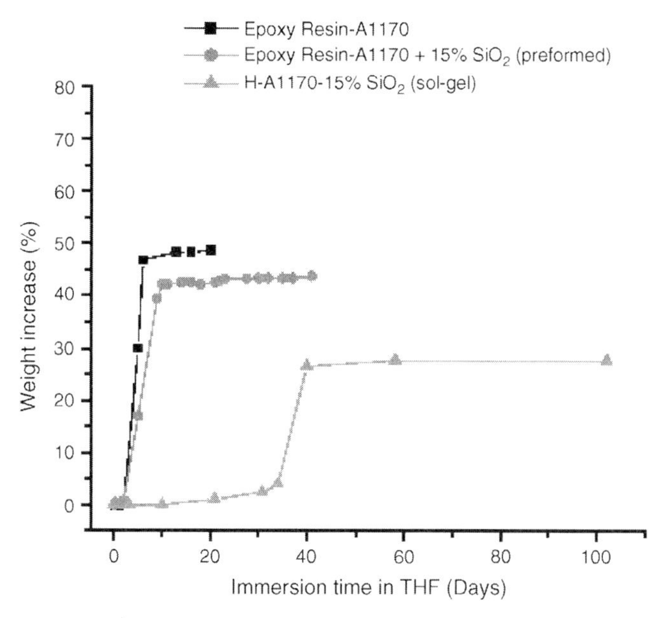

Figure 24.6 Comparison of room-temperature THF absorption as a function of immersion time of an epoxy resin control with epoxy/silica particulate nanocomposite and epoxy/silica hybrid (bicontinuous nanocomposite).

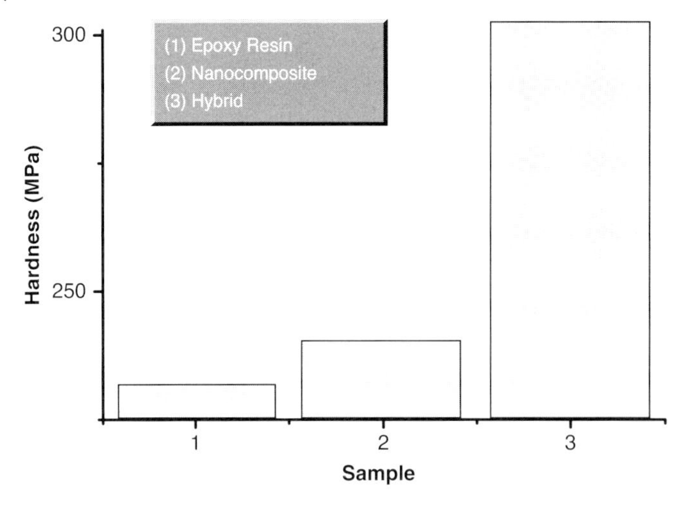

Figure 24.7 Comparison of effects of phase bicontinuity on nanocomposites. Nanocomposite: particulate nanocomposite. Hybrid: Bicontinuous nanocomposite (without molybdate dopant).

exhibiting a slow release of corrosion inhibitors has only recently been considered [23]. In this work the functionalization of the epoxy resin was carried out using suitable trimethoxysilane compounds, containing amine and mercaptan reactive groups. The molybdate anions were added as "bound" species in the form of salts produced from prereactions with a fractional amount of the cycloaliphatic amine hardener, and as "free" molybdic acid. The cocontinuous silica domains within the epoxy networks were produced by the incorporation of prehydrolyzed mixtures of tetraethoxysilane and GOTMS into the precursor solution used for the production of epoxy/silica hybrids.

The addition of a source of molybdate anions to the hybrid precursor solution gave rise to the formation of a denser siloxane network and to a substantial increase in T_g of the organic phase. This also resulted in a large increase in the rubberyplateau modulus and in a substantial reduction in both the rate of solvent uptake and the equilibrium amount of solvent absorption. These effects are illustrated in Figures 24.8 and 24.9.

In this study it was found that the presence of organic–inorganic hybrid networks allows the "entrapped" molybdate anions to be released at a slow and controlled rate. Furthermore, when the precursor solution mixtures for the formation of such doped networks are used as coatings for ferrous metal substrates, the molybdate anions slowly migrate to the interface and provide an enhanced mechanism for the passivation of the metal substrate. This is illustrated in Figure 24.10 where it is shown that the hybrid formulation doped with molybdate anions exhibit a much higher corrosion potential than the corresponding formulations that do not contain the molybdate anions. It is also noteworthy that the epoxy–silica hybrid without molybdate doping does not show any improvements over the neat epoxy resin. The migration of molybdate anions and the passivation effect of the iron corrosion were also confirmed by chemical analysis.

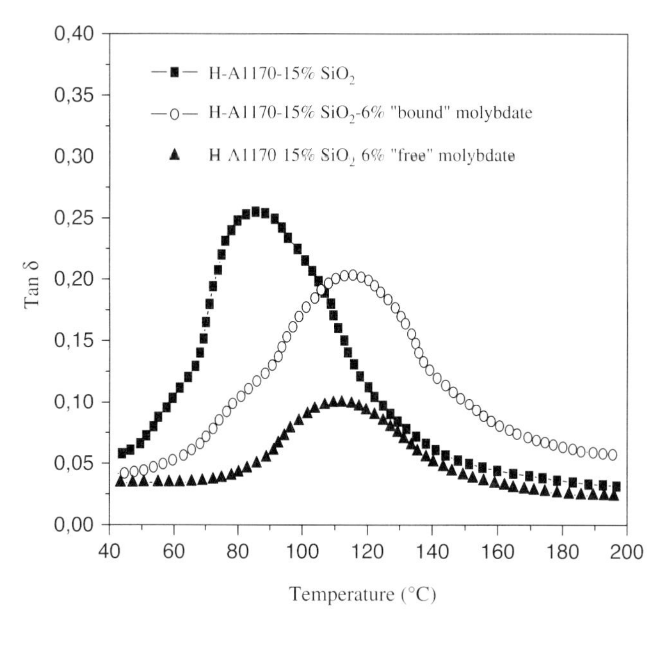

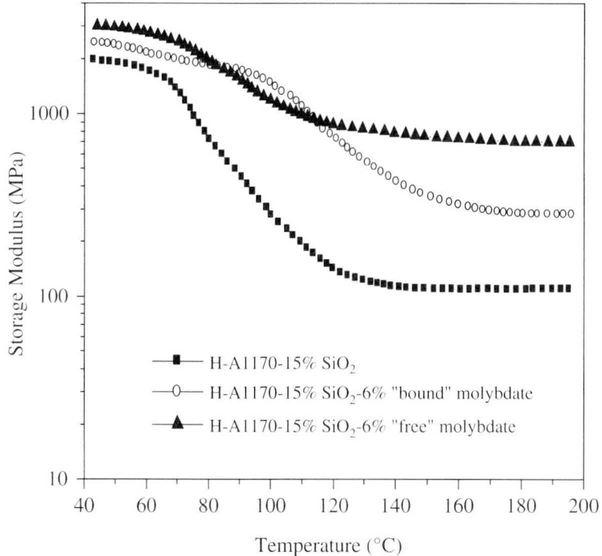

Figure 24.8 Effects of type of molybdate dopant on the dynamic mechanical properties of epoxy/silica bicontinuous nanocomposites.

The enhancement of the corrosion protection capability of the coatings was found to be greater when the molybdate anions were added in "bound" form, as amine molybdate, than as "free" molybdic acid as shown in Figure 24.11 [24].

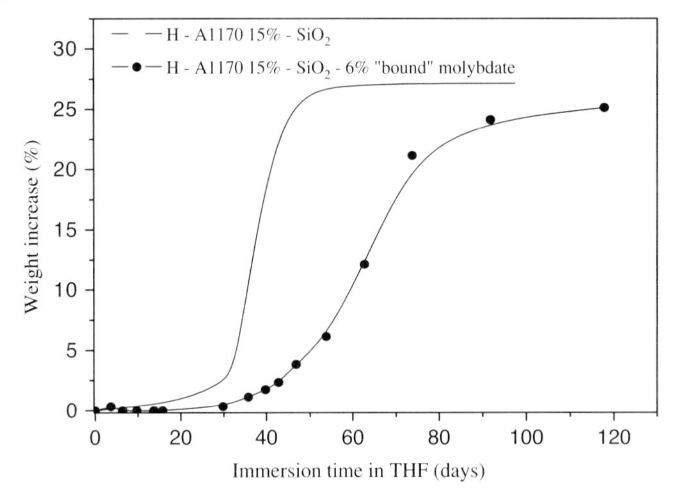

Figure 24.9 Effect of molybdate dopant on the solvent (THF) absorption characteristics of bicontinuous nanocomposites.

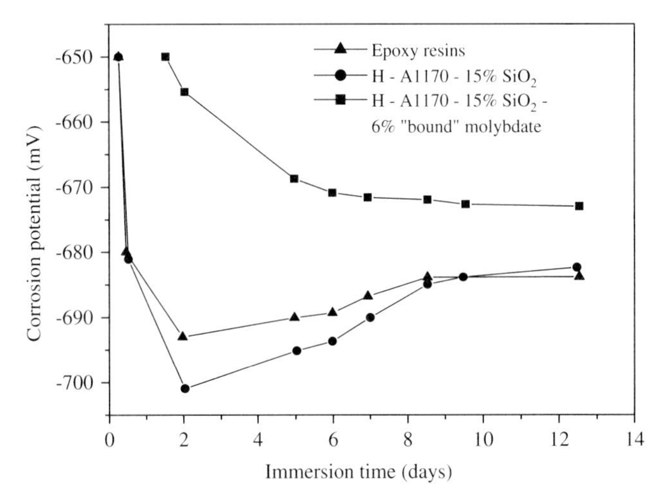

Figure 24.10 Effects of molybdate dopant on the corrosion protection characteristics of epoxy/silica bicontinuous nanocomposites.

24.4.2 Use of Polyimide/Silica Systems in Polymer Composites

24.4.2.1 General

Polyimides are probably the most appropriate class of polymers for the production of bicontinuous nanocomposites, in so far as they represent types of polymers that have been widely used in applications requiring stringent performance at high temperatures. The incorporation of continuous silica domains, therefore, is expected to enhance the high-temperature properties to bridge the gap between polymers and

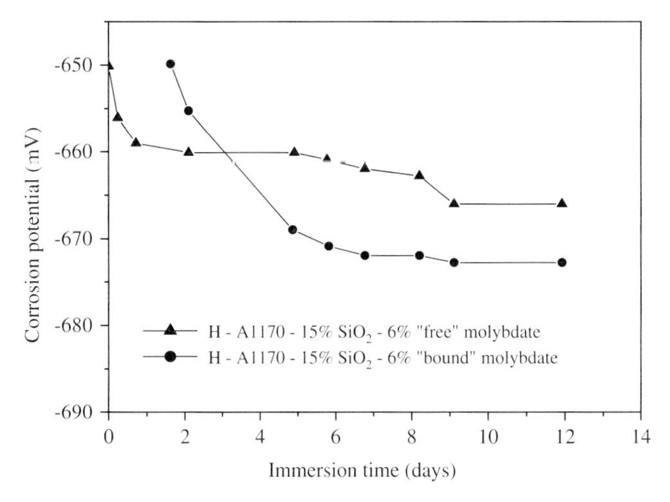

Figure 24.11 Comparison of corrosion protection characteristics of nanocomposites prepared with the addition of "free" molybdic acid and "bound" molybdate as a mono salt of the diamine hardener.

ceramics, particularly for use in microelectronics and electrical insulation. In both applications the polymer is frequently in contact with a metal substrate, hence a mismatch in thermal expansion and Young's modulus between the two components would produce stress concentrations leading to mechanical failure.

At the same time, polyimides are particularly suitable for the production of nanocomposites by the sol–gel method as they are derived from polyamic acid solutions in hydrophilic solvents, which are miscible with inorganic alkoxide precursor solutions for the production of ceramics. Furthermore, since the condensation reaction to convert polyamic acid to polyimide is an intramolecular process (see reaction scheme in Figure 24.12), there will be no steric hindrance by the surrounding inorganic oxide network after gelation at low temperatures. The incorporation of silica domains through the formation of hybrids represents a natural approach for further extending the use of polymers in high-temperature applications.

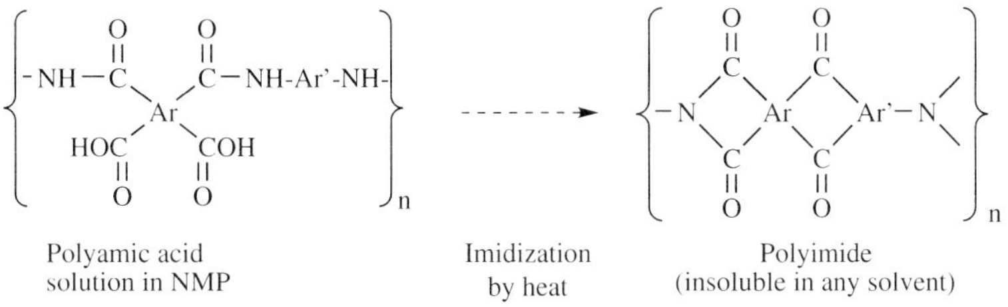

Polyamic acid solution in NMP — Imidization by heat → Polyimide (insoluble in any solvent)

Figure 24.12 Reaction scheme for the intramolecular condensation reactions in the conversion of polyamic acid to polyimide.

There are two main areas where polyimide nanocomposites have been investigated by the author as a means of enhancing the performance of composites:

(a) Fiber coatings for continuous fiber/epoxy composites, particularly for glass fibers.
(b) Matrices for continuous carbon-fiber composites.

The chemical structures of the materials used to prepare the hybrids are shown in Figure 24.13. In both cases, the hybrid material is a polyimide/silica system derived by the sol–gel processing of a polyamic acid and an appropriate alkoxysilane solution

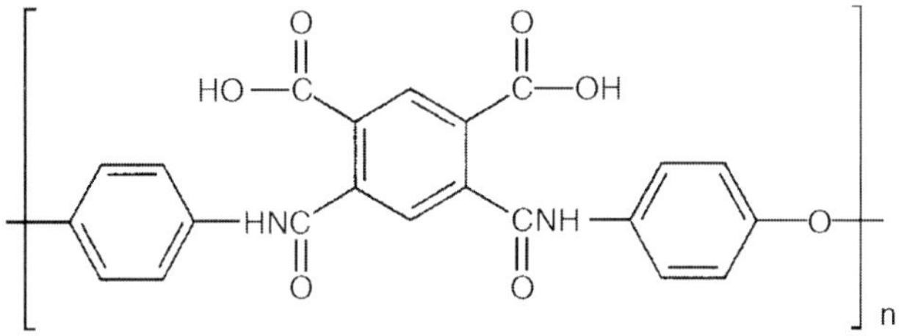

(a) PYRE ML-RK602 (DuPont de Nemours) (Polyamic acid - Polyimide precursor)

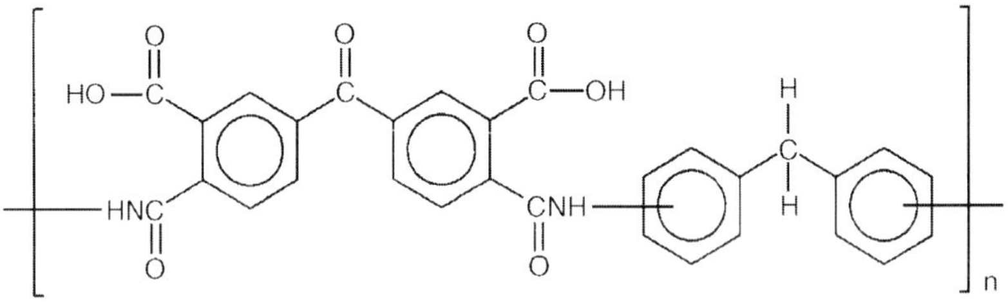

(b) SKYBOND 703 (Monsanto) (Polyamic acid - Polyimide precursor)

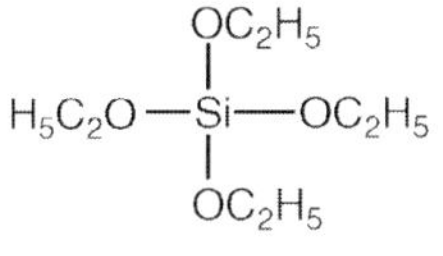

(c) Tetraethoxysilane "TEOS" (Supramolecular silica network former)

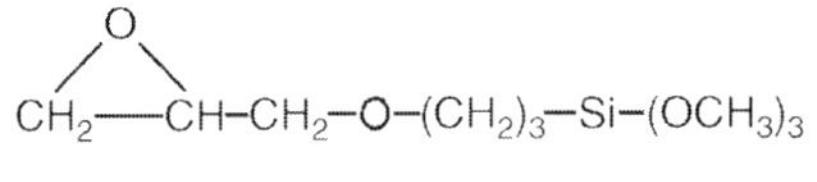

(d) γ-Glycidyloxypropyltrimethoxysilane (Coupling agent - Compatibilizer)

Figure 24.13 Chemical structure of the components of the polyimide/silica nanocomposites.

for the silica phase, using GOTMS as dispersing/coupling agent. A high MW polyamic acid (Pyre ML-RK602, DuPont) was used to prepare the precursor solution in *N*-methylpyrrolidone for the coating of glass fibers and a low MW polyamic acid (Skybond 703, Monsanto) was used for the production of the hybrid matrix of carbon-fiber composites. The silica content of the polyimide/silica hybrids used was about 25% by weight for the coating formulation and about 8 and 16% by weight for hybrids used as the matrix for carbon-fiber composites.

The SEM micrographs in Figure 24.14 [25] show the morphology of typical polyimide/silica hybrids based, respectively, on Skybond 703 and Pyre ML-RK692. Both micrographs reveal the presence of cocontinuous domains. The hybrids based on the low MW polyimide (Skybond) display fractal topography, which indicates the

(a) Skybond based

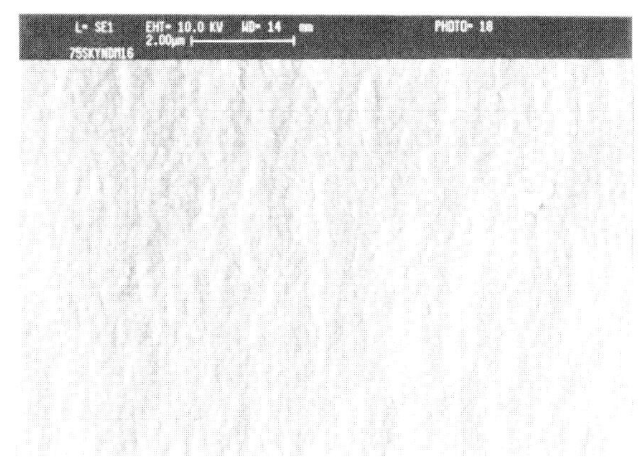

(b) Pyre ML based

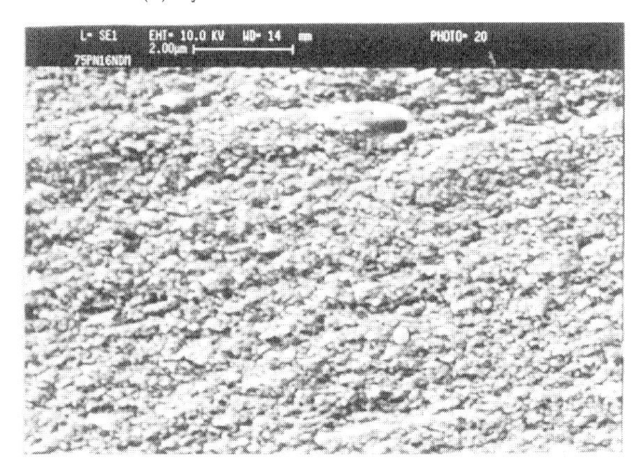

Figure 24.14 SEM micrographs of fractured surfaces of polyimide/silica hybrids (25% w/w SiO_2): (a) systems based on low MW polyamic acid (Skybond 703) and (b) systems based on high MW polyamic acid (Pyre ML).

existence of highly diffused nanostructured phases. The hybrids based on the high MW polyimide (Pyre ML), on the other hand, have a coarser and more nodular structure, but the films are very transparent. TEM micrographs of the latter systems (see below) show that the dispersed phase consists of cocontinuous domains of the two components, probably richer in silica than the surrounding areas [26]. Not only the high MW but also the higher chain rigidity of Pyre ML, relative to Skybond, may be responsible for the formation of a coarser morphology.

Polyimides were selected for these systems as the manufacturing processes for their applications are solution based, and the solvent used for the polyamic acid (precursor for polyimides) is miscible with water. Both aspects are compatible with the processes used for sol–gel applications.

24.4.2.2 Coatings for Glass Fibers

Commercially available glass fibers, for use in epoxy resin composites, were washed with hot xylene to remove the commercial "size" and subsequently recoated with the hybrid precursor solution. After curing, the coating thickness was in the region of 0.5 μm. Films were cast from similar solutions using different amounts of alkoxysilane precursor and their properties were evaluated as a basis for selecting the most appropriate composition for the fiber coatings. The drying and curing schedule was the same as for the fiber coatings. Unidirectional composites were then prepared by winding and impregnating the fibers on a frame that was then fitted in a "leaky mold" and pressed at the appropriate temperature to cure the bisphenol-A epoxy resin matrix with hexahydrophthalic anhydride. Glass concentration was 65 wt%.

The force–displacement curves, shown in Figure 24.15 [27], were obtained in a 3-point bending at a 5 : 1 span to thickness ratio; such short spans are normally used to measure the interlaminar shear strength of composites. The results indicate that the use of the polyimide–silica coating maintains the rigidity and the interlaminar shear strength achieved with commercial sizes. The ductility associated with interlaminar shear failures, however, is increased substantially over that obtained with conventional sizes due to a change in failure mechanism from a rapid growth of a few large cracks in the central planes to slow growth of multiple fine cracks formed in several parallel planes in the middle regions of the specimen. The polyimide/silica hybrid at the interphase between glass fibers and epoxy resin matrix acts as a "crumbling" zone, which arrests the propagation of the cracks in the planes along the fibers.

The methods that have been studied in the past for increasing the interlaminar shear ductility are based on coating fibers with rubbery polymers or very ductile thermoplastics from solutions. In these cases, the enhancement in ductility occurs at the expense of the interlaminar shear rigidity and strength [25, 27].

Figure 24.16 shows the variation in interlaminar shear rigidity with temperature, measured by dynamic mechanical tests at 1 Hz, using a cantilever bending mode with loading distance equal to half the span for the corresponding interlaminar shear tests in the 3-point bending. A comparison of the two curves indicates that the interlaminar shear rigidity becomes increasingly higher than for conventional sized

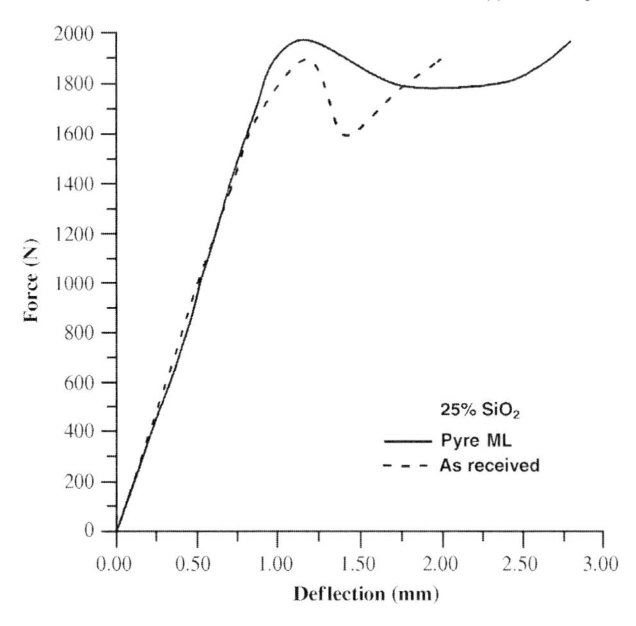

Figure 24.15 Comparison of force–deflection curves recorded in interlaminar shear strength tests between composites made with commercial glass fibers (as received) and fibers coated with the Pyre ML-RK692 (PI)–silica (25 wt%) hybrid.

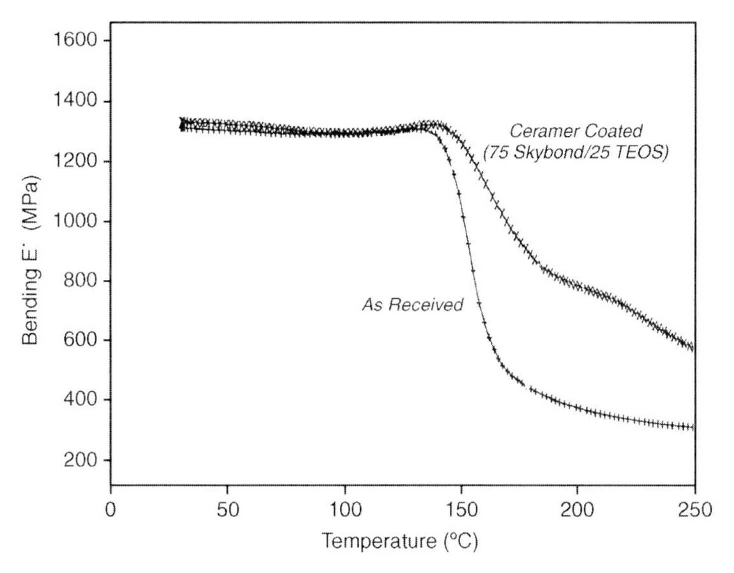

Figure 24.16 Effect of replacing a commercial sizing with a polyimide–silica hybrid coating on the interlaminar shear rigidity as a function of temperature of glass fiber/epoxy composites.

fiber systems upon traversing T_g of the polymeric matrix due to the much higher T_g of the interphase material (280–300 °C) compared to that of the matrix (around 150 °C) [27]. The higher interlaminar shear rigidity above T_g of the matrix suggests that the use of polyimide/silica hybrids as coatings for fibers can provide better safety margins for the designers against possible excursions of the product to high temperatures.

24.4.2.3 **Polyimide/Silica Matrix**

The precursor solution for these polyimide–silica hybrids was prepared in the same way as for coating formulations. The only difference was the type and molecular weight of the polyamic acid (see Figure 24.4) and the amounts of alkoxysilane used for the formation of silica domains. Films were produced using the same curing

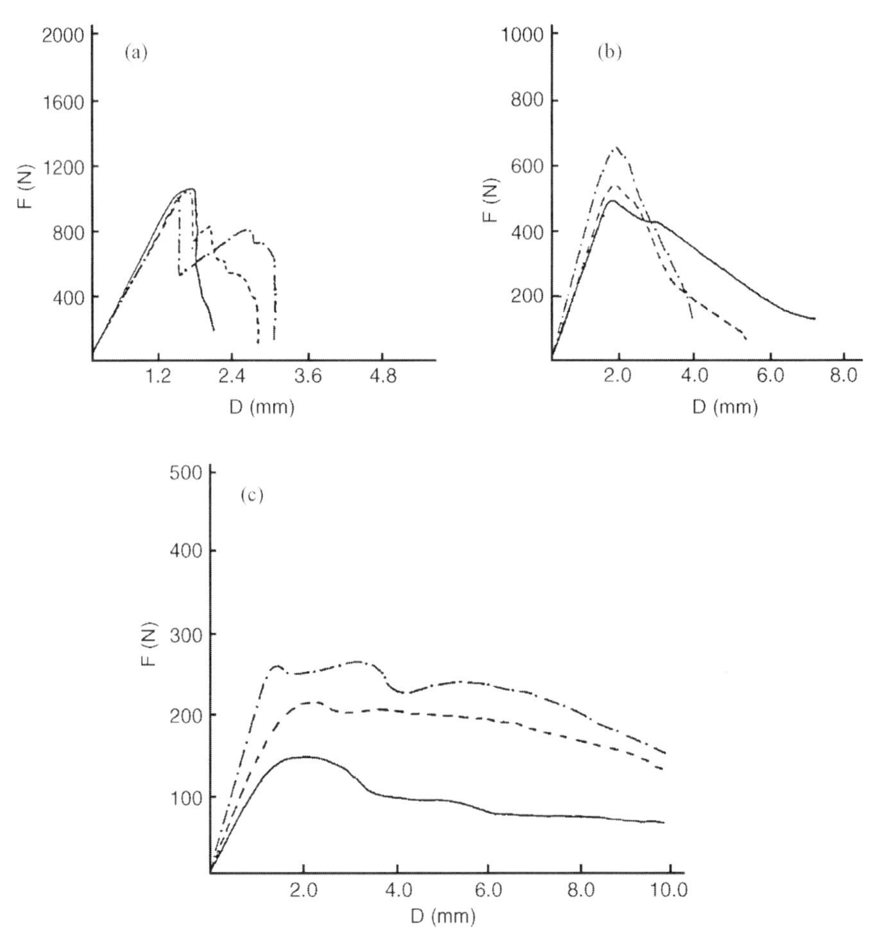

Figure 24.17 Flexural load–deflection curves for carbon-fiber composites measured at (a) room temperature, (b) 150 °C, and (c) 280 °C. (—) Polyimide; (– – –) hybrid with 7.5% SiO_2; (–·–·–) hybrid with 15% SiO_2.

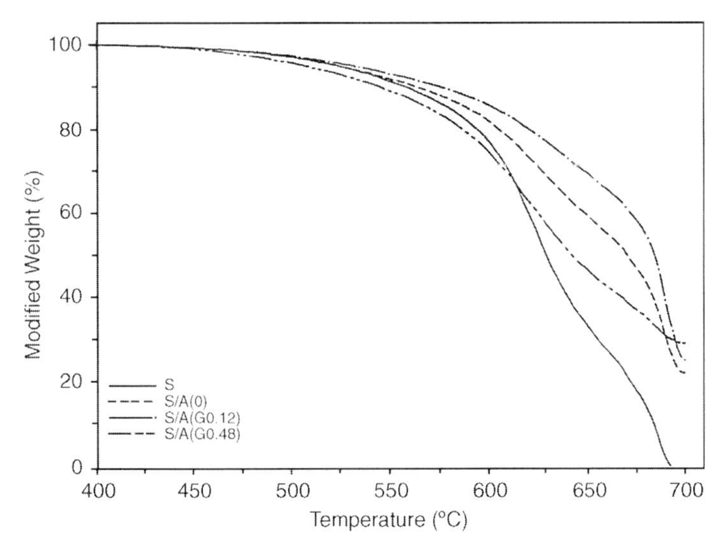

Figure 24.18 TGA thermograms for polyimide/silica (Skybond based) containing various amounts of coupling agent. S: polyimide; S/A(0): polymide/silica hybrid (no coupling agent). S/A(G0.12): polymide/silica hybrid (0.12 molar ratio coupling agent/TEOS). S/A(G0.48): polymide/silica hybrid (0.48 molar ratio coupling agent/TEOS).

schedule described for the coatings. Composites were prepared as previously described using commercial carbon fibers.

The SEM micrographs of fractured surfaces revealed a fractal topography, which suggests that the two phases are not segregated but exhibit a graded supramolecular structure. The mechanical properties were evaluated both at room temperature and at high temperatures, close to T_g of the polyimide. The load–deflection curves obtained from flexural tests (span:depth ratio = 20 : 1) are compared in Figure 24.17 [28]. These show that for systems in which a polymeric matrix is replaced by its corresponding silica hybrid, the resulting increase in modulus at higher temperatures is even more dramatic.

Finally, the thermograms in Figure 24.18 [29] show that the presence of the cocontinuous silica domains within the polyimide matrix increases the thermal oxidative stability by shifting to higher temperatures the curve for the weight loss. This is associated with the improved barrier properties provided by the silica phase, both for the infusion of oxygen and the outer diffusion of gases resulting from pyrolysis of the polymer chains. However, when large amounts of coupling agents (above the optimum) are used, the resulting increase in the content of aliphatic matter brings about deterioration in thermal stability.

Acknowledgments

The author is grateful to many of his colleagues who have contributed to the contents of this chapter and in particular Drs H. Demirer, C. Xenopoulos, and L. Prezzi.

References

1 Huang, H.H., Orler, B., and Wilkes, G.L. (1987) Structure–property behavior of new hybrid materials incorporating oligomeric species into sol–gel glasses. *Macromolecules*, **20**, 1322.

2 Novak, B.M. (1993) Hybrid nanocomposite materials – between inorganic glasses and organic polymers. *Adv. Mater.*, **5**, 43.

3 Brinker, C.J. (1990) Chapter 3, in *Sol–Gel Science: The Physics and Chemistry of Sol–Gel Processing*, London Academic Press, London, UK.

4 Schmidt, H. and Seiferling, B. (1986) Chemistry and applications of inorganic–organic polymers (organically modified silicates). *Proc. Mater. Res. Soc. Symp.*, **73**, 739.

5 Sanchez, C., Julián, B., Belleville, P., and Popall, M. (2005) Applications of hybrid organic–inorganic nanocomposites. *J. Mater. Chem.*, **15**, 3559.

6 Ravaine, D., Seminel, A., Charbouillot, Y., and Vincent, M. (1986) A new family of organically modified silicates prepared from gels. *J. Non-Cryst. Solids*, **82**, 210.

7 Mark, J.E., Lee, Y.-C., and Bianconi, P.A. (eds) (1995) *Hybrids Organic–Inorganic Composites*, ACS Symposium Series 585 Washington, D.C.

8 Mascia, L. (1995) Developments in organic–inorganic polymeric hybrids: ceramers. *Trends Polym. Sci.*, **3**, 61.

9 Mascia, L., Prezzi, L., and Haworth, B. (2006) Substantiating the role of phase bicontinuity and interfacial bonding in epoxy–silica nanocomposites. *J. Mater. Sci.*, **41**, 1145.

10 Wang, B., Wilkes, G.L., Hedrick, J.C., Liptak, S.C., and McGrath, J.E. (1991) New high-refractive-index organic/inorganic hybrid materials from sol–gel processing. *Macromolecules*, **24**, 3449.

11 Huang, H.-H., Glaser, R.H., and Wilkes, G.L. (1987) Structure–property behavior of new hybrid materials incorporating organic oligomeric species into sol–gel glasses. IV. Characterization of structure and extent of reaction. *Proc. PMSE ACS Div.*, **56**, 85812.

12 Matejka, L., Prestil, J., and Dusek, K. (1998) Structure evolution in epoxy-silica hybrids: sol–gel process. *J. Non-Cryst. Solids*, **226**, 114.

13 Prezzi, L. and Mascia, L. (2005) Network density control in epoxy–silica hybrids by selective silane functionalization of precursors. *Adv. Polym. Technol.*, **24**, 13.

14 Mascia, L. and Tang, T. (1998) Ceramers based on crosslinked epoxy resins–silica hybrids: low surface energy systems. *J. Sol-Gel. Sci. Technol.*, **13**, 405.

15 Prosek, T. and Thierry, D. (2004) A model for the release of chromate from organic coatings. *Prog. Org. Coat.*, **49**, 209.

16 Buchheit, R.G., Mamidipally, S.B., Schmutz, P., and Guan, H. (2002) Active corrosion protection in Ce-modified hydrotalcite conversion coatings. *Corrosion*, **58**, 3.

17 Cabot, B. and Foissy, A. (1998) Reversal of surface charges of mineral powder: application to electrophoretic deposition of silica for anticorrosion coatings. *J. Mater. Sci. Lett.*, **33**, 3945.

18 El Din Shams, A.M., Mohamed, R.A., and Haggag, H.H. (1997) Corrosion inhibition by molybdate/polymaleate mixtures. *Desalination*, **114**, 85.

19 Jabeera, B., Shibli, S.M.A., and Anirudhan, T.S. (2001) The synergistic effect of molybdate with zinc for the effective inhibition of corrosion of mild steel. *Corros. Prevent. Contr.*, **48**, 65.

20 El Din Shams, A.M. and Wang, L. (1996) Mechanism of corrosion by sodium molybdate. *Desalination*, **107**, 29.

21 Treacy, G.M., Wilcox, G.D., and Richardson, M.O.W. (1999) Behaviour of passivated zinc coated steel exposed to corrosive chloride environments. *J. Appl. Electrochem.*, **29**, 647.

22 Kumary, V.A., Sreevalsan, K., and Shimli, S.M.A. (2001) Inhibitive effects of calcium gluconate. *Corros. Prevent. Contr.*, **48**, 83.

23 Sousa, J.S., Falcao, A.N., Carrapico, M., Marcaca, F.M.A., Carvalho, F.G., Miranda Salvado, I.M., and Teixeira, J. (2003) SANS study of zirconia–silica and titania–silica hybrid materials. *J. Sol-Gel Sci. Technol.*, **26**, 345.

24 Mascia, L., Prezzi, L., Wilcox, G.D., and Lavorgna, M. (2006) Molybdate-doping of networks in epoxy–silica hybrids: domain structuring and corrosion inhibition. *Prog. Org. Coat.*, **56**, 13.

25 Mascia, L. (1998) Materiali ibridi organici-inorganici (ceramers o nanocompositi): aspetti strutturali e proprieta. *La Chimica e L' Industria*, **80**, 623.

26 Mascia, L. and Kioul, A. (1994) Compatibility of polyimide-silica hybrids induced by alkoxysilane coupling agents. *J. Cryst. Solids*, **175**, 169.

27 Demirer, H. (1999) Ph.D. Thesis, Loughborough University, UK.

28 Mascia, L., Zhang, Z., and Shaw, S.J. (1996) Carbon fibres composites based on polyimide/silica ceramers: aspects of structure-properties relationship. *Composites, Part A*, **27A**, 1211.

29 Xenopoulos, C., Mascia, L., and Shaw, S.J. (2001) Optimisation of morphology of polyimide–silica hybrids in the production of matrices for carbon fibre. *High Perform. Polym.*, **13**, 1.

Index

Functional Fillers for Plastics: Second, updated and enlarged edition. Edited by Marino Xanthos

ISBN: 978-3-527-32361-6

d

n

u

v